Lecture Notes in Computer Science 10504

Commenced Publication in 1973
Founding and Former Series Editors:
Gerhard Goos, Juris Hartmanis, and Jan van Leeuwen

Advanced Research in Computing and Software Science
Subline of Lecture Notes in Computer Science

More information about this series at http://www.springer.com/series/7409

Vittorio Bilò · Michele Flammini (Eds.)

Algorithmic Game Theory

10th International Symposium, SAGT 2017
L'Aquila, Italy, September 12–14, 2017
Proceedings

 Springer

Editors
Vittorio Bilò
University of Salento
Lecce
Italy

Michele Flammini
University of L'Aquila
L'Aquila
Italy

ISSN 0302-9743   ISSN 1611-3349   (electronic)
Lecture Notes in Computer Science
ISBN 978-3-319-66699-0   ISBN 978-3-319-66700-3   (eBook)
DOI 10.1007/978-3-319-66700-3

Library of Congress Control Number: 2017950086

LNCS Sublibrary: SL3 – Information Systems and Applications, incl. Internet/Web, and HCI

Printed on acid-free paper

This Springer imprint is published by Springer Nature
The registered company is Springer International Publishing AG
The registered company address is: Gewerbestrasse 11, 6330 Cham, Switzerland

Preface

This volume contains the papers presented at the 10th International Symposium on Algorithmic Game Theory (SAGT 2017), which was held on September 12–14, 2017, in L'Aquila, Italy.

This year, we received a record number of 66 submissions. Each submission was reviewed by at least three Program Committee members. After a careful reviewing process, the committee decided to accept 30 papers. The program also included three invited talks by distinguished researchers in Algorithmic Game Theory: Michal Feldman (Tel-Aviv University), Martin Hoefer (Goethe University), and Nicole Immorlica (Microsoft Research).

To accommodate the publishing traditions of different fields, the authors of accepted papers could request that only a one-page abstract of the paper appears in the proceedings. Among the 30 accepted papers, the authors of 4 papers opted to publish a one-page abstract. The accepted submissions cover various important aspects of algorithmic game theory, such as auctions, computational aspects of games, congestion games, network and opinion formation, mechanism design, incentives and regret minimization, and resource allocation. The best paper award, generously supported by Springer, has been shared between the papers *Tight Welfare Guarantees for Pure Nash Equilibria of the Uniform Price Auction* by Georgios Birmpas, Evangelos Markakis, Orestis Telelis and Artem Tsikiridis, and *Online Random Sampling for Budgeted Settings* by Alon Eden, Michal Feldman and Adi Vardi.

We would like to thank all authors who submitted their research work, the Program Committee members, and the external reviewers who assisted them, for their wonderful work. We are indebted to the Gran Sasso Science Institute of L'Aquila, Fondazione Cassa di Risparmio della Provincia dell'Aquila, the Department of Mathematics and Physics "Ennio De Giorgi" of the University of Salento, EATCS, and Springer for their generous support. We thank Anna Kramer and Alfred Hoffmann at Springer for helping with the proceedings. We are grateful for the use of the EasyChair paper management system.

July 2017

Vittorio Bilò
Michele Flammini

Organization

Program Committee

Elliot Anshelevich	Rensselaer Polytechnic Institute (RPI), USA
Vittorio Bilò (Co-chair)	University of Salento, Italy
Ioannis Caragiannis	University of Patras, Greece
George Christodoulou	University of Liverpool, UK
Xiaotie Deng	Shanghai Jiao Tong University, China
Angelo Fanelli	CNRS, France
Michele Flammini (Co-chair)	University of L'Aquila and Gran Sasso Science Institute (GSSI), Italy
Dimitris Fotakis	National Technical University of Athens, Greece
Martin Gairing	University of Liverpool, UK
Paul Goldberg	University of Oxford, UK
Tobias Harks	Augsburg University, Germany
Thomas Kesselheim	TU Dortmund, Germany
Elias Koutsoupias	University of Oxford, UK
Stefano Leonardi	University of Rome "La Sapienza", Italy
Marios Mavronicolas	University of Cyprus, Cyprus
Gianpiero Monaco	University of L'Aquila, Italy
Luca Moscardelli	University of Chieti-Pescara, Italy
Giuseppe Persiano	University of Salerno, Italy
Guido Proietti	University of L'Aquila and Institute for Systems Analysis and Computer Science (IASI-CNR), Italy
Amin Saberi	Stanford University, USA
Rahul Savani	University of Liverpool, UK
Guido Schaefer	CWI Amsterdam, The Netherlands
Alexander Skopalik	Heinz Nixdorf Institute and Paderborn University, Germany
Nicolas Stier-Moses	Torcuato Di Tella University, Argentina
Chaitanya Swamy	University of Waterloo, Canada
Vasilis Syrgkanis	Microsoft Research, USA
Marc Uetz	University of Twente, The Netherlands
Adrian Vetta	McGill University, Canada

Steering Committee

Elias Koutsoupias	University of Oxford, UK
Marios Mavronicolas	University of Cyprus, Cyprus
Dov Monderer	Technion, Israel
Burkhard Monien	Paderborn University, Germany
Christos Papadimitriou	UC Berkeley, USA

Giuseppe Persiano   University of Salerno, Italy
Paul Spirakis (Chair)   University of Liverpool, UK

Organizing Committee

Vittorio Bilò (Co-chair)   University of Salento, Italy
Michele Flammini   University of L'Aquila and Gran Sasso Science Institute
  (Co-chair)   (GSSI), Italy
Gianlorenzo D'Angelo   Gran Sasso Science Institute (GSSI), Italy
Mattia D'Emidio   Gran Sasso Science Institute (GSSI), Italy
Gianpiero Monaco   University of L'Aquila, Italy
Luca Moscardelli   University of Chieti-Pescara, Italy
Cosimo Vinci   Gran Sasso Science Institute (GSSI), Italy

Additional Reviewers

Amanatidis, Georgios
Baumeister, Dorothea
Bhaskar, Umang
Bilò, Davide
Birmpas, Georgios
Biro, Peter
Carosi, Raffaello
Colini Baldeschi,
  Riccardo
Cord-Landwehr, Andreas
Cseh, Ágnes
de Jong, Jasper
De Keijzer, Bart
Deligkas, Argyrios
Drees, Maximilian
Duetting, Paul
Eden, Alon
Epitropou, Markos
Fearnley, John
Feldkord, Björn
Feldotto, Matthias
Ferraioli, Diodato
Filos-Ratsikas, Aris

Fischer, Felix
Freeman, Rupert
Giannakopoulos, Yiannis
Greco, Gianluigi
Gualà, Luciano
Gur, Yonatan
Hajiaghayi,
  Mohammadtaghi
Hamada, Koki
Hoeksma, Ruben
Igarashi, Ayumi
Jabbari, Shahin
Kern, Walter
Kliemann, Lasse
Kodric, Bojana
Krimpas, George
Krysta, Piotr
Kyropoulou, Maria
Lazos, Philip
Lenzner, Pascal
Leucci, Stefano
Lianeas, Thanasis
Liu, Zhengyang

Lykouris, Thodoris
Maillé, Patrick
Malladi, Suraj
Mandal, Debmalya
Manlove, David
Markakis, Evangelos
Peters, Dominik
Schroder, Marc
Schwiegelshohn, Chris
Sgouritsa, Alkmini
Shameli, Ali
Skochdopoloe, Nolan
Tuffin, Bruno
Tönnis, Andreas
Vegh, Laszlo
Ventre, Carmine
von Stengel, Bernhard
Voudouris, Alexandros
Xiao, Tao
Yukun, Cheng
Zick, Yair

Abstracts

Position Ranking and Auctions for Online Marketplaces

Leon Yang Chu, Hamid Nazerzadeh, and Heng Zhang

Data Sciences and Operations Department, University of Southern California,
Los Angeles, CA, USA, 90089
{leonyzhu,hamidnz,Heng.Zhang.2019}@marshall.usc.edu

Abstract. E-commerce platforms such as Amazon, Ebay, Taobao, and Google Shopping connect thousands of sellers and consumers everyday. When a consumer enters a search keyword related to a product of interest, the platform's search engine returns a list. Typically, the consumer looks for a desired item by searching downward in the list. With a large volume of returned results, consumers rarely consider all of the items because examining each option is costly. Therefore, the ranking of items is an important decision that determines the welfare of sellers, consumers and the platform. In this work, we study how such platforms should rank products displayed, and utilize the top and most salient slots. Building on the optimal sequential search theory, we present a model that considers consumers' search costs and the externalities sellers impose on each other. This model allows us to study a multi-objective optimization, whose objective includes consumer and seller surplus, as well as the sales revenue, and derive the optimal ranking decision. One of the challenges in obtaining a satisfactory solution in practice is information asymmetry. The platform may be unaware of sellers' private benefits of each consumer purchase, for example, profits, brand effects, and so on. We show that an uninformed decision, one in which sellers' private valuations are unknown to the platform, can lead to an arbitrary loss of average welfare. We propose selling the platform's top slots to extract private information, using the *surplus-ordered ranking* (SOR) mechanism. This mechanism is motivated in part by Amazon's sponsored search program. We study this mechanism in a mechanism design framework, and show that it is a near-optimal solution. In addition, when the platform sells all slots, we show that our mechanism can be implemented as a Nash equilibrium in a modified generalized second price (GSP) auction.

Full paper available online at: https://papers.ssrn.com/sol3/papers.cfm?abstract_id=2926176.

Asymptotic Existence of Fair Divisions
for Groups

Pasin Manurangsi[1] and Warut Suksompong[2]

[1] Department of Computer Science, UC Berkeley
253 Cory Hall, Berkeley, CA 94720, USA
pasin@berkeley.edu

[2] Department of Computer Science, Stanford University, 353 Serra Mall,
Stanford, CA 94305, USA
warut@cs.stanford.edu

The problem of dividing resources fairly occurs in many practical situations and is therefore an important topic of study in economics. In this paper, we investigate envy-free divisions in the setting where there are multiple players in each party. While all players in a party share the same set of resources, each player has her own preferences. In this generalized setting, we consider a division to be envy-free if every player values the set of items assigned to her group at least as much as that assigned to any other group.

We show that under additive valuations drawn randomly from probability distributions, when all groups contain an equal number of players, an envy-free division is likely to exist if the number of goods exceeds the total number of players by a logarithmic factor, no matter whether the players are distributed into several groups of small size or few groups of large size. In particular, any allocation that maximizes social welfare is likely to be envy-free. A similar result holds when there are two groups with possibly unequal numbers of players and the distribution on the valuation of each item is symmetric.

To complement our existence results, we show on the other hand that we cannot get away with a much lower number of items and still have an envy-free division with high probability. In particular, if the number of items is less than the total number of players by a superconstant factor, or if the number of items is less than the total number of players and the number of groups is large, the probability that an envy-free division exists is low. This leaves the gap between asymptotic existence and non-existence of envy-free divisions at a mere logarithmic factor.

Finally, we tackle the issue of truthfulness and show that a simple truthful mechanism, namely the random assignment mechanism, is α-*approximate envy-free* with high probability for any constant $\alpha \in [0,1)$. Approximate envy-freeness means that even though a player may envy another player in the resulting division, the values of the

---

The full version of this paper is available at http://arxiv.org/abs/1706.08219.

player for her own allocation and for the other player's allocation differ by no more than a multiplicative factor of α. Our result shows that it is possible to achieve truthfulness and approximate envy-freeness simultaneously in a wide range of random instances, and improves upon the previous result for the setting with one player per group in several ways.

Approximate Maximin Shares for Groups of Agents

Warut Suksompong

Department of Computer Science, Stanford University, 353 Serra Mall, Stanford,
CA 94305, USA
warut@cs.stanford.edu

We consider the problem of fairly allocating indivisible goods to interested agents. Several notions of fairness have been proposed, including envy-freeness and proportionality. However, the existence of allocations satisfying these notions, or even a multiplicative approximation of them, cannot be guaranteed. A notion that was designed to fix this problem and has been a subject of much interest in the last few years is the maximin share.

In this paper, we apply the concept of maximin share to a more general setting of fair division in which goods are allocated not to individual agents, but rather to groups of agents who can have varying preferences on the goods. Several practical situations involving fair division fit into this model. For instance, an outcome of a negotiation between countries may have to be approved by members of the cabinets of each country who have different opinions on the outcome. Another example is a large company or university that needs to divide its resources among competing groups of agents (e.g., departments in a university). The agents in each group have different and possibly misaligned interests; the professors who perform theoretical research may prefer more whiteboards and open space in the department building, while those who engage in experimental work are more likely to prefer laboratories.

We extend the maximin share to groups in a natural way by calculating the maximin share for each agent using the number of groups instead of the number of agents. When there are two groups, we completely characterize the cardinality of agents in the groups for which it is possible to approximate the maximin share within a constant factor regardless of the number of goods. In particular, an approximation is possible when one of the groups contain a single agent, when both groups contain two agents, or when the groups contain three and two agents respectively. In all other cases, no approximation is possible in a strong sense: There exists an instance with only four goods in which some agent with positive maximin share necessarily gets zero utility.

We then generalize to the setting with several groups of agents. On the positive side, we show that a constant factor approximation is possible if only one group contains more than a single agent. On the other hand, we show on the negative side that when all groups contain at least two agents and one group contains at least five agents, it is possible that some agent with positive maximin share will be forced to get zero utility, which means that there is no hope of obtaining an approximation in this case.

The full version of this paper is available at http://arxiv.org/abs/1706.09869.

On Black-Box Transformations
in Downward-Closed Environments

Warut Suksompong

Department of Computer Science, Stanford University, 353 Serra Mall, Stanford,
CA 94305, USA
warut@cs.stanford.edu

A major line of work in algorithmic mechanism design involves taking a setting where the optimization problem is computationally intractable, and designing computationally tractable mechanisms that yield a good global outcome and such that the agents have a truth-telling incentive. The widespread success of designing such mechanisms has raised the question of whether there exists a "black-box transformation" for transforming any computationally tractable algorithm into a computationally tractable mechanism without degrading the approximation guarantee. Chawla et al. showed that no fully general black-box transformation exists for single-parameter environments.

Despite this negative result, it is still conceivable that there are transformations that work for certain large subclasses of single-parameter environments. One important subclass is that of downward-closed environments, which occur in a wide variety of settings in mechanism design. In this paper, we consider such settings and assume, crucially, that the black-box transformation is aware that the feasible set is downward-closed. We investigate the potentials and limits of black-box transformations when they are endowed with this extra power.

We begin by showing the limits of black-box transformations in downward-closed environments. We prove that such transformations cannot preserve the full welfare at every input, even when the private valuations can take on only two arbitrary values. Preserving a constant fraction of the welfare pointwise is impossible if the ratio between the two values $l < h$ is sublinear, i.e., $h/l \in O(n^\alpha)$ for $\alpha \in [0,1)$, where n is the number of agents, while preserving the approximation ratio is also impossible if the values are within a constant factor of each other and the transformation is restricted to querying inputs of Hamming distance $o(n)$ away from its input.

Next, we show the powers of black-box transformations in downward-closed environments. We prove that when the private valuations can take on only a constant number of values, each pair of values separated by a ratio of $\Omega(n)$, it becomes possible for a transformation to preserve a constant fraction of the welfare pointwise, and therefore of the approximation ratio as well. The same is also true if the private valuations are all within a constant factor of each other. Combined with the negative results, this gives us a complete picture of constant-fraction welfare-preserving transformations for multiple input values. Not only are these results interesting in their own right, but they also demonstrate the borders of the negative results that we can hope to prove.

The full version of this paper is available at http://arxiv.org/abs/1707.00230.

Contents

Congestion Games, Network and Opinion Formation Games

Mechanism Design, Incentives and Regret Minimization

Resource Allocation

Auctions

Liquid Price of Anarchy

Yossi Azar[1], Michal Feldman[1], Nick Gravin[2], and Alan Roytman[3(✉)]

[1] Tel Aviv University, Tel Aviv, Israel
{azar,mfeldman}@tau.ac.il
[2] Massachusetts Institute of Technology, Cambridge, USA
ngravin@mit.edu
[3] University of Copenhagen, Copenhagen, Denmark
alanr@di.ku.dk

Abstract. Incorporating budget constraints into the analysis of auctions has become increasingly important, as they model practical settings more accurately. The social welfare function, which is the standard measure of efficiency in auctions, is inadequate for settings with budgets, since there may be a large disconnect between the value a bidder derives from obtaining an item and what can be liquidated from her. The *Liquid Welfare* objective function has been suggested as a natural alternative for settings with budgets. Simple auctions, like simultaneous item auctions, are evaluated by their performance at equilibrium using the Price of Anarchy (PoA) measure – the ratio of the objective function value of the optimal outcome to the worst equilibrium. Accordingly, we evaluate the performance of simultaneous item auctions in budgeted settings by the *Liquid Price of Anarchy* (LPoA) measure – the ratio of the optimal Liquid Welfare to the Liquid Welfare obtained in the worst equilibrium.

For pure Nash equilibria of simultaneous first price auctions, we obtain a bound of 2 on the LPoA for additive buyers. Our results easily extend to the larger class of fractionally-subadditive valuations. Next we show that the LPoA of mixed Nash equilibria for first price auctions with additive bidders is bounded by a constant. Our proofs are robust, and can be extended to achieve similar bounds for Bayesian Nash equilibria. To derive our results, we develop a new technique in which some bidders deviate (surprisingly) toward a non-optimal solution. In particular, this technique goes beyond the smoothness-based approach.

Y. Azar—Supported in part by the Israel Science Foundation (grant No. 1506/16), by the I-CORE program (Center No. 4/11), and by the Blavatnik Fund.

M. Feldman—This work was partially supported by the European Research Council under the European Union's Seventh Framework Programme (FP7/2007-2013)/ERC grant agreement number 337122.

A. Roytman—This work was partially supported by the European Research Council under the European Union's Seventh Framework Programme (FP7/2007-2013)/ERC grant agreement number 337122, by Thorup's Advanced Grant DFF-0602-02499B from the Danish Council for Independent Research, by grant number 822/10 from the Israel Science Foundation, and by the Israeli Centers for Research Excellence (ICORE) program.

V. Bilò and M. Flammini (Eds.): SAGT 2017, LNCS 10504, pp. 3–15, 2017.
DOI: 10.1007/978-3-319-66700-3_1

1   Introduction

Budget constraints have become an important practical consideration in most existing auctions, as reflected in recent literature (see, e.g., [4,6,20,37]), because they model reality more accurately. The issue of limited liquidity of buyers arises when transaction amounts are large and may exhaust bidders' liquid assets, as is the case for privatization auctions in Eastern Europe and FCC spectrum auctions in the U.S. (see, e.g., [5]). As another example, advertisers in Google Adword auctions are instructed to specify their budget even before specifying their bids and keywords. Many other massive electronic marketplaces have a large number of participants with limited liquidity, which impose budget constraints. Buyers would not borrow money from a bank to partake in multiple auctions on eBay, and even with available credit, they only have a limited amount of attention, so that in aggregate they cannot spend too much money by participating in every auction online. Finally, budget constraints also arise in small scale systems, such as the reality TV show Storage Wars, where people participate in cash-only auctions to win the content of an expired storage locker with an unknown asset.

Maximizing social welfare is a classic objective function that has been extensively studied within the context of resource allocation problems, and auctions in particular. The *social welfare* of an allocation is the sum of agents' valuations for their allocated bundles. Unfortunately, in settings where agents have limited budgets (hereafter, *budgeted settings*), the social welfare objective fails to accurately capture what happens in practice. Consider, for example, an auction in which there are two bidders and one item to be allocated among the bidders. One bidder has a high value but a very small budget, while the second bidder has a medium value along with a medium budget. In this case, a high social welfare is achieved by allocating the item to the bidder who values the item highly. In contrast, most Internet advertising and electronic marketplaces (such as Google and eBay) would allocate the item in the opposite way, namely to the bidder with a medium value and budget. Indeed, it seems reasonable to favor participants with substantial investments and engagement in the economical system to maintain a healthy economy regardless of the marketplace intermediary's personal gains. Hence, the social welfare objective is a poor model for how auctions are executed in reality.

In this work, we study the efficiency of simultaneous first price auctions in budgeted settings. Following Dobzinski and Leme [21] (see also [11,23,32, 38]), we measure the efficiency of outcomes in budgeted settings according to their *Liquid Welfare* objective, motivated as follows. In the mechanism design literature, a buyer i with additive values for items v_{ij} (where v_{ij} denotes buyer i's value for item j) and a budget cap B_i is usually modeled with *budget additive* valuations $v_i(S) = \min(B_i, \sum_{j \in S} v_{ij})$, where S is the set of items that player i receives (see, e.g., Lehmann et al. [31] and many follow-up works). Budget additive valuations are convenient to work with (they form a simple subclass of submodular valuations) and, from the designer's perspective, are a natural proxy for the contribution of each bidder to the economical system.

However, in reality such valuations do not capture the real preferences of the buyers, since each buyer usually prefers to get as many items of high value as possible $v_i(S) = \sum_{j \in S} v_{ij}$, with the only concern being that her total payments for the received set of items S, denoted by $p_i(S)$, should not exceed her budget constraint $p_i(S) \leq B_i$. To reconcile this discrepancy, Dobzinski and Leme [21] proposed to evaluate the welfare of buyers in budgeted settings according to their *admissibility-to-pay*; that is, the minimum between the buyer's value for the allocated bundle and the buyer's budget. The aggregate welfare according to this definition is termed the *Liquid Welfare* (LW). Hence, the Liquid Welfare objective can be seen as a natural analogue to social welfare in budgeted settings, as it simultaneously captures the health of an economic system while still modeling buyers as preferring items of high value, despite budget constraints.

For simultaneous first price auctions, we use the following natural *item-clearing mechanism* for each individual item. Each player submits a bid they are willing to pay for the whole item, along with the maximal fraction of the item they are willing to purchase. Then, in decreasing order of the bids and as long as some fraction of the item remains to be allocated, each buyer receives their requested fraction of the item (or whatever remains), and pays their bid multiplied by the fraction they received. In the context of additive values, we model players' utilities for each item as their value for the item minus their submitted bid (both of which are scaled by the fraction of the item they receive).

Our model is closely related to a prominent simultaneous item auction format with heterogeneous items, which has been extensively studied recently. In such auctions, buyers submit bids simultaneously on all items, and the allocation and prices are determined separately for each individual item, based only on the bids submitted for that item. This format is similar to auctions used in practice (e.g., eBay auctions). The standard measure for quantifying efficiency in such settings is the *Price of Anarchy* (PoA) [29,35,38], defined as the ratio of the optimal social welfare to the social welfare of the worst equilibrium. In budgeted settings, it is thus natural to quantify the efficiency of such auctions by the *Liquid Price of Anarchy* (LPoA), defined as the ratio of the optimal Liquid Welfare to the Liquid Welfare of the worst equilibrium.

New Techniques. The most common framework for analyzing the Price of Anarchy of games and auctions is the *smoothness* framework (see, e.g., [35,38]). Such techniques usually involve a thought experiment in which each player deviates toward some strategy related to the optimal solution, and hence the total utility of all players can be bounded appropriately. One important and necessary condition for applying the smoothness framework is that the objective function must dominate the sum of utilities (which holds for social welfare). However, this technique falls short in the case of Liquid Welfare, since a bidder's utility can be arbitrarily higher than their admissibility-to-pay, and in aggregate, bidders may achieve a total utility that is much larger than the Liquid Welfare at equilibrium. To overcome this issue, we develop new techniques to bound the LPoA in budgeted settings. Our techniques include a novel type of hypothetical deviation that is used to *upper bound* the aggregate utility of bidders (in addition to the traditional deviation that is used to lower bound it), and the consideration of a special set of carefully chosen bidders to engage in these hypothetical deviations (see more details in the full version of our paper [3]). To the best of our knowledge, most prior techniques, including those that depart from the smoothness framework (e.g., [25]), examine the utility derived when *every* player deviates toward the optimal solution.

With our new techniques at hand, we address the following question: *What is the Liquid Price of Anarchy of simultaneous first price item auctions in settings with budgets?*

Clarifying Remarks and Examples

Settings where agents have additive valuations and are constrained by budgets (as in our setting) should not be confused with settings with *budget additive* valuations. The latter assumes quasilinear utilities, while the former does not[1]. The class of budget additive valuations is a proper subclass of submodular valuations, a setting for which the Price of Anarchy of simultaneous combinatorial auctions is well understood, and known to be bounded by a constant[2]. However, these results do not apply to the budgeted setting, since a bidder's perspective and consequently their behavior at equilibrium is very different from the budget additive setting (see the full version of our paper [3] for an example and a more detailed discussion).

The budgeted simultaneous item bidding setting has also been studied by [38], where a different approach was taken. They measured the social welfare at equilibrium against the optimal Liquid Welfare. Note that according to their measure, the benchmark (i.e., optimal Liquid Welfare) may be lower than the measured welfare. Please see the full version of our paper [3] for an example illustrating the difference between their measure and our LPoA measure.

Our Contributions

We show that simultaneous first price item auctions achieve nearly optimal performance, i.e., a constant Liquid Price of Anarchy. Our main result concerns the case in which agent valuations are additive (i.e., agent i's value for item j is v_{ij} and the value for a set of items is the sum of the individual valuations, each of which is scaled by the corresponding fraction received).

Main Theorem: For simultaneous first price auctions with additive bidders and divisible items, the LPoA with respect to mixed Nash equilibria and Bayesian Nash equilibria is constant.

We also show that for pure Nash equilibria in simultaneous first price auctions, our results hold for more general settings.

---

[1] The difference is also pointed out in the literature on the design of truthful combinatorial auctions [20,21].

[2] In particular, there are tight PoA bounds of $\frac{e}{e-1}$ for submodular bidders, and 2 for subadditive bidders.

Theorem: For fractionally-subadditive bidders, the LPoA of pure Nash equilibria in simultaneous first price auctions is 2. Moreover, this bound is tight.

The following remarks are in order:

1. In settings without budgets, simultaneous first price item auctions for additive bidders reduce to m independent auctions (where m is the number of items). In contrast, when agents have budget constraints, the separate auctions exhibit non-trivial dependencies even under additive valuations.

2. Since fractionally-subadditive valuations are not typically defined over divisible items, we discretize the bidding space so that requested fractions of items can only be multiples of a fixed small size in our fractionally-subadditive results. This essentially induces an indivisible setting with discrete items, and hence fractionally-subadditive valuations are well-defined.

Related Work

There is a vast literature in algorithmic game theory that incorporates budgets into the design of incentive compatible mechanisms. The paper of [6] showed that, in the case of one divisible good, the adaptive clinching auction is incentive compatible under some assumptions. Moreover, the work of [37] initiated the design of incentive compatible mechanisms in the context of reverse auctions, where the payments of the auctioneer cannot exceed a hard budget constraint (follow-up works include [1, 4, 12, 15, 22]). A great deal of work focused on designing incentive compatible mechanisms that approximately maximize the auctioneer's revenue in various settings with budget-constrained bidders [9, 13, 30, 33, 34]. Some works analyzed how budgets affect markets and non-truthful mechanisms [5, 14].

Earlier work on multi-unit auctions with budgets deals with designing incentive compatible mechanisms that always produce Pareto-optimal allocations [20]. The results in this line of work are mostly negative with a notable exception of mechanisms based on Ausubel's adaptive clinching auction framework [2].

Some recent results concern the design of incentive compatible mechanisms with respect to the Liquid Welfare objective, introduced by [21]. They gave a constant approximation for the auction that sells a single divisible good to additive buyers with budgets. In a follow-up work, [32] gave an $O(1)$-approximation for bidders with general valuations in the single-item setting. The work of [23] extended the notion of a combinatorial Walrasian equilibrium (see [26]) to settings with budgets. They showed that their generalization, termed a lottery pricing equilibrium, achieves high Liquid Welfare. They also argued how to efficiently compute randomized allocations that have near-optimal Liquid Welfare for large classes of valuation functions (including subadditive valuations).

A large body of literature is concerned with simultaneous item bidding auctions. These simple auctions have been studied from a computational perspective [10, 19]. There is also extensive work addressing the Price of Anarchy of such simple auctions (see [36] for more general Price of Anarchy results). The work of [16] initiated the study of simultaneous item auctions within the Price of Anarchy framework. The authors showed that, for second price auctions, the social welfare of every Bayesian Nash equilibrium is a 2-approximation to the optimal social welfare, even for players with fractionally-subadditive valuation functions. A large amount of follow-up work [7, 8, 17, 24, 25, 28, 38] made significant progress in understanding simultaneous item auctions, but all of these works measure inefficiency only with respect to the social welfare objective.

Much less is known about the Price of Anarchy in auctions for objectives other than social welfare. In fact, we are aware of only one such work [27], which studies the revenue of simultaneous auctions with reserve prices for *single-parameter* bidders with regular distributions. This work essentially reduces the revenue maximization problem to the welfare maximization problem for virtual values in single-parameter settings and then employs smoothness analysis to bound virtual value welfare. We note that this approach fails for multi-parameter settings such as simultaneous multi-item auctions with additive valuations.

The work of [38] considered Liquid Welfare when measuring the inefficiency of equilibria. They gave various Price of Anarchy results, developed a smoothness framework for broad solution concepts such as correlated and Bayesian Nash equilibria, and explored composition properties of various mechanisms. They extended their results to the setting where players are budget-constrained, achieving similar approximation guarantees when comparing the *social welfare achieved at equilibrium* to the *optimal Liquid Welfare*. In particular, their results imply an $\frac{e}{e-1}$-approximation for simultaneous first price auctions, and a 2-approximation for all-pay auctions and simultaneous second price auctions under the no-overbidding assumption. While [38] show that the social welfare at equilibrium cannot be much worse than the optimal Liquid Welfare, one should note that the social welfare at equilibrium can be arbitrarily better than the optimal Liquid Welfare (e.g., if all budgets are small, the optimal Liquid Welfare is small). It is useful to note that, in general, the ratio between the Liquid Welfare at equilibrium and the social welfare at equilibrium can be arbitrarily bad (if all budgets are small, then the Liquid Welfare of any allocation is small, while players' values for received goods can be arbitrarily large).

The works of [11, 18] also considered the setting where players have budgets and studied the same ratio we consider, namely the *Liquid Welfare at equilibrium* to the *optimal Liquid Welfare*. In [11], they studied the proportional allocation mechanism, which concerns auctioning off one divisible item proportionally according to the bids that players submit. They showed that, assuming players have concave non-decreasing valuation functions, the Liquid Welfare at coarse-correlated equilibria and Bayesian Nash equilibria achieve at least a constant fraction of the optimal Liquid Welfare. It should be noted that, for random allocations, they measure the benchmark at equilibrium ex-ante over the randomness of the allocation, i.e., $\sum_{i=1}^{n} \min\{\mathbb{E}_{\mathbf{v}_{-i}, B_{-i}}[v_i(x_i)], B_i\}$, where v_i is player i's valuation, B_i is player i's budget, and x_i denotes the allocation player i receives. In contrast, for random allocations, we use the stronger ex-post measure of the expected Liquid Welfare at equilibrium given by $\sum_{i=1}^{n} \mathbb{E}[\min\{v_i(x_i), B_i\}]$. The work of [18] studied a similar setting, except that multiple divisible items were considered. They gave improved bounds for coarse-correlated equilibria even with multiple items and also an improved bound for Bayesian Nash equilibria with one item. In addition, they studied the polyhedral environment and showed an exact bound of 2 for pure Nash equilibria when agents have subadditive valuations.

2 Model and Preliminaries

We consider *simultaneous first price item auctions*, in which m heterogeneous items are sold to n bidders (or players) in m independent auctions. We first describe our notation in the context of indivisible items, and then describe the divisible model. A bidder's *strategy* is a bid vector $b_i \in \mathbb{R}^m_{\geq 0}$, where b_{ij} represents player i's bid for item j. We use $\mathbf{b}$ to denote the bid profile $\mathbf{b} = (b_1, \ldots, b_n)$, and we will often use the notation $\mathbf{b} = (b_i, \mathbf{b}_{-i})$ to denote the strategy profile where player i bids b_i and the remaining players bid according to $\mathbf{b}_{-i} = (b_1, \ldots, b_{i-1}, b_{i+1}, \ldots, b_n)$.

The outcome of an auction consists of an allocation rule $\mathbf{x}$ and payment rule $\mathbf{p}$. The allocation rule $\mathbf{x}$ maps bid profiles to an allocation vector for each bidder i, where $x_i(\mathbf{b}) = (x_{i1}, \ldots, x_{im})$ denotes the set of items won by player i ($x_{ij} = 1 \Leftrightarrow$ player i wins item j). In a *simultaneous first price auction*, each item is allocated to the highest bidder (breaking ties according to some rule) and the winner pays their bid. The total payment of bidder i is $p_i(\mathbf{b}) = \sum_{j \in x_i(\mathbf{b})} b_{ij}$.

Each player i has a *valuation function* v_i, which maps sets of items to $\mathbb{R}_{\geq 0}$ (v_i captures how much player i values item bundles), and a budget B_i. We assume that all valuations are normalized and monotone, i.e., $v_i(\emptyset) = 0$ and $v_i(S) \leq v_i(T)$ for any $i \in [n]$ and $S \subseteq T \subseteq [m]$. We mostly consider bidders with additive valuations, i.e., $v_i(S) = \sum_{j \in S} v_{ij}$ (where v_{ij} denotes agent i's value for item j). The *utility* $u_i(x_i(\mathbf{b}))$ of each player i is $v_i(x_i(\mathbf{b})) - p_i(\mathbf{b}) = \sum_j v_{ij} \cdot x_{ij} - p_i(\mathbf{b})$ if $p_i(\mathbf{b}) \leq B_i$; and $u_i(x_i(\mathbf{b})) = -\infty$ if $p_i(\mathbf{b}) > B_i$. Buyers select their bids strategically in order to maximize utility.

For divisible items, agent's i bid $\mathbf{b}_{ij}$ consists of two parameters for each item j: a price b_{ij} and a desired fraction δ_{ij}. For each item j, in decreasing order of b_{ij}, and as long as some fraction of item j remains, each buyer i receives a fraction of item j given by $x_{ij} = \delta_{ij}$ (or whatever remains). If the agent receives an x_{ij}-fraction of item j, then their value is given by $v_i(x_{ij}) = v_{ij} \cdot x_{ij}$ and they pay $b_{ij} \cdot x_{ij}$. We write all individual bids $\mathbf{b}_{ij}$ on item j as a vector $\mathbf{b}_{.j}$ and bids for all items as $\mathbf{b}$.

Definition 1 (Pure Nash Equilibrium). *A bid profile $\mathbf{b}$ is a Pure Nash Equilibrium if, for any player i and any deviating bid b'_i: $u_i(b_i, \mathbf{b}_{-i}) \geq u_i(b'_i, \mathbf{b}_{-i})$.*

A mixed Nash equilibrium is defined similarly, except that bidding strategies can be randomized $b_i \sim s_i$ and utility is measured in expectation over the joint bid distribution $\mathbf{s} = s_1 \times \cdots \times s_n$.

Definition 2 (Mixed Nash Equilibrium). *A bid profile $\mathbf{s}$ is a mixed Nash equilibrium if, for any player i and any deviating bid b'_i: $\mathbb{E}_{\mathbf{b} \sim \mathbf{s}}[u_i(b_i, \mathbf{b}_{-i})] \geq \mathbb{E}_{\mathbf{b}_{-i} \sim \mathbf{s}_{-i}}[u_i(b'_i, \mathbf{b}_{-i})]$.*

Note that, in general, we assume the bidding space is discretized (i.e., each player can only bid in multiples of a sufficiently small value ε). This is done to ensure that there always exists a mixed Nash equilibrium, as otherwise we do not have a finite game.

Definition 3 (Liquid Welfare). *The Liquid Welfare, denoted by* LW, *of an allocation* $\mathbf{x}$ *is given by* $\mathrm{LW}(\mathbf{x}) = \sum_{i \in [n]} \min\{v_i(x_i), B_i\}$. *For random allocations, we use the measure given by* $\mathrm{LW}(\mathbf{x}) = \sum_{i \in [n]} \mathbb{E}[\min\{v_i(x_i), B_i\}]$.

For a given vector of valuations $\mathbf{v} = (v_1, \ldots, v_n)$, we use $\mathrm{OPT}(\mathbf{v})$ to denote the Liquid Welfare of an optimal outcome given by the expression $\mathrm{OPT}(\mathbf{v}) = \max_{x_1,\ldots,x_n} \sum_i \min\{v_i(x_i), B_i\}$, where the bundles x_i form a fractional partition of $[m]$ (i.e., $\sum_i x_{ij} = 1$ for any j). We often use OPT instead of $\mathrm{OPT}(\mathbf{v})$ when the context is clear.

Definition 4 (Liquid Price of Anarchy). *Given a fixed valuation profile* $\mathbf{v}$, *the* Liquid Price of Anarchy *(LPoA) is the worst-case ratio between the optimal Liquid Welfare and the expected Liquid Welfare at an equilibrium (pure, mixed, or Bayesian Nash) and is given by* $\mathrm{LPoA}(\mathbf{v}) = \sup_{\mathbf{s}} \left\{ \frac{\mathrm{OPT}(\mathbf{v})}{\mathrm{LW}(\mathbf{s}(\mathbf{v}))} \mid \mathbf{s} \in Equilibria \right\}$.

3 Simultaneous First Price Auctions

In this section, we prove our main theorem that, for mixed Nash equilibria, the Liquid Price of Anarchy of simultaneous first price auctions is constant. In what follows we build up notation and intuition toward the proof. Recall that agents have additive valuations and submit separate bids on each item. We assume that the buyers bid according to a mixed Nash equilibrium $\mathbf{b} \sim \mathbf{s}$. For all items we can define an "expected price per item" at equilibrium or just a "price per item" as $\mathbf{p} = (p_1, \ldots, p_m)$, where $p_j = \sum_{i=1}^{n} \mathbb{E}_{\mathbf{b} \sim \mathbf{s}}[b_{ij} \cdot x_{ij}(\mathbf{b}_{\cdot j})]$.

Each bidder i has a good chance of winning a particular item if they bid above the expected price of this item. To this end we define boosted prices $\overline{\mathbf{p}} = (\overline{p}_1, \ldots, \overline{p}_m)$, where $\overline{p}_j = \alpha p_j$ for some $\alpha > 1$ ($\alpha = 2$ will be sufficient for us). One simple observation about $\overline{\mathbf{p}}$ is the following:

Observation 1. *Revenue is related to prices, namely:* $\mathrm{REV}(\mathbf{s}) = \frac{1}{\alpha} \sum_{j=1}^{m} \overline{p}_j$, *where* $\mathrm{REV}(\mathbf{s})$ *denotes the expected revenue at the equilibrium profile* $\mathbf{s}$.

We next show that if players bid on a fraction of an item j according to $\overline{p}_j$, then they win a large fraction of j in expectation. The proof is in the full version of our paper [3].

Proposition 1. *For any item* j, *if a player bids on a δ-fraction of item j at price* $\overline{p}_j$, *then the player receives in expectation at least a* $\delta \cdot \left(1 - \frac{1}{\alpha}\right)$-*fraction of item* j.

When relating prices to Liquid Welfare we notice that

Observation 2. *Revenue is bounded by the Liquid Welfare:* $\mathrm{REV}(\mathbf{s}) \leq \mathrm{LW}(\mathbf{s})$, *where* $\mathrm{LW}(\mathbf{s})$ *denotes the expected Liquid Welfare at the equilibrium profile* $\mathbf{s}$.

We consider the following Linear Program for the allocation problem with the goal of optimizing Liquid Welfare.

$$\text{Maximize} \sum_{i=1}^{n} \sum_{j=1}^{m} v_{ij} \cdot z_{ij} \qquad\qquad \text{Liquid linear program (LLP)}$$

$$\text{Subject to} \sum_{j} v_{ij} \cdot z_{ij} \leq B_i \quad \forall i; \qquad \sum_{i} z_{ij} \leq 1 \quad \forall j; \qquad z_{ij} \geq 0 \quad \forall i,j.$$

We denote by $\mathbf{y} = (y_{ij})$ the optimal solution to LLP. Notice that the solution for the Liquid Welfare never benefits from allocating a set of items to a player such that their value for the set exceeds their budget. Thus

Observation 3. *The solution to LLP is equal to the optimal allocation, namely:*
$$\sum_{i=1}^{n} \sum_{j=1}^{m} v_{ij} \cdot y_{ij} = \mathrm{OPT}.$$

We now define some notation that will be useful in order to obtain our result. We let $q_{ij} \overset{\text{def}}{=} \mathbb{E}_{\mathbf{b}\sim\mathbf{s}}[x_{ij}(\mathbf{b}_{\cdot j})]$ be the expected fraction of item j that player i receives at an equilibrium strategy $\mathbf{s}$. In addition, for each agent i, we consider a set of high value items $J_i \overset{\text{def}}{=} \{j \mid v_{ij} \geq \bar{p}_j\}$. In particular, in our deviations, we ensure that each player i only bids on items in J_i, since this guarantees that they attain nonnegative utility if they win such items. We further define $Q_i \overset{\text{def}}{=} \mathbf{Pr}_{\mathbf{b}\sim\mathbf{s}}[v_i(x_i(\mathbf{b})) \geq B_i]$ to be the probability that $v_i(x_i) \geq B_i$ at equilibrium (recall that x_i denotes the random set that player i receives in the mixed Nash equilibrium). We also define two sets of bidders, the first one is for budget feasibility reasons and the second is for bidders that often fall under their budget in equilibrium (these sets need not be disjoint). In particular, for a fixed parameter $\gamma > 1$ ($\gamma = 4$ will be sufficient for us), we define sets $\mathcal{I}_1$ and $\mathcal{I}_2$:
$$\mathcal{I}_1 \overset{\text{def}}{=} \left\{ i \ \Big| \ \gamma \sum_{j \in J_i} \bar{p}_j \cdot q_{ij} \leq B_i \right\} \text{ and } \mathcal{I}_2 \overset{\text{def}}{=} \left\{ i \ \Big| \ Q_i \leq \tfrac{1}{2\gamma} \right\}.$$

Throughout our proof, we focus on bidders in the set $\mathcal{I} \overset{\text{def}}{=} \mathcal{I}_1 \cap \mathcal{I}_2$. As mentioned, the way we achieve our main result is to consider two types of deviations for all players in $\mathcal{I}$, the first of which is a deviation towards an optimal solution, and the second of which is a γ-boosting deviation (where players essentially bid on a larger fraction of items by a factor γ relative to what they receive at equilibrium, namely $\gamma \cdot q_{ij}$). We only consider deviations for players in set $\mathcal{I}$ for the following reasons. Players in $\mathcal{I}_1$ are guaranteed to respect their budgetary constraints in our γ-boosting deviation. Players that do not belong to $\mathcal{I}_2$ have the property that the value they receive at equilibrium often exceeds their budgets, and hence such players already contribute a lot to the Liquid Welfare at equilibrium. In particular, whenever a player receives a value that exceeds their budget, their contribution to the Liquid Welfare at equilibrium is at least as much as their contribution in the optimal Liquid Welfare, which is always bounded above by their budget. Hence, we need only consider players in $\mathcal{I}_2$ as far as deviations are concerned. We define sets $\overline{\mathcal{I}}_1 \overset{\text{def}}{=} [n] \setminus \mathcal{I}_1$, $\overline{\mathcal{I}}_2 \overset{\text{def}}{=} [n] \setminus \mathcal{I}_2$, and $\overline{\mathcal{I}} \overset{\text{def}}{=} [n] \setminus \mathcal{I}$. To this end, we relate the total budget of bidders outside of the set $\mathcal{I}$ to the revenue at equilibrium $\mathbf{s}$. The proof is in the full version of our paper [3].

Proposition 2. *The total budget of players in $\overline{\mathcal{I}}$ is small:* $\sum_{i \in \overline{\mathcal{I}}} B_i < \alpha \cdot \gamma \cdot$ REV$(\mathbf{s}) + \sum_{i \in \overline{\mathcal{I}}_2} B_i$.

To achieve our result, we essentially consider two main ideas for player deviations in set $\mathcal{I}$. The first idea is to use the solution to LLP as guidance to claim that players can extract a large amount of value relative to the optimal solution. Define the first LLP deviation to be $\mathbf{b}_1 = (b_i', \mathbf{b}_{-i})$, where in b_i' buyer i bids on a y_{ij}-fraction of each item $j \in J_i$ with price $\overline{p}_j$. We note that the LLP deviation $\mathbf{b}_1$ is feasible, since $v_{ij} \geq \overline{p}_j$ for every $j \in J_i$, and $\sum_j v_{ij} \cdot y_{ij} \leq B_i$ as $\mathbf{y}$ is a solution to LLP.

Lemma 1 (LLP deviations). *Buyers in $\mathcal{I}$ at equilibrium $\mathbf{s}$ derive large value:*

$$\sum_{i \in \mathcal{I}} \sum_j v_{ij} \cdot q_{ij} \geq \left(1 - \frac{1}{\alpha}\right) \left(\text{OPT} - \alpha \left(1 + \gamma\right) \text{REV}(\mathbf{s}) - \sum_{i \in \overline{\mathcal{I}}_2} B_i\right).$$

We defer the proof of Lemma 1 to the full version of our paper [3]. We now turn to our second type of deviation, but we need to further restrict the set of items that players bid on. In particular, we let $\Gamma_i = \left\{j \;\middle|\; q_{ij} \leq \frac{1}{\gamma}\right\}$, and define $G_i = J_i \cap \Gamma_i$. The set Γ_i is defined to ensure that each player i that deviates never bids on a fraction of an item j that exceeds 1. We now define the γ-boosting deviation as $\mathbf{b}_2 = (b_i', \mathbf{b}_{-i})$, where in b_i' buyer i bids on a $\gamma \cdot q_{ij}$-fraction of each item $j \in G_i$ with price $\overline{p}_j$, where $\gamma > 1$ is a constant. Note that each $\mathbf{b}_2$ deviation for every $i \in \mathcal{I}$ is feasible since $\mathcal{I} \subseteq \mathcal{I}_1$.

Lemma 2 (γ-boosting deviation). *The value derived by buyers in $\mathcal{I}$ is comparable to the Liquid Welfare obtained at equilibrium:*

$$\left(1 - \frac{\alpha}{\gamma(\alpha - 1)}\right) \sum_{i \in \mathcal{I}} \sum_j v_{ij} \cdot q_{ij} \leq \alpha \cdot \text{REV}(\mathbf{s}) + 2 \cdot \text{LW}(\mathbf{s}) - \frac{1}{\gamma} \sum_{i \in \overline{\mathcal{I}}_2} B_i.$$

We defer the proof of Lemma 2 to the full version of our paper [3]. Now we have all necessary components to conclude the proof of our main theorem and show that the Liquid Price of Anarchy of any mixed Nash equilibrium is bounded. The proof is given in the full version of our paper [3].

Theorem 1. *For mixed Nash equilibria, the Liquid Price of Anarchy of simultaneous first price auctions is constant (at most 26).*

The above reasoning also extends to Bayesian Nash equilibria with the same LPoA bound.

Theorem 2. *For Bayesian Nash equilibria, the Liquid Price of Anarchy of simultaneous first price auctions is constant (at most 26).*

We omit the proof as it is very similar to the proof of Theorem 1. We also study pure Nash equilibria of simultaneous first price auctions. The proof of the next theorem is given in the full version of our paper [3].

Theorem 3. *Consider a simultaneous first price auction where budgeted bidders have fractionally-subadditive valuations[3]. If* **b** *is a pure Nash equilibrium, then* $\mathrm{LW}(\mathbf{b}) \geq \frac{\mathrm{OPT}}{2}$.

A complementary tightness result for Theorem 3 is given in the full version of our paper [3]. Unfortunately, this result is not quite satisfying compared to mixed Nash equilibria, as pure Nash equilibria might not even exist (see the full version of our paper [3]).

References

1. Anari, N., Goel, G., Nikzad, A.: Mechanism design for crowdsourcing: an optimal 1–1/e competitive budget-feasible mechanism for large markets. In: FOCS, pp. 266–275 (2014)
2. Ausubel, L.M.: An efficient ascending-bid auction for multiple objects. Am. Econ. Rev. **94**(5), 1452–1475 (2004)
3. Azar, Y., Feldman, M., Gravin, N., Roytman, A.: Liquid price of anarchy. CoRR abs/1511.01132 (2015)
4. Bei, X., Chen, N., Gravin, N., Lu, P.: Budget feasible mechanism design: From prior-free to bayesian. In: STOC, pp. 449–458 (2012)
5. Benoit, J.P., Krishna, V.: Multiple-object auctions with budget constrained bidders. Rev. Econ. Stud. **68**(1), 155–179 (2001)
6. Bhattacharya, S., Conitzer, V., Munagala, K., Xia, L.: Incentive compatible budget elicitation in multi-unit auctions. In: SODA, pp. 554–572 (2010)
7. Bhawalkar, K., Roughgarden, T.: Welfare guarantees for combinatorial auctions with item bidding. In: SODA, pp. 700–709 (2011)
8. Bhawalkar, K., Roughgarden, T.: Simultaneous single-item auctions. In: Goldberg, P.W. (ed.) WINE 2012. LNCS, vol. 7695, pp. 337–349. Springer, Heidelberg (2012). doi:10.1007/978-3-642-35311-6_25
9. Borgs, C., Chayes, J., Immorlica, N., Mahdian, M., Saberi, A.: Multi-unit auctions with budget-constrained bidders. In: EC, pp. 44–51 (2005)
10. Cai, Y., Papadimitriou, C.: Simultaneous Bayesian auctions and computational complexity. In: EC, pp. 895–910 (2014)
11. Caragiannis, I., Voudouris, A.A.: Welfare guarantees for proportional allocations. In: Lavi, R. (ed.) SAGT 2014. LNCS, vol. 8768, pp. 206–217. Springer, Heidelberg (2014). doi:10.1007/978-3-662-44803-8_18
12. Chan, H., Chen, J.: Truthful multi-unit procurements with budgets. In: Liu, T.-Y., Qi, Q., Ye, Y. (eds.) WINE 2014. LNCS, vol. 8877, pp. 89–105. Springer, Cham (2014). doi:10.1007/978-3-319-13129-0_7
13. Chawla, S., Malec, D., Malekian, A.: Bayesian mechanism design for budget-constrained agents. In: EC, pp. 253–262 (2011)

---

[3] Valuation v is fractionally-subadditive or equivalently XOS if there is a set of additive valuations $A = \{a_1, \ldots, a_\ell\}$ such that $v_i(S) = \max_{a \in A} a(S)$ for every $S \subseteq [m]$. XOS is a super class of submodular and additive valuations.

14. Che, Y.K., Gale, I.: Standard auctions with financially constrained bidders. Rev. Econ. Stud. **65**(1), 1–21 (1998)
15. Chen, N., Gravin, N., Lu, P.: On the approximability of budget feasible mechanisms. In: SODA, pp. 685–699 (2011)
16. Christodoulou, G., Kovács, A., Schapira, M.: Bayesian combinatorial auctions. In: Aceto, L., Damgård, I., Goldberg, L.A., Halldórsson, M.M., Ingólfsdóttir, A., Walukiewicz, I. (eds.) ICALP 2008. LNCS, vol. 5125, pp. 820–832. Springer, Heidelberg (2008). doi:10.1007/978-3-540-70575-8_67
17. Christodoulou, G., Kovács, A., Sgouritsa, A., Tang, B.: Tight bounds for the price of anarchy of simultaneous first price auctions. CoRR abs/1312.2371 (2013)
18. Christodoulou, G., Sgouritsa, A., Tang, B.: On the efficiency of the proportional allocation mechanism for divisible resources. In: Hoefer, M. (ed.) SAGT 2015. LNCS, vol. 9347, pp. 165–177. Springer, Heidelberg (2015). doi:10.1007/978-3-662-48433-3_13
19. Dobzinski, S., Fu, H., Kleinberg, R.: On the complexity of computing an equilibrium in combinatorial auctions. In: SODA, pp. 110–122 (2015)
20. Dobzinski, S., Lavi, R., Nisan, N.: Multi-unit auctions with budget limits. In: FOCS, pp. 260–269 (2008)
21. Dobzinski, S., Leme, R.P.: Efficiency guarantees in auctions with budgets. In: Esparza, J., Fraigniaud, P., Husfeldt, T., Koutsoupias, E. (eds.) ICALP 2014. LNCS, vol. 8572, pp. 392–404. Springer, Heidelberg (2014). doi:10.1007/978-3-662-43948-7_33
22. Dobzinski, S., Papadimitriou, C., Singer, Y.: Mechanisms for complement-free procurement. In: EC, pp. 273–282 (2011)
23. Dughmi, S., Eden, A., Feldman, M., Fiat, A., Leonardi, S.: Lottery pricing equilibria. In: EC, pp. 401–418 (2016)
24. Dütting, P., Henzinger, M., Starnberger, M.: Valuation compressions in VCG-based combinatorial auctions. In: Chen, Y., Immorlica, N. (eds.) WINE 2013. LNCS, vol. 8289, pp. 146–159. Springer, Heidelberg (2013). doi:10.1007/978-3-642-45046-4_13
25. Feldman, M., Fu, H., Gravin, N., Lucier, B.: Simultaneous auctions are (almost) efficient. In: STOC, pp. 201–210 (2013)
26. Feldman, M., Gravin, N., Lucier, B.: Combinatorial walrasian equilibrium. In: STOC, pp. 61–70 (2013)
27. Hartline, J., Hoy, D., Taggart, S.: Price of anarchy for auction revenue. In: EC, pp. 693–710 (2014)
28. Hassidim, A., Kaplan, H., Mansour, Y., Nisan, N.: Non-price equilibria in markets of discrete goods. In: EC, pp. 295–296 (2011)
29. Koutsoupias, E., Papadimitriou, C.: Worst-case equilibria. In: Meinel, C., Tison, S. (eds.) STACS 1999. LNCS, vol. 1563, pp. 404–413. Springer, Heidelberg (1999). doi:10.1007/3-540-49116-3_38
30. Laffont, J.J., Robert, J.: Optimal auction with financially constrained buyers. Econ. Lett. **52**(2), 181–186 (1996)
31. Lehmann, B., Lehmann, D., Nisan, N.: Combinatorial auctions with decreasing marginal utilities. Games Econ. Behav. **55**(2), 270–296 (2006)
32. Lu, P., Xiao, T.: Improved efficiency guarantees in auctions with budgets. In: EC, pp. 397–413 (2015)
33. Malakhov, A., Vohra, R.V.: Optimal auctions for asymmetrically budget constrained bidders. Rev. Econ. Des. **12**(4), 245–257 (2008)
34. Pai, M., Vohra, R.V.: Optimal auctions with financially constrained bidders. Discussion papers, Northwestern University, Center for Mathematical Studies in Economics and Management Science, August 2008

35. Roughgarden, T.: Intrinsic robustness of the price of anarchy. In: STOC, pp. 513–522 (2009)
36. Roughgarden, T., Tardos, E.: Introduction to the inefficiency of equilibria. In: Nisan, N., Roughgarden, T., Tardos, E., Vazirani, V.V. (eds.) Algorithmic Game Theory. Cambridge University Press, New York (2007)
37. Singer, Y.: Budget feasible mechanisms. In: FOCS, pp. 765–774 (2010)
38. Syrgkanis, V., Tardos, E.: Composable and efficient mechanisms. In: STOC, pp. 211–220 (2013)

Tight Welfare Guarantees for Pure Nash Equilibria of the Uniform Price Auction

Georgios Birmpas[1], Evangelos Markakis[1], Orestis Telelis[2(✉)], and Artem Tsikiridis[1]

[1] Department of Informatics, Athens University of Economics and Business, Athens, Greece
{gebirbas,markakis,tsikiridis15}@aueb.gr
[2] Department of Digital Systems, University of Piraeus, Piraeus, Greece
telelis@gmail.com

Abstract. We revisit the inefficiency of the uniform price auction, one of the standard multi-unit auction formats, for allocating multiple units of a single good. In the uniform price auction, each bidder submits a sequence of non-increasing marginal bids, one for each additional unit. The per unit price is then set to be the highest losing bid. We focus on the pure Nash equilibria of such auctions, for bidders with submodular valuation functions. Our result is a tight upper and lower bound on the inefficiency of equilibria, showing that the Price of Anarchy is bounded by 2.188. This resolves one of the open questions posed in previous works on multi-unit auctions.

1 Introduction

The Uniform Price Auction is one of the standard multi-unit auction formats, for allocating multiple units of a single good, at a uniform price per unit. Multi-unit auctions have been in use for a long time, with important applications, such as the auctions offered by the U.S. and U.K. Treasuries for selling bonds to investors. They are also being deployed in various platforms, including several online brokers [9,10]. In the literature, multi-unit auctions have been a subject of study ever since the seminal work of Vickrey [13], and some formats were conceived even earlier, by Friedman [6]. The quantification of their inefficiency at equilibrium has been the subject of recent works [4,7,8,12], via derivation of upper and lower bounds on the Price of Anarchy, in the full and incomplete information models. The outcomes of these works are quite encouraging, as they establish that the inefficiency is bounded by a small constant.

In this work we derive tight bounds for the Price of Anarchy of pure Nash equilibria of the uniform price auction. For such an auction on k units, each bidder is required to submit a sequence of non-increasing bids, one for each additional unit. Among all submitted bids the k highest win the auction and

This work has been partly supported by the University of Piraeus Research Center, and by a research program of the Athens University of Economics and Business.

V. Bilò and M. Flammini (Eds.): SAGT 2017, LNCS 10504, pp. 16–28, 2017.
DOI: 10.1007/978-3-319-66700-3_2

each bidder receives as many units of the good as the number of his winning bids. The highest losing bid is then chosen as the uniform price that each bidder pays, per unit won. The simplicity of this auction format is counterbalanced however by the fact that it does not support truthful bidding in dominant strategies, thus encouraging strategic behavior. The underlying strategic game induced by the auction is known to possess pure Nash equilibria, and a polynomial time algorithm for computing such an equilibrium is developed in [8]. However, [8] derives a tight bound for the Price of Anarchy of only a strict subset of such equilibria (in undominated strategies). For the full set of pure Nash equilibria, the results of [7] show that the Price of Anarchy is between $2 - \frac{1}{k}$ and 3.146. The source of inefficiency in the uniform price auction is partly due to a *demand reduction* effect discussed in [1], where bidders may have incentives to understate their demand, so as to receive less units at a lower price per unit, and partly due to the use of dominated strategies. These effects motivate the further study and quantification of the inefficiency for this auction format.

Contribution. We focus on the pure Nash equilibria of the uniform price auction, for bidders with submodular valuation functions. Our results are tight upper and lower bounds on the inefficiency of pure equilibria for submodular bidders, showing that the Price of Anarchy is bounded by 2.188. This resolves one of the questions left open from [7,8]. The proof of the upper bound is based on carefully analyzing the performance of equilibria with respect to bidders who receive less units than in the optimal assignment. Our lower bound is obtained by an explicit construction, which matches our upper bound as the number of units becomes large enough.

2 Model and Definitions

We consider a multi-unit auction, involving the allocation of k units of a single item, to a set $\mathcal{N}$ of n bidders, $\mathcal{N} = \{1, \ldots, n\}$. Each bidder $i \in \mathcal{N}$ has a private valuation function $v_i : \{0, 1, \ldots, k\} \mapsto \mathbb{R}^+$, defined over the quantity of units that he receives, with $v_i(0) = 0$. In this work, we assume that each function v_i is a non-decreasing submodular function.

Definition 1. *A valuation function* $f : \{0, 1, \ldots, k\} \mapsto \mathbb{R}^+$ *is called* **submodular***, if for every* $x < y$, $f(x) - f(x-1) \geqslant f(y) - f(y-1)$.

A valuation function can also be specified through a sequence of *marginal values*, corresponding to the value that each additional unit yields for the bidder. For the j-th additional unit, the bidder obtains marginal value $v_i(j) - v_i(j-1)$, which we denote by m_{ij}. Then, the function v_i can be determined by the vector $\mathbf{m}_i = (m_{i1}, \ldots, m_{ik})$. For submodular functions, $m_{i1} \geqslant \ldots \geqslant m_{ik}$, by definition. We will often use the representation of v_i by $\mathbf{m}_i$ in the sequel. We will also use the following well known properties of submodular functions:

Proposition 1. *Given* $x, y \in \{0, 1, \ldots, k\}$ *with* $x \leqslant y$, *any non-decreasing submodular function* f, *with* $f(0) = 0$, *satisfies* $y f(x) \geqslant x f(y)$. *Moreover, when* $x < y$, *for any* $j = 1, \ldots, y - x$ *the function* f *satisfies:* $(f(x + j) - f(x))/j \geqslant (f(y) - f(x))/(y - x)$.

The standard Uniform Price Auction requires that bidders submit their non-increasing marginal value for each additional unit; every bidder i is asked to declare his valuation curve, as a *bid vector* $\mathbf{b}_i = (b_{i1}, b_{i2}, \ldots, b_{ik})$, satisfying $b_{i1} \geqslant b_{i2} \geqslant \ldots \geqslant b_{ik}$. Thus, b_{ij} is the *declared* marginal bid of i, for obtaining the j-th unit of the item. Note that each b_{ij} may differ from the bidder's actual marginal value, m_{ij}. Given a *bidding profile* $\mathbf{b} = (\mathbf{b}_1, \ldots, \mathbf{b}_n)$, the auction allocates the k units to the k highest marginal bids. We denote this allocation by $\mathbf{x}(\mathbf{b}) = (x_1(\mathbf{b}), x_2(\mathbf{b}), \ldots, x_n(\mathbf{b}))$, where $x_i(\mathbf{b})$ is the number of units allocated to bidder i. Each bidder pays a uniform price $p(\mathbf{b})$ *per received unit*, which equals the highest rejected marginal bid, i.e., the $(k+1)$-th highest marginal bid. The total payment of bidder i is then $x_i(\mathbf{b}) \cdot p(\mathbf{b})$, and his utility for the allocation is: $u_i(\mathbf{b}) = v_i(x_i(\mathbf{b})) - x_i(\mathbf{b}) \cdot p(\mathbf{b})$.

The (utilitarian) *Social Welfare* achieved under a bidding profile $\mathbf{b}$ is defined as the sum of utilities of all interacting parties, inclusively of the auctioneer's revenue. This sum equals the sum of the bidders' values for their allocations:

$$SW(\mathbf{b}) = \sum_{i=1}^{n} v_i(x_i(\mathbf{b}))$$

Our goal is to derive upper and lower bounds on the Price of Anarchy (PoA) of pure Nash equilibria of the Uniform Price Auction. This is the worst-case ratio of the optimal welfare, over the welfare achieved at a pure Nash equilibrium. If $\mathbf{x}^*$ denotes an optimal allocation, then

$$PoA = \sup_{\mathbf{b}} \frac{SW(\mathbf{x}^*)}{SW(\mathbf{b})}$$

where the supremum is taken over pure equilibrium profiles.

Finally, following previous works on equilibrium analysis of auctions, e.g., [2, 3, 8], we focus on *non-overbidding* profiles $\mathbf{b}$, wherein no bidder ever outbids his value, for any number of units. That is, for any $\ell \leqslant k$, we assume $\sum_{j=1}^{\ell} b_{ij} \leqslant v_i(\ell)$. Note that, this *does not* necessarily imply $b_{ij} \leqslant m_{ij}$, except for when $j = 1$: i.e., $b_{i1} \leqslant m_{i1} = v_i(1)$. In our analysis, we refer to non-overbidding vectors, $\mathbf{b}_i$, and profiles, $\mathbf{b}$, as *feasible*.

3   Inefficiency Upper Bound

In this section we develop tight welfare guarantees for feasible (non-overbidding) pure Nash equilibrium profiles of the uniform price auction, when the bidders have submodular valuation functions. By the results of [7], it is already known that for submodular valuations on k units, $2 - \frac{1}{k} \leqslant PoA \leqslant 3.146$. We show that:

Theorem 1. *The Price of Anarchy of non-overbidding pure Nash equilibria of the Uniform Price Auction with submodular bidders is at most:*

$$\frac{2 + \mathcal{W}_0(-e^{-2})}{1 + \mathcal{W}_0(-e^{-2})} \approx 2.188$$

where $\mathcal{W}_0$ is the first branch of the Lambert W function.

The Lambert W function is the multi-valued inverse function of $f(W) = We^W$. For more on the properties of this function, see [5].

Before we continue, we introduce first some notation to be used throughout the section. Let $\mathbf{b}$ denote a feasible bidding profile. We denote the k winning marginal bids under $\mathbf{b}$ by $\beta_j(\mathbf{b})$, $j = 1, \ldots, k$, so that $\beta_j(\mathbf{b})$ is the j-th lowest winning bid under $\mathbf{b}$, thus, $\beta_1(\mathbf{b}) \leqslant \beta_2(\mathbf{b}) \leqslant \ldots \leqslant \beta_k(\mathbf{b})$. We will often apply this notation to profiles of the form $\mathbf{b}_{-i}$, for some bidder $i \in \mathcal{N}$. For a profile of valuation functions $(v_1, v_2, \ldots, v_n)$, we denote the *socially optimal* – i.e., welfare maximizing – allocation by $\mathbf{x}^* = (x_1^*, \ldots, x_n^*)$. If there are multiple such allocations, we fix one for the remainder of the analysis. We define a partition of the set of bidders, $\mathcal{N}$, with reference to $\mathbf{x}^*$ and any arbitrary allocation $\mathbf{x}$, into two subsets, $\mathcal{O}$ and $\mathcal{U}$, as follows:

$$\mathcal{N} = \mathcal{O} \cup \mathcal{U}, \quad \mathcal{O} = \{i \in \mathcal{N} : x_i \geqslant x_i^*\}, \quad \mathcal{U} = \{i \in \mathcal{N} : x_i < x_i^*\}.$$

The set $\mathcal{O}$ contains the *"overwinners"*, i.e., bidders that receive in $\mathbf{x}$ at least as many units as in $\mathbf{x}^*$. The set $\mathcal{U}$ contains respectively the *"underwinners"*. In our analysis, the allocations we refer to are determined by some profile $\mathbf{b}$, i.e., $\mathbf{x} \equiv \mathbf{x}(\mathbf{b})$. Consequently, the sets $\mathcal{O}$ and $\mathcal{U}$ will depend on $\mathbf{b}$; for simplicity, we omit this dependence from our notation. The following lemma states that, under a feasible bidding profile $\mathbf{b}$, every bidder $i \in \mathcal{O}$ retains value at least equal to a convex combination of her socially optimal value, $v_i(x_i^*)$, and of the sum of her winning bids.

Lemma 1. *Let $\mathbf{b}$ be a feasible bidding profile, and let $\mathcal{O}$ be the set of overwinners with respect to the allocation $\mathbf{x}(\mathbf{b})$. For every $\lambda \in [0,1]$, and every bidder $i \in \mathcal{O}$:*

$$v_i(x_i(\mathbf{b})) \geqslant \lambda \cdot v_i(x_i^*) + (1 - \lambda) \cdot \sum_{j=1}^{x_i(\mathbf{b})} b_{ij} \tag{1}$$

Proof. Indeed, for every $i \in \mathcal{O}$: $v_i(x_i(\mathbf{b})) = \lambda v_i(x_i(\mathbf{b})) + (1 - \lambda)v_i(x_i(\mathbf{b}))$, which is at least equal to $\lambda v_i(x_i^*) + (1 - \lambda)v_i(x_i(\mathbf{b}))$, by definition of $\mathcal{O}$. Then, (1) follows by our no-overbidding assumption on $\mathbf{b}$. □

By definition, each overwinner is capable of *"covering"* her socially optimal value. Conversely, the underwinners are the cause of social inefficiency. We will bound the total inefficiency by transforming the leftover fractions of winning bids of bidders in $\mathcal{O}$, i.e., the term $(1 - \lambda) \cdot \sum_{j=1}^{x_i(\mathbf{b})} b_{ij}$ for each bidder $i \in \mathcal{O}$ in (1), into fractions of the value attained by bidders in $\mathcal{U}$ in the optimal allocation. In this manner, we will quantify the value that the underwinners are missing (due to their strategic bidding), and determine the worst-case scenario that can arise at a pure Nash equilibrium. The following claim can be inferred from [8], and will be used to facilitate this transformation. We present the proof for completeness.

Claim 1. *Let* **b** *be any bidding profile. Then it holds that:*

$$\sum_{i \in \mathcal{U}} \sum_{j=1}^{x_i^* - x_i(\mathbf{b})} \beta_j(\mathbf{b}) \leqslant \sum_{i \in \mathcal{O}} \sum_{j=x_i^*+1}^{x_i(\mathbf{b})} b_{ij}. \tag{2}$$

Proof. For every unit missed under **b** by any bidder $i \in \mathcal{U}$ (with respect to the units won by i in the optimal allocation), there must exist some bidder $\ell \in \mathcal{O}$ that obtains this unit. If i missed $x_i^* - x_i(\mathbf{b}) > 0$ units under **b**, there are at least as many bids issued by bidders in $\mathcal{O}$ who obtained collectively these units. The sum of these bids cannot be less than the sum $\sum_{j=1}^{x_i^*-x_i(\mathbf{b})} \beta_j(\mathbf{b})$ of the $x_i^* - x_i(\mathbf{b})$ lowest winning bids in **b**. Summing over $i \in \mathcal{U}$ yields the desired inequality. $\square$

Next, we develop a characterization of upper bounds on the Price of Anarchy. To this end, let us first define the following set, $\Lambda(\mathbf{b})$, for any bidding profile **b**.

$$\Lambda(\mathbf{b}) = \left\{ \lambda \in [0,1] : v_i(x_i(\mathbf{b})) + (1 - \lambda) \sum_{j=1}^{x_i^* - x_i(\mathbf{b})} \beta_j(\mathbf{b}) \geqslant \lambda v_i(x_i^*), \forall i \in \mathcal{U} \right\} \tag{3}$$

Notice that, for every **b**, $\Lambda(\mathbf{b}) \neq \emptyset$, because $\lambda = 0 \in \Lambda(\mathbf{b})$. The following lemma helps us understand how one can obtain upper bounds on the Price of Anarchy.

Lemma 2. *If there exists* $\lambda \in [0,1]$ *such that* $\lambda \in \Lambda(\mathbf{b})$, *for every feasible pure Nash equilibrium profile* **b** *of the Uniform Price Auction, then the Price of Anarchy of feasible pure Nash equilibria is at most* λ^{-1}.

Proof. Fix a feasible pure Nash equilibrium profile **b** and consider any $\lambda \in \Lambda(\mathbf{b})$. Then, we can apply consecutively the partition $\mathcal{N} = \mathcal{O} \cup \mathcal{U}$ with respect to **b**, Lemma 1, Claim 1 and, finally, the definition of $\Lambda(\mathbf{b})$, to obtain:

$$SW(\mathbf{b}) = \sum_{i \in \mathcal{O}} v_i(x_i(\mathbf{b})) + \sum_{i \in \mathcal{U}} v_i(x_i(\mathbf{b}))$$

$$\geqslant \lambda \sum_{i \in \mathcal{O}} v_i(x_i^*) + (1 - \lambda) \sum_{i \in \mathcal{O}} \sum_{j=1}^{x_i(\mathbf{b})} b_{ij} + \sum_{i \in \mathcal{U}} v_i(x_i(\mathbf{b}))$$

$$\geqslant \lambda \sum_{i \in \mathcal{O}} v_i(x_i^*) + (1 - \lambda) \sum_{i \in \mathcal{O}} \sum_{j=x_i^*+1}^{x_i(\mathbf{b})} b_{ij} + \sum_{i \in \mathcal{U}} v_i(x_i(\mathbf{b}))$$

$$\geqslant \lambda \sum_{i \in \mathcal{O}} v_i(x_i^*) + \sum_{i \in \mathcal{U}} \left((1 - \lambda) \sum_{j=1}^{x_i^* - x_i(\mathbf{b})} \beta_j(\mathbf{b}) + v_i(x_i(\mathbf{b})) \right)$$

$$\geqslant \lambda \sum_{i \in \mathcal{O}} v_i(x_i^*) + \sum_{i \in \mathcal{U}} \lambda \cdot v_i(x_i^*) = \lambda \cdot SW(\mathbf{x}^*)$$

$\square$

Using $\lambda = 0$ with Lemma 2, yields the trivial upper bound of ∞. To obtain better upper bounds, Lemma 2 shows that we need to understand better the sets $\Lambda(\mathbf{b})$, and whether underwinners can extract at equilibrium a good fraction of their value under the optimal assignment. By the definition of these sets, the next step towards this is to derive lower bounds on every $\beta_\ell(\mathbf{b})$ for each underwinner $i \in \mathcal{U}$, and every value $\ell = 1, \ldots, x_i^* - x_i(\mathbf{b})$. The lower bound that we will use is formally expressed below.

Lemma 3. *Let* $\mathbf{b}$ *be a pure Nash equilibrium of the Uniform Price Auction and* $\mathbf{x}^*$ *be a socially optimal allocation. For every underwinning bidder* $i \in \mathcal{U}$ *under* $\mathbf{b}$ *and for every* $\ell = 1, \ldots, x_i^* - x_i(\mathbf{b})$:

$$\beta_\ell(\mathbf{b}) \geqslant \frac{1}{x_i(\mathbf{b}) + \ell} \cdot \Big(v_i(x_i(\mathbf{b}) + \ell) - v_i(x_i(\mathbf{b})) \Big) \tag{4}$$

We defer the proof of this statement, in order to explain first how – along with Lemma 2 – it leads to the proof of Theorem 1.

Proof (of Theorem 1). Using Lemma 2, we identify values of λ that belong to every $\Lambda(\mathbf{b})$. Fix any feasible pure Nash equilibrium profile $\mathbf{b}$ and, for every bidder $i \in \mathcal{U}$, let $q_i(\mathbf{b}) = x_i^* - x_i(\mathbf{b})$. To simplify the notation, we use hereafter x_i for $x_i(\mathbf{b})$, p for $p(\mathbf{b})$, q_i for $q_i(\mathbf{b})$, and β_j for $\beta_j(\mathbf{b})$, (always with respect to the Nash equilibrium $\mathbf{b}$).

Consider an arbitrary $\lambda \in [0, 1]$ and, keeping everything else fixed, define $h(\lambda) = v_i(x_i) + (1 - \lambda) \cdot \sum_{j=1}^{q_i} \beta_j$. We can now have the following implications.

$$h(\lambda) = v_i(x_i) + (1 - \lambda) \cdot \sum_{j=1}^{q_i} \beta_j$$

$$\geqslant v_i(x_i) + (1 - \lambda) \cdot \sum_{j=1}^{q_i} \frac{1}{j + x_i} \cdot \Big(v_i(x_i + j) - v_i(x_i) \Big) \tag{5}$$

$$= v_i(x_i) + (1 - \lambda) \cdot \sum_{j=1}^{q_i} \left(\frac{j}{j + x_i} \cdot \frac{v_i(x_i + j) - v_i(x_i)}{j} \right)$$

$$\geqslant v_i(x_i) + (1 - \lambda) \cdot \frac{v_i(x_i^*) - v_i(x_i)}{x_i^* - x_i} \cdot \sum_{j=1}^{q_i} \frac{j}{j + x_i}. \tag{6}$$

In the derivation above, inequality (5) follows by applying (4) from Lemma 3, for every β_j, $j = 1, \ldots, q_i$. Inequality (6) follows by application of the second statement of Proposition 1, which yields $\frac{v_i(x_i+j) - v_i(x_i)}{j} \geqslant \frac{v_i(x_i^*) - v_i(x_i)}{x_i^* - x_i}$, for any $j = 1, \ldots, q_i$.

Suppose now that under the equilibrium $\mathbf{b}$, there exists $i \in \mathcal{U}$ such that $x_i = 0$. In order for some λ to belong to $\Lambda(\mathbf{b})$, we would need to have $h(\lambda) \geqslant \lambda v_i(x_i^*)$. Using (6), for the underwinners with $x_i = 0$, and substituting $v_i(x_i) = 0$, we obtain: $h(\lambda) \geqslant (1 - \lambda)v_i(x_i^*)$. If we now impose that $(1 - \lambda)v_i(x_i^*) \geqslant \lambda v_i(x_i^*)$, we obtain $\lambda \leqslant 1/2$. Thus, any value of λ in $[0, 1/2]$ satisfies the constraint in the definition of $\Lambda(\mathbf{b})$ for bidders in $\mathcal{U}$ with $x_i = 0$. It remains to consider the more interesting case, which is for bidders in $\mathcal{U}$ with $x_i > 0$. We continue from (6) to bound $h(\lambda)$ as follows:

$$h(\lambda) \geqslant \lambda v_i(x_i) + (1 - \lambda) \cdot \left(v_i(x_i) + \frac{v_i(x_i^*) - v_i(x_i)}{x_i^* - x_i} \cdot \sum_{j=1}^{q_i} \frac{j}{j + x_i} \right)$$

$$\geqslant \lambda \cdot v_i(x_i) + (1 - \lambda) \cdot \left(\sum_{j=x_i+1}^{x_i^*} m_{ij} \right) \cdot \left(1 + \frac{x_i}{x_i^* - x_i} \cdot \left(1 - \sum_{j=1}^{q_i} \frac{1}{j + x_i} \right) \right)$$

$$\geqslant \lambda \cdot v_i(x_i) + (1 - \lambda) \cdot \left(\sum_{j=x_i+1}^{x_i^*} m_{ij} \right) \cdot \left(1 + \frac{x_i}{x_i^* - x_i} \cdot \left(1 - \int_{x_i}^{x_i^*} \frac{1}{y} dy \right) \right)$$

$$\geqslant \lambda \cdot v_i(x_i) + (1 - \lambda) \cdot \left(\sum_{j=x_i+1}^{x_i^*} m_{ij} \right) \cdot \left(1 + \frac{x_i}{x_i^* - x_i} \cdot \left(1 + \ln \frac{x_i}{x_i^*} \right) \right)$$

$$= \lambda \cdot v_i(x_i) + (1 - \lambda) \cdot \left(\sum_{j=x_i+1}^{x_i^*} m_{ij} \right) \cdot \left(1 + \frac{\frac{x_i}{x_i^*}}{1 - \frac{x_i}{x_i^*}} \cdot \left(1 + \ln \frac{x_i}{x_i^*} \right) \right)$$

The second inequality follows from the fact that $v_i(x_i(\mathbf{b})) \geqslant \frac{x_i}{x_i^* - x_i} \cdot \sum_{j=x_i+1}^{x_i^*} m_{ij}$, which is an implication of the first statement of Proposition 1. We have bounded the sum of harmonic terms by using $\sum_{k=m}^{n} f(k) \leqslant \int_{m-1}^{n} f(x)dx$, which holds for any monotonically decreasing positive function.

To continue, we minimize the function $f(y) = 1 + \frac{y}{1-y} \cdot (1 + \ln y)$ over $(0, 1)$, since x_i/x_i^* belongs to this interval.

Fact 1. *The minimum of the function* $f(y) = 1 + \frac{y}{1-y} \cdot (1 + \ln y)$ *over* $(0, 1)$, *is achieved at* $y = -\mathcal{W}_0(-e^{-2})$, *where* $\mathcal{W}_0$ *is the first branch of the Lambert W function.*

By substituting, we obtain a new lower bound on $h(\lambda)$ as follows:

$$h(\lambda) \geqslant \lambda \cdot v_i(x_i(\mathbf{b})) + (1 - \lambda) \cdot \left(\sum_{j=x_i+1}^{x_i^*} m_{ij} \right) \cdot \left(1 + \mathcal{W}_0(-e^{-2}) \right)$$

If we now set the right hand side of the above to be greater than or equal to $\lambda v_i(x_i^*)$, we can check which values of λ can belong to $\Lambda(\mathbf{b})$. In particular, we notice that by using $\lambda^* = (1 + \mathcal{W}_0(-e^{-2}))/(2 + \mathcal{W}_0(-e^{-2})) \approx 0.457$, we have that $h(\lambda^*) \geqslant \lambda^* v_i(x_i^*)$ for every bidder $i \in \mathcal{U}$ with $x_i > 0$. Since for bidders with $x_i = 0$, we found earlier that $\lambda \leqslant 1/2$ suffices, and since $\lambda^* < 1/2$, we conclude that $\lambda^* \in \Lambda(\mathbf{b})$. Hence, the theorem follows by Lemma 2. $\square$

To complete our analysis, we provide the proof of Lemma 3.

Proof (of Lemma 3). Let $\mathbf{b}$ denote a feasible pure Nash equilibrium profile and $p(\mathbf{b})$ be the uniform price under $\mathbf{b}$. Fix an underwinning bidder $i \in \mathcal{U}$. We explore whether i is able to deviate from $\mathbf{b}$ feasibly and unilaterally to obtain ℓ additional units for $\ell = 1, \ldots, x_i^* - x_i(\mathbf{b})$. Consider the following deviation $\mathbf{b}_i'$, for bidder i, and for any such ℓ.

$$\mathbf{b}_i' = \Big(\underbrace{b_{i1}, \cdots, b_{ir}}_{r \text{ bids}}, \underbrace{\beta_\ell(\mathbf{b}_{-i}) + \epsilon, \beta_\ell(\mathbf{b}_{-i}) + \epsilon, \ldots, \beta_\ell(\mathbf{b}_{-i}) + \epsilon}_{x_i(\mathbf{b}) + \ell - r \text{ bids}}, 0, 0, \ldots, 0 \Big),$$

where $0 \leqslant r \leqslant x_i(\mathbf{b})$ is the index of the last bid in $\mathbf{b}_i$, up to position $x_i(\mathbf{b})$, that is strictly greater than $\beta_\ell(\mathbf{b}_{-i}) + \epsilon$, and $\epsilon > 0$ is any *sufficiently small* positive constant. The last index of $\mathbf{b}_i'$ with a value of $\beta_\ell(\mathbf{b}_{-i}) + \epsilon$ is the $(x_i + \ell)$-th bid. All subsequent bids are set to 0. Observe that such a bidding vector (should it be feasible) would grant bidder i exactly $x_i(\mathbf{b}) + \ell$ units in total in the profile $(\mathbf{b}_i', \mathbf{b}_{-i})$: the first r bids of $\mathbf{b}_i'$ were already winning bids in $\mathbf{b}$ and each of the next $x_i(\mathbf{b}) + \ell - r$ bids exceed the ℓ-th lowest winning bid of the other bidders, $\beta_\ell(\mathbf{b}_{-i})$. Moreover, the price at $(\mathbf{b}_i', \mathbf{b}_{-i})$ would be $\beta_\ell(\mathbf{b}_{-i})$; this is now the highest losing bid (issued by some other bidder in the auction).

Note that $\mathbf{b}_i'$ may not always be a feasible deviation, since it may not obey the no-overbidding assumption. We continue by examining two cases separately.

Case 1: The bidding vector $\mathbf{b}_i'$ is a feasible deviation. Then bidder i obtains ℓ additional units by deviating. But since $\mathbf{b}$ is a pure Nash equilibrium, the utility of the bidder at $(\mathbf{b}_i', \mathbf{b}_{-i})$ cannot be higher than the utility obtained by the bidder at $\mathbf{b}$, i.e.:

$$v_i(x_i(\mathbf{b}) + \ell) - (x_i(\mathbf{b}) + \ell) \cdot \beta_\ell(\mathbf{b}_{-i}) \leqslant v_i(x_i(\mathbf{b})) - x_i(\mathbf{b}) \cdot p(\mathbf{b})$$

Thus, for a bidder that may feasibly perform such a deviation, a lower bound for β_ℓ is, for $\ell = 1, \ldots, x_i^* - x_i(\mathbf{b})$:

$$\beta_\ell(\mathbf{b}_{-i}) \geqslant \frac{1}{\ell + x_i(\mathbf{b})} \cdot \Big(v_i(x_i(\mathbf{b}) + \ell) - v_i(x_i(\mathbf{b})) + x_i(\mathbf{b}) \cdot p(\mathbf{b}) \Big)$$

By dropping the non-negative term $x_i(\mathbf{b}) \cdot p(\mathbf{b})$ and since $\beta_\ell(\mathbf{b}) \geqslant \beta_\ell(\mathbf{b}_{-i})$ for every $\ell = 1, \ldots, x_i^* - x_i(\mathbf{b})$, we obtain (4).

Before continuing to examine the second case, we identify first a useful inequality pertaining to the feasibility of $\mathbf{b}_i'$, given the initial feasible profile $\mathbf{b}$.

Claim 2. *For $\ell = 1, \ldots, x_i^* - x_i(\mathbf{b})$, the vector $\mathbf{b}_i'$ is feasible if and only if*

$$v_i(x_i(\mathbf{b}) + \ell) \geqslant \sum_{j=1}^{x_i(\mathbf{b}) + \ell} b_{ij}'$$

Proof. If $\mathbf{b}'_i$ is a feasible deviation, the inequality holds, by definition of no-overbidding. For the reverse direction, we will show that if $\mathbf{b}'_i$ is not feasible, i.e., it violates the no-overbidding assumption, then $v_i(x_i(\mathbf{b}) + \ell) < \sum_{j=1}^{x_i(\mathbf{b})+\ell} b'_{ij}$. When $\mathbf{b}'_i$ is not feasible, we know there exists an index $t \leqslant x_i(\mathbf{b}) + \ell$, such that $v_i(t) < \sum_{j=1}^{t} b'_{ij}$. Note also that $t > r$, because $b'_{ij} = b_{ij}$ for $j \leqslant r$ and $\mathbf{b}$ is a feasible bidding vector. Assume that $t < x_i(\mathbf{b}) + \ell$ since, otherwise, we are done. We can decompose the sum of bids in our inequality as:

$$v_i(t) < \sum_{j=1}^{t} b'_{ij} = \sum_{j=1}^{r} b'_{ij} + \sum_{j=r+1}^{t} b'_{ij} = \sum_{j=1}^{r} b_{ij} + (t - r)(\beta_\ell(\mathbf{b}_{-i}) + \epsilon)$$

By rearranging the terms we obtain:

$$(t - r)(\beta_\ell(\mathbf{b}_{-i}) + \epsilon) > v_i(t) - \sum_{j=1}^{r} b_{ij} = v_i(t) - v_i(r) + v_i(r) - \sum_{j=1}^{r} b_{ij}$$

$$\geqslant v_i(t) - v_i(r) = \sum_{j=r+1}^{t} m_i(j)$$

This means that there exists an index $s \in \{r + 1, \ldots, t\}$ such that $m_{is} < \beta_\ell(\mathbf{b}_{-i}) + \epsilon$. Then, by definition of $\mathbf{b}'_i$ and by the non-increasing marginal values of the submodular valuation function, we derive that $m_{ij} < b'_{ij}$, for $j = s + 1, \ldots, x_i(\mathbf{b}) + \ell$. Hence, since the no-overbidding assumption was violated at index t, it will continue to be violated if we include all the non-zero bids of $\mathbf{b}'_i$, up until the index $x_i(\mathbf{b}) + \ell$. Thus, $v_i(x_i(\mathbf{b}) + \ell) < \sum_{j=1}^{x_i(\mathbf{b})+\ell} b'_{ij}$, as claimed. □

Case 2: Suppose $\mathbf{b}'_i$ is not feasible, i.e., it involves overbidding. Then we can still infer a lower bound on $\beta_\ell(\mathbf{b})$, by exploiting Claim 2, as follows:

$$v_i(x_i(\mathbf{b}) + \ell) < \sum_{j=1}^{x_i(\mathbf{b})+\ell} b'_{ij} = \sum_{j=1}^{r} b_{ij} + (x_i(\mathbf{b}) + \ell - r) \cdot \left(\beta_\ell(\mathbf{b}_{-i}) + \epsilon\right)$$

$$\leqslant \sum_{j=1}^{x_i(\mathbf{b})} b_{ij} + (x_i(\mathbf{b}) + \ell) \cdot \left(\beta_\ell(\mathbf{b}_{-i}) + \epsilon\right)$$

$$\leqslant v_i(x_i(\mathbf{b})) + (x_i(\mathbf{b}) + \ell) \cdot \left(\beta_\ell(\mathbf{b}_{-i}) + \epsilon\right)$$

where the last inequality holds because $\mathbf{b}$ is feasible. By rearranging, we obtain:

$$\beta_\ell(\mathbf{b}_{-i}) > \frac{1}{\ell + x_i(\mathbf{b})} \cdot \left(v_i(x_i(\mathbf{b}) + \ell) - v_i(x_i(\mathbf{b}))\right) - \epsilon.$$

Observe that the above *strict* inequality holds for any sufficiently small constant $\epsilon > 0$. Since also $\beta_\ell(\mathbf{b}_{-i}) \leqslant \beta_\ell(\mathbf{b})$, inequality (4) follows. □

4   A Matching Lower Bound

We now present a lower bound, establishing that our upper bound is tight[1].

Theorem 2. *For any $k \geqslant 8$, the Price of Anarchy of the Uniform Price Auction for pure Nash equilibria and submodular bidders is at least*

$$1 + \frac{(1 - \frac{1}{k})(1 + \mathcal{W}_0(-e^{-2}))}{\frac{1}{k-1} + 1 + (-(1 - \frac{1}{k})\mathcal{W}_0(-e^{-2}) - \frac{1}{k})\ln(-\mathcal{W}_0(-e^{-2}) + \frac{1}{k})}$$

and approaches $(2 + \mathcal{W}_0(-e^{-2}))/(1 + \mathcal{W}_0(-e^{-2})) \approx 2.188$ as k grows.

Proof. We construct an instance of the auction with two bidders and $k \geqslant 8$ units. Let $x \in \{1, 2, \dots, k - 2\}$ be a parameter that we will set later on. The valuation function of bidder 1 assigns value only for the first unit and equals

$$m_{11} = \frac{k - 1 - x}{k - 1} + \sum_{i=1}^{k-1-x} \frac{i}{x + i}$$

For the remaining units, we have $m_{1j} = 0$, for any $j \geqslant 2$.

The valuation function of bidder 2 is given by the following marginal values:

$$m_{2j} = \begin{cases} 1, & j = 1, \dots, k - 1 \\ 0, & j = k \end{cases}$$

Hence, the optimal allocation is for bidder 1 to obtain 1 unit and for bidder 2 to obtain $k - 1$ units. Consider a bidding profile $\mathbf{b} = (\mathbf{b}_1, \mathbf{b}_2)$ defined as follows:

$$b_{1j} = \begin{cases} 1 - \frac{x}{k-1}, & j = 1 \\ 1 - \frac{x}{k-j+1}, & j = 2, \dots, k - x \\ 0, & \text{otherwise} \end{cases} \quad \text{and} \quad b_{2j} = \begin{cases} \epsilon, & j = 1, \dots, x \\ 0, & j > x \end{cases}.$$

Here, $\epsilon > 0$ is any small positive quantity no larger than 1.

We will see that this construction yields better lower bounds than the previously known bound of $2 - \frac{1}{k}$, when $k \geqslant 11$. For example, for $k = 11$ and $x = 2$ we obtain the following instance:

$$\mathbf{m}_1 = (5.942, 0, 0, \dots, 0, 0, 0), \quad \mathbf{b}_1 = \left(\frac{8}{10}, \frac{8}{10}, \frac{7}{9}, \frac{6}{8}, \frac{5}{7}, \dots, \frac{1}{3}, 0, 0\right)$$
$$\mathbf{m}_2 = (1, 1, \dots, 1, 1, 0), \quad \mathbf{b}_2 = (\epsilon, \epsilon, 0, 0, \dots, 0, 0)$$

It is easy to verify that this instance already yields a lower bound of 2.007, on the Price of Anarchy. Coming back now to the analysis for general k and x, we will first ensure that both bidding vectors $\mathbf{b}_1, \mathbf{b}_2$ adhere to no-overbidding. For the vector $\mathbf{b}_1$, it suffices to note that

$$\sum_{j=1}^{k-x} b_{1j} = \frac{k - 1 - x}{k - 1} + \sum_{j=2}^{k-x} \frac{k - j + 1 - x}{k - j + 1} = \frac{k - 1 - x}{k - 1} + \sum_{i=1}^{k-1-x} \frac{i}{x + i} = m_{11}$$

---

[1] We note that Theorem 2 holds for any deterministic tie-breaking rule.

where the last equality holds by changing indices and setting $i = k - j + 1 - x$. Therefore, we have that $\sum_{j=1}^{k-x} b_{1j} = v_1(k - x)$. And this directly implies that for any $\ell < k - x$, we have $\sum_{j=1}^{\ell} b_{1j} < v_1(\ell)$. It is also straightforward that for $\ell > k - x$, the no-overbidding assumption cannot be violated. Similarly, for the vector $\mathbf{b}_2$, it is easy to check that it complies to no-overbidding.

Under $\mathbf{b}$, bidder 1 obtains $k - x$ units and bidder 2 obtains x units. Notice that in this profile the uniform price is 0, as there is no contest for any unit; bidder 1 bids for exactly $k - x$ units, while bidder 2 bids for x units. All other bids are 0. We also note that $\mathbf{b}_1$ is a weakly dominated strategy.

We now argue that $\mathbf{b}$ is a pure Nash equilibrium. Bidder 1 clearly has no incentive to deviate. She is interested only in the first unit, and there is no incentive to win more units, since that would increase the price. She is also not interested in deviating to receive less units than in the current profile. Such a deviation, would either grant her zero units, and thus no utility, or would grant her at least one unit with the same utility as in $\mathbf{b}$.

Let us examine the case of bidder 2. Since bidder 2 is not interested in the last unit, we can consider only deviation vectors $\mathbf{b}_2'$ with $b_{2k}' = 0$. Note that under $\mathbf{b}$, $u_2(\mathbf{b}) = x$. Hence, bidder 2 does not have an incentive to obtain less than x units, since the price will then still remain 0, and she will only have a lower utility. It therefore suffices to consider what happens when she tries to obtain ℓ additional units, where $\ell = 1, \ldots, k - x - 1$. To do so, bidder 2 must outbid some of the winning bids of $\mathbf{b}_1$. In particular, to obtain ℓ additional units at the minimum possible price, she must outbid the bid b_{1t} of bidder 1, where t is the index $t = k - x - (\ell - 1)$. If she issues a bid $\mathbf{b}_2'$, where the first $x + \ell$ coordinates outbid b_{1t} and the remaining bids are 0, then she will obtain exactly $x + \ell$ units, and the new price (i.e., the new highest losing bid) will be precisely b_{1t}. However, any such attempt will grant bidder 2 utility equal to $u_2(\mathbf{b})$, since

$$u_2(\mathbf{b}_1, \mathbf{b}_2') = v(x + \ell) - (x + \ell) \cdot b_{1t}$$
$$= x + \ell - (x + \ell) \cdot \left(1 - \frac{x}{x + \ell}\right) = x = u_2(\mathbf{b}).$$

We conclude that the profile $\mathbf{b}$ is a pure Nash Equilibrium. The ratio of the optimal Social Welfare to the one in $\mathbf{b}$ is at least:

$$\frac{SW(\mathbf{x}^*)}{SW(\mathbf{b})} = \frac{v_1(1) + v_2(k - 1)}{v_1(k - x) + v_2(x)}$$

$$= 1 + \frac{k - 1 - x}{\frac{k-1-x}{k-1} + \sum_{i=1}^{k-1-x} \frac{i}{x+i} + x} = 1 + \frac{k - 1 - x}{\frac{k-1-x}{k-1} + k - 1 - x \sum_{i=x+1}^{k-1} \frac{1}{i}}$$

$$\geqslant 1 + \frac{k - 1 - x}{\frac{k-1-x}{k-1} + k - 1 - x \int_{x+1}^{k} \frac{1}{y} dy} \geqslant 1 + \frac{k - 1 - x}{\frac{k-1-x}{k-1} + k - 1 - x \ln \frac{k}{x+1}} \quad (7)$$

At this point we set $x = \lfloor -(k - 1)\mathcal{W}_0(-e^{-2}) \rfloor$, where $\mathcal{W}_0$ is the first branch of the Lambert W function. To continue from (7), we will need to ensure that $-\mathcal{W}_0(-e^{-2})(k-1) - 1 > 0$, which holds for $k \geqslant 8$. Substitution of x in (7) and standard algebraic manipulations yield a lower bound for (7) equal to:

$$f(k) = 1 + \frac{(1 - \frac{1}{k})(1 + \mathcal{W}_0(-e^{-2}))}{\frac{1}{k-1} + 1 + (-(1-\frac{1}{k})\mathcal{W}_0(-e^{-2}) - \frac{1}{k})\ln(-\mathcal{W}_0(-e^{-2}) + \frac{1}{k})}$$

So far, we have shown that $SW(\mathbf{x}^*)/SW(\mathbf{b}) \geqslant f(k)$. The theorem follows by observing that, as k goes to ∞:

$$\lim_{k \to \infty} f(k) = 1 + \frac{1 + \mathcal{W}_0(-e^{-2})}{1 - \mathcal{W}_0(-e^{-2}) \cdot \ln(-\mathcal{W}_0(-e^{-2}))} = \frac{2 + \mathcal{W}_0(-e^{-2})}{1 + \mathcal{W}_0(-e^{-2})},$$

where we use that $\ln(-\mathcal{W}_0(y)) = -\mathcal{W}_0(y) + \ln(-y)$, for $y \in [-e^{-1}, 0)$. This stems from the definition of the Lambert W function, particularly of $\mathcal{W}_0$ [5].   □

5 Conclusions

We presented a tight bound for the Price of Anarchy of pure Nash equilibria and for bidders with submodular valuation functions. There are still several intriguing open questions for future research in multi-unit auctions. First, it is not clear to us if our proof can be recast into the smoothness framework of [11,12]. Second, when going beyond submodular valuations to the superclass of subadditive functions, the bounds are not tight. It still remains elusive to produce lower bounds tailored for subadditive functions, and the best known upper bound is 4, due to [7]. Finally, a major open problem is to tighten the known gaps for the set of Bayes-Nash equilibria.

References

1. Ausubel, L., Cramton, P.: Demand reduction and inefficiency in multi-unit auctions. Techincal report, University of Maryland (2002)
2. Bhawalkar, K., Roughgarden, T.: Welfare guarantees for combinatorial auctions with item bidding. In: Proceedings of the Twenty-Second Annual ACM-SIAM Symposium on Discrete Algorithms (SODA), pp. 700–709 (2011)
3. Christodoulou, G., Kovács, A., Schapira, M.: Bayesian combinatorial auctions. J. ACM 63(2), 11:1–11:19 (2016). (Preliminary version appeared at ICALP(1) 2008:820–832)
4. Christodoulou, G., Kovács, A., Sgouritsa, A., Tang, B.: Tight bounds for the price of anarchy of simultaneous first-price auctions. ACM Trans. Econ. Comput. 4(2), 9:1–9:33 (2016)
5. Corless, R.M., Gonnet, G.H., Hare, D.E.G., Jeffrey, D.J., Knuth, D.E.: On the Lambert W function. Adv. Comput. Math. 5, 329–359 (1996)
6. Friedman, M.: A Program for Monetary Stability. Fordham University Press, New York (1960)
7. de Keijzer, B., Markakis, E., Schäfer, G., Telelis, O.: Inefficiency of standard multi-unit auctions. In: Bodlaender, H.L., Italiano, G.F. (eds.) ESA 2013. LNCS, vol. 8125, pp. 385–396. Springer, Heidelberg (2013). doi:10.1007/978-3-642-40450-4_33. (Extended version available at arXiv:1303.1646)

8. Markakis, E., Telelis, O.: Uniform price auctions: equilibria and effciency. Theor. Comput. Syst. **57**(3), 549–575 (2015). (Preliminary version appeared at SAGT2012:227–238)
9. Milgrom, P.: Putting Auction Theory to Work. Cambridge University Press, Cambridge (2004)
10. Ockenfels, A., Reiley, D.H., Sadrieh, A.: Economics and Information Systems. Handbooks in Information Systems, vol. 1, chap. 12, pp. 571–628. Online Auctions, Emerald Group Publishing Limited (2006)
11. Roughgarden, T.: Intrinsic robustness of the price of anarchy. J. ACM **62**(5), 32:1–32:42 (2015). (Preliminary version appeared at ACM STOC 2009:513–522)
12. Syrgkanis, V., Tardos, E.: Composable and efficient mechanisms. In: Proceedings of the 45th ACM Symposium on the Theory of Computing (STOC 2013), pp. 211–220 (2013). (Extended version available at arXiv:1213.1325)
13. Vickrey, W.: Counterspeculation, auctions, and competitive sealed tenders. J. Finan. **16**(1), 8–37 (1961)

Online Random Sampling for Budgeted Settings

Alon Eden[1(✉)], Michal Feldman[1,2], and Adi Vardi[1]

[1] Computer Science Department, Tel Aviv University, Tel Aviv-Yafo, Israel
alonarden@gmail.com, michal.feldman@cs.tau.ac.il, adi.vardi@gmail.com
[2] Microsoft Research Israel, Herzliya, Israel

Abstract. We study online multi-unit auctions in which each agent's private type consists of the agent's arrival and departure times, valuation function and budget. Similarly to secretary settings, the different attributes of the agents' types are determined by an adversary, but the arrival process is random. We establish a general framework for devising truthful random sampling mechanisms for online multi-unit settings with budgeted agents. We demonstrate the applicability of our framework by applying it to different objective functions (revenue and liquid welfare), and a range of assumptions about the agents' valuations (additive or general) and the items' nature (divisible or indivisible). Our main result is the design of mechanisms for additive bidders with budget constraints that extract a constant fraction of the optimal revenue, for divisible and indivisible items (under a standard large market assumption). We also show a mechanism that extracts a constant fraction of the optimal liquid welfare for general valuations over divisible items.

1 Introduction

In a typical setting of sales of online ad slots, advertisers arrive at different times, each with her own preferences. The auctioneer (e.g., cnn.com) decides about the allocation of the ad slots to the advertisers and how much to charge them. Scenario of this type have inspired the study of mechanism design in online settings, and random sampling has been proposed as a useful approach for the design of truthful mechanisms in online settings [19].

The random sampling framework was first introduced in the context of selling identical items in offline settings [18]. The basic idea of random sampling is to divide the agents into two sets of roughly equal size. Then, sell half of the items to each set at a price calculated according to the counterpart set. Random sampling mechanisms differ from each other in the allocation and pricing functions they apply, but they all operate according to the principle described above. The use of random sampling became widespread due to its desired properties: It is simple, trivially truthful and achieves good guarantees for a wide variety of settings. Our

This work was partially supported by the European Research Council under the European Unions Seventh Framework Programme (FP7/2007-2013)/ERC grant agreement number 337122.

V. Bilò and M. Flammini (Eds.): SAGT 2017, LNCS 10504, pp. 29–40, 2017.
DOI: 10.1007/978-3-319-66700-3_3

goal in this paper is to generalize the random sampling approach to settings in which agents arrive in an online fashion and have budgets.

Random sampling mechanisms have been previously proposed for online settings without budgets and for offline settings with budgets. Hajiaghayi et al. [19] applied the random sampling approach to online settings without budgets, and devised truthful mechanisms that approximate the auctioneer's revenue and the social welfare up to constant factors. While this was a major progress in the applicability of random sampling, their techniques are restricted to unit-demand valuations and quasi-linear utilities.

Borgs et al. [7] applied the random sampling approach to offline settings with budgets. Budget constraints impose major challenges on auction design since utilities are no longer quasi-linear. For example, the seminal VCG mechanism cannot work in non quasi-linear settings. In fact, it was shown that no truthful mechanism can give a non-trivial approximation to the social welfare in such settings [13].[1] Borgs et al. [7] were able to overcome these challenges in offline settings, and designed a truthful mechanism that gives constant approximation to revenue in offline settings with budgets (under a standard large market assumption).

The scenarios described above address either the budget constraints or the online nature of arrival. It might seem that we already have all the necessary ingredients to address the combination of both. Not surprisingly, however, it is not at all clear how to combine the existing techniques for our setting. First, some of the techniques above are tailored to unit-demand valuations, whereas we are interested in mechanisms for additive (or even general) valuations. Like in other problems in the literature on revenue approximation, the transition from unit-demand valuations to additive valuations requires entirely new techniques (e.g., pricing mechanisms for revenue maximization, [4,8,10]). Second, the combination of online arrival and budgets imposes new challenges that require new ideas. In what follows, we present our model followed by our results and techniques.

1.1   Model

We consider a setting with a set $N = \{1, \ldots, n\}$ of n agents, who arrive in an online fashion. Each agent i has a private type, represented by the following parameters: (i) a_i and d_i: The respective arrival and departure times of agent i (clearly, $d_i \geq a_i$). The interval $[a_i, d_i]$ is referred to as agent i's *time frame*. (ii) $t_i = (v_i, b_i)$: The utility type of the agent, which contains the agent's valuation function $v_i : \mathbb{R}^{\geq 0} \to \mathbb{R}^{\geq 0}$ (which maps a given amount of items to a non-negative value) and the agent's budget, b_i. There are m identical items (either divisible or indivisible). An allocation function determines an allocation x_i for agent i. The utility of agent i for obtaining an allocation x_i within her *real* time frame for a payment of p_i is:

$$u_i(x_i, p_i) = \begin{cases} v_i(x_i) - p_i & p_i \leq b_i \\ -\infty & p_i > b_i \end{cases}. \tag{1}$$

---

[1] This impossibility holds even if budgets are public.

We consider an online auction with a secretary "flavor." An adversary states a vector of time frames $([a_1, d_1], [a_2, d_2], \ldots, [a_n, d_n])$ such that $a_i < a_j$ for every $i < j$, and a vector of utility types $(t_1, t_2, \ldots, t_n)$. A random permutation is used to match time frames with utility types. That is, a permutation $\pi : N \mapsto N$ is sampled uniformly at random and agent i's type is given by the tuple $(a_i, d_i, t_{\pi(i)})$. As in [19][2], we assume that arrival times are distinct, but the results also extend to non-distinct arrival times if agents cannot bid before their real arrival times[3].

Agents report their type upon arrival, and can manipulate any component of it. In particular, they can report earlier or later arrival and departure times, and arbitrary utility types. We consider mechanisms that satisfy the following properties: (i) Feasibility: The mechanism does not sell more items than are available. (ii) Ex-ante Individual Rationality: An agent's expected utility from an allocation and payment is non-negative. (iii) Incentive Compatibility: An agent's expected utility is maximized when she reports her true type.

We consider two natural objective for budgeted settings: revenue and liquid welfare. The revenue obtained by the mechanism is simply the sum of payments of the agents to the mechanism.

The liquid welfare objective is defined as follows. Since the items are identical and divisible, it is without loss of generality to consider a single divisible item. Each agent has a non-decreasing valuation function $v_i : [0, 1] \to \mathbb{R}^+$ and a budget b_i. The *liquid welfare* of an agent i from allocation x_i is defined as the minimum between her valuation for x_i and her budget, and is denoted by $\overline{v}_i(x_i) = \min(v_i(x_i), b_i)$. The liquid welfare of an allocation vector x is $\overline{W}(x) = \sum_i \overline{v}_i(x_i)$. Let $OPT = \max_x \overline{W}(x)$ be the optimal liquid welfare. In the full version we design a mechanism for the allocation of divisible items to agents with hard budget constraints and monotonically non-decreasing valuation functions with the goal of approximating the optimal liquid welfare.

1.2   Previous Techniques and Their Limitations

Hajiaghayi et al. [19] suggested the following scheme for online settings with unit-demand valuations (and no budget constraints): Set the first (roughly) half of the arriving agents as the sampling set, and the remaining ones as the performance set. All the revenue guarantees are obtained from the agents in the performance set; the agents in the sampling set are used merely for learning the valuations in the market. Induce the agents in the sampling set to reveal their true type by applying the VCG mechanism[4], and use this information to extract revenue

---

[2] Based on personal communication with the authors, this is essentially what is assumed for the correctness of Mechanism RM_k in Sect. 6 in [19].

[3] A description of the tie-breaking rule for this case appears in the full version.

[4] While VCG is defined and analyzed for offline settings, it is shown in [19] that it can also be applied in online settings by invoking it at the time where last agent arrives, serving only the agents that haven't departed yet. While this method does not give any revenue guarantees, it is only used to extract truthful information from agents in the sampling set.

from the agents in the performance set. The proposed mechanism is truthful and obtains a constant factor approximation to revenue for unit-demand agents. Borgs et al. [7] devised allocation and payment functions that, when applied to a random sampling scheme, give good approximation to the revenue obtained in offline settings with budgets.

A natural approach for devising mechanisms for our setting is to combine the techniques of Hajiaghayi et al. [19] and Borgs et al. [7]. In particular, use the scheme of the former (to address the online nature of arrival) with the allocation and payments of the latter (to handle budgets). However, there are several obstacles to such an approach. First, the scheme proposed by [19] is restricted to agents with unit-demand valuations. In particular, agents cannot benefit from reporting a later arrival time (thereby being considered in the performance set) since the scheme ensures that the price in the sampling set never exceeds the price in the performance set. This is indeed sufficient for unit-demand agents, but does not extend beyond this class of valuations (in particular, to additive valuations). Second, while VCG can be used to encourage truthfulness in settings without budgets, it is well known that VCG is limited to quasi-linear settings, thus cannot work for settings with budgets.

1.3 Our Results and Techniques

We establish a general framework for devising truthful random sampling mechanisms for online multi-unit settings with budgeted agents. We demonstrate the applicability of our framework by applying it to different objective functions (revenue and liquid welfare), and a range of assumptions about the agents' valuations (additive or general) and the items' nature (divisible or indivisible).

Our framework splits the agents into sampling and performance sets of roughly the same size, based on the arrival time. In order to induce agents in the sampling set to report their type truthfully, we invoke random sampling at the time where the last agent in the sampling set arrives (recall VCG cannot be used due to budget constraints). In order to ensure that an agent cannot gain by delaying her arrival time, we use an *impersonation technique*. Namely, we treat an agent in the performance set *as if* she were in the sampling set. In particular, an agent in the performance set receives at most the allocation she would have received had she been in the sampling set, and at the same price. The impersonation techniques guarantees truthfulness; it remains to design the allocation and payment functions so that they do not lead to a high loss in revenue. We devise such functions for two cases of interest. Our main result is the following:

Theorem: We devise truthful mechanisms for the allocation of divisible and indivisible items among budgeted agents with additive valuations in online settings. These mechanisms give a constant approximation to the optimal revenue, under a large market assumption.

The proposed mechanism combines our general framework with the allocation and payment functions of [7]. We first show that the impersonation technique does not lead to a high loss in revenue. In particular, we have to prove that when the allocation of an agent from the performance set is limited by the allocation she would have gotten in the sampling set, the approximation guarantee is not significantly hurt. An additional challenge that arises in the case of indivisible items is the need to sell items in their entirety. The use of lotteries is limited since the realization must be carried out in an online manner. To address this challenge, we provide an online rounding procedure which rounds the allocation in a way that loses only a constant factor in the approximation. As in [7], we apply a large market assumption. Without this assumption, no truthful mechanism can achieve a constant approximation to the optimal revenue. In addition to the setting above, we demonstrate the applicability of our framework in a different setting with budgets:

Theorem: We devise a truthful mechanism for the allocation of divisible items among budgeted agents with general (monotone) valuation functions in online settings. This mechanism gives a constant approximation to the optimal liquid welfare.

This setting imposes new challenges, since the valuations are general and no large market assumption is invoked. While we apply our framework to settings with budgets, it is general enough to be applied to other non quasi-linear settings.

1.4 Additional Related Work

Online auctions have been the subject of a vast body of work. [22] introduced an online model in which agents have decreasing marginal valuations over identical indivisible items. While in there model the agent's value is private, her arrival time is public. A wide variety of additional online auction settings, such as digital goods [5,6] and combinatorial auctions [2], have been studied under the assumption of private values and public arrival times.

Similarly to our model and the model of [19], online settings in which agents arrive in a random order were also considered in [3,21]. Friedman and Parkes [16] considered the case where an agent's arrival time is also part of her private information, and thus can also be manipulated. Additional auction settings were studied under the assumption that an agent's arrival and departure times are private [20,23,26].

Offline mechanisms for budgeted agents have been considered, both for revenue maximization and welfare maximization. In the context of revenue maximization, Abrams [1] also considered the model of Borgs et al. [7] where indivisible and identical items are sold to additive agents with hard budget constrains. Abrams establishes the relation between the revenue achieved by an optimal uniform-price mechanism and an optimal mechanism with heterogeneous prices. Maximizing revenue was also considered in Bayesian settings [9] and with the goal of approximating the optimal envy free revenue [11,14].

As stated above, in general budgeted settings it is impossible to approximate the optimal social welfare using a truthful mechanism. Therefore, previous works that seeks to maximizing efficiency ensure that the outcome is Pareto Optimal [12,15,17]. The most related works to our model are [7,13,19,24], which are discussed in great detail throughout the paper.

2   Online Implementation of Random Sampling Based Mechanisms

In this section we present a general framework for adjusting offline random sampling based mechanisms to online settings. Proofs of this section are deferred to the full version. Our template mechanism **Online-RS** is specified in Fig. 2. It receives as input a set of agents N, number of items m, an **offline** pricing function $\mathcal{P}$, and an **offline** allocation function *Allocation*.

The price function $\mathcal{P}$ receives a set of agents S and computes a per-item price. The allocation function *Allocation* receives a set of agents S, a per-item price p, total number of items k, and a cap per agent ℓ (where the cap per agent limits the number of items an individual agent can get). It outputs an allocation vector, where $Allocation_i$ denotes the number of items allocated to agent i.

Before we give a formal definition of Mechanism **Online-RS**, we provide a non-formal description.

Step (a) [splitting]: We split the agents into two sets A and B, such that roughly the first $n/2$ arriving agents are placed in set A and the rest in set B (see Fig. 1a).

Step (b) [sampling set]: We apply an (offline) random sampling mechanism on set A. In particular, once the last agent of set A arrives, we divide the agents into two sets, A_1 and A_2, uniformly at random. We set a per-item price p_1 by applying $\mathcal{P}$ on set A_1 and sell $m/4$ items to agents in A_2. We apply an analogous procedure to agents in set A_2 (see Fig. 1b).

Step (c) [performance set]: Upon the arrival of an agent in set B, she is placed in one of sets B_1 or B_2 uniformly at random. An agent i in B_1 is treated as if she were in A_1, with the additional limitation determined by the actual number of remaining items. Therefore, the price is calculated based on A_2. An agent in B_2 is treated analogously, with allocation and payments according to sets A_1 and A_2, respectively (see Fig. 1c).

The repositioning of an agent from set B in set A pushes the last agent in A toward B, thereby altering the set A. This issue is addressed by a careful specification of the sets by which the allocation and price are determined to agents from B. Let j denote the last agent to arrive to A (as defined in **Online-RS**) and let $N' = N - \{j\}$. The following observation is useful in our analysis.

Observation 1. *Sets A_1, $A_2 - j$, B_1 or B_2 form a uniform partition of the agents in N'.*

Recall that while a_i and d_i are the real arrival and departure time of agent i, an agent may misreport its type. We denote the reported arrival and departure time of agent i by $\hat{a}_i$, $\hat{d}_i$, respectively. A formal definition of Mechanism **Online-RS** is provided in Fig. 2.

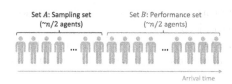

(a) Agents are split into sets A and B.

(b) Upon arrival of the last agent in A, random sampling with $m/2$ items is applied on set A.

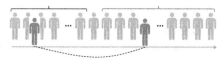

(c) $m/2$ items are sold to agents in B, upon arrival, where each agent in B is treated as if she were in set A.

Fig. 1. The different steps of **Online-RS**.

Online-RS $(N, m, \mathcal{P}, Allocation)$

Partition:

1. Rename agents according to a permutation $\pi : [n] \to [n]$ chosen uniformly at random.
2. $j \leftarrow Bin(n - 1, 1/2) + 1$, $A \leftarrow$ first j arriving agents, $B \leftarrow N \setminus A$, $t_0 \leftarrow \hat{a}_j$.

Sampling phase: (Set A)

1. At time t_0: Partition the agents in A into sets A_1, A_2, uniformly at random. (without loss of generality, $j \in A_2$).
2. $x^1 \leftarrow Allocation(A_1, \mathcal{P}(A_2), \frac{m}{4}, \infty)$, $x^2 \leftarrow Allocation(A_2, \mathcal{P}(A_1), \frac{m}{4}, \infty)$.
3. For every agent $i \in A_1$: if $\hat{d}_i \geq t_0$, agent i is allocated x_i^1 items and pays $\mathcal{P}(A_2)$ per item. Apply the analogous rules for agents in A_2.

Performance phase: (Set B)
$\alpha_1, \alpha_2 \leftarrow \frac{m}{4}$. Upon arrival of agent $i \in B$:

1. with probability $1/2$, $i \in B_1$, agent i is allocated
   $x_i \leftarrow Allocation_i \left(\{A_1 \cup i\}, \mathcal{P}(A_2 - j), \frac{m}{4}, \alpha_1\right)$, and pays $\mathcal{P}(A_2 - j)$ per item.
   $\alpha_1 \leftarrow \alpha_1 - x_i$.
2. with probability $1/2$, $i \in B_2$, agent i is allocated
   $x_i \leftarrow Allocation_i \left(\{A_2 \cup i\} - j, \mathcal{P}(A_1), \frac{m}{4}, \alpha_2\right)$, and pays $\mathcal{P}(A_1)$ per item.
   $\alpha_2 \leftarrow \alpha_2 - x_i$.

Fig. 2. A template online random sampling mechanism.

2.1 Truthfulness and Feasibility

An allocation is said to be *feasible* if the total number of allocated items does not exceed the limitation and the number of items allocated to every individual agent does not exceed the cap per agent. Let's consider the offline mechanism **M**, that determines allocation based on *Allocation* and payment based on price per item $\mathcal{P}(\cdot)$.

The main result in this section is that mechanism **Online-RS** is truthful and feasible as long as mechanism **M** satisfies incentive compatibility (IC), individual rationality (IR), feasibility and cap monotonicity: The expected utility of an agent is non-decreasing in the cap per agent.

Theorem 1. *Mechanism **Online-RS** is truthful and feasible if: (i) Mechanism **M** is IC, IR, feasible and cap monotone. (ii) Agent's valuations are monotonically non-decreasing.*

3 Revenue Maximization for Additive Agents with Budgets

In this section we study a mechanism for the allocation of *identical* and *divisible* items for agents with hard budget constraints and additive valuation functions. The utility of agent i for obtaining an allocation x_i within her time frame for a payment of p_i is:

$$u_i(x_i, p_i) = \begin{cases} v_i \cdot x_i - p_i & p_i \leq b_i \\ -\infty & p_i > b_i \end{cases}. \tag{2}$$

Our goal in this section is to devise a truthful mechanism that approximates the optimal revenue. Some of the proofs of this section are omitted and appear in the full version.

In the full version we extend our results and analysis to the case of indivisible items. The design of mechanisms for the indivisible case is similar, but it requires new ideas for transforming fractional allocations into integral ones. To this end, we devise an online rounding procedure, which leads to a constant loss in the approximation ratio.

Let S be a set of agents and k a number of items. We define $P_k(S)$ to be the price that maximizes the revenue of selling at most k items to agents in S at a uniform price: $P_k(S) = \operatorname{argmax}_p \min \left(\sum_{i \in S, v_i \geq p} \frac{b_i}{p}, k \right) \cdot p$. Let $OPT(S, k)$ be the optimal revenue from selling at most k items to agent in S at a uniform price: $OPT(S, k) = \min \left(\sum_{i \in S, v_i \geq P_k(S)} \frac{b_i}{p}, k \right) \cdot P_k(S)$. Let $OPT^*(S, k)$ be the optimal revenue from selling at most k items to agents in S at non-uniform prices (*i.e.*, where the seller can discriminate between the agents).

The following theorem implies that it is sufficient to bound the performance of our mechanism with respect to $OPT^*(S, k)$.

Theorem 2 [1]. *For every set of agents S and k items, $OPT(S, k) \geq \frac{OPT^*(S,k)}{2}$.*

3.1 The Mechanism

Our mechanism, **Rev-RS** is an implementation of mechanism **Online-RS** with the following allocation and payment functions. The allocation function, **Rev-Allocation** (see Fig. 3) approaches the agents sequentially, allocates the items in a greedy manner, and verifies that none of the agents exceeds the cap limitation. The payment function is $P_{m/4}(\cdot)$.

Rev-Allocation (S, p, k, ℓ)

1. Initialize $k' \leftarrow k$, $x_i \leftarrow 0$ for every $i \in S$.
2. For each $i \in S$ approached in the order of the random renaming (see first line in **Online-RS**):
   - If $v_i \geq p$:
     $$x_i \leftarrow \min\{b_i/p, k'\}$$
     $$k' \leftarrow k' - x_i$$
3. For every i, if $x_i > \ell$, set $x_i \leftarrow \ell$.
4. Return x.

Fig. 3. A procedure for the allocation of divisible items.

Recall that without a large market assumption, no truthful mechanism can achieve a constant approximation to the revenue [7]. Adopting the parameter used in Borgs et al. [7], we define $\epsilon(S, k) = \frac{\max\{b_i\}_{i \in S}}{OPT(S, k)}$. Intuitively, a smaller ϵ implies a larger market. The main result of this section is:

Theorem 3. Rev-RS *is a truthful mechanism that allocates divisible items and gives an 8-approximation to the optimal revenue as ϵ tends to 0.*

3.2 A Constant Loss Due to the Impersonation Technique

Our mechanism uses an impersonation technique. Specifically, it calculates the allocation for each agent in set B_1 as if she were in a random permutation in set A_1, and for each agent in B_2 as if she were in $A_2 - j$. According to Observation 1, the sets A_1 and B_1 are equally distributed. Therefore, one can view the impersonation technique as follows. We are given a set $N_1 = A_1 \cup B_1$. Then, the sets are partitioned uniformly at random into sets A_1 and B_1, and each agent in B_1 gets the same price and allocation as if she were placed in set A_1. The same holds for $N_2 = (A_2 - j) \cup B_2$. With this interpretation in mind, we prove that we lose at most a factor 2 in the revenue due to the impersonation technique.

Let **Offline-Sale**(S, p, k) be a mechanism that allocates k items to agents in set S according to the allocation returned by **Rev-Allocation**(S, p, k, ∞), and charges a price p per item. **Impersonation-Sale**(S, p, k) (see Fig. 4) is a mechanism that sells according to the impersonation technique described above.

I.e., it divides S into sets S_1 and S_2 uniformly at random. It then sells items to agents in set S_1 in a random order at a fixed price p per item. The allocation for each agent is calculated as if the agent was in a random permutation in set S_2.

Impersonation-Sale (S, p, k)

1. Initialize $\ell \leftarrow k$.
2. For each agent $i \in S$: With probability of $\frac{1}{2}$ place i in set S_1 and with probability of $\frac{1}{2}$ in set S_2.
3. Let $\pi_{S_1} : \{1, 2, \dots, |S_1|\} \rightarrow S_1$ be a random permutation of the agents of S_1.
4. For $i = 1, 2, \dots, |S_1|$:
   (a) Let $j \leftarrow \pi_{S_1}(i)$, and let x be the allocation returned by **Rev-Allocation** $(S_2 \cup \{j\}, p, k, \ell)$.
   (b) Allocate to agent j x_j items at a price of p per item.
   (c) Update $\ell \leftarrow \ell - x_j$.

Fig. 4. A mechanism defined for the analysis of the impersonation technique.

Observation 2. *Let T and S be two sets of agents such that $T \subseteq S$ and let k be a number of items. For every agent $i \in T$, $\mathrm{E}[\textbf{Rev-Allocation}_i(T, p, k, \infty)] \geq \mathrm{E}[\textbf{Rev-Allocation}_i(S, p, k, \infty)]$.*

Lemma 1. *For every set S, price p and a fraction of items k, the expected revenue from mechanism **Impersonation-Sale**(S, p, k) is at least half of the expected revenue from mechanism **Offline-Sale**(S, p, k).*

Proof. Let r be the expected revenue from mechanism **Offline-Sale**(S, p, k). For the sake of the analysis, assume that **Offline-Sale** begins by placing each agent $i \in S$ in S_1 and S_2 uniformly at random. Since each agent has a probability of $\frac{1}{2}$ to be placed in S_1, the expected revenue extracted from agents in S_1 is $\frac{r}{2}$. Therefore, it is sufficient to show that for each partition of S into S_1 and S_2, the expected revenue **Impersonation-Sale**(S, p, k) extracts from agents in S_1 is at least the expected revenue **Offline-Sale**(S, p, k) extracts from agents in S_1. There are two cases. If **Impersonation-Sale**(S, p, k) allocates the entire fraction of the item, then it collects the maximal revenue possible from price p from agents in S_1 and the claim is obviously true.

Otherwise, for each agent $i \in S_1$, $x_i < \ell$, where x_i is the allocation agent i receives in **Impersonation-Sale**. Therefore, the allocation of every agent $i \in S_1$ is not bounded by ℓ in **Rev-Allocation**$(S_2 \cup \{i\}, p, k, \ell)$, *i.e.*, the agent receives the same allocation she would receive in **Rev-Allocation**$(S_2 \cup \{i\}, p, k, \infty)$. Let $\bar{x}$ be the allocation returned by **Rev-Allocation**(S, p, k, ∞). According to Observation 2, $\mathrm{E}[x_i] \geq \mathrm{E}[\bar{x}_i]$. Since the revenue extracted out of agent i in **Impersonation-Sale** is $x_i \cdot p$ while the revenue extracted out of agent i in **Offline-Sale** is $\bar{x}_i \cdot p$, we get the desired result. ∎

Therefore, it is sufficient to bound the performance of **Offline-Sale**(S, p, k).

3.3   Analysis of Our Mechanism

Theorem 4. *Mechanism* **Rev-RS** *is a truthful and feasible mechanism.*

In order to prove that our mechanism approximates the optimal revenue, we relate the revenue of our online mechanism to an offline sale **Offline-Sale**(S, p, k) using Lemma 1. Then, we use Theorems from Borgs et al. [7] in order to prove the offline sale performs well. Specifically, they are used to prove that the random partitions of agents to sets, as done by our mechanism, do not incur a high loss for the revenue guarantees of our mechanism. Due to lack of space, the description of the specific theorems we use from [7] appears in the full version.

Recall that $N' = N - j$ and let $OPT = OPT(N, m)$, $OPT' = OPT(N', m)$.

Lemma 2. $\Pr\left[OPT' \geq OPT/2\right] \geq 1 - 1/n$.

The following is the main technical lemma of this section, and its proof appears in the full version.

Lemma 3. *For every* $\delta \in [0, \frac{1}{3}]$, *the probability that the expected revenue obtained from mechanism* **Rev-RS**(N, m) *is greater than* $\frac{(1-3\delta)OPT'}{8}$ *is at least* $\left(1 - 2e^{-1/16\epsilon'}\right)\left(1 - 4e^{-\delta^2/(16\epsilon')}\right)$.

Let $OPT^* = OPT^*(N, m)$, $\epsilon = \epsilon(N, m)$ we now use Theorem 2 and Lemma 2, to bound the expected revenue from mechanism **Rev-RS** with respect to OPT^*.

Theorem 5. *For every* $\delta \in [0, \frac{1}{3}]$, *the probability that the expected revenue obtained from mechanism* **Rev-RS**(N, m) *is greater than* $\frac{(1-3\delta)OPT^*}{32}$ *is at least* $(1 - 1/n)\left(1 - 2e^{-1/32\epsilon}\right)\left(1 - 4e^{-\delta^2/(32\epsilon)}\right)$.

The proof of Theorem 3 now follows directly:

Proof of Theorem 3: By definition, when ϵ tends to 0, OPT' approaches OPT. Therefore, the bound from Lemma 3 becomes: for every $\delta \in [0, \frac{1}{3}]$, the probability that the expected revenue obtained from mechanism **Rev-RS**(N, m) is greater than $\frac{(1-3\delta)OPT^*}{16}$ is at least $\left(1 - 2e^{-1/32\epsilon}\right)\left(1 - 4e^{-\delta^2/(32\epsilon)}\right)$. When ϵ tends to 0, for every δ this probability tends to 1. Therefore, the current analysis gives a 16-approximation.

In the full version we show how to improve the approximation factor to be arbitrarily close to 8. ∎

References

1. Abrams, Z.: Revenue maximization when bidders have budgets. In: SODA, pp. 1074–1082. SIAM (2006)
2. Awerbuch, B., Azar, Y., Meyerson, A.: Reducing truth-telling online mechanisms to online optimization. In: STOC, pp. 503–510. ACM (2003)
3. Babaioff, M., Immorlica, N., Kleinberg, R.: Matroids, secretary problems, and online mechanisms. In: SODA, pp. 434–443. SIAM (2007)

4. Babaioff, M., Immorlica, N., Lucier, Weinberg, S.M.: A simple and approximately optimal mechanism for an additive buyer. In: FOCS (2014)
5. Bar-Yossef, Z., Hildrum, K., Wu, F.: Incentive-compatible online auctions for digital goods. In: SODA, pp. 964–970. SIAM (2002)
6. Blum, A., Hartline, J.D.: Near-optimal online auctions. In: SODA, pp. 1156–1163. SIAM (2005)
7. Borgs, C., Chayes, J., Immorlica, N., Mahdian, M., Saberi, A.: Multi-unit auctions with budget-constrained bidders. In: EC, pp. 44–51. ACM (2005)
8. Chawla, S., Hartline, J.D., Malec, D.L., Sivan, B.: Multi-parameter mechanism design and sequential posted pricing. In: STOC. ACM (2010)
9. Chawla, S., Malec, D.L., Malekian, A.: Bayesian mechanism design for budget-constrained agents. In: EC, pp. 253–262. ACM (2011)
10. Chawla, S., Malec, D.L., Sivan, B.: The power of randomness in Bayesian optimal mechanism design. Games Econ. Behav. **91**, 297–317 (2015)
11. Devanur, N.R., Ha, B.Q., Hartline, J.D.: Prior-free auctions for budgeted agents. In: EC, pp. 287–304. ACM (2013)
12. Dobzinski, S., Lavi, R., Nisan, N.: Multi-unit auctions with budget limits. Games Econ. Behav. **74**(2), 486–503 (2012)
13. Dobzinski, S., Leme, R.P.: Efficiency guarantees in auctions with budgets. In: Esparza, J., Fraigniaud, P., Husfeldt, T., Koutsoupias, E. (eds.) ICALP 2014. LNCS, vol. 8572, pp. 392–404. Springer, Heidelberg (2014). doi:10.1007/978-3-662-43948-7_33
14. Feldman, M., Fiat, A., Leonardi, S., Sankowski, P.: Revenue maximizing envy-free multi-unit auctions with budgets. In: EC, pp. 532–549. ACM (2012)
15. Fiat, A., Leonardi, S., Saia, J., Sankowski, P.: Single valued combinatorial auctions with budgets. In: EC, pp. 223–232. ACM (2011)
16. Friedman, E.J., Parkes, D.C.: Pricing WiFi at starbucks: issues in online mechanism design. In: EC, pp. 240–241. ACM (2003)
17. Goel, G., Mirrokni, V., Leme, R.P.: Clinching auctions with online supply. In: SODA, pp. 605–619. SIAM (2013)
18. Goldberg, A.V., Hartline, J.D., Karlin, A.R., Saks, M., Wright, A.: Competitive auctions. Games Econ. Behav. **55**, 242–269 (2006). Elsevier
19. Hajiaghayi, M.T., Kleinberg, R., Parkes, D.C.: Adaptive limited-supply online auctions. In: EC, pp. 71–80. ACM (2004)
20. Hajiaghayi, M.T., Kleinberg, R.D., Mahdian, M., Parkes, D.C.: Online auctions with re-usable goods. In: EC, pp. 165–174. ACM (2005)
21. Kleinberg, R.: A multiple-choice secretary algorithm with applications to online auctions. In: SODA, pp. 630–631. SIAM (2005)
22. Lavi, R., Nisan, N.: Competitive analysis of incentive compatible on-line auctions. In: EC, pp. 233–241. ACM (2000)
23. Lavi, R., Nisan, N.: Online ascending auctions for gradually expiring items. In: SODA, pp. 1146–1155. SIAM (2005)
24. Lu, P., Xiao, T.: Improved efficiency guarantees in auctions with budgets. In: EC 2015, Portland, OR, USA, 15–19 June 2015, pp. 397–413 (2015)
25. Parkes, D.C.: Online Mechanisms (2007)
26. Parkes, D.C., Singh, S.P.: An MDP-based approach to online mechanism design. In: NIPS (2003)

Liquid Welfare Maximization in Auctions with Multiple Items

Pinyan Lu[1]([✉]) and Tao Xiao[2]

[1] ITCS, Shanghai University of Finance and Economics, Shanghai, China
lu.pinyan@mail.shufe.edu.cn
[2] Department of Computer Science, Shanghai Jiaotong University, Shanghai, China
xt_1992@sjtu.edu.cn

Abstract. Liquid welfare is an alternative efficiency measure for auctions with budget constrained agents. Previous studies focused on auctions of a single (type of) good. In this paper, we initiate the study of general multi-item auctions, obtaining a truthful budget feasible auction with constant approximation ratio of liquid welfare under the assumption of large market.

Our main technique is random sampling. Previously, random sampling was usually used in the setting of single-parameter auctions. When it comes to multi-dimensional settings, this technique meets a number of obstacles and difficulties. In this work, we develop a series of analysis tools and frameworks to overcome these. These tools and frameworks are quite general and they may find applications in other scenarios.

1 Introduction

Let us consider the following auction environment: there is one auctioneer, who has m heterogeneous divisible items and wants to distribute them among n agents. Since the items are divisible, W.L.O.G we assume that each item is of one unit. Each agent i has a value per unit v_{ij} for item j. Each agent i is also constrained by a budget B_i, which is the maximum amount of money i is able to pay during the auction. An allocation rule $A = (x_{ij})_{n \times m}$ specifies the fraction of items everyone is allocated in an auction, where x_{ij} denotes that i is allocated x_{ij} fraction of item j. We say an allocation is feasible if for each item j, $\sum_i x_{ij} \leq 1$. A feasible payment rule $p = (p_1, \ldots, p_n)$ specifies the amount of money each agent needs to pay, while satisfying budget constraint $p_i \leq B_i$. Basically, an auction is an algorithm that takes all agents' bids as inputs, and outputs a feasible allocation and payment rule. We say an auction is *truthful* if it is every agent's dominant strategy to bid her/his true private profile (here it means value and budget). We say an auction is *universally truthful* if this auction is a distribution over deterministic truthful auctions. Put it more precisely, agent i's utility is $x_{ij}v_{ij} - p_i$ if $p_i \leq B_i$ and $-\infty$ if $p_i \geq B_i$. We also assume that the agents are risk-neutral.

Liquid Welfare. Due to the budget constraints, it is impossible to get any reasonable guarantee for the social welfare objective, even in the simplest setting

© Springer International Publishing AG 2017
V. Bilò and M. Flammini (Eds.): SAGT 2017, LNCS 10504, pp. 41–52, 2017.
DOI: 10.1007/978-3-319-66700-3_4

of single item auction. The main obstacle is that we cannot allocate item(s) to an agent with very high value but low budget truthfully. To overcome this, an alternative measure called liquid welfare was proposed in [16]. Basically, each agent's contribution to the liquid welfare is her/his valuation for the allocated bundle, capped by her/his budget. A precise definition is given as follows.

Definition 1. *The liquid welfare of an assignment* $A = (x_{ij})_{n \times m}$ *in the multi-item setting is*

$$LW(A) = \sum_{i=1}^{n} \min\{\sum_{j=1}^{m} v_{ij}x_{ij}, B_i\}.$$

Just like social welfare is the maximum amount of money an omniscient auctioneer can obtain in a budget-free setting, the liquid welfare measure is actually the maximum amount of money an omniscient auctioneer would be able to extract from agents in the budget setting. Therefore, this measure is a quite reasonable efficiency measure in the budget setting. More justification about the liquid welfare can be found in [16]. Our goal is to design a universally truthful, budget feasible mechanism that guarantees some good approximation towards this liquid welfare objective.

For the simplest setting of single item environment, the problem was first studied in [16], where an $O(\log n)$ approximation mechanism was obtained. In a previous work [25], we improved the result to $O(1)$ approximation. Nothing was previously known for multi-item setting. Although the valuation for each item is additive, the total budget for each agent is shared by different items. This fact makes the multi-item setting much more complicated and challenging than single item setting.

Large Market. Generally speaking, the large market assumption says that a single agent's contribution (power) to the total market is very small. There is a number of recent works which are based on this assumption [2,21]. From practical point of view, this is a very realistic assumption especially in the age of internet economy; from theoretical point of view, this assumption is a very interesting mathematical framework to overcome some impossibility results or get better results than general setting.

In this paper, we study the above liquid welfare maximization problem also with the assumption of large market. It is crucial to give a good characterization of this large market assumption. There are a number of alternative definitions characterizing this. We choose the following one:

$$\forall i, B_i \leq \frac{OPT}{m \cdot c},$$

where OPT is the liquid welfare for an optimal allocation and c is some large constant. The quantity of $\frac{OPT}{m}$ represents the average contribution of each item to the total market. Basically, the above assumption says that each agent does not have enough budget to make a significant interference to a typical item in the market.

Results and Techniques. We get the first constant approximation budget feasible truthful mechanism for liquid welfare maximization problem under the large market assumption. Notice that the liquid welfare is an upper bound for revenue obtained in any individual rational auction. From our proof of liquid welfare guarantee by our auction, what we indeed prove is that the revenue from our auction is a constant fraction of the optimal liquid welfare. As a corollary, our mechanism also guarantees a constant approximation in terms of revenue.

The main technique used in designing our auction is random sampling. Random sampling is a very powerful tool in designing truthful mechanisms, which is widely applied in various of different settings [4–7,23,24]. A typical random sampling mechanism follows the following routine: first divides the agents into two groups randomly, then gathers information from one group and uses this information as a guide to design mechanism for the agents in other group. This approach is usually seen to be applied on single item setting. However, for the multi-item setting, there could be a number of equally optimal solutions for the sample set of agents, but the allocations in these different optimal solutions can be quite different for the same item. Such fragility of optimal solutions brings in difficulty in directly applying random sampling: from the sampling set, one can get good estimation of the total welfare of the set, but does not necessary give stable and useful information for individual items. To overcome this, we use a greedy solution rather than the optimal solution as the guidance. The greedy solution has certain robustness and monotonicity properties which are very helpful to get useful information for every single item.

To argue that a random sampling algorithm does give a good guarantee to some objective, the analysis usually has two steps. In the first step, one proves that with a constant probability, the sampling set is a good estimation of the remaining set. Then, in the second step, one proves that under the condition that it is a good sampling, one can get a good allocation from the remaining set. But in the multi-item setting, there are obstacles in proving both steps. With the large market assumption, one can prove that for a single item, with a constant probability, it *is* a good sampling. But to show that a sampling indeed gives a good estimation for a constant fraction of items, it is not sufficient to apply union bound and it is not clear if applying any other tool from probability theory would work, since the correlation between different items could be very complicated. We still do not know how to prove that this is true. Instead, we are able to bypass this with a very subtle and direct estimation of the performance without conditioning that the sampling is a good one. Our analysis has some similarities with that in [12].

Related Work. As budget is becoming an important issue that cannot be neglected in practice, many theoretical investigations have been devoted to analyzing auctions for budget constrained agents. One of the important directions leads to optimal auction design which tries to maximize revenue for the auctioneer [1,8,10,14,20]. Another direction focus on maximizing social efficiency. In particular, there are a number of previous works focusing on a solution concept of Pareto Efficiency [15,22]. Note that the liquid welfare is not the only quantifiable measure for efficiency for budget constrained agents. There are similar alternatives for this measure, studied in [14,27], but for different solution concepts.

Beyond designing auctions that maximize liquid welfare itself, there are other interesting works that follow this liquid welfare notion. For example [3,9,13,18, 27] focus on the liquid welfare guarantee at equilibrium. [19] focused on an online version of auctions with budget constraints.

Another line of research is devoted to study budget feasible mechanism design for reversal auction, in which the budget constrained buyer becomes the auctioneer rather than bidder. This model was first proposed and studied by Singer [26]. Since then, several improvements have been obtained [7,11,17].

For random sampling technique applied on mechanism design, there are also a long line of research focusing on it [4–7,23–25]. Most of them are for single item setting. Some of them [4–6] also applied random sampling techniques on multi-item setting. But unlike our setting, they have constraints on solution space, number of agents and value profile.

Open Problems and Discussions. Here, we consider the simplest valuation function, which is linear for each item and additive across different items. It is natural to extend them to more complicated ones. We conjecture that a similar mechanism can be applied to concave (for each item) and sub-modular (across items) functions and leave it to future work.

Theoretically, the most important and interesting open question is whether we can remove the large market assumption and obtain a constant approximation mechanism in general multi item setting. It is easy to see that one can combine the random sampling mechanism with the modified ground bundle second price auction to get an $O(m)$ approximation mechanism. So, it is a constant approximation when the number of items is a constant. But if m is not a constant, the problem remains open.

2   Greedy Algorithm

If all valuations and budgets are common knowledge, the off line liquid welfare maximization problem can be solved by a simple linear program. However, due to the dedication of linear programming, we do not really have much structural understanding or nice properties about this optimal solution. This is in contrast to the single item setting where the optimal solution can be obtained by a simple greedy algorithm.

To overcome this, we propose the following natural greedy algorithm for the multi-item setting. A high level idea of this algorithm is the following: traverse entry (i, j) in decreasing order of value (per unit). At entry (i, j), let agent i buy some fraction of item j at price v_{ij} per unit, so that this fraction is constrained by remaining supply and budget. A detailed formulation can be referred in the following.

Unlike in the single item case, this greedy algorithm is not necessarily optimal but gives a good guarantee towards the optimal. We shall prove that the

---

Algorithm 1. Greedy Algorithm

---

input : n agents with valuations $(v_{ij})_{n \times m}$ and corresponding budgets
$\quad\quad\quad B_1, \ldots, B_n$

output: An allocation $(x_{ij})_{n \times m}$

begin

   for *each $i \in [n]$* **do**
      $C_i \leftarrow B_i$;

   for *each $j \in [m]$* **do**
      $s_j \leftarrow 1$;

   for *each $i \in [n]$ and $j \in [m]$* **do**
      $x_{ij} \leftarrow 0$;

   for *each $v_{ij} > 0$ in decreasing order* **do**
      if $C_i > v_{ij}s_j$ **then**
         $x_{ij} \leftarrow s_j$;
         $C_i \leftarrow C_i - v_{ij}s_j$;
         $s_j \leftarrow 0$;

      else
         $x_{ij} \leftarrow \frac{C_i}{v_{ij}}$;
         $s_j \leftarrow s_j - \frac{C_i}{v_{ij}}$;
         $C_i \leftarrow 0$;

---

greedy solution is a 2-approximation to optimal liquid welfare, implying that this solution is good enough to serve as a reference to design mechanism. Most importantly, this greedy solution enjoys a number of nice monotonicity properties which are very essential for the analysis of our mechanism in Sect. 3.

In the algorithm, if there are ties among different v_{ij}s, we break them arbitrary but in fixed order (a simple way is to break ties by the index of agents and items). The tie breaking rule gives a total order on v_{ij}'s, thus making the algorithm outputs solution deterministically.

Before we analyze the properties of the algorithm, we introduce a few more necessary notations. Let $A = (x_{ij})_{n \times m}$ be some allocation. We say an allocation is budget compatible if for every i we have $\sum_{j=1}^{m} v_{ij}x_{ij} \leq B_i$. It is obvious that the allocation derived from the above greedy algorithm is budget compatible. For a feasible allocation that is not budget compatible, we can get a new allocation that is budget compatible while achieving the same liquid welfare by just cutting off some fraction of items given to this agent in order to make the value equals to the budget. Thus we can also assume that the optimal allocation given by linear programming is budget compatible. For budget compatible allocations, the liquid welfare is the same as social welfare. For convenience we denote $v_i(A) = \sum_{j=1}^{m} v_{ij}x_{ij}$, $v(A_j) = \sum_{i=1}^{n} v_{ij}x_{ij}$ and $v(A) = \sum_i \sum_j v_{ij}x_{ij}$ respectively. In this paper, unless otherwise specified, we always use $A = (x_{ij})_{n \times m}$ to denote the allocation outputted by the above greedy algorithm. For any subset $T \subseteq [n]$, we use A^T to denote the allocation when running the greedy algorithm only on the subset of agents in T. We use $A^* = (x_{ij}^*)_{n \times m}$ to denote a budget compatible optimal allocation.

We first prove that greedy algorithm with full information guarantees at least half of optimal liquid welfare.

Lemma 1. $v(A) \geq \frac{1}{2} OPT$.

Proof. In the greedy algorithm, by decreasing order of v_{ij}s, we always allocate fraction of item j to agent i until agent i's budget is exhausted or item j is sold out. Up to the termination of the algorithm, we denote by $D \subseteq [n]$ the subset of agents who exhaust their budgets ($C_i = 0$), and by $F \subseteq [m]$ the subset of items which are sold out ($s_j = 0$). It is clear that we have $v_{ij} = 0$ if $i \notin D$ and $j \notin F$.

A lower bound of greedy algorithm's liquid welfare is as follows:

$$2v(A) \geq \sum_{i \in D} v_i(A) + \sum_{j \in F} v(A_j) = \sum_{i \in D} B_i + \sum_{j \in F} v(A_j)$$

For i, j such that $i \notin D$ and $j \in F$, we can see that in greedy algorithm, after the algorithm go through this entry (i, j), item j is already sold out. This implies $v_{ij} \leq v(A_j)$ since every fraction of item j is sold at a price of at least v_{ij}.

To bound optimal liquid welfare, we also divide all the agents into two groups: D and the rest. We note that these sets D and F are defined with respect to the greedy solution rather than the optimal solution. We have

$$OPT = v(A^*) = \sum_{i \in D} v_{ij} x_{ij}^* + \sum_{i \notin D} v_{ij} x_{ij}^* = \sum_{i \in D} v_{ij} x_{ij}^* + \sum_{i \notin D} \sum_{j \in F} v_{ij} x_{ij}^*,$$

where the last equality uses the fact that $v_{ij} = 0$ for $i \notin D$ and $j \notin F$. We can further bound this by

$$OPT = \sum_{i \in D} v_{ij} x_{ij}^* + \sum_{i \notin D} \sum_{j \in F} v_{ij} x_{ij}^* \leq \sum_{i \in D} B_i + \sum_{i \notin D} \sum_{j \in F} v_{ij} x_{ij}^*$$

$$\leq \sum_{i \in D} B_i + \sum_{j \in F} v(A_j) \sum_{i \notin D} x_{ij}^*$$

$$\leq \sum_{i \in D} B_i + \sum_{j \in F} v(A_j) \leq 2v(A),$$

where the first inequality is from the budget compatibility of optimal allocation, the second inequality uses the fact that $v_{ij} \leq v(A_j)$ for $i \notin D$ and $j \in F$ while the third inequality uses that fact that $\sum_{i \notin D} x_{ij}^* \leq 1$ since it is a feasible allocation. This completes the proof. □

Not only greedy algorithm is a good approximation, we shall also prove that it has a nice monotonicity property when running on a subset of the agents. This is crucial for our random sampling mechanism to work.

Lemma 2. (monotonicity of greedy). *Let $T \subseteq [n]$ be a subset of agents, A and A^T be the greedy solutions running on the total set $[n]$ and its subset T respectively. Then $\forall i \in T$ and j, we have $v_i(A^T) \geq v_i(A)$ and $v(A_j^T) \leq v(A_j)$.*

The intuition is clear that when there are less agents, each remaining agent can get more and each item generates less welfare. Notice that this property does not necessary hold for every single agent and item if we use optimal solution rather than greedy solution.

Proof. We prove this by coupling every step of greedy algorithm for the inputs $[n]$ and T. When generating assignments A and A^T, the entries (i,j)s traversed in the algorithm keep the order in v_{ij}, except for $[n]$ it experiences some extra entries (i,j) when $i \notin T$. For these cases, we couple them with empty steps.

For $i \in T$, we denote the remaining budget for agent i by C_i and C_i^T respectively. We also denote remaining supply for item j by s_j and s_j^T respectively. We inductively prove that after each step, $\forall i \in T$ we have $C_i \geq C_i^T$, and $\forall j \in [m]$ we have $s_j \leq s_j^T$.

Initially, $C_i = B_i = C_i^T$ and $s_j = 1 = s_j^T$. Now, we assume that the property holds before the algorithm processes entry (i,j). Notice that after going through an entry (i,j), $\forall k \in T \backslash \{i\}$ both C_k^T and C_k remain the same. Similarly, $\forall l \in [m] \backslash \{j\}$ both s_l^T and s_l remain the same. Thus we only need to consider the changes in C_i, C_i^T, s_j and s_j^T at step (i,j). There are three cases.

$i \notin T$. In this case, the only possible change is that s_j may decrease by some certain amount. So, the monotonicity property remains to hold.

$i \in T$ and $C_i^T \geq s_j^T v_{ij}$. In this case, $C_i \geq C_i^T \geq v_{ij}s_j^T \geq v_{ij}s_j$, thus after step (i,j), $\hat{s}_j^T = \hat{s}_j = 0$, and $\hat{C}_i^T = C_i^T - s_j^T v_{ij} \leq C_i - s_j v_{ij} = \hat{C}_i$, which shows that after this step, the two properties still hold.

$i \in T$ and $C_i^T < s_j^T v_{ij}$. In this case $\hat{C}_i^T = 0 \leq \hat{C}_i$. Also $\hat{s}_j^T = s_j^T - \frac{C_i^T}{v_{ij}} \geq \max\{s_j - \frac{C_i}{v_{ij}}, 0\} = \hat{s}_j$, which shows that after this step, the two properties still hold.

From the above argument, when both algorithms terminate, $\forall i \in T$, $C_i \geq C_i^T$, thus $\forall i \in T$, $v_i(A^T) = B_i - C_i^T \geq B_i - C_i \geq v_i(A)$.

We further prove that $v(A_j^T) \leq v(A_j)$. Since $\forall j$, $s_j \leq s_j^T$ at every step of greedy algorithm, which indicates that each fraction of unit in A is allocated with a price no less than in A^T (if not allocated the price of that fraction is 0). Thus $v(A_j^T) \leq v(A_j)$. $\square$

3   The Random Sampling Mechanism

The greedy algorithm has some nice properties but is obviously not truthful. To design truthful mechanism that has good guarantee, we combine this greedy algorithm with random sampling. The idea is very simple, we randomly divide all the agents into two groups T and R. For agents in T, they do not get any allocation in the auction. We run the greedy algorithm on set T, and use this result as a guide for pricing for agents in R. From the solution of the greedy algorithm, we have a rough idea of how to set the price for each item. Then, we simply sell the items to agents in R at fixed prices which are determined by the output of greedy algorithm. A formal description of the auction is as follows.

We present our main theorem in the following.

Algorithm 2. Random Sampling Mechanism

input : n agents with valuations $(v_{ij})_{n \times m}$ and corresponding budgets
$B_1, \ldots, B_n$
output: An allocation and payments

begin
| Randomly divide all agents with equal probability into group T and R
| $A^T \leftarrow$ the greedy solution running on group T.
| **for** $j \in [m]$ **do**
| $\lfloor \ p_j = \frac{1}{6} v(A_j^T);$
| Each agent $i \in R$ comes in a given fixed order and buy the most profitable
| part with respect to price vector $\{p_j\}$ under budget feasibility and unit item
| supply constraint.

Theorem 1. *The random sampling mechanism is a universal truthful budget feasible mechanism which guarantees a constant fraction of the liquid welfare under the large market assumption.*

The truthfulness and budget feasibility of this auction is obvious. In the following two subsections, we analyse its approximation ratio. Before that, we introduce one more notion: for a subset of agents $T \subseteq [n]$, denote $v(A_j \cap T) = \sum_{i \in T} v_{ij} x_{ij}$.

3.1 Random Sampling and Large Market

We divide all the items into two groups. Let H be the set of easily samplable items consists of item j that satisfies condition $Pr_T(\frac{1}{3} v(A_j) \leq v(A_j \cap T) \leq \frac{2}{3} v(A_j)) \geq \frac{3}{4}$. We also denote the remaining set as G.

Firstly, we provide a simple technical concentration lemma.

Lemma 3. [12] *Let $a_1 \geq a_2 \geq \cdots \geq a_l$ be positive real numbers such that the sum $a = \sum_{i=1}^{l} a_i$ satisfies $a > 36a_1$. We select each number $a_1, \cdots, a_l$ independently at random with probability $1/2$ each and let b be a random variable representing the sum of these selected numbers. Then*

$$Pr[\frac{a}{3} < b < \frac{2a}{3}] \geq \frac{3}{4}.$$

The key lemma in this subsection is as follows. It basically says that the items in group G do not contribute too much in the greedy solution. This is also the only place we use the assumption of large market throughout this paper.

Lemma 4. $\sum_{j \in G} v(A_j) \leq \frac{1}{6} v(A)$

Proof. Lemma 3 provides a sufficient condition for an item to be in H, namely, $B_i \leq \frac{v(A_j)}{36}, \forall i \in [n]$. So, for $j \in G$, there exist $i \in [n]$ such that $B_i > \frac{v(A_j)}{36}$. By the large market assumption, we have that $B_i \leq \frac{OPT}{m \cdot c}$. As a result, we have that $v(A_j) < \frac{36 OPT}{m \cdot c} \leq \frac{72 v(A)}{m \cdot c}$ for all $j \in G$. Since there are at most m items in G, we get $\sum_{j \in G} v(A_j) \leq \frac{72}{c} v(A)$. By choosing $c = 432$ in the large market assumption, we get the claimed result. □

3.2   Approximation Ratio

In the above section, we already show that items in G do not contribute much. Thus, if our auction do get a good guarantee on items in H, then we are done. In this subsection we will prove this. Before that we introduce a few more necessary definitions. For an item $j \in H$, we denote by Π_j the set of T such that for $T \in \Pi_j$, $\frac{1}{3}v(A_j) \leq v(A_j \cap T) \leq \frac{2}{3}v(A_j)$. Then, from the definition, we know that for $j \in H$, $Pr(T \in \Pi_j) \geq \frac{3}{4}$. For convenience, we also abuse the notation Π_j to denote the conditional distribution of T over the subset Π_j. We use Π to denote the distribution of T in the mechanism.

The following lemma shows that even if we restrict to the agents in T and only count these T in Π_j, the contribution from items in H is still significant.

Lemma 5.
$$\sum_{j \in H} Pr(T \in \Pi_j) E_{T \sim \Pi_j} v(A_j^T) \geq \frac{1}{8}v(A).$$

Proof. We give both lower bound and upper bound for the term $\sum_j E_{T \sim \Pi} v(A_j^T)$. On one hand, we have

$$\sum_j E_{T \sim \Pi} v(A_j^T) = \sum_i E_{T \sim \Pi} v_i(A^T) \geq \sum_i Pr(i \in T) v_i(A) = \frac{1}{2} \sum_i v_i(A) = \frac{1}{2}v(A),$$

where the inequality uses the fact that $v_i(A^T) \geq v_i(A)$ for any subset T and $i \in T$.

On the other hand, we have

$$\sum_j E_{T \sim \Pi} v(A_j^T) = \sum_{j \in G} E_{T \sim \Pi} v(A_j^T) + \sum_{j \in H} E_{T \sim \Pi} v(A_j^T)$$

$$= \sum_{j \in G} E_{T \sim \Pi} v(A_j^T) + \sum_{j \in H} [Pr(T \in \Pi_j) E_{T \sim \Pi_j} v(A_j^T)$$

$$+ Pr(T \notin \Pi_j) E_{T \sim \Pi \backslash \Pi_j} v(A_j^T)]$$

$$\leq \sum_{j \in G} E_{T \sim \Pi} v(A_j) + \sum_{j \in H} [Pr(T \in \Pi_j) E_{T \sim \Pi_j} v(A_j^T) + \frac{1}{4}v(A_j)]$$

$$= \frac{1}{4}v(A) + \frac{3}{4} \sum_{j \in G} v(A_j) + \sum_{j \in H} Pr(T \in \Pi_j) E_{T \sim \Pi_j} v(A_j^T)$$

$$\leq \frac{1}{4}v(A) + \frac{1}{8}v(A) + \sum_{j \in H} Pr(T \in \Pi_j) E_{T \sim \Pi_j} v(A_j^T)$$

$$= \frac{3}{8}v(A) + \sum_{j \in H} Pr(T \in \Pi_j) E_{T \sim \Pi_j} v(A_j^T)$$

where the first inequality uses the fact that $Pr(T \notin \Pi_j) \leq \frac{1}{4}$ and $v(A_j^T) \leq v(A_j)$ for all item j, while the last inequality uses Lemma 4.

Connecting the lower and upper bounds for $\sum_j E_{T \sim \Pi} v(A_j^T)$, we have that

$$\sum_{j \in H} Pr(T \in \Pi_j) E_{T \sim \Pi_j} v(A_j^T) \geq \frac{1}{8} v(A).$$

$\square$

Up to this point, we have not talked about the allocation of our random sampling algorithm but only the property of greedy solution under random sampling. The following lemma gives the last piece of the analysis which connects liquid welfare of our mechanism to the above quantity.

Lemma 6. *The liquid welfare obtained from the random sampling algorithm is at least*

$$\frac{1}{12} \sum_{j \in H} Pr(T \in \Pi_j) E_{T \sim \Pi_j} v(A_j^T).$$

Proof. We note that the allocation outputted by our mechanism may not be budget compatible. However, the liquid welfare is always lower bounded by the revenue obtained by a truthful auction (note that the payment of each agent is also bounded by both value and budget), so we only need to bound the revenue obtained by our mechanism.

In our auction, we denote by $D \subseteq R$ the subset of agents who exhaust their budgets, and by $F \subseteq [m]$ the subset of items which are sold out. Both sets are random which depend on the random set T. One key observation is that for all $j \notin F$ and $i \in R \setminus D$, we have $v_{ij} \leq \frac{1}{6} v(A_j^T)$. For $j \notin F$ and $i \in R \setminus D$, agent i did not exhaust his/her budget and item j is not sold out. The only possible reason why agent i did not buy item j is that the price of $\frac{1}{6} v(A_j^T)$ is higher than his/her value v_{ij}, thus we have $v_{ij} \leq \frac{1}{6} v(A_j^T)$.

On one hand, the revenue (and thus the liquid welfare) is bounded by $\sum_{i \in D} B_i$. We further have that

$$\sum_{i \in D} B_i \geq \sum_{i \in D} \sum_{j \notin F} v_{ij} x_{ij} = \sum_{j \notin F} \sum_{i \in D} v_{ij} x_{ij} = \sum_{j \notin F} \left(\sum_{i \in R} v_{ij} x_{ij} - \sum_{i \in R \setminus D} v_{ij} x_{ij} \right)$$

$$\geq \sum_{j \notin F} \max \left\{ 0, (v(A_j \cap R) - \frac{1}{6} v(A_j^T)) \right\}.$$

We note that the allocation A and x_{ij} in the above calculation are from the greedy solution rather than the allocation given by the random sampling mechanism. According to the above argument, this quantity does give a lower bound for the liquid welfare of the random sampling mechanism.

On the other hand, we can also bound the revenue (and thus the liquid welfare) from the item point of view. It is bounded by $\sum_{j \in F} \frac{1}{6} v(A_j^T)$ as the item $j \in F$ is sold out at a price $\frac{1}{6} v(A_j^T)$ per unit.

Let Y_j be a random variable that

$$Y_j := \mathbb{1}_{j \in F} \frac{v(A_j^T)}{6} + \mathbb{1}_{j \notin F} \max \left\{ 0, (v(A_j \cap R) - \frac{1}{6} v(A_j^T)) \right\}.$$

Then the above argument showed that the expected liquid welfare of our mechanism is bounded by $\frac{1}{2}\sum_j E_{T\sim\Pi}Y_j$. We further have that

$$\sum_j E_{T\sim\Pi}Y_j \geq \sum_{j\in H} E_{T\sim\Pi}Y_j \geq \sum_{j\in H} Pr(T\in\Pi_j)E_{T\sim\Pi_j}Y_j,$$

where the inequalities use the simple fact that $Y_j \geq 0$ and one simply throw away some terms in the summation for computing expectation.

For $j \in H$ and $T \in \Pi_j$, we have a better bound for Y_j. If $j \in F$, we directly have $Y_j \geq \frac{1}{6}v(A_j^T)$. For $j \notin F$ we have $Y_j \geq v(A_j \cap R) - \frac{1}{6}v(A_j^T)$. Since $j \in H$ and $T \in \Pi_j$, we have

$$v(A_j \cap R) - \frac{1}{6}v(A_j^T) \geq \frac{1}{3}v(A_j) - \frac{1}{6}v(A_j^T) \geq \frac{1}{3}v(A_j^T) - \frac{1}{6}v(A_j^T) = \frac{1}{6}v(A_j^T).$$

Thus $\forall j \in H$ and $T \in \Pi_j$, we have $Y_j \geq \frac{1}{6}v(A_j^T)$. Therefore, the expected liquid welfare obtained by our mechanism is at least $\frac{1}{12}\sum_{j\in H} Pr(T\in\Pi_j)E_{T\sim\Pi_j}v(A_j^T)$. This completes the proof.   □

Put Lemmas 1, 5 and 6 together, we know that the liquid welfare of random sampling mechanism is at least $\frac{1}{192}$ of the optimal one. This completes the proof of the main theorem.

References

1. Abrams, Z.: Revenue maximization when bidders have budgets. In: Proceedings of the Seventeenth Annual ACM-SIAM Symposium on Discrete Algorithm, pp. 1074–1082. Society for Industrial and Applied Mathematics (2006)
2. Anari, N., Goel, G., Nikzad, A.: Mechanism design for crowdsourcing: an optimal 1-1/e competitive budget-feasible mechanism for large markets. In: 2014 IEEE 55th Annual Symposium on Foundations of Computer Science (FOCS), pp. 266–275. IEEE (2014)
3. Azar, Y., Feldman, M., Gravin, M., Roytman, A.: Liquid price of anarchy. arXiv preprint arXiv:1511.01132 (2015)
4. Balcan, M.-F., Blum, A., Hartline, J.D., Mansour, Y.: Mechanism design via machine learning. In 46th Annual IEEE Symposium on Foundations of Computer Science, FOCS 2005, pp. 605–614. IEEE (2005)
5. Balcan, M.-F., Blum, A., Hartline, J.D., Mansour, Y.: Reducing mechanism design to algorithm design via machine learning. J. Comput. Syst. Sci. **74**(8), 1245–1270 (2008)
6. Balcan, M.-F., Devanur, N., Hartline, J.D., Talwar, K.: Random sampling auctions for limited supply. Manuscript (2007, submitted)
7. Bei, X., Chen, N., Gravin, N., Lu, P.: Budget feasible mechanism design: from prior-free to Bayesian. In: STOC, pp. 449–458 (2012)
8. Borgs, C., Chayes, J.T., Immorlica, N., Mahdian, M., Saberi, A.: Multi-unit auctions with budget-constrained bidders. In: EC, pp. 44–51 (2005)
9. Caragiannis, I., Voudouris, A.A.: Welfare guarantees for proportional allocations. Theory Comput. Syst. **59**(4), 581–599 (2016)

10. Chawla, S., Malec, D.L., Malekian, A.: Bayesian mechanism design for budget-constrained agents. In: EC, pp. 253–262 (2011)
11. Chen, N., Gravin, N., Lu, P.: On the approximability of budget feasible mechanisms. In: SODA, pp. 685–699 (2011)
12. Chen, N., Gravin, N., Lu, P.: Truthful generalized assignments via stable matching. Math. Oper. Res. **39**(3), 722–736 (2013)
13. Christodoulou, G., Sgouritsa, A., Tang, B.: On the efficiency of the proportional allocation mechanism for divisible resources. Theory Comput. Syst. **59**(4), 600–618 (2016)
14. Devanur, N.R., Ha, B.Q., Hartline, D.: Prior-free auctions for budgeted agents. In: EC, pp. 287–304 (2013)
15. Dobzinski, S., Lavi, R., Nisan, N.: Multi-unit auctions with budget limits. Games Econ. Behav. **74**(2), 486–503 (2012)
16. Dobzinski, S., Leme, R.P.: Efficiency guarantees in auctions with budgets. In: Esparza, J., Fraigniaud, P., Husfeldt, T., Koutsoupias, E. (eds.) ICALP 2014. LNCS, vol. 8572, pp. 392–404. Springer, Heidelberg (2014). doi:10.1007/978-3-662-43948-7_33
17. Dobzinski, S., Papadimitriou, C.H., Singer, Y.: Mechanisms for complement-free procurement. In: EC, pp. 273–282 (2011)
18. Dughmi, S., Eden, A., Feldman, M., Fiat, A., Leonardi, S.: Lottery pricing equilibria. In: Proceedings of the 2016 ACM Conference on Economics and Computation, pp. 401–418. ACM (2016)
19. Eden, A., Feldman, M., Vardi, A.: Truthful secretaries with budgets. arXiv preprint arXiv:1504.03625 (2015)
20. Feldman, M., Fiat, A., Leonardi, S., Sankowski, P.: Revenue maximizing envy-free multi-unit auctions with budgets. In: EC, pp. 532–549 (2012)
21. Feldman, M., Immorlica, M., Lucier, B., Roughgarden, T., Syrgkanis, V.: The price of anarchy in large games. In: Proceedings of the 48th Annual ACM SIGACT Symposium on Theory of Computing, pp. 963–976. ACM (2016)
22. Fiat, A., Leonardi, S., Saia, J., Sankowski, P.: Single valued combinatorial auctions with budgets. In: EC, pp. 223–232 (2011)
23. Goldberg, A.V., Hartline, J.D., Karlin, A.R., Saks, M., Wright, A.: Competitive auctions. Games Econ. Behav. **55**(2), 242–269 (2006)
24. Gravin, N., Lu, P.: Competitive auctions for markets with positive externalities. In: Fomin, F.V., Freivalds, R., Kwiatkowska, M., Peleg, D. (eds.) ICALP 2013. LNCS, vol. 7966, pp. 569–580. Springer, Heidelberg (2013). doi:10.1007/978-3-642-39212-2_50
25. Lu, P., Xiao, T.: Improved efficiency guarantees in auctions with budgets. In: Proceedings of the Sixteenth ACM Conference on Economics and Computation, pp. 397–413. ACM (2015)
26. Singer, Y.: Budget feasible mechanisms. In: FOCS, pp. 765–774 (2010)
27. Syrgkanis, V., Tardos, É.: Composable and efficient mechanisms. In: STOC, pp. 211–220 (2013)

Computational Aspects of Games

On the Nucleolus of Shortest Path Games

Mourad Baïou[1] and Francisco Barahona[2($\boxtimes$)]

[1] CNRS and Université Clermont II,
Campus des Cézeaux BP 125, 63173 Aubière Cedex, France
[2] IBM T.J. Watson Research Center, Yorktown Heights, NY 10589, USA
barahon@us.ibm.com

Abstract. We study a type of cooperative games introduced in [8] called shortest path games. They arise on a network that has two special nodes s and t. A coalition corresponds to a set of arcs and it receives a reward if it can connect s and t. A coalition also incurs a cost for each arc that it uses to connect s and t, thus the coalition must choose a path of minimum cost among all the arcs that it controls. These games are relevant to logistics, communication, or supply-chain networks. We give a polynomial combinatorial algorithm to compute the nucleolus. This vector reflects the relative importance of each arc to ensure the connectivity between s and t. Our development is done on a directed graph, but it can be extended to undirected graphs and to similar games defined on the nodes of a graph.

Keywords: Cooperative games · Shortest path games · Nucleolus

1 Introduction

Shortest path games arise on a network, a coalition corresponds to a set of arcs and it receives a reward if it can connect two fixed nodes s and t. A coalition also incurs a cost for each arc that it uses to connect s and t, thus the coalition must choose a path of minimum cost among all the arcs that it controls. Shortest path games have been introduced in [8], this type of games is useful to determine the most critical links to ensure connectivity between two distinguished nodes s and t in a network. This analysis is relevant to logistics, communication, or supply-chain networks.

Our Results. We give a polynomial combinatorial algorithm to compute the nucleolus of a shortest path game. This vector reflects the relative importance of the different arcs in the network that ensure the connectivity between s and t. Our development is done on a directed graph, but it can be extended to undirected graphs and to similar games defined on the nodes of a graph.

Related Work. The core and the Shapley Value of shortest path games were studied in [8], also the least core was studied in [2]. Flow games were introduced in [12]. Polynomial combinatorial algorithms for computing the nucleolus of simple flow games were given in [4,16]. Algorithms for computing the nucleolus of

© Springer International Publishing AG 2017
V. Bilò and M. Flammini (Eds.): SAGT 2017, LNCS 10504, pp. 55–66, 2017.
DOI: 10.1007/978-3-319-66700-3_5

several other combinatorial games have been found, see [7] for path cooperative games, see [15] for cost allocation games in a tree, [18] for assignment games, [13] for cardinality matching games, [5] for weighted voting games. On the other hand computing the nucleolus is NP-Hard for minimum spanning tree games [6], and general flow games [4].

This paper is organized as follows. In Sect. 2 we give some basic definitions and mention some basic results of Network Flows and Linear Programming. In Sect. 3 we study the core. Section 4 contains the basics for the computation of the nucleolus. In Sect. 5 we give an algorithm to compute the nucleolus when the core is non-empty. In Sect. 6 we extend the algorithm to the case when the core is empty. In Sect. 7 we study the time complexity of the algorithm.

2   Preliminaries

Here we give some definitions and mention some basic results on Network Flows and Linear Programming. Throughout this paper we assume that we are working with a directed graph $G = (V, A)$. We use n to denote $|V|$, and m to denote $|A|$. Given two distinguished nodes s and t. An st-path is a sequence of arcs $(u_1, v_1), (u_2, v_2), \ldots, (u_k, v_k)$, where $s = u_1$, $v_k = t$, and $v_i = u_{i+1}$ for $i = 1, \ldots, k - 1$. A cycle is a sequence of arcs $(u_1, v_1), (u_2, v_2), \ldots, (u_k, v_k)$, where $u_1 = v_k$, and $v_i = u_{i+1}$ for $i = 1, \ldots, k - 1$. Consider a partition of V into U and $V \setminus U$, with $s \in U$ and $t \in V \setminus U$. Then the arc-set $\{(u, v) \mid u \in U, v \in V \setminus U\}$ is called an st-cut. For a function $w : A \to \mathbb{R}$, and $S \subseteq A$, we use $w(S)$ to denote $w(S) = \sum_{a \in S} w(a)$. For $S \subseteq A$, we use $V(S)$ to denote the set of nodes covered by S, i.e., $V(S) = \cup_{(u,v) \in S} \{u, v\}$.

Shortest path games have been introduced in [8], and they are defined as follows. The graph has two distinguished vertices s and t, and there is a cost function $c : A \to \mathbb{R}_+$, that gives the costs for using different arcs. There is also a reward r obtained if s and t are connected by a path, then for a coalition $S \subseteq A$ the value function is

$$\mathbf{v}(S) = \begin{cases} r - m(S) & \text{if } m(S) < \infty, \\ 0 & \text{otherwise,} \end{cases} \tag{1}$$

where $m(S) = \min \{c(P) \mid P \subseteq S, P \text{ is an } st\text{-path and } c(P) \leq r\}$.

2.1   Shortest Paths

For a cost function $c : A \to \mathbb{R}$, a shortest st-path is an st-path of minimum cost. The cost of a path is the sum of the costs of the arcs in the path. If the costs are non-negative, a shortest path can be found in $O(m + n \log n)$ time with Dijkstra's algorithm, see [1].

2.2   Minimum Ratio Cycles

Assume that there is a cost function $c : A \to \mathbb{R}$, and a "time" function $\tau : A \to \mathbb{Z}_+$, then the *cost to time ratio* of a cycle D is $c(D)/\tau(D)$. An algorithm to find a cycle that minimizes the cost to time ratio was given in [11]. In our case we require the function τ to take the values 0 or 1, so the algorithm of [11] takes $O(nm + n^2 \log n)$ time.

2.3   Network Flows and Linear Programming

A reference for Minimum Cost Network Flows and Maximum Cost Circulations is [1]. We just mention that the algorithm of [10] runs in $O(n^2 m^3 \log n)$ time.

Consider the following network flow problem.

$$\max \sum_{(u,v)\in A} c(u,v)x(u,v)$$

$$\sum_{(u,v)\in A} x(u,v) - \sum_{(v,u)\in A} x(v,u) = 0, \quad \text{for } v \in V,$$

$$x(u,v) \geq 0, \quad \text{for all } (u,v) \in A.$$

The dual problem is

$$\min \sum_{v\in V} 0 \cdot \pi(v); \quad \pi(v) - \pi(u) \geq c(u,v) \quad \text{for all } (u,v) \in A.$$

Here we have a variable π for each node in V. The dual problem has a solution if and only if there is no cycle with positive cost. The cost of a cycle is the sum of the costs of its arcs.

In the following sections we use the following basic result on linear programming. Let $g(\epsilon)$ be the value of the linear program $\min\{cx \mid Ax = b + \epsilon d, \, x \geq 0\}$, then g is a convex piecewise linear function of ϵ, cf. [3].

3   The Core

Here we study some basic properties of the core. A vector $x : A \to \mathbb{R}$ is called an *allocation* if $x(A) = \mathbf{v}(A)$. Here $x(a)$ represents the amount paid to the player a. The core is a concept introduced in [9], it is based on the following stability condition: No subgroup of players does better if they break away from the joint decision of all players to form their own coalition. Thus an allocation x is in the core if $x(S) \geq \mathbf{v}(S)$ for each coalition $S \subseteq A$. Therefore the core is the polytope below.

$$\{x \in \mathbb{R}^A \mid x(A) = \mathbf{v}(A), \, x(S) \geq \mathbf{v}(S), \text{ for } S \subseteq A\}. \tag{2}$$

Let λ be the cost of a shortest st-path in G. If $\lambda \geq r$ then the core consists of only the vector $x = 0$. If $\lambda < r$, the following lemma gives a description of the core that is useful for our purposes.

Lemma 1. *If $\lambda < r$, the core is also defined by*

$$x(A) = r - \lambda, \tag{3}$$
$$x(P) \geq r - c(P) \quad \text{for each st-path } P \text{ with } c(P) \leq r, \tag{4}$$
$$x \geq 0. \tag{5}$$

Proof. Consider an arc a, if a does not go from s to t, we have $x(a) \geq 0$. If a goes from s to t then we have $x(a) \geq \max\{0, r - c(a)\}$. Thus either we have an inequality (4) or an inequality (5).

If P is an st-path with $c(P) \leq r$, then its associated inequality (4) is among the inequalities in (2).

Now assume that $S \subseteq A$ is not an st-path, but it contains an st-path with cost at most r. We have $m(S) = c(\bar{P})$ where $\bar{P} \subseteq S$ is an st-path. Then $x(S) \geq \mathbf{v}(S)$ can be written as $x(S \setminus \bar{P}) + x(\bar{P}) \geq r - c(\bar{P})$. This last inequality is implied by $x(a) \geq 0$ for $a \in S \setminus \bar{P}$, and $x(\bar{P}) \geq r - c(\bar{P})$ that is one of the inequalities (4).

Finally if $S \subseteq A$ does not contain an st-path with cost at most r, then $x(S) \geq 0$ is implied by $x(a) \geq 0$ for $a \in S$.

To understand the core, we look at the set of solutions of the linear program below.

$$\min x(A), \quad x(P) \geq r - c(P), \text{ for each st-path } P, \quad x \geq 0. \tag{6}$$

Notice that the minimum above is at least $\max\{0, r-\lambda\}$. If it is exactly this value, then the core is nonempty, and it corresponds to the set of optimal solutions of this linear program. If we look at r as a parameter, we have that the optimal value is a convex piecewise linear function of r. Then there is a value $\bar{r}$ such that the core is non-empty if and only if $r \leq \bar{r}$. See Fig. 1.

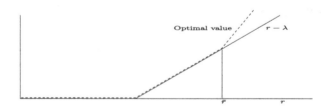

Fig. 1. Optimal value.

Consider now the dual of (6), this is

$$\max \sum_P (r - c(P)) y_P; \quad \sum \{y_P \mid a \in P\} \leq 1, \quad \text{for each arc } a; \ y \geq 0.$$

There is a variable y_P for each st-path P, and its associated cost is $r - c(P)$. Each arc has capacity one, and the sum of the variables y for the paths using a particular arc is at most one. We can reduce this to a network flow problem as follows. We add an artificial arc from t to s with cost coefficient r and infinite capacity. To every other arc a we give cost $-c(a)$ and capacity one. Then we look for a circulation of maximum cost.

Recall that λ is the value of a shortest st-path. We need the optimal value to be $r - \lambda$, for this we should have one optimal solution consisting of exactly one shortest path. The value $\bar{r}$ is the maximum value of r so that there is an optimal solution consisting of exactly one path.

Notice that the graph could contain more than one st-path with cost equal to λ. Let A' be the set of arcs that belong to a shortest path, and assume that the core is non-empty. Then the graph should not contain two arc-disjoint shortest paths. Thus the subgraph $G' = (V, A')$ contains an st-cut consisting of exactly one arc. Such an arc is called a *veto-player*, because every shortest path contains it. Let P_1 and P_2 be two shortest st-paths. If x is an element of the core, we have $x(P_1) = x(P_2) = r - \lambda$, and since $x(A) = r - \lambda$, we have $x(P_1 \setminus P_2) = x(P_2 \setminus P_1) = 0$. Thus if A'' is the set of veto-players, we have $x(a) = 0$, for $a \in A \setminus A''$. This had already been established in [8], using different techniques.

4   The Nucleolus

For a coalition S and a vector $x \in \mathbb{R}^A$, their *excess* is $e(x, S) = x(S) - \mathbf{v}(S)$. The nucleolus has been introduced in [17] trying to minimize dissatisfaction of players, more precisely, the nucleolus is the allocation that lexicographically maximizes the vector of non-decreasingly ordered excesses, cf. [17]. Thus in some sense, it is the fairest allocation. The nucleolus can be computed with a sequence of linear programs as follows, cf. [14]. First solve

$$\max \epsilon$$
$$x(S) \geq \mathbf{v}(S) + \epsilon, \quad \forall S \subset A$$
$$x(A) = \mathbf{v}(A).$$

Let ϵ_1 be the optimal value of this, and $P_1(\epsilon_1)$ be the polytope defined above, with $\epsilon = \epsilon_1$, i.e., $P_1(\epsilon_1)$ is the set of optimal solutions of the linear program above. For a polytope $P \subset \mathbb{R}^A$ let

$$\mathcal{F}(P) = \{S \subseteq A \mid x(S) = y(S), \forall x, y \in P\} \tag{7}$$

denote the set of coalitions fixed for P. In general given ϵ_{r-1} we have to solve

$$\max \epsilon \tag{8}$$
$$x(S) \geq \mathbf{v}(S) + \epsilon, \forall S \notin \mathcal{F}(P_{r-1}(\epsilon_{r-1})) \tag{9}$$
$$x \in P_{r-1}(\epsilon_{r-1}). \tag{10}$$

We denote by ϵ_r the optimal value of this, and $P_r(\epsilon_r)$ the polytope above with $\epsilon = \epsilon_r$. We continue for $r = 2, ..., |A|$, or until $P_r(\epsilon_r)$ is a singleton. Notice that each time the dimension of $P_r(\epsilon_r)$ decreases by at least one, so it takes at most $|A|$ steps for $P_r(\epsilon_r)$ to be a singleton.

In general the difficulty in computing the nucleolus resides in having to solve a sequence of linear programs with an exponential number of inequalities. In our case we shall see that most of the inequalities are redundant, and only a polynomial number of them is needed. Moreover these linear programs can be solved in a combinatorial way.

5   The Nucleolus When the Core is Non-empty

In this section we study the nucleolus under the assumption that the core is non-empty. In that case the dual of (6) should have an optimal solution consisting of exactly one path. Thus if there are several shortest st-paths, no two of them should be arc-disjoint. The only arcs a for which $x(a)$ can take a non-zero value, are the arcs that belong to the intersection of all shortest paths. To compute the nucleolus, we have to solve the sequence of linear programs described in Sect. 4, for that we need some changes of variables as follows.

5.1   An Alternative Description of the Core

For x in the core, define $z = x + c$, where c is the vectors of costs. We fix one shortest st-path $\mathcal{P}$, then the following should be satisfied: $z(a) = c(a)$, if $a \notin \mathcal{P}$; $z(a) \geq c(a)$, if $a \in \mathcal{P}$; $z(\mathcal{P}) = r$; $z(P) \geq r$, if P is an st-path; $z(P) = c(P) \geq r$, if P is an st-path not containing arcs in $\mathcal{P}$.

Now we use a change of variables similar to the one used in [16]. For two nodes a and b in $\mathcal{P}$ we denote by $\mathcal{P}_{ab}$ the sub-path of $\mathcal{P}$ going from a to b. Given z as defined above, for each node u in $\mathcal{P}$, let $p_z(u) = z(\mathcal{P}_{su})$. Therefore we have

$$p_z(s) = 0, \quad p_z(t) = r, \tag{11}$$
$$p_z(v) - p_z(u) \geq c(u,v), \quad \text{for } (u,v) \in \mathcal{P}. \tag{12}$$

For two nodes u and v in $\mathcal{P}$, a *jump* J_{uv} is a path from u to v, such that all nodes in J_{uv} different from u and v are not in $\mathcal{P}$, this notion was used in [16]. If there is an arc $a = (u,v)$ with $a \notin \mathcal{P}$, then we have a jump consisting of one arc. Consider an st-path P consisting of the sub-path $\mathcal{P}_{su}$ from s to u, then a jump J_{uv} from u to v, and a sub-path $\mathcal{P}_{vt}$ from v to t. Then the inequality $z(P) \geq r$ can be written as $p_z(u) + c(J_{uv}) + r - p_z(v) \geq r$, or

$$p_z(u) - p_z(v) \geq -c(J_{uv}). \tag{13}$$

Now we show that (11), (12) and (13) are sufficient to describe the core. We need the lemma below.

Lemma 2. *Inequalities associated with paths with more than one jump are redundant.*

Proof. Consider an st-path P with two jumps. Thus assume that P consists of the following segments: a sub-path $\mathcal{P}_{sa}$ from s to a, a jump J_{ab} from a to b, a sub-path $\mathcal{P}_{bd}$ from b to d, a jump J_{de} from d to e, and a sub-path of $\mathcal{P}_{et}$ from e to t. Then the inequality $z(P) \geq r$, can be written as $z(P) = p_z(a) + c(J_{ab}) + p_z(d) - p_z(b) + c(J_{de}) + r - p_z(e) \geq r$, or $p_z(a) - p_z(b) + c(J_{ab}) + p_z(d) - p_z(e) + c(J_{de}) \geq 0$. This is implied by $p_z(a) - p_z(b) + c(J_{ab}) \geq 0$ and $p_z(d) - p_z(e) + c(J_{de}) \geq 0$. These two inequalities correspond to st-paths with one jump. Paths with more than two jumps can be treated in a similar way.

On the other hand, let $\bar{V} = V(\mathcal{P})$, the set of nodes spanned by $\mathcal{P}$. Consider any function $p : \bar{V} \to \mathbb{R}$ satisfying (11), (12) and (13), we can define $\bar{z}(u, v) = p(v) - p(u)$ for each arc $(u, v) \in \mathcal{P}$, $\bar{z}(u, v) = c(u, v)$ for each arc $(u, v) \in A \setminus \mathcal{P}$, and $\bar{x} = \bar{z} - c$. It is easy to see that $\bar{x}$ is an element of the core. Thus there is a bijection between the vectors in the core and the functions $p : \bar{V} \to \mathbb{R}$ satisfying (11), (12) and (13).

5.2 The Nucleolus

To compute the nucleolus we have to solve the sequence of linear programs defined in Sect. 4. The development above suggests the following procedure.

We create an auxiliary graph $G' = (\bar{V}, A')$ as follows. First we include in A' each arc $(u, v) \in \mathcal{P}$ with cost $d(u, v) = c(u, v)$. Then for every pair of nodes u and v in $\bar{V}$, we find the cost of a shortest path in G from u to v using only arcs not in $\mathcal{P}$ and going only through nodes in $V \setminus \bar{V}$. Let ρ be the value of this shortest path, we add an arc (v, u) to A' with cost $d(v, u) = -\rho$. Finally we add the arcs (s, t) and (t, s) to A', with costs $d(s, t) = r$ and $d(t, s) = -r$. Then we impose the inequalities

$$p(v) - p(u) \geq d(u, v) \quad \text{for all } (u, v) \in A'. \tag{14}$$

These inequalities come from the following conditions.

- For an arc $(u, v) \in \mathcal{P}$ they correspond to $x(u, v) \geq 0$.
- For an arc $(u, v) \notin \mathcal{P}$, $(u, v) \neq (s, t), (t, s)$ they correspond to $x(P) \geq r - c(P)$, where P is the path consisting of $\mathcal{P}_{sv}$, a jump J_{vu}, and $\mathcal{P}_{ut}$.
- The inequalities for (s, t) and (t, s) imply $x(\mathcal{P}) + c(\mathcal{P}) = p(t) - p(s) = r$.

Thus from a vector p satisfying (14) we can derive a vector in the core. Moreover, as mentioned in Subsect. 2.3, the system (14) has a solution if and only if the graph G' has no cycle of positive weight.

Now we can describe the computation of the nucleolus. We set $k = 0$, $\bar{\epsilon} = 0$, and we call the arcs (s, t) and (t, s) *fixed*. We have to solve

$$\max \mu \tag{15}$$

$$p(v) - p(u) \geq d(u, v) \quad \text{if } (u, v) \in A' \text{ is fixed}, \tag{16}$$

$$p(v) - p(u) \geq d(u, v) + \mu \quad \text{if } (u, v) \in A' \text{ is not fixed}, \tag{17}$$

$$\mu \geq 0. \tag{18}$$

Inequalities (17) come from $x(u,v) \geq \mu$ for $(u,v) \in \mathcal{P}$, or $x(P) \geq r - c(P) + \mu$ if P is an st-path with one jump. To prove that these inequalities are sufficient we need the following lemma, its proof is similar to the one of Lemma 1.

Lemma 3. *The inequalities $x(S) \geq \mathbf{v}(S) + \epsilon$, for $S \subseteq A$, are implied by*

$$x(P) \geq r - c(P) + \epsilon, \quad \text{for each } st\text{-path } P; \ x(a) \geq \epsilon; \ \epsilon \geq 0.$$

As seen in Subsect. 2.3, when solving (15)–(18) we are looking for the maximum value of μ so that when we increase the costs of the non-fixed arcs by this amount, the graph has no positive cycle. Thus we need $d(C) + \mu\, n(C) \leq 0$, for each cycle C, where $n(C)$ is the number of non-fixed arcs in the cycle C. Then we have to compute

$$\bar{\mu} = \min_{C} \frac{-d(C)}{n(C)}, \tag{19}$$

where the minimum is taken over all cycles in G'. Here we are looking for a cycle that minimizes a "cost to time" ratio. As mentioned in Subsect. 2.2, this can be solved with the algorithm of [11]. Once the value $\bar{\mu}$ is obtained in (19), we update the arcs costs as $d(u,v) \leftarrow d(u,v) + \bar{\mu}$, for each non-fixed arc $(u,v) \in A'$. Let $\bar{C}$ be a cycle giving the minimum in (19), we declare *fixed* all arcs in $\bar{C}$. We update $\bar{\epsilon} \leftarrow \bar{\epsilon} + \bar{\mu}$, $k \leftarrow k+1$, and set $\epsilon_k = \bar{\epsilon}$. As long as there is an arc in A' that is not fixed we solve (15)–(18) and continue. Since at each iteration at least one new arc in A' becomes fixed, this procedure takes at most $|A'|$ iterations.

6   The Nucleolus When the Core Is Empty

Here we assume that the core is empty, thus for the first linear program defined in Sect. 4, we have $\epsilon_1 < 0$. First we have to prove that its set of optimal solutions $P_1(\epsilon_1)$, is in the non-negative orthant.

Lemma 4. $P_1(\epsilon_1) \subset \mathbb{R}_+^A$.

Proof. Assume that $(\bar{x}, \epsilon_1)$ is a solution of

$$\max \epsilon$$
$$x(A) = \mathbf{v}(A)$$
$$x(S) \geq \mathbf{v}(S) + \epsilon, \quad \text{for } S \subset A,$$

and $\bar{x}(a_0) < 0$ for some arc $a_0 \in A$.

First we prove that if $\bar{x}(S) = \mathbf{v}(S) + \epsilon_1$, then $a_0 \in S$. Suppose $a_0 \notin S$. We have two cases:

– $S \cup \{a_0\} \neq A$. Then $\bar{x}(S \cup \{a_0\}) < \bar{x}(S) = \mathbf{v}(S) + \epsilon_1 \leq \mathbf{v}(S \cup \{a_0\}) + \epsilon_1$. This contradicts the feasibility of $(\bar{x}, \epsilon_1)$.

– $S \cup \{a_0\} = A$. Then $\bar{x}(A) = \bar{x}(S) + \bar{x}(a_0) = v(S) + \epsilon_1 + \bar{x}(a_0) < v(S) \leq v(A)$. Again this contradicts the feasibility of $(\bar{x}, \epsilon_1)$.

Then we can define $x'(a_0) = \bar{x}(a_0) + \beta$, and $x'(a) = \bar{x}(a) - \beta/(m-1)$, for $a \in A \setminus \{a_0\}$, for a small number $\beta > 0$, $m = |A|$. Since x' is a better solution we have a contradiction.

As in the previous section, we have to see that when computing the nucleolus most inequalities are redundant. This is in the lemma below, its proof is similar to the one of Lemma 1.

Lemma 5. *Inequalities $x(S) \geq v(S) + \epsilon$, for $S \subset A$, are implied by $x(P) \geq r - c(P) + \epsilon$, for each st-path P, and $x \geq 0$.*

Then to compute ϵ_1, we can look for the maximum value of the parameter $\epsilon < 0$, so that the value of the linear program below is $r - \lambda$.

$$\min x(A); \; x(P) \geq r + \epsilon - c(P), \text{ for each } st\text{-path } P; \; x \geq 0. \tag{20}$$

The dual of (20), is

$$\max \sum_P (r + \epsilon - c(P))y_P; \; \sum \{y_P \mid a \in P\} \leq 1, \text{ for each arc } a; \; y \geq 0. \tag{21}$$

As before, we can reduce this to a network flow problem. We add an artificial arc from t to s with cost coefficient $r + \epsilon$ and infinite capacity. To every other arc a we give cost $-c(a)$ and capacity one. Then we look for a circulation of maximum cost. Notice that a solution of this flow problem corresponds to a set of arc-disjoint st-paths of minimum cost. Recall that $m = |A|$. For a non-negative integer k, $0 \leq k \leq m$, let $f(k)$ be the value of a minimum cost set of k arc-disjoint st-paths. Denote by $g(\epsilon)$ the value of (20), then

$$g(\epsilon) = \max_k \{k(r + \epsilon) - f(k)\}. \tag{22}$$

Here the maximum is taken over all possible values of k so that G has k arc-disjoint st-paths. The function $g(\cdot)$ is convex and piece-wise linear. One evaluation of $g(\cdot)$ is done by solving a network flow problem. In what follows we discuss how to find the value ϵ_1 so that $g(\epsilon_1) = r - \lambda$, this is done with the algorithm below.

Step 0. We set $\epsilon^- = -r$ and $\epsilon^+ = 0$, thus $g(\epsilon^-) < r - \lambda$ and $g(\epsilon^+) > r - \lambda$. We denote by $k(\epsilon)$ a value of k giving the maximum in (22).

Step 1. Let $k^- = k(\epsilon^-)$, $k^+ = k(\epsilon^+)$. If $k^- = k^+$, then ϵ_1 is the solution of $g(\epsilon^-) + k^- (\epsilon - \epsilon^-) = r - \lambda$, and we Stop. Otherwise let $\tilde{\epsilon}$ be the solution of $g(\epsilon^-) + k^- (\epsilon - \epsilon^-) = g(\epsilon^+) + k^+ (\epsilon - \epsilon^+)$.

Step 2. If $g(\tilde{\epsilon}) < r - \lambda$ set $\epsilon^- = \tilde{\epsilon}$. Otherwise set $\epsilon^+ = \tilde{\epsilon}$ and go to Step 1.

Notice that at each iteration, either k^- increases or k^+ decreases, thus this algorithm takes at most $2m$ iterations.

Now we assume that the value ϵ_1 has been found. An optimal solution of (21) corresponds to a set of k arc-disjoint st-paths $\mathcal{S} = \{\mathcal{P}^1, \ldots, \mathcal{P}^k\}$. We set $\mathcal{P} = \cup_i \mathcal{P}^i$. Let $\bar{x}$ be a solution of (20), then the complementary slackness conditions imply $\bar{x}(a) = 0$ if $a \in A \setminus \mathcal{P}$, and $\bar{x}(\mathcal{P}^i) = r + \epsilon_1 - c(\mathcal{P}^i)$, $i = 1, \ldots, k$.

Let $\bar{z}(a) = \bar{x}(a) + c(a)$, for each $a \in A$, then $\bar{z}(a) \geq c(a)$, for $a \in \mathcal{P}$; $\bar{z}(a) = c(a)$, if $a \in A \setminus \mathcal{P}$; $\bar{z}(\mathcal{P}^i) = r + \epsilon_1$, $i = 1, \ldots, k$; $\bar{z}(P) \geq r + \epsilon_1$, for every st-path P. Let $\bar{V} = V(\mathcal{P})$. Before the next change of variables, we need the lemma below.

Lemma 6. *Let $v \in \bar{V}$, and assume that there are two paths $\mathcal{P}^i$ and $\mathcal{P}^j$ going through v. Then $\bar{z}(\mathcal{P}^i_{sv}) = \bar{z}(\mathcal{P}^j_{sv})$.*

Proof. We have $\bar{z}(\mathcal{P}^i_{sv}) + z(\mathcal{P}^i_{vt}) = \bar{z}(\mathcal{P}^j_{sv}) + \bar{z}(\mathcal{P}^j_{vt}) = r + \epsilon_1$. If $\bar{z}(\mathcal{P}^i_{sv}) < \bar{z}(\mathcal{P}^j_{sv})$, then $\bar{z}(\mathcal{P}^i_{sv}) + z(\mathcal{P}^j_{vt}) < r + \epsilon_1$. This leads to a contradiction because for the st-path $\mathcal{P}^i_{sv} \cup \mathcal{P}^j_{vt}$ we should have $\bar{z}(\mathcal{P}^i_{sv}) + z(\mathcal{P}^j_{vt}) \geq r + \epsilon_1$.

Based on Lemma 6, for any $u \in \bar{V}$ we can define $p_z(u) = z(\mathcal{P}^i_{su})$, where $\mathcal{P}^i$ is any path in $\mathcal{S}$ going through u. We have

$$p_z(s) = 0, \quad p_z(t) = r, \tag{23}$$
$$p_z(v) - p_z(u) \geq c(u, v), \quad \text{for } (u, v) \in \mathcal{P}. \tag{24}$$

Now we use the same notion of a jump used in the last section. Consider an st-path P consisting of the sub-path $\mathcal{P}^i_{su}$ from s to u, then a jump J_{uv} from u to v, and a sub-path $\mathcal{P}^j_{vt}$ from v to t. Then the inequality $z(P) \geq r + \epsilon_1$ can be written as $p_z(u) + c(J_{uv}) + r + \epsilon_1 - p_z(v) \geq r + \epsilon_1$, or $p_z(u) - p_z(v) \geq -c(J_{uv})$.

Below we have the analogue of Lemma 2, its proof is similar.

Lemma 7. *Inequalities associated with paths with more than one jump are redundant.*

As in Sect. 5, we create an auxiliary graph $G' = (\bar{V}, A')$ as follows. First we include in A' each arc $(u, v) \in \mathcal{P}$ with cost $d(u, v) = c(u, v)$. Then for every pair of nodes u and v in $\bar{V}$, we find the cost of a shortest path in G from u to v using arcs not in $\mathcal{P}$ and going only through nodes in $V \setminus \bar{V}$. Let ρ be the value of this shortest path, we add an arc (v, u) to A' with cost $d(v, u) = -\rho$. Finally we add the arcs (s, t) and (t, s) to A', with costs $d(s, t) = r + \epsilon_1$ and $d(t, s) = -r - \epsilon_1$. Then we impose the inequalities

$$p(v) - p(u) \geq d(u, v) \quad \text{for all } (u, v) \in A'. \tag{25}$$

For a vector p satisfying (25) we can define $z(u, v) = p(v) - p(u)$ for $(u, v) \in \mathcal{P}$, and $z(u, v) = c(u, v)$ for $(u, v) \in A \setminus \mathcal{P}$. Then $x = z - c$ satisfies $x(A) = r - \lambda$ and the inequalities of (20). As before, there is a bijection between vectors x satisfying $x(A) = r - \lambda$ and the inequalities in (20), and vectors p satisfying $p(s) = 0$ and (25). Now we can proceed to the computation of the nucleolus.

6.1   Computation of the Nucleolus

Recall that $\epsilon_1 < 0$. As seen in Sect. 4, we have to solve a sequence of linear programs where we maximize a parameter ϵ. We divide this into two phases. The case when $\epsilon \leq 0$ is treated first, and then we continue with the case when $\epsilon > 0$.

The Case When $\epsilon \leq 0$. We set $\bar{\epsilon} = \epsilon_1$, $k = 0$, we call *fixed* the arcs (s,t) and (t,s), and we have to solve

$$\max \mu \tag{26}$$
$$p(v) - p(u) \geq d(u,v) \quad \text{if } (u,v) \in A' \text{ is fixed or } (u,v) \in \mathcal{P} \tag{27}$$
$$p(v) - p(u) \geq d(u,v) + \mu \quad \text{if } (u,v) \in A' \setminus \mathcal{P}, \text{ and } (u,v) \text{ is not fixed,} \tag{28}$$
$$0 \leq \mu \leq -\bar{\epsilon}. \tag{29}$$

After solving this, the new value for ϵ will be $\bar{\epsilon} + \mu$. For $(u,v) \in \mathcal{P}$ inequalities (27) correspond to $x(u,v) \geq 0$. Inequalities (28) correspond to $x(P) \geq r - c(P) + \bar{\epsilon} + \mu$, if P is an st-path with one jump. We need the inequality $\mu \leq -\bar{\epsilon}$ in (29) because the new value of ϵ should be non-positive.

When solving (26)–(29) we are looking for the maximum value of μ so that the graph has no positive cycle, if we increase the costs of the non-fixed arcs in $A' \setminus \mathcal{P}$ by this amount. Thus we need $d(C) + \mu n(C) \leq 0$ for each cycle C. Here $n(C)$ is the number of arcs in $A' \setminus \mathcal{P}$ that are non-fixed, in the cycle C. Then we have to compute

$$\alpha = \min_{C} \frac{-d(C)}{n(C)}. \tag{30}$$

Here the minimum is taken over all cycles in G'. As before this can be found with the algorithm of [11].

Once the value α is obtained in (30), we set $\mu = \min\{\alpha, -\bar{\epsilon}\}$. Then we update the arcs costs as

$$d(u,v) \leftarrow d(u,v) + \mu,$$

for each non-fixed arc $(u,v) \in A' \setminus \mathcal{P}$. If $\mu < -\bar{\epsilon}$, let $\bar{C}$ be a cycle giving the minimum in (30). We declare *fixed* all arcs in $\bar{C} \cap (A' \setminus \mathcal{P})$. We also update $\bar{\epsilon} \leftarrow \bar{\epsilon} + \mu$, $k \leftarrow k + 1$, $\epsilon_k = \bar{\epsilon}$. Notice that at the first iteration we obtain $\mu = 0$, because at this point $\bar{\epsilon} = \epsilon_1$. So this iteration just gives a set of arcs that should be fixed.

If $\bar{\epsilon} < 0$ and there is an arc in $A' \setminus \mathcal{P}$ that is not fixed we solve (26)–(29) and continue. Otherwise $\bar{\epsilon} = 0$, or all arcs in $A' \setminus \mathcal{P}$ are fixed, in this case we should have $\epsilon \geq 0$, this is treated below.

The Case When $\epsilon \geq 0$. Here the arcs that have been fixed remain fixed. We have to impose $x(a) \geq \epsilon$ for $a \in \mathcal{P}$, this corresponds to inequalities (33) for $a \in \mathcal{P}$. Thus we have to solve

$$\max \mu \tag{31}$$
$$p(v) - p(u) \geq d(u,v) \quad \text{if } (u,v) \in A' \text{ is fixed,} \tag{32}$$
$$p(v) - p(u) \geq d(u,v) + \mu \quad \text{if } (u,v) \in A' \text{ is not fixed,} \tag{33}$$
$$\mu \geq 0. \tag{34}$$

Then we proceed as in Subsect. 5.2.

7   Complexity

For space reasons we do not discuss the complexity of the algorithms given in each section. We just mention that the time complexity is dominated by the computation of ϵ_1 in Sect. 6, that requires solving $2\,m$ network flow problems. Then we have the theorem below.

Theorem 8. *The nucleolus of a shortest path game can be computed in $O(n^2 m^4 \log n)$ time.*

All the development was done on a directed graph, undirected graphs can be treated in an analogous way. Similar games defined on the nodes of a graph can also be treated with this approach.

References

1. Ahuja, R.K., Magnanti, T.L., Orlin, J.B.: Network Flows: Theory, Algorithms, and Applications. Prentice hall, Upper Saddle River (1993)
2. Aziz, H., Sørensen, T.B.: Path coalitional games, arXiv preprint arXiv:1103.3310 (2011)
3. Chvatal, V.: Linear Programming. Macmillan, London (1983)
4. Deng, X., Fang, Q., Sun, X.: Finding nucleolus of flow game. J. Comb. Optim. **18**, 64–86 (2009)
5. Elkind, E., Pasechnik, D.: Computing the nucleolus of weighted voting games. In: Proceedings of the Twentieth Annual ACM-SIAM Symposium on Discrete Algorithms, Society for Industrial and Applied Mathematics, pp. 327–335 (2009)
6. Faigle, U., Kern, W., Kuipers, J.: Note computing the nucleolus of min-cost spanning tree games is NP-hard. Int. J. Game Theory **27**, 443–450 (1998)
7. Fang, Q., Li, B., Shan, X., Sun, X.: The least-core and nucleolus of path cooperative games. In: Xu, D., Du, D., Du, D. (eds.) COCOON 2015. LNCS, vol. 9198, pp. 70–82. Springer, Cham (2015). doi:10.1007/978-3-319-21398-9_6
8. Fragnelli, V., Garcia-Jurado, I., Mendez-Naya, L.: On shortest path games. Math. Methods Oper. Res. **52**, 251–264 (2000)
9. Gillies, D.B.: Solutions to general non-zero-sum games. Contrib. Theory Games **4**, 47–85 (1959)
10. Goldberg, A.V., Tarjan, R.E.: Finding minimum-cost circulations by canceling negative cycles. J. ACM (JACM) **36**, 873–886 (1989)
11. Hartmann, M., Orlin, J.B.: Finding minimum cost to time ratio cycles with small integral transit times. Networks **23**, 567–574 (1993)
12. Kalai, E., Zemel, E.: Generalized network problems yielding totally balanced games. Oper. Res. **30**, 998–1008 (1982)
13. Kern, W., Paulusma, D.: Matching games: the least core and the nucleolus. Math. Oper. Res. **28**, 294–308 (2003)
14. Kopelowitz, A.: Computation of the kernels of simple games and the nucleolus of n-person games. Technical report, DTIC Document (1967)
15. Megiddo, N.: Computational complexity of the game theory approach to cost allocation for a tree. Math. Oper. Res. **3**, 189–196 (1978)
16. Potters, J., Reijnierse, H., Biswas, A.: The nucleolus of balanced simple flow networks. Games Econ. Behav. **54**, 205–225 (2006)
17. Schmeidler, D.: The nucleolus of a characteristic function game. SIAM J. Appl. Math. **17**, 1163–1170 (1969)
18. Solymosi, T., Raghavan, T.E.: An algorithm for finding the nucleolus of assignment games. Int. J. Game Theory **23**, 119–143 (1994)

Earning Limits in Fisher Markets with Spending-Constraint Utilities

Xiaohui Bei[1], Jugal Garg[2], Martin Hoefer[3]([✉]), and Kurt Mehlhorn[4]

[1] Nanyang Technological University, Singapore, Singapore
`xhbei@ntu.edu.sg`
[2] University of Illinois at Urbana-Champaign, Champaign, USA
`jugal@illinois.edu`
[3] Goethe University Frankfurt, Frankfurt, Germany
`mhoefer@cs.uni-frankfurt.de`
[4] MPI Informatik, Saarbrücken, Germany
`mehlhorn@mpi-inf.mpg.de`

Abstract. Earning limits are an interesting novel aspect in the classic Fisher market model. Here sellers have bounds on their income and can decide to lower the supply they bring to the market if income exceeds the limit. Beyond several applications, in which earning limits are natural, equilibria of such markets are a central concept in the allocation of indivisible items to maximize Nash social welfare.

In this paper, we analyze earning limits in Fisher markets with linear and spending-constraint utilities. We show a variety of structural and computational results about market equilibria. The equilibrium price vectors form a lattice, and the spending of buyers is unique in non-degenerate markets. We provide a scaling-based algorithm that computes an equilibrium in time $O(n^3 \ell \log(\ell + nU))$, where n is the number of agents, $\ell \geq n$ a bound on the segments in the utility functions, and U the largest integer in the market representation. Moreover, we show how to refine any equilibrium in polynomial time to one with minimal prices, or one with maximal prices (if it exists). Finally, we discuss how our algorithm can be used to obtain in polynomial time a 2-approximation for Nash social welfare in multi-unit markets with indivisible items that come in multiple copies.

1 Introduction

Fisher markets are a fundamental model to study competitive allocation of goods among rational agents. In a Fisher market, there is a set B of buyers and a set G of divisible goods. Each buyer $i \in B$ has a budget $m_i > 0$ of money and a utility function u_i that maps any bundle of goods to a non-negative utility value. Each good $j \in G$ is assumed to come in unit supply and to be sold by a separate seller. A *competitive* or *market equilibrium* is an allocation vector of goods and a vector of prices, such that (1) every buyer spends his budget to buy an optimal bundle of goods, and (2) supply equals demand.

© Springer International Publishing AG 2017
V. Bilò and M. Flammini (Eds.): SAGT 2017, LNCS 10504, pp. 67–79, 2017.
DOI: 10.1007/978-3-319-66700-3_6

Fisher markets have been studied intensively in algorithmic game theory. For many strictly increasing and concave utility functions, market equilibria can be described by convex programs [10,15]. There are a variety of algorithms for computing market equilibria [7,8,12,19]. For linear markets, there are algorithms that run in strongly polynomial time [14,18], and proportional response dynamics that converge to equilibrium quickly [4,20].

A common assumption in all this work is that utility functions are non-satiated, that is, the utility of every buyer i strictly increases with amount of good allocated, and the utility of every seller j strictly increases with the money earned. Consequently, when buyers and sellers are price-taking agents, it is in their best interest to spend their entire budget and bring all supply to the market, resp. In this paper, we study new variants of linear Fisher markets with satiated utility functions recently proposed in [5]. In these markets, each seller has an earning limit, which gives him an incentive to possibly reduce the supply that he brings to the market. This is a natural property in many domains, e.g., when sellers have revenue targets. Many properties of such markets are not well-understood.

Interestingly, equilibria in Fisher markets with earning limits also relate closely to *fair allocations of indivisible* items. There has been a surge of interest in allocating indivisible items to maximize Nash social welfare. Very recent work [1,6] has provided the first constant-factor approximation algorithms for this important problem. The algorithms first compute and then cleverly round a market equilibrium of a Fisher market with earning limits. The tools and techniques for computing market equilibria are a key component in this approach.

In this paper, we consider algorithmic and structural properties of markets with earning limits and spending-constraint utilities. Spending-constraint utilities are a natural generalization of linear utilities with many additional applications [8,16]. We show structural properties of equilibria and provide new and improved polynomial-time algorithms for computation. Moreover, we show how these algorithms can be used to approximate Nash social welfare in markets where each item j is provided in d_j copies (where d_j is a given integer). We obtain the first polynomial-time approximation algorithms for multi-unit markets.

Contribution and Outline. After formal discussion of the market model, we discuss some preliminaries in Sect. 2, including a formal condition for existence of equilibrium. In Sect. 3, we show that the set of equilibrium price vectors forms a lattice. While there always exists an equilibrium with pointwise smallest prices, an equilibrium with largest prices might not exist. Moreover, in non-degenerate markets (for a formal definition see Sect. 2) the spending of buyers in every equilibrium is unique.

In Sect. 4 we outline a novel algorithm to compute an equilibrium in time $O(n^3 \ell \log(\ell + nU))$, where n is the total number of agents, ℓ is the maximum number of segments in the description of the utility functions that is incident to any buyer or any good, and U is the largest integer in the representation of utilities, budgets, and earning limits. For linear markets, the running time simplifies to $O(n^4 \log nU))$. Our algorithm uses a scaling technique with decreasing prices and maintains assignments in which buyers overspend their money. A technical challenge is to maintain rounded versions of the spending restrictions in the utility functions. The algorithm runs until the maximum overspending of all buyers becomes tiny and then rounds the outcome to an exact equilibrium. Given an arbitrary equilibrium, we show how to find in polynomial time an equilibrium with smallest prices, or one with largest prices (if it exists).

Finally, in Sect. 5 we round a market equilibrium in linear markets with earning caps to an allocation in indivisible multi-unit markets to approximate the Nash social welfare. In these markets, the representation is given by the set of items and for each item j a number d_j of the available copies. The direct application of existing algorithms [1,6] would require pseudo-polynomial time. Instead, we show how to adjust the rounding procedure in [6] to run in strongly polynomial time. The resulting algorithm yields a 2-approximation and runs in time $O(n^4 \log(nU))$, which is polynomial in the input size.

Due to spatial constraints, all missing details and proofs are deferred to the full version.

Related Work. For Fisher markets we focus on some directly related work about computation of market equilibria. For markets with linear utilities a number of polynomial-time algorithms have been derived [7,12,19], including ones that run in strongly polynomial time [14,18]. For spending-constraint utilities in exchange markets [8] a polynomial-time algorithm was recently obtained [2]. For Fisher markets with spending-constraint utilities, the algorithm by Végh [18] runs in strongly polynomial time.

Linear markets with either utility or earning limits were studied only recently [3,5]. The equilibria solve standard convex programs. The Shmyrev program for earning limits also applies to spending-constraint utilities. Our paper complements our previous results [3] on linear markets with utility limits, where we proved that (1) equilibria form a lattice, (2) an equilibrium with maximum prices can be computed in time $O(n^8 \log(nU))$, (3) it can be refined in polynomial time to an equilibrium with minimum prices, and (4) several related problem variants are NP- or PPAD-hard. The framework of [17] provides an (arbitrary) equilibrium in time $O(n^5 \log(nU))$. For earning limits, our algorithm runs in time $O(n^3 \ell \log(\ell + nU))$ for spending-constraint and $O(n^4 \log(nU))$ for linear utilities. It computes an approximate solution that can be rounded to an exact equilibrium. An approximate solution could also be obtained with classic algorithms for separable convex optimization [11,13]. These algorithms seem slower – the algorithm of [13] obtains the required precision only in time $O(n^3 \ell^2 \log(\ell) \log(\ell + nU))$.

An interesting open problem are strongly polynomial-time algorithms for arbitrary earning limits. A non-trivial challenge in adjusting [14] is the precision of intermediate prices. For the framework of [18] it appears challenging to generalize the ERROR-method to markets with earning limits.

Approximating optimal allocations of indivisible items that maximize Nash social welfare has been studied recently for markets with additive [5,6] and separable concave valuations [1]. Here equilibria of markets with earning limits can be rounded to yield a 2-approximation. We extend this approach to markets with multi-unit items, where each item j comes in d_j copies (and the input includes d_j in standard logarithmic coding). In contrast to the direct, pseudo-polynomial extensions of previous work, we show how to obtain a 2-approximation in polynomial time.

2   Preliminaries

In a spending-constraint Fisher market with earning limits, there is a set B of buyers and a set G of goods. Every buyer $i \in B$ has a *budget* $m_i > 0$ of money. The utility of buyer i is a *spending-constraint function* given by non-empty sets of *segments* $K_{ij} = \{(i,j,k) \mid 1 \le k \le \ell_{ij}\}$ for each good $j \in G$. Each segment $(i,j,k) \in K_{ij}$ comes with a *utility value* u_{ijk} and a *spending limit* $c_{ijk} > 0$. We assume that the utility function is piecewise linear and concave, i.e., $u_{ijk} > u_{ij,k+1} > 0$ for all $\ell_{ij} - 2 \ge k \ge 1$. W.l.o.g. we assume that the last segment has $u_{ij\ell_{ij}} = 0$ and $c_{ijk} = \infty$.

Buyer i can spend at most an amount of c_{ijk} of money on segment (i,j,k). We use $\mathbf{f} = (f_{ijk})_{(i,j,k) \in K_{ij}}$ to denote the spending of money on segments. $\mathbf{f}$ is termed *money flow*. A segment is *closed* if $f_{ijk} \ge c_{ijk}$, otherwise *open*. For notational convenience, we let $f_{ij} = \sum_{(i,j,k) \in K_{ij}} f_{ijk}$.

Given a vector $\mathbf{p} = (p_j)_{j \in G}$ of strictly positive *prices* for goods, a money flow results in an allocation $x_{ij} = \sum_k f_{ijk}/p_j$ of good j. The *bang-per-buck ratio* of segment (i,j,k) is $\alpha_{ijk} = u_{ijk}/p_j$. To maximize his utility, buyer i spends his budget m_i on segments in non-increasing order of bang-per-buck ratio, while respecting the spending limits. A bundle $\mathbf{x}_i = (x_{ij})_{j \in G}$ that results from this approach is termed a *demand bundle* and denoted by $\mathbf{x}_i^*$. The corresponding money flow on the segments is termed *demand flow* $\mathbf{f}_i^*$.

Demand bundles and flows might not be unique, but they differ only on the allocated segments with smallest bang-per-buck ratio. This smallest ratio is termed *maximum bang-per-back (MBB) ratio* and denoted by α_i. Note that α_i is unique given $\mathbf{p}$. All segments with $\alpha_{ijk} \ge \alpha_i$ are termed *MBB segments*. The segments with $\alpha_{ijk} = \alpha_i$ are termed *active segments*. We assume w.l.o.g. $m_i \le \sum_{j,k:u_{ijk}>0} c_{ijk}$, since no buyer would spend more, and we can assume there is no allocation on segments with $u_{ijk} = 0$. Therefore, we assume buyers always spend all their money.

In this paper, we study a natural condition on seller supplies. Each good is owned by a different seller, and the seller has a maximum endowment of 1. Seller j comes with an *earning limit* d_j. He only brings a supply $e_j \le 1$ that suffices to reach this earning limit under the given prices. Intuitively, while each seller has utility $\min\{d_j, e_j p_j\}$, we also assume that he has tiny utility for unsold parts of his good. Hence, he only brings a supply to earn d_j.

More formally, the *active price* of good j is given by $p_j^a = \min(d_j, p_j)$. His good is *capped* if $p_j^a = d_j$ and *uncapped* otherwise. A *thrifty supply* is $e_j = p_j^a / p_j$, which guarantees $e_j p_j \le d_j$, i.e., the earning limit holds when market clears.
The goal is to find a *thrifty equilibrium*.

Definition 1. *A pair* $(\mathbf{f}, \mathbf{p})$ *is a thrifty equilibrium if (1)* $\mathbf{f}_i$ *is a demand flow for prices* $\mathbf{p}$ *for every* $i \in B$*, and (2)* $\sum_{i,k} f_{ijk} = p_j^a$*, for every* $j \in G$*.*

Proposition 1. *Across all thrifty equilibria: (1) the seller incomes are unique; (2) there is a unique set of uncapped goods, and their prices are unique; and (3) uncapped goods are available in full supply, capped goods in thrifty supply.*

These uniqueness properties are a direct consequence of the fact [5] that thrifty equilibria are the solutions of the following convex program.

$$\text{Max.} \sum_{i \in B} \sum_{j \in G} \sum_{(i,j,k) \in K_{ij}} f_{ijk} \log u_{ijk} - \sum_{j \in G} (q_j \log q_j - q_j)$$

$$\text{s.t.} \sum_{j \in G} \sum_{(i,j,k) \in K_{ij}} f_{ijk} = m_i \quad \forall i \in B$$

$$\sum_{i \in B} \sum_{(i,j,k) \in K_{ij}} f_{ijk} = q_j \quad \forall j \in G \qquad (1)$$

$$f_{ijk} \le c_{ijk} \quad \forall (i,j,k) \in K_{ij}$$

$$q_j \le d_j \quad \forall j \in G$$

$$f_{ijk} \ge 0 \quad \forall i \in B, \, j \in G, (i,j,k) \in K_{ij}$$

The incomes of sellers and, consequently, the sets of capped and uncapped goods are unique in all thrifty equilibria. The money flow, allocation, and prices of capped goods might not be unique.

Buyers always spend all their budget, but this can be impossible when every seller must not earn more than its limit[1]. Then a thrifty equilibrium does not exist. This, however, turns out to be the only obstruction to nonexistence.

Let $\hat{B} \subseteq B$ be a set of buyers, and $N(\hat{B}) = \{ j \in G \mid u_{ij1} > 0, i \in \hat{B} \}$ be the set of goods such that there is at least one buyer in $\hat{B}$ with positive utility on its first segment for the good. The following *money clearing* condition states that buyers can spend their money without violating the earning limits.

Definition 2 (Money Clearing). *A market is* money clearing *if for every subset of buyers* $\hat{B} \subseteq B$ *there is a flow* f *such that*

$$f_{ij} \le \sum_{k=1}^{k^+} c_{ijk}, \quad \forall i \in \hat{B}, \forall j \in N(\hat{B}), k^+ = \max\{k \mid u_{ijk} > 0\}$$

$$\sum_{i \in \hat{B}} f_{ij} \le d_j, \quad \forall j \in N(\hat{B}) \quad \text{and} \quad \sum_{j \in N(\hat{B})} f_{ij} \ge m_i, \quad \forall i \in \hat{B} \, . \qquad (\text{MC})$$

---

[1] Consider the example of a linear market with one buyer and one good. The utility is $u_{11} > 0$, the buyer has a budget $m_1 = 2$, the good has an earning limit $d_1 = 1$.

Money clearing is clearly necessary for the existence of a thrifty equilibrium. It is also sufficient since, e.g., our algorithm in Sect. 4 will successfully compute an equilibrium iff money clearing holds. Alternatively, it can be verified that this is the unique necessary and sufficient feasibility condition for convex program (1). It is easy to check condition (MC) by a max-flow computation. We therefore assume that our market instance satisfies it.

Lemma 1. *A thrifty equilibrium exists iff the market is money clearing.*

Let us define some more useful concepts for the analysis. For any pair $(\mathbf{f}, \mathbf{p})$ the *surplus of buyer* i is given by $s(i) = \sum_{j \in G} f_{ij} - m_i$, and the *surplus of good* j is $s(j) = p_j^a - \sum_{i \in B} f_{ij}$. The *active-segment graph* $G(\mathbf{p})$ is a bipartite graph $(B \cup G, E)$ which contains edge $\{i, j\}$ iff there is some active segment (i, j, k). Note that there can be at most one active segment (i, j, k) for an (i, j). A market is called *non-degenerate* if the active segment graph for any non-zero $\mathbf{p}$ is a forest.

3   Structure of Thrifty Equilibria

Some Intuition. We start by providing some intuition for the structural results in the case where all utility functions are linear, i.e., with a single segment in every K_{ij}. Consider a thrifty equilibrium $(\mathbf{f}, \mathbf{p})$. Call an edge (i, j) $\mathbf{p}$-MBB if $u_{ij}/p_j = \alpha_i$. The active-segment graph here simplifies to an *MBB graph* $G(\mathbf{p})$.

Let C be any connected component of the MBB graph. The buyers in C spend all budget on the goods in C, and no other buyer spends money on the goods in C. Thus

$$\sum_{i \in C \cap B} m_i = \sum_{j \in C \cap G} p_j^a = \sum_{j \in C \cap G_u} p_j + \sum_{j \in C \cap G_c} d_j,$$

where G_c and G_u are the sets of capped and uncapped goods, resp. First, assume all goods in C are capped. Let r be a positive real and consider the pair $(\mathbf{f}, \mathbf{p}')$, where $p_j' = r \cdot p_j$ if $j \in C \cap G_c$ and $p_j' = p_j$ otherwise.

Note that the allocations for any good $j \in C \cap G_c$ are scaled by $1/r$. The pair $(\mathbf{f}, \mathbf{p}')$ is an equilibrium provided that all edges with positive allocation are also $\mathbf{p}'$-MBB and $p_j' \geq d_j$ for all $j \in C \cap G_c$. This certainly holds for $r > 1$ and $r - 1$ sufficiently small. If $p_j > d_j$ for all $j \in C$ this also holds for $r < 1$ and $1 - r$ sufficiently small. Thus, there is some freedom in choosing the prices in components containing only capped goods even for a fixed MBB graph. For non-degenerate instances, the money flow is unique (but not the allocation).

Now assume that there is at least one uncapped good in C, and let j_u be such an uncapped good. The price of any other good j in the component is linearly related to the price j_u, i.e., $p_j = \gamma_j p_{j_u}$, where γ_j is a rational number whose numerator and denominator is a product of utilities. Thus,

$$\sum_{i \in C \cap B} m_i = \sum_{j \in C \cap G} p_j^a = \sum_{j \in C \cap G_u} \gamma_j p_{j_u} + \sum_{j \in C \cap G_c} d_j,$$

and the reference price is uniquely determined. All prices in the component are uniquely determined. For a non-degenerate instance the money flow and allocation are also uniquely determined.

Suppose in a component C containing only capped goods we increase the prices by a common factor $r > 1$. We raise r continuously until a new MBB edge arises. If we can raise r indefinitely, no buyer in the component is interested in any good outside the component. Otherwise, a new MBB edge arises, and then C is united with some other component. At this moment, the money flow over the new MBB edge is zero. If the newly formed component contains an uncapped good, prices in the component are fixed and money flow is exactly as in the moment of joining the components. Otherwise, we raise all prices in the newly formed component, and so on. If the market is non-degenerate, then money flow is unique, and money will never flow on the new MBB edge.

If the component contains only capped goods j with $p_j > d_j$, we can decrease prices continuously by a common factor $r < 1$ until a new MBB edge arises. If no MBB edge ever arises, no buyer outside the component is interested in any good in the component, which allows to argue as above.

We have so far described how the prices in a component of the MBB graph of an equilibrium are determined if at least one good is uncapped, and how the prices can be scaled by a common factor if all goods are capped. We have also discussed how components are merged and that the new MBB edge arising in a merge will never carry nonzero flow. Components can also be split if they contain an edge with zero flow.

Consider an equilibrium $(\mathbf{f}, \mathbf{p})$ and assume $f_{ij} = 0$ for some edge (i, j) of the MBB graph w.r.t. $\mathbf{p}$. Let C be the component containing (i, j) and let C_1 and C_2 be the components of $C \setminus \{i, j\}$. Let the instance be non-degenerate. Hence, the MBB graph is a forest. If we want to retain all MBB edges within C_1 and C_2 and only drop (i, j), we have to either increase all prices in the subcomponent containing j or decrease all prices in the subcomponent containing i. Both options are infeasible if both components contain a good with price strictly below its earning limit. The first option is feasible if the component containing j contains only goods with prices at least their earning limits. The latter option is feasible if the component containing i contains only goods with prices strictly larger than their earning limits. The split does not affect the money flow.

If the above described changes allow to change any equilibrium into any other equilibrium, then money flow should be unique across all equilibria. Moreover, the set of edges carrying flow should be the same in all equilibria. The MBB graph for an equilibrium contains these edges, and maybe some more edges that do not carry flow. Next, we prove that this intuition captures the truth, even for the general case of spending-constraint utility functions.

Lattice Structure. We characterize the set of price vectors of thrifty equilibria, which we denote by $\mathcal{P} = \{\mathbf{p} \mid \exists \mathbf{f} \text{ s.t. } (\mathbf{f}, \mathbf{p}) \text{ is a thrifty equilibrium}\}$. For money clearing markets, we establish two results: (1) the set of equilibrium price vectors forms a lattice, and (2) the money flow is unique in nondegenerate markets. For the first result, we consider the coordinate-wise comparison, i.e., $\mathbf{p} \geq \mathbf{p}'$ iff $p_j \geq p'_j, \forall j \in G$.

Theorem 1. *The pair $(\mathcal{P}, \geq)$ is a lattice.*

The proof relies on the following structural properties. Given $\mathbf{p}$ and $\mathbf{p}'$, we partition the set of goods into sets $S_r = \{j \mid p'_j = r \cdot p_j\}$, for $r > 0$. For a price vector $\mathbf{p}$, let segment (i, j, k) be $\mathbf{p}$-MBB if $u_{ijk}/p_j \geq \alpha_i$, and $\mathbf{p}$-active if $u_{ijk}/p_j = \alpha_i$. For a set T of goods and an equilibrium $(\mathbf{f}, \mathbf{p})$, let

$$K(T, \mathbf{p}) = \{(i, j, k) \mid \text{segment is } \mathbf{p}\text{-MBB for some } j \in T\},$$
$$K_a(T, \mathbf{p}) = \{(i, j, k) \mid f_{ijk} > 0 \text{ for some } j \in T \text{ and some equilibrium } (\mathbf{f}, \mathbf{p})\},$$

where the sets denote the set of $\mathbf{p}$-MBB segments for goods in T and the ones on which some good in T is allocated. Note that $K_a(T, \mathbf{p}) \subseteq K(T, \mathbf{p})$.

Lemma 2. *For any two thrifty equilibria $E = (\mathbf{f}, \mathbf{p})$ and $E' = (\mathbf{f}', \mathbf{p}')$:*

1. $K_a(S_r, \mathbf{p}) = K_a(S_r, \mathbf{p}')$ *for every $r > 0$, i.e., the goods in S_r are allocated on the same set of segments in both equilibria.*
2. $K_a(S_r, \mathbf{p}) = K_a(S_r, \mathbf{p}') \subseteq K(S_r, \mathbf{p}') \subseteq K(S_r, \mathbf{p})$ *for $r > 1$. Similarly, $K_a(S_r, \mathbf{p}') = K_a(S_r, \mathbf{p}) \subseteq K(S_r, \mathbf{p}) \subseteq K(S_r, \mathbf{p}')$ for $r < 1$. For $r = 1$, $K_a(S_r, \mathbf{p}') = K_a(S_r, \mathbf{p})$.*
3. *If $f_{ijk} > 0$ for $(i, j, k) \in K_a(S_r, \mathbf{p})$ with $r > 1$, then (i, j, k) is $\mathbf{p}'$-MBB. If $f'_{ijk} > 0$ for $(i, j, k) \in K_a(S_r, \mathbf{p}')$ with $r < 1$, then (i, j, k) is $\mathbf{p}$-MBB.*

Corollary 1. *There exists a thrifty equilibrium with coordinate-wise lowest prices. Among all thrifty equilibria, it yields the largest supply in the market and the maximum utility for every buyer.*

Theorem 2. *In a non-degenerate market, all thrifty equilibria have the same money flow.*

The convex program implies that there is a unique income for each seller. This is consistent with our observation that a good can have different prices in two equilibria only when income equals its earning limit.

While existence of an equilibrium with smallest prices is guaranteed, we might or might not have an equilibrium with coordinate-wise largest prices (e.g., when all goods are capped in equilibrium, prices can be raised indefinitely).

4   Algorithms to Compute Thrifty Equilibria

Scaling Algorithm. We first propose and discuss a polynomial-time scaling algorithm to compute a thrifty equilibrium. We begin with defining some useful tools and concepts. The *active-segment network* $N(\mathbf{p}) = (\{s, t\} \cup B \cup G, E)$ contains a node for each buyer and each good, along with two additional nodes s and t. It contains every edge (s, i) for $i \in B$ with capacity $m_i - c_i^c$, where $c_i^c =$

$\sum_{(i,j,k) \text{ closed}} c_{ijk}$. Also, it contains every (j,t) for $j \in G$ with capacity $p_j^a - c_j^c$, where $c_j^c = \sum_{(i,j,k) \text{ closed}} c_{ijk}$. It contains edge (i,j) with infinite capacity iff there is some active segment (i,j,k). Finally, the *active-residual network* $G_r(\mathbf{f}, \mathbf{p})$ contains a node for each buyer and each good. It contains forward edge (i,j) iff there is some active segment (i,j,k) with $f_{ijk} < c_{ijk}$ and contains backward edge (j,i) iff there is some active segment (i,j,k) with $f_{ijk} > 0$. Moreover, $G_r(\mathbf{f}, \mathbf{p}, i)$ is the subgraph of $G_r(\mathbf{f}, \mathbf{p})$ induced by the set of all buyers $i' \in G_r(\mathbf{f}, \mathbf{p})$ such that there is an augmenting path from i' to i.

Our algorithm uses Δ-*discrete capacities* $\hat{c}_{ijk} = \lceil c_{ijk}/\Delta \rceil \cdot \Delta$ for all $i \in B, j \in G$ and $(i,j,k) \in K_{ij}$, where we iteratively decrease Δ. Initially, the algorithm overestimates the budget of buyer i, where it assumes the buyer has $r\Delta$ money and every segment has Δ-discrete capacities. Then $\mathbf{f}_i$ is a (Δ, r)-*discrete demand* for buyer i iff it is a demand flow for buyer i under these conditions.

We also adjust the definitions of MBB ratio, active segments, active-segment graph, network, and residual network to the case of Δ-discrete capacities. We denote these adjusted versions by $\hat{\alpha}$, $\hat{\mathcal{G}}(\mathbf{p})$, $\hat{N}(\mathbf{p})$, $\hat{\mathcal{G}}_r(\mathbf{f}, \mathbf{p})$ and $\hat{\mathcal{G}}_r(\mathbf{f}, \mathbf{p}, j)$ resp.

Finally, we make a number of assumptions to simplify the stated bound on the running time. We assume w.l.o.g. that $|B| = |G|$ (by adding dummy buyers and/or goods) and define $n = |B| + |G|$. Moreover, we let $K_i = \bigcup_{j \in G} K_{ij}$ and $K_j = \bigcup_{i \in B} K_{ij}$ and assume w.l.o.g. that $\ell = |K_i| = |K_j| \geq n$ for every buyer i and every good j (by adding dummy segments with 0 utility).

Algorithm 1 computes a thrifty equilibrium in polynomial time. It uses descending prices and maintains a money flow on closed and open MBB segments with increasing precision and decreasing surplus. We call a run of the outer while-loop a Δ-*phase*. The algorithm runs until the precision parameter Δ is decreased to exponentially small size. A final rounding procedure (similar to [9] and deferred to the full version) then rounds the solution to an exact equilibrium.

For the analysis, we use the following notion of Δ-feasible solution.

Definition 3. *Given a value $\Delta > 0$, a pair $(\mathbf{f}, \mathbf{p})$ of flow and prices with $\mathbf{p} \geq 0$ and $\mathbf{f} \geq 0$ is a Δ-feasible solution if*

- $\ell\Delta \leq s(i) \leq (\ell+1)\Delta, \forall i \in B$.
- $\forall j \in G$: If $p_j < p_j^0$, then $0 \leq s(j) \leq \Delta$. If $p_j = p_j^0$, then $-\infty < s(j) \leq \Delta$.
- $\mathbf{f}$ *is Δ-integral, and $f_{ijk} > 0$ only if (i,j,k) is a closed or open MBB segment w.r.t. Δ-discretized capacities.*

For the running time, note that prices are non-increasing. Once a capped good becomes uncapped, it remains uncapped. We refer to an execution of the repeat loop in Algorithm 1 as an iteration. After the initialization, there may be goods j for which d_j is smaller than the initial value of Δ and which receive flow from some buyer. As long as their surplus is negative, these goods keep their initial price. The following observations are useful to prove a bound on the running time. We also observe that the precision of prices and flow values is always bounded.

Algorithm 1. Scaling Algorithm for Markets $\mathcal{M}^s$ with Earning Limits

---

Input : Fisher market $\mathcal{M}$ with spending constraint utilities and earning limits
Budget m_i, earning limits d_j, and parameters u_{ijk}, c_{ijk}

Output: Thrifty equilibrium $(\mathbf{f}, \mathbf{p})$

1   $\Delta \leftarrow U^{n+1} \sum_{i \in B} m_i; \ p_j^0 \leftarrow n(\ell+1)\Delta, \forall j \in G; \mathbf{p} \leftarrow \mathbf{p}^0$

2   $f_i \leftarrow (\Delta, \ell+1)$-discrete demand for buyer i

3   **while** $\Delta > 1/(2\ell(2nU)^{4n})$ **do**

4      $\Delta \leftarrow \Delta/2;$

5   **for** *each closed segment* (i,j,k) **do** $f_{ijk} \leftarrow \lceil c_{ijk}/\Delta \rceil \cdot \Delta$

6   **for** *each* $i \in B$ *with* $s(i) > (\ell+1)\Delta$ **do**

7   Pick any active segment (i,j,k) with $f_{ijk} > 0$ and set $f_{ijk} \leftarrow f_{ijk} - \Delta$

8   **while** *there is a good* j' *with* $s(j') > \Delta$ **do**   // Δ-phase

9   **repeat**   // iteration

10          $(\hat{B}, \hat{G}) \leftarrow$ Set of (buyers, goods) in $\hat{\mathcal{G}}_r(\mathbf{f}, \mathbf{p}, j')$

11          $x \leftarrow 1$; Define $p_j \leftarrow x p_j, \forall j \in \hat{G}$ // active prices & surpluses

12   change, too

13   Decrease x continuously down from 1 until one of the following

14   events occurs

15   **Event 1:** $s(j') = \Delta$

16   **Event 2:** $s(j) \le 0$ for a $j \in \hat{G}$

17          $P \leftarrow$ path from j to j' in $\hat{\mathcal{G}}_r(\mathbf{f}, \mathbf{p}, j')$ // Δ-augmentation

18   Update $\mathbf{f} : f_{ijk} = \begin{cases} f_{ijk} + \Delta \text{ if } (i,j) \text{ is a forward arc in } P \\ f_{ijk} - \Delta \text{ if } (i,j) \text{ is a backward arc in } P \\ f_{ijk} \quad \text{otherwise} \end{cases}$

19   **Event 3:** A capped good becomes uncapped

20   **Event 4:** New active segment (i,j,k) with $i \notin \hat{B}, \ j \in \hat{G}, \ f_{ijk} < \hat{c}_{ijk}$

21   **until** *Event 1 or 2 occurs*

22 $(\mathbf{f}, \mathbf{p}) \leftarrow$ Rounding$(\mathbf{f}, \mathbf{p})$

Lemma 3. *Once the surplus of a good is non-negative, it stays non-negative. If the surplus of a good is negative, its price is the initial price.*

Lemma 4. *The first run of the outer while-loop in Algorithm 1 takes $O(n^3\ell)$ time, every subsequent one takes $O(n^2\ell)$ time. At the end of each Δ-phase, the pair $(\mathbf{f}, \mathbf{p})$ is a Δ-feasible solution.*

Lemma 5. *If all budgets, earning limits and utility values are integers bounded by U, then all flow values and prices at the end of each iteration are rational numbers whose denominators are at most $poly(1/\Delta, n, U^n)$.*

Finally, for correctness of the algorithm, it maintains the following condition resulting from (MC) for active prices.

Lemma 6. *Let $\hat{B} \subseteq B$ be a set of buyers and let $N(\hat{B})$ be the goods having positive utility for some buyer in $\hat{B}$. At all times $\sum_{j \in N(\hat{B})} p_j^a - \sum_{i \in \hat{B}} m_i \ge 0$.*

Algorithm 2. MinPrices

Input   : Market parameters and any thrifty equilibrium $(\mathbf{f}, \mathbf{p})$
Output: Thrifty equilibrium with smallest prices

1 $E(\mathbf{f}) \leftarrow \{(i,j,k) \mid f_{ijk} > 0\}$; $G_c \leftarrow$ Set of capped goods at $(\mathbf{f}, \mathbf{p})$

2 Solve an LP in q_j and λ_i:
$$
\begin{array}{l}
\min \sum_i \lambda_i \\
q_j \leq u_{ijk}\lambda_i, \text{ for segment } (i,j,k) \in E(\mathbf{f}) \\
q_j = p_j, \qquad \forall j \in G \setminus G_c \\
q_j \geq d_j, \qquad \forall j \in G_c \\
\lambda_i, q_j \geq 0 \quad \forall i \in B, j \in G
\end{array}
$$

return $(\mathbf{f}, \mathbf{q})$

Algorithm 3. MaxPrices

Input   : Market parameters and any thrifty equilibrium $(\mathbf{f}, \mathbf{p})$
Output: Thrifty equilibrium with largest prices

1 Initialize active price $p_j^a \leftarrow \min\{d_j, p_j\}$ for every good j
2 $S \leftarrow \{j \mid p_j > 0$ and j is not connected to any uncapped good in $G(\boldsymbol{p})\}$
3 **while** $S \neq \emptyset$ **do**
4 $\quad$ $x \leftarrow 1$; Set prices $p_j \leftarrow xp_j, \ \forall j \in S$
5 $\quad$ Increase x continuously from 1 until a new active segment appears
6 $\quad$ Recompute S
7 **return** $(\mathbf{f}, \mathbf{p})$

Lemma 7. *Let $(\mathbf{f}, \mathbf{p})$ be the flow and price vector computed by the outer while-loop in Algorithm 1. The pair is Δ-feasible for $\Delta = 1/(2\ell(2nU)^{4n})$ and $-n(\ell+1) \Delta \leq s(j) \leq \Delta$ for all $j \in G$.*

Theorem 3. *Algorithm 1 computes a thrifty equilibrium for money-clearing markets $\mathcal{M}^s$ with earning limits in $O(n^3\ell \log(\ell + nU))$ time.*

Extremal Prices. Given an arbitrary thrifty equilibrium, Algorithm 2 computes a thrifty equilibrium with smallest prices. Algorithm 3 computes a thrifty equilibrium with largest prices if it exists. Otherwise, it yields a set S of goods for which prices can be raised indefinitely.

Theorem 4. *Algorithm 2 computes a thrifty equilibrium with smallest prices.*

Theorem 5. *Algorithm 3 computes a thrifty equilibrium with largest prices if it exists.*

5   Nash Social Welfare in Additive Multi-unit Markets

Using our algorithm to compute a thrifty equilibrium in linear markets with earning limits, we can approximate the optimal Nash social welfare for additive valuations, indivisible items, and multiple copies for each item. Here there are n

agents and m items. For item j, there are $d_j \in \mathbb{N}$ copies. The valuation of agent i for an assignment x of goods is $v_i(x) = \sum_j v_{ij} x_{ij}$, where x_{ij} denotes the number of copies of item j that agent i receives. The goal is to find an assignment such that the Nash social welfare $(\prod_i v_i(x))^{1/n}$ is maximized.

When all $d_j = 1$, the algorithm of [6] provides a 2-approximation [5]. It finds an equilibrium for a linear market, where each agent i is a buyer with $m_i = 1$, and each item j is a good with earning limit $d_j = 1$. Then it rounds the allocation to an integral assignment. The direct adjustment to handle $d_j \geq 1$ copies is to represent each copy of item j by a separate auxiliary item with unit supply (all valued exactly the same way as item j) and run the algorithm from [6]. A similar approach is used by [1] to provide a 2-approximation for separable concave utilities. This, however, yields a running time polynomial in $\max_j d_j$, which is only pseudo-polynomial for multi-unit markets (due to standard logarithmic coding of d_j's). We make the algorithm efficient.

Proposition 2. *There is a polynomial-time 2-approximation algorithm for maximizing Nash social welfare in multi-unit markets with additive valuations.*

References

1. Anari, N., Mai, T., Gharan, S.O., Vazirani, V.: Nash social welfare for indivisible items under separable, piecewise-linear concave utilities (2016). CoRR abs/1612.05191
2. Bei, X., Garg, J., Hoefer, M.: Ascending-price algorithms for unknown markets. In: Proceedings of 17th Conference Economics and Computation (EC), p. 699 (2016)
3. Bei, X., Garg, J., Hoefer, M., Mehlhorn, K.: Computing equilibria in markets with budget-additive utilities. In: Proceedings of 24th European Symposium Algorithms (ESA), pp. 8:1–8:14 (2016)
4. Birnbaum, B., Devanur, N., Xiao, L.: Distributed algorithms via gradient descent for Fisher markets. In: Proceedings of 12th Conference Electronic Commerce (EC), pp. 127–136 (2011)
5. Cole, R., Devanur, N., Gkatzelis, V., Jain, K., Mai, T., Vazirani, V., Yazdanbod, S.: Convex program duality, Fisher markets, and Nash social welfare. In: Proceedings of 18th Conference Economics and Computation (EC) (2017, to appear)
6. Cole, R., Gkatzelis, V.: Approximating the Nash social welfare with indivisible items. In: Proceedings of 47th Symposium Theory of Computing (STOC), pp. 371–380 (2015)
7. Devanur, N., Papadimitriou, C., Saberi, A., Vazirani, V.: Market equilibrium via a primal-dual algorithm for a convex program. J. ACM **55**(5), 22:1–22:18 (2008)
8. Devanur, N., Vazirani, V.: The spending constraint model for market equilibrium: algorithmic, existence and uniqueness results. In: Proceedings of 36th Symposium Theory of Computing (STOC), pp. 519–528 (2004)
9. Duan, R., Mehlhorn, K.: A combinatorial polynomial algorithm for the linear Arrow-Debreu market. Inf. Comput. **243**, 112–132 (2015)
10. Eisenberg, E., Gale, D.: Consensus of subjective probabilities: the Pari-Mutuel method. Ann. Math. Stat. **30**(1), 165–168 (1959)
11. Hochbaum, D., Shanthikumar, G.: Convex separable optimization is not much harder than linear optimization. J. ACM **37**(4), 843–862 (1990)

12. Jain, K.: A polynomial time algorithm for computing the Arrow-Debreu market equilibrium for linear utilities. SIAM J. Comput. **37**(1), 306–318 (2007)
13. Karzanov, A., McCormick, T.: Polynomial methods for separable convex optimization in unimodular linear spaces with applications. SIAM J. Comput. **26**(4), 1245–1275 (1997)
14. Orlin, J.: Improved algorithms for computing Fisher's market clearing prices. In: Proceedings of 42nd Symposium Theory of Computing (STOC), pp. 291–300 (2010)
15. Shmyrev, V.: An algorithm for finding equilibrium in the linear exchange model with fixed budgets. J. Appl. Indust. Math. **3**(4), 505–518 (2009)
16. Vazirani, V.: Spending constraint utilities with applications to the adwords market. Math. Oper. Res. **35**(2), 458–478 (2010)
17. Végh, L.: Concave generalized flows with applications to market equilibria. Math. Oper. Res. **39**(2), 573–596 (2014)
18. Végh, L.: Strongly polynomial algorithm for a class of minimum-cost flow problems with separable convex objectives. SIAM J. Comput. **45**(5), 1729–1761 (2016)
19. Ye, Y.: A path to the Arrow-Debreu competitive market equilibrium. Math. Prog. **111**(1–2), 315–348 (2008)
20. Zhang, L.: Proportional response dynamics in the Fisher market. Theoret. Comput. Sci. **412**(24), 2691–2698 (2011)

Robustness Among Multiwinner Voting Rules

Robert Bredereck[1], Piotr Faliszewski[2], Andrzej Kaczmarczyk[3($\boxtimes$)],
Rolf Niedermeier[3], Piotr Skowron[3], and Nimrod Talmon[4]

[1] University of Oxford, Oxford, UK
[2] AGH University, Krakow, Poland
[3] TU Berlin, Berlin, Germany
a.kaczmarczyk@tu-berlin.de
[4] Weizmann Institute of Science, Rehovot, Israel

Abstract. We investigate how robust are results of committee elections to small changes in the input preference orders, depending on the voting rules used. We find that for typical rules the effect of making a single swap of adjacent candidates in a single preference order is either that (1) at most one committee member can be replaced, or (2) it is possible that the whole committee can be replaced. We also show that the problem of computing the smallest number of swaps that lead to changing the election outcome is typically NP-hard, but there are natural FPT algorithms. Finally, for a number of rules we assess experimentally the average number of random swaps necessary to change the election result.

1 Introduction

We study how multiwinner voting rules (i.e., procedures used to select fixed-size committees of candidates) react to (small) changes in the input votes. We are interested both in the complexity of computing the smallest modification of the votes that affects the election outcome, and in the extent of the possible changes. We start by discussing our ideas informally in the following example.

Consider a research-funding agency that needs to choose which of the submitted project proposals to support. The agency asks a group of experts to evaluate the proposals and to rank them from the best to the worst one. Then, the agency uses some formal process—here modeled as a multiwinner voting rule—to aggregate these rankings and to select k projects to be funded. Let us imagine that one of the experts realized that instead of ranking some proposal A as better than B, he or she should have given the opposite opinion. What are the consequences of such a "mistake" of the expert? It may not affect the results at all, or it may cause only a minor change: Perhaps proposal A would be dropped (to the benefit of B or some other proposal) or B would be selected (at the expense of A or some other proposal). We show that while this indeed would be the case under a number of multiwinner voting rules (e.g., under the k-Borda rule; see Sect. 2 for definitions), there exist other rules (e.g., STV or the Chamberlin–Courant rule) for which such a single swap could lead to selecting a completely disjoint set of proposals. The agency would prefer to avoid situations

© Springer International Publishing AG 2017
V. Bilò and M. Flammini (Eds.): SAGT 2017, LNCS 10504, pp. 80–92, 2017.
DOI: 10.1007/978-3-319-66700-3_7

where small changes in the experts' opinions lead to (possibly large) changes in the outcomes; so the agency would want to be able to compute the smallest number of swaps that would change the result. In cases where this number is too small, the agency might invite more experts to gain confidence in the results.

More formally, a multiwinner voting rule is a function that, given a set of rankings of the candidates and an integer k, outputs a family of size-k subsets of the candidates (the winning committees). We consider the following three issues (for simplicity, below we assume to always have a unique winning committee):

1. We say that a multiwinner rule $\mathcal{R}$ is ℓ-*robust* if (1) swapping two adjacent candidates in a single vote can lead to replacing no more than ℓ candidates in the winning committee, and (2) there are examples where exactly ℓ candidates are indeed replaced; we refer to ℓ as the *robustness level* of $\mathcal{R}$.[1] Notably, the robustness level is between 1 and k, with 1-robust being the strongest form of robustness one could ask for. We ask for the robustness levels of several multiwinner rules.
2. We say that the *robustness radius* of an election E (for committee size k) under a multiwinner rule $\mathcal{R}$ is the smallest number of swaps (of adjacent candidates) which are necessary to change the election outcome. We ask for the complexity of computing the robustness radius (referred to as the ROBUSTNESS RADIUS problem) under a number of multiwinner rules (this problem is strongly related to the MARGIN OF VICTORY [4,7,20,26] and DESTRUCTIVE SWAP BRIBERY problems [13,25]; in particular, it follows up on the study of robustness of single-winner rules of Shiryaev et al. [25])
3. We ask how many random swaps of adjacent candidates are necessary, on average, to move from a randomly generated election to one with a different outcome. Doing experiments, we assess the practical robustness of our rules.

There is quite a number of multiwinner rules. We consider only several of them, selected to represent a varied set of ideas from the literature (ranging from variants of scoring rules, through rules inspired by the Condorcet criterion, to the elimination-based STV rule). We find that all these rules are either 1-robust (a single swap can replace at most one committee member) or are k-robust (a single swap can replace the whole committee of size k).[2] Somewhat surprisingly, this phenomenon is deeply connected to the complexity of winner determination. Specifically, we show (under mild assumptions) that if a rule has a constant robustness level, then it has a polynomial-time computable refinement (that is, it is possible to compute *one* of its outcomes in polynomial time). Since for many rules the problem of computing such a refinement is NP-hard, we get a quick way of finding out that such rules have non-constant robustness levels.

The ROBUSTNESS RADIUS problem tends to be NP-hard (sometimes even for a single swap) and, thus, we seek fixed-parameter tractability (FPT) results. For example, we find several FPT algorithms parametrized by the number of voters

---

[1] Indeed, the formal definition is more complex due to taking care of ties.

[2] We also construct somewhat artificial rules with robustness levels between 1 and k.

(useful, e.g., for scenarios with few experts, such as our introductory example). See Table 1 for an overview on our theoretical results.

We complement our work with an experimental evaluation of how robust are our rules with respect to random swaps. On the average, to change the outcome of an election, one needs to make the most swaps under the k-Borda rule. All the omitted proofs are present in the long version of our paper [5].

Table 1. Summary of our results. Together with each rule, we provide the complexity of its winner determination. The parameters m, n, and B mean, respectively, the number of candidates, voters, and the robustness radius; NP-hard(B) means NP-hard even for constant B. (♣) For STV there is a polynomial-time algorithm for computing a single winning committee but deciding if a given committee wins is NP-hard [10].

Voting rule	Robustness level	Complexity of ROBUSTNESS RADIUS
SNTV, Bloc, k-Borda (P)	1	P
k-Copeland (P)	1	NP-hard, FPT(m)
NED (NP-hard [1])	k	NP-hard, FPT(m)
STV (P) (♣)	k	NP-hard(B), FPT(m), FPT(n)
β-CC (NP-hard [3, 19, 23])	k	NP-hard(B), FPT(m), FPT(n)

2 Preliminaries

Elections. An election $E = (C, V)$ consists of a set of candidates $C = \{c_1, \ldots, c_m\}$ and of a collection of voters $V = (v_1, \ldots, v_n)$. We consider the ordinal election model, where each voter v is associated with a preference order $\succ_v$, that is, with a ranking of the candidates from the most to the least desirable one (according to this voter). A multiwinner voting rule $\mathcal{R}$ is a function that, given an election $E = (C, V)$ and a committee size k, outputs a set $\mathcal{R}(E, k)$ of size-k subsets of C, referred to as the winning committees (each of these committees ties for victory).

(Committee) Scoring Rules. Given a voter v and a candidate c, by $\text{pos}_v(c)$ we denote the position of c in v's preference order (the top-ranked candidate has position 1 and the following candidate has position 2, and so on). A scoring function for m candidates is a function $\gamma_m \colon [m] \to \mathbb{R}$ that associates each candidate-position with a score. Examples of scoring functions include (1) the Borda scoring functions, $\beta_m(i) = m - i$; and (2) the t-Approval scoring functions, $\alpha_t(i)$ defined so that $\alpha_t(i) = 1$ if $i \leq t$ and $\alpha_t(i) = 0$ otherwise (α_1 is typically referred to as the Plurality scoring function). For a scoring function γ_m, the γ_m-score of a candidate c in an m-candidate election $E = (C, V)$ is defined as $\gamma_m\text{-score}_E(c) = \sum_{v \in V} \gamma_m(\text{pos}_v(c))$.

For a given election E and a committee size k, the SNTV score of a size-k committee S is defined as the sum of the Plurality scores of its members. SNTV outputs the committee(s) with the highest score (i.e., the rule outputs the committees that consist of k candidates with the highest plurality scores; there may be more than one such committee due to ties). Bloc and k-Borda rules are defined analogously, but using k-Approval and Borda scoring functions, respectively. The Chamberlin–Courant rule [8] (abbreviated as β-CC) also outputs the committees with the highest score, but computes these scores in a different way: The score of committee S in a vote v is the Borda score of the highest-ranked member of S (the score of a committee is the sum of the scores from all voters).

SNTV, Bloc, k-Borda, and β-CC are examples of committee scoring rules [12, 14]. However, while the first three rules are polynomial-time computable, winner determination for β-CC is well-known to be NP-hard [3,19,23].

Condorcet-Inspired Rules. A candidate c is a Condorcet winner (resp. a weak Condorcet winner) if for each candidate d, more than (at least) half of the voters prefer c to d. In the multiwinner case, a committee is *Gehrlein strongly-stable* (resp. *weakly-stable*) if every committee member is preferred to every non-member by more than (at least) half of the voters [15], and a multiwinner rule is Gehrlein strongly-stable (resp. weakly-stable) if it outputs exactly the Gehrlein strongly-stable (weakly-stable) committees whenever they exist. For example, let the NED score of a committee S be the number of pairs (c, d) such that (i) c is a candidate in S, (ii) d is a candidate outside of S, and (iii) at least half of the voters prefer c to d. Then, the NED rule [9], defined to output the committees with the highest NED score, is Gehrlein weakly-stable. In contrast, the k-Copeland^0 rule is Gehrlein strongly-stable but not weakly-stable (the Copeland$^\alpha$ score of a candidate c, where $\alpha \in [0,1]$, is the number of candidates d such that a majority of the voters prefer c to d, plus α times the number of candidates e such that exactly half of the voters prefer c to e; winning k-Copeland$^\alpha$ committees consist of k candidates with the highest scores). Detailed studies of Gehrlein stability mostly focused on the weak variant of the notion [2,17]. Very recent findings, as well as results from this paper, suggest that the strong variant is more appealing [1,24]; for example, all Gehrlein weakly-stable rules are NP-hard to compute [1], whereas there are strongly-stable rules (such as k-Copeland^0) that are polynomial-time computable.

STV. For an election with m candidates, the STV rule executes up to m rounds as follows. In a single round, it checks whether there is a candidate c who is ranked first by at least $q = \lfloor \frac{n}{k+1} \rfloor + 1$ voters and, if so, then it (i) includes c into the winning committee, (ii) removes exactly q voters that rank c first from the election, and (iii) removes c from the remaining preference orders. If such a candidate does not exist, then a candidate d that is ranked first by the fewest voters is removed. Note that this description does not specify *which* q voters to remove or *which* candidate to remove if there is more than one that is ranked first by the fewest voters. We adopt the parallel-universes tie-breaking model and we say that a committee wins under STV if there is any way of breaking such internal ties that leads to him or her being elected [10].

We can compute *some* STV winning committee by breaking the internal ties in some arbitrary way, but it is NP-hard to decide if a given committee wins [10].

Parametrized Complexity. We assume familiarity with basic notions of parametrized complexity, such as parametrized problems, FPT-algorithms, and $W[1]$-hardness. For details, we refer to the textbook of Cygan et al. [11].

3 Robustness Levels of Multiwinner Rules

In this section we identify the robustness levels of our multiwinner rules. We start by defining this notion formally; note that the definition below has to take into account that a voting rule can output several tied committees.

Definition 1. *The* robustness level *of a multiwinner rule $\mathcal{R}$ for elections with m candidates and committee size k is the smallest value ℓ such that for each election $E = (C, V)$ with $|C| = m$, each election E' obtained from E by making a single swap of adjacent candidates in a single vote, and each committee $W \in \mathcal{R}(E, k)$, there exists a committee $W' \in \mathcal{R}(E', k)$ such that $|W \cap W'| \geq k - \ell$.*

All rules that we consider belong to one of two extremes: Either they are 1-robust (i.e., they are very robust) or they are k-robust (i.e., they are possibly very non-robust). We start with a large class of rules that are 1-robust.

Proposition 1. *Let $\mathcal{R}$ be a voting rule that assigns points to candidates and selects those with the highest scores. If a single swap in an election affects the scores of at most two candidates (decreases the score of one and increases the score of the other), then the robustness level of $\mathcal{R}$ is equal to one.*

The proof uses the observation that, after a single swap, either the candidate whose score increases can push out a single (lowest-scoring) member of the winning committee W, or a member of W who loses score can be replaced by the highest-scoring candidate outside W. This suffices to deal with four of our rules.

Corollary 1. *SNTV, Bloc, k-Borda, and k-Copeland$^\alpha$ (for each α) are 1-robust.*

In contrast, Gehrlein weakly-stable rules are k-robust.

Proposition 2. *The robustness level of each Gehrlein weakly-stable rule is k.*

Proof. Consider the following election, described through its majority graph (in a majority graph, each candidate is a vertex and there is a directed arc from candidate u to candidate v if more than half of the voters prefer u to v; the classic McGarvey's theorem says that each majority graph can be implemented with polynomially many votes [22]). We form an election with candidate set $C = A \cup B \cup \{c\}$, where $A = \{a_1, \ldots, a_k\}$ and $B = \{b_1, \ldots, b_k\}$, and with the following majority graph: The candidates in A form one cycle, the candidates in B form another cycle, and there are no other arcs (i.e., for all other pairs of candidates (x, y) the same number of voters prefers x to y as the other way round). We further assume that there is a vote, call it v, where c is ranked directly below a_1 (McGarvey's theorem easily accommodates this need).

In the constructed election, there are two Gehrlein weakly-stable committees, A and B. To see this, note that if a Gehrlein weakly-stable contains some a_i then it must also contain all other members of A (otherwise there would be a candidate outside of the committee that is preferred by a majority of the voters to a committee member). An analogous argument holds for B.

If we push c ahead of a_1 in vote v, then a majority of the voters prefers c to a_1. Thus, A is no longer Gehrlein weakly-stable and B becomes the unique winning committee. Since (1) A and B are disjoint, (2) A is among the winning committees prior to the swap, and (3) B is the unique winning committee after the swap; we have that every Gehrlein weakly-stable rule is k-robust.   □

To conclude this section we show that the robustness levels of β-CC and STV are k. Such negative results seem unavoidable among rules that—like β-CC and STV—provide diversity or proportionality (we discuss this further in Sect. 4).

Proposition 3. *Both β-CC and STV are k-robust.*

One may wonder whether there exist any voting rules with robustness level between 1 and k. Although we could not identify any classical rules with this property, we found natural hybrid multi-stage rules which satisfy it. For example, the rule which first elects half of the committee as k-Borda does and then the other half as β-CC does has robustness level of roughly $k/2$.

Proposition 4. *For each ℓ, there is an ℓ-robust rule.*

4   Computing Refinements of Robust Rules

It turns out that the dichotomy between 1-robust and k-robust rules is strongly connected to the one between polynomial-time computable rules and those that are NP-hard. To make this claim formal, we need the following definition.

Definition 2. *A multiwinner rule $\mathcal{R}$ is scoring-efficient if the following holds:*

1. *For each three positive integers n, m, and k ($k \leq m$) there is a polynomial-time computable election E with n voters and m candidates, such that at least one member of $\mathcal{R}(E, k)$ can be computed in polynomial time.*
2. *There is a polynomial-time computable function $f_{\mathcal{R}}$ that for each election E, committee size k, and committee S, associates score $f_{\mathcal{R}}(E, k, S)$ with S, so that $\mathcal{R}(E, k)$ consists exactly of the committees with the highest $f_{\mathcal{R}}$-score.*

The first condition from Definition 2 is satisfied, e.g., by *weakly unanimous* rules.

Definition 3 (Elkind et al. [12]). *A rule $\mathcal{R}$ is weakly unanimous if for each election $E = (C, V)$ and each committee size k, if each voter ranks the same set W of k candidates on top (possibly in different order), then $W \in \mathcal{R}(E, k)$.*

All voting rules which we consider in this paper are weakly unanimous (indeed, voting rules which are not weakly unanimous are somewhat "suspicious"). Further, all our rules, except STV, satisfy the second condition from Definition 2. For example, while winner determination for β-CC is indeed NP-hard, computing the score of a given committee can be done in polynomial time. We are ready to state and prove the main theorem of this section.

Theorem 1. *Let $\mathcal{R}$ be a 1-robust scoring-efficient multiwinner rule. Then there is a polynomial-time computable rule $\mathcal{R}'$ such that for each election E and committee size k we have $\mathcal{R}'(E, k) \subseteq \mathcal{R}(E, k)$.*

Proof. We will show a polynomial-time algorithm that, given an election E and committee size k, finds a committee $W \in \mathcal{R}(E, k)$; we let $\mathcal{R}'(E, k)$ output $\{W\}$.

Let $E = (C, V)$ be our input election and let k be the size of the desired committee. Let $E' = (C, V')$ be an election with $|V'| = |V|$, whose existence is guaranteed by the first condition of Definition 2, and let S' be a size-k $\mathcal{R}$-winning committee for this election, also guaranteed by Definition 2.

Let $E_0, E_1 \ldots, E_t$ be a sequence of elections such that $E_0 = E'$, $E_t = E$, and for each integer $i \in [t]$, we obtain E_i from E_{i-1} by (i) finding a voter v and two candidates c and d such that in E_{i-1} voter v ranks c right ahead of d, but in E voter v ranks d ahead of c (although not necessarily right ahead of c), and (ii) swapping c and d in v's preference order. We note that at most $|C||V|^2$ swaps suffice to transform E' into E (i.e., $t \leq |C||V|^2$).

For each $i \in \{0, 1, \ldots, t\}$, we find a committee $S_i \in \mathcal{R}(E_i, k)$. We start with $S_0 = S'$ (which satisfies our condition) and for each $i \in [t]$, we obtain S_i from S_{i-1} as follows: Since $\mathcal{R}$ is 1-robust, we know that at least one committee S'' from the set $\{S'' \mid |S_{i-1} \cap S''| \geq k - 1\}$ is winning in E_i. We try each committee S'' from this set (there are polynomially many such committees) and we compute the $f_{\mathcal{R}}$-score of each of them (recall Condition 2 of Definition 2). The committee with the highest $f_{\mathcal{R}}$-score must be winning in E_i and we set S_i to be this committee (by Definition 2, computing the $f_{\mathcal{R}}$-scores is a polynomial-time task).

Finally, we output S_t. By our arguments, we have that $S_t \in \mathcal{R}(E, k)$. $\square$

Theorem 1 generalizes to the case of r-robust rules for constant r: Our algorithm simply has to try more (but still polynomially many) committees S''.

Let us note how Theorem 1 relates to single-winner rules (that can be seen as multiwinner rules for $k = 1$). All such rules are 1-robust, but for those with NP-hard winner determination, even computing the candidates' scores is NP-hard (see, e.g., the survey of Caragiannis et al. [6]), so Theorem 1 does not apply.

5   Complexity of Computing the Robustness Radius

In the ROBUSTNESS RADIUS problem we ask if it is possible to change the election result by performing a given number of swaps of adjacent candidates. Intuitively, the more swaps are necessary, the more robust a particular election is.

Definition 4. *Let $\mathcal{R}$ be a multiwinner rule. In the $\mathcal{R}$ ROBUSTNESS RADIUS problem we are given an election $E = (C, V)$, a committee size k, and an integer B. We ask if it is possible to obtain an election E' by making at most B swaps of adjacent candidates within the rankings in E so that $\mathcal{R}(E', k) \neq \mathcal{R}(E, k)$.*

This problem is strongly connected to some other problems studied in the literature. Specifically, in the DESTRUCTIVE SWAP BRIBERY problem [13,16,25] (DSB for short) we ask if it is possible to preclude a particular candidate from winning by making a given number of swaps. DSB was already used to study robustness of single-winner election rules by Shiryaev et al. [25]. We decided to give our problem a different name, and not to refer to it as a multiwinner variant of DSB, because we feel that in the latter the goal should be to preclude a given candidate from being a member of any of the winning committees, instead of changing the outcome in any arbitrary way. In this sense, our problem is very similar to the MARGIN OF VICTORY problem [4,7,20,26], which has the same goal, but instead of counting single swaps, counts how many votes are changed.

We find that ROBUSTNESS RADIUS tends to be computationally challenging. Indeed, we find polynomial-time algorithms only for the following simple rules.

Theorem 2. *ROBUSTNESS RADIUS is computable in polynomial time for SNTV, Bloc, and k-Borda.*

The rules in Theorem 2 are all 1-robust, but not all 1-robust rules have efficient ROBUSTNESS RADIUS algorithms. In particular, a simple modification of a proof of Kaczmarczyk and Faliszewski [16, Theorem 3] shows that for k-Copeland$^\alpha$ rules (which are 1-robust) we obtain NP-hardness. We also obtain a general NP-hardness for all Gehrlein weakly-stable rules.

Corollary 2. *k-Copeland ROBUSTNESS RADIUS is NP-hard.*

Theorem 3. *ROBUSTNESS RADIUS is NP-hard for Gehrlein weakly-stable rules.*

Without much surprise, we find that ROBUSTNESS RADIUS is also NP-hard for STV and for β-CC. For these rules, however, the hardness results are, in fact, significantly stronger. In both cases it is NP-hard to decide if the election outcome changes after a single swap, and for STV the result holds even for committees of size one (β-CC with committees of size one is simply the single-winner Borda rule, for which the problem is polynomial-time solvable [25]).

Theorem 4. *ROBUSTNESS RADIUS is NP-hard both for STV and for β-CC, even if we can perform only a single swap; for STV this holds even for committees of size 1. For β-CC, the problem is W[1]-hard with respect to the committee size.*

For the case of β-CC, the proof of Theorem 4 gives much more than stated in the theorem. Indeed, our construction shows that the problem remains NP-hard even if we are given a current winning committee as part of the input. Further, the same construction gives the following corollary (whose first part is sometimes taken for granted in the literature, but has not been shown formally yet).

Corollary 3. *The problem of deciding if a given candidate belongs to some β-CC winning committee (for a given election and committee size) is both NP-hard and coNP-hard.*

We conclude this section by providing FPT algorithms for ROBUSTNESS RADIUS. An FPT algorithm for a given parameter (e.g., the number of candidates or the number of voters) must have running time of the form $f(k)|I|^{O(1)}$, where k is the value of the parameter and $|I|$ is the length of the encoding of the input instance. Using the standard approach of formulating integer linear programs and invoking the algorithm of Lenstra [18], we find that ROBUSTNESS RADIUS is in FPT when parametrized by the number of candidates.

Proposition 5. ROBUSTNESS RADIUS *for k-Copeland, NED, STV, and β-CC is in FPT (parametrized by the number of candidates).*

For STV and β-CC we also get algorithms parametrized by the number of voters. For the case of STV, we assume that the value of k is such that we never need to "delete non-existent voters" and we refer to committee sizes, k, where such deleting is not necessary as *normal*. For example, k is not normal if $k > n$. Another example is to take $n = 12$ and $k = 5$: We need to delete $q = \lfloor \frac{12}{5+1} \rfloor + 1 = 3$ voters for each committee member, which requires deleting 15 out of 12 voters.

Theorem 5. *STV* ROBUSTNESS RADIUS *is in FPT when parametrized by the number n of voters (for normal committee sizes).*

Proof. Let $E = (C, V)$ be an input election and k be the size of the desired committee. For each candidate c, we defnie "rank of c" as $\text{rank}(c) = \min_{v \in V}(\text{pos}_v(c))$.

First, we prove that a candidate with a rank higher than n cannot be a member of a winning committee. Let us assume towards a contradiction that there exists a candidate c with $\text{rank}(c) > n$ who is a member of some winning committee W. When STV adds some candidate to the committee (this happens when the number of voters who rank such a candidate first exceeds the quota $\lfloor \frac{n}{k+1} \rfloor + 1$), it removes this candidate and at least one voter from the election. Thus, before c were included in W, STV must have removed some candidate c' from the election without adding it to W (since c had to be ranked first by some voter to be included in the committee; for c to be ranked first, STV had to delete at least n candidates, so by the assumption that the committee size is normal, not all of them could have been included in the committee). Since STV always removes a candidate with the lowest Plurality score, and at the moment when c' was removed the Plurality score of c was equal to zero, the Plurality score of c' also must have been zero. Thus, removing c' from the election did not affect the top preferences of the voters, and STV, right after removing c', removed another candidate with zero Plurality score. By repeating this argument sufficiently many times, we conclude that c must have been eventually eliminated, and so could not have been added to W. This gives a contradiction and proves our claim.

The same reasoning also shows that the number of committees winning according to STV is bounded by a function of n: In each step either one of at most n voters is removed, or all candidates who are not ranked first by any voter are removed from the election (which leaves at most n candidates in the election).

Second, we observe that the robustness radius for our election is at most n^2. Indeed, we can take any winning candidate, and with at most n^2 swaps we can push him or her to have rank at least $n + 1$. Such a candidate will no longer be a member of W and, so, the outcome of the election will change.

Third, we observe that in order to change the outcome of an election, we should only swap such candidates that at least one of them has rank at most $n^2 + n + 1$. Indeed, consider a candidate c with $\text{rank}(c) > n^2 + n + 1$. After n^2 swaps, the rank of this candidate would still be above n, so he or she still would not belong to any winning committee. Thus, a swap of any two candidates with ranks higher than $n^2 + n + 1$ does not belong to any of the sequences of at most n^2 swaps that change the election result (the exact positions of these two candidates would have no influence on the STV outcome).

As a result, it suffices to focus on the candidates with ranks at most n^2+n+1. There are at most $n(n^2+n+1)$ such candidates. Consequently, there are at most $(2n^3 + 2n^2 + 2n)^{n^2}$ possible n^2-long sequences of swaps which we need to check in order to find the optimal one. This completes the proof. $\square$

The algorithm for the case of β-CC is more involved. Briefly put, it relies on finding either the unique winning committee or two committees tied for victory. In the former case, it combines brute-force search with dynamic programming, and in the latter either a single swap or a clever greedy algorithm suffice.

Theorem 6. β-CC ROBUSTNESS RADIUS *is in* FPT *for the number of voters.*

6   Beyond the Worst Case: An Experimental Evaluation

In this section we present results of experiments, in which we measure how many randomly-selected swaps are necessary to change election results.[3]

We perform a series of experiments using four distributions of rankings obtained from PrefLib [21], a library that contains both real-life preference data and tools for generating synthetic elections. We use three artificial distributions: (i) Impartial Culture (IC), (ii) Mallows Model with parameter ϕ between 0 and 1 drawn uniformly, and (iii) the mixture of two Mallows Models with separate parameters ϕ chosen identically to the previous model. (Intuitively, in the Mallows model there is a single most probable preference order r_0, and the more swaps are necessary to modify a given order r to become r_0, the less probable it is to draw r; in the IC model, drawing each preference order is equally likely.) Additionally, we use one dataset describing real-life preferences over sushi sets (we treat this dataset as a distribution selecting votes from this dataset uniformly

---

[3] We found STV to be computationally too intensive for our experiments, so we used a simplified variant where all internal ties are broken lexicographically. We omit NED for similar reasons (but we expect the results to be similar as for k-Copeland).

at random). For each of these four distributions, and for every investigated rule (for k-Copeland we took $\alpha = 0.5$), we perform 200 simulations. In each simulation we draw an election containing 10 candidates and 50 voters from the given distribution. Then we repeatedly draw a pair of adjacent candidates uniformly at random and perform a swap, until the outcome of the election changes or 1000 swaps are done (taking 1000 for the following computations). The average number of swaps required to change the outcome of an election for different rules and for different datasets is depicted in Fig. 1. We present the results for committee size $k = 3$; simulations for $k = 5$ led to analogous conclusions.

Among our rules, k-Borda is the most robust one (k-Copeland holds second place), whereas rules that achieve either diversity (β-CC and, to some extent, SNTV) or proportionality (STV) are more vulnerable to small changes in the input. This is aligned with what we have seen in the theoretical part of the paper (with a minor exception of SNTV). As expected, the robustness radius decreases with the increase of randomness in the voters' preferences. Indeed, one needs to make more swaps to change the outcome of elections which are highly biased towards the resulting winners. Thus we conclude that preferences in the sushi datasets are not highly centered, as the radiuses of these elections are small.

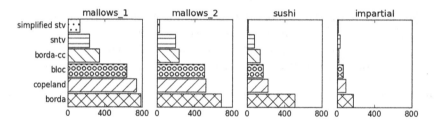

Fig. 1. Experimental results showing the average number of swaps needed to change the outcome of random elections obtained according to the description in Sect. 6.

7   Conclusions

We formalized the notion of robustness of multiwinner rules and studied the complexity of assessing the robustness/confidence of collective multiwinner decisions. Our theoretical and experimental analysis indicates that k-Borda is the most robust among our rules, and that proportional rules, such as STV and the Chamberlin–Courant rule, are on the other end of the spectrum.

Acknowledgments. We are grateful to anonymous SAGT reviewers for their useful comments. R. Bredereck was supported by the DFG fellowship BR 5207/2. P. Faliszewski was supported by the NCN, Poland, under project 2016/21/B/ST6/01509. A. Kaczmarczyk was supported by the DFG project AFFA (BR 5207/1 and NI 369/15). P. Skowron was supported by a Humboldt Fellowship. N. Talmon was supported by an I-CORE ALGO fellowship.

References

1. Aziz, H., Elkind, E., Faliszewski, P., Lackner, M., Skowron, P.: The Condorcet principle for multiwinner elections: from shortlisting to proportionality. arXiv preprint arXiv:1701.08023 (2017)
2. Barberà, S., Coelho, D.: How to choose a non-controversial list with k names. Soc. Choice Welf. **31**(1), 79–96 (2008)
3. Betzler, N., Slinko, A., Uhlmann, J.: On the computation of fully proportional representation. J. Artif. Intell. Res. **47**, 475–519 (2013)
4. Blom, M., Stuckey, P., Teague, V.: Towards computing victory margins in STV elections. arXiv preprint arXiv:1703.03511 (2017)
5. Bredereck, R., Faliszewski, P., Kaczmarczyk, A., Niedermeier, R., Skowron, P., Talmon, N.: Robustness among multiwinner voting rules. arXiv preprint arXiv:1707.01417 (2017)
6. Caragiannis, I., Hemaspaandra, E., Hemaspaandra, L.: Dodgson's rule and Young's rule. In: Brandt, F., Conitzer, V., Endriss, U., Lang, J., Procaccia, A.D. (eds.) Handbook of Computational Social Choice. Cambridge University Press, Cambridge (2016)
7. Cary, D.: Estimating the margin of victory for instant-runoff voting. Presented at EVT/WOTE-2011, August 2011
8. Chamberlin, B., Courant, P.: Representative deliberations and representative decisions: proportional representation and the Borda rule. Am. Polit. Sci. Rev. **77**(3), 718–733 (1983)
9. Coelho, D.: Understanding, evaluating and selecting voting rules through games and axioms. Ph.D. thesis, Universitat Autònoma de Barcelona (2004)
10. Conitzer, V., Rognlie, M., Xia, L.: Preference functions that score rankings and maximum likelihood estimation. In: Proceedings of IJCAI-2009, pp. 109–115, July 2009
11. Cygan, M., Fomin, F.V., Kowalik, Ł., Lokshtanov, D., Marx, D., Pilipczuk, M., Pilipczuk, M., Saurabh, S.: Parameterized Algorithms. Springer, Heidelberg (2015)
12. Elkind, E., Faliszewski, P., Skowron, P., Slinko, A.: Properties of multiwinner voting rules. Social Choice Welf. **48**(3), 599–632 (2017)
13. Elkind, E., Faliszewski, P., Slinko, A.: Swap bribery. In: Mavronicolas, M., Papadopoulou, V.G. (eds.) SAGT 2009. LNCS, vol. 5814, pp. 299–310. Springer, Heidelberg (2009). doi:10.1007/978-3-642-04645-2_27
14. Faliszewski, P., Skowron, P., Slinko, A., Talmon, N.: Committee scoring rules: axiomatic classification and hierarchy. In: Proceedings of IJCAI-2016, pp. 250–256 (2016)
15. Gehrlein, W.: The Condorcet criterion and committee selection. Math. Soc. Sci. **10**(3), 199–209 (1985)
16. Kaczmarczyk, A., Faliszewski, P.: Algorithms for destructive shift bribery. In: Proceedings of AAMAS-2016, pp. 305–313 (2016)
17. Kamwa, E.: On stable voting rules for selecting committees. J. Math. Econ. **70**, 36–44 (2017)
18. Lenstra Jr., H.: Integer programming with a fixed number of variables. Math. Oper. Res. **8**(4), 538–548 (1983)
19. Lu, T., Boutilier, C.: Budgeted social choice: from consensus to personalized decision making. In: Proceedings of IJCAI-2011, pp. 280–286 (2011)
20. Magrino, T., Rivest, R., Shen, E., Wagner, D.: Computing the margin of victory in IRV elections. Presented at EVT/WOTE-2011, August 2011

21. Mattei, N., Walsh, T.: Preflib: a library for preferences. In: Proceedings of the 3rd International Conference on Algorithmic Decision Theory, pp. 259–270 (2013)
22. McGarvey, D.: A theorem on the construction of voting paradoxes. Econometrica **21**(4), 608–610 (1953)
23. Procaccia, A., Rosenschein, J., Zohar, A.: On the complexity of achieving proportional representation. Soc. Choice Welf. **30**(3), 353–362 (2008)
24. Sekar, S. Sikdar., Xia, L.: Condorcet consistent bundling with social choice. In: Proceedings of AAMAS-2017, May 2017
25. Shiryaev, D., Yu, L., Elkind, E.: On elections with robust winners. In: Proceedings of AAMAS-2013, pp. 415–422 (2013)
26. Xia, L.: Computing the margin of victory for various voting rules. In: Proceedings of EC-2012, pp. 982–999, June 2012

Computing Constrained Approximate Equilibria in Polymatrix Games

Argyrios Deligkas[1(✉)], John Fearnley[2], and Rahul Savani[2]

[1] Technion, Haifa, Israel
argyris@technion.ac.il
[2] University of Liverpool, Liverpool, UK

Abstract. This paper studies *constrained* approximate Nash equilibria in polymatrix games. We show that is NP-hard to decide if a polymatrix game has a constrained approximate equilibrium for 9 natural constraints and *any* non-trivial ϵ. We then provide a QPTAS for polymatrix games with bounded treewidth and logarithmically many actions per player that finds constrained approximate equilibria for a wide family of constraints.

1 Introduction

In this paper we study *polymatrix games*, which provide a succinct representation of a many-player game. In these games, each player is a vertex in a graph, and each edge of the graph is a bimatrix game. Every player chooses a single strategy and plays it in *all* of the bimatrix games that he is involved in, and his payoff is the *sum* of the payoffs that he obtains from each individual edge game.

A fundamental problem in algorithmic game theory is to design efficient algorithms for computing *Nash equilibria*. Unfortunately, even in bimatrix games, this is PPAD-complete [12,17], which probably rules out efficient algorithms. Thus, attention has shifted to *approximate* equilibria. There are two natural notions of an approximate equilibrium. An ϵ-*Nash equilibrium* (ϵ-NE) requires that each player has an expected payoff that is within ϵ of their best response payoff. An ϵ-*well-supported Nash equilibrium* (ϵ-WSNE) requires that all players only play pure strategies whose payoff is within ϵ of the best response payoff.

Constrained Approximate Equilibria. Sometimes, it is not enough to find an approximate NE, but instead we want to find one that satisfies certain constraints, such as having high social welfare. For bimatrix games, the algorithm of Lipton, Markakis, and Mehta (henceforth LMM) can be adapted to provide a quasi-polynomial time approximation scheme (QPTAS) for this task [31]: we can find in $m^{O(\frac{\ln m}{\epsilon^2})}$ time an ϵ-NE whose social welfare is at least as good as any ϵ'-NE where $\epsilon' < \epsilon$.

A sequence of papers [1,11,21,29] has shown that polynomial time algorithms for finding ϵ-NEs with good social welfare are unlikely to exist, subject to various

---

This work was supported by ISF grant 2021296 and EPSRC grant EP/P020909/1.

V. Bilò and M. Flammini (Eds.): SAGT 2017, LNCS 10504, pp. 93–105, 2017.
DOI: 10.1007/978-3-319-66700-3_8

hardness assumptions such as ETH. These hardness results carry over to a range of other properties, and apply for all $\epsilon < \frac{1}{8}$ [21].

Our Contribution. We show that deciding whether there is an ϵ-NE with good social welfare in a polymatrix game is NP-complete for all $\epsilon \in [0, 1]$. We then study a variety of further constraints (Table 1). For each one, we show that deciding whether there is an ϵ-WSNE that satisfies the constraint is NP-complete for all $\epsilon \in (0, 1)$. Our results hold even when the game is a planar bipartite graph with degree at most 3, and each player has at most 7 actions.

To put these results into context, let us contrast them with the known lower bounds for bimatrix games, which also apply directly to polymatrix games. Those results [1,11,21,29] imply that one cannot hope to find an algorithm that is better than a QPTAS for polymatrix games when $\epsilon < \frac{1}{8}$. In comparison, our results show a stronger, NP-hardness, result, apply to all ϵ in the range $(0, 1)$, and hold even when the players have constantly many actions.

We then study the problem of computing constrained approximate equilibria in polymatrix games with restricted graphs. Although our hardness results apply to a broad class of graphs, bounded treewidth graphs do not fall within their scope. A recent result of Ortiz and Irfan [33,34] provides a QPTAS for finding ϵ-NEs in polymatrix games with bounded treewidth where every player has at most logarithmically many actions. We devise a dynamic programming algorithm for finding approximate equilibria in polymatrix games with bounded treewidth. Much like the algorithm in [33], we discretize both the strategy and payoff spaces, and obtain a complexity result that matches theirs. However, our algorithm works directly on the game, avoiding the reduction to a CSP used in their result.

The main benefit is that this algorithm can be adapted to provide a QPTAS for constrained approximate Nash equilibria. We introduce *one variable decomposable (OVD)* constraints, which are a broad class of optimization constraints, covering many of the problems listed in Table 1. We show that our algorithm can be adapted to find good approximate equilibria relative to an OVD constraint. Initially, we do this for the restricted class of *k-uniform* strategies: we can find a k-uniform 1.5ϵ-NE whose value is better than any k-uniform $\epsilon/4$-NE. Note that this is similar to the guarantee given by the LMM technique in bimatrix games. We extend this beyond the class of k-uniform strategies for constraints that are defined by a linear combination of the payoffs, such as social welfare. In this case, we find a 1.5ϵ-NE whose value is within $O(\epsilon)$ of *any* $\epsilon/8$-NE.

Related Work. Barman et al. [4] have provided a randomised QPTAS for polymatrix games played on trees. Their algorithm is also a dynamic programming algorithm that discretizes the strategy space using the notion of a k-uniform strategy. Their algorithm is a QPTAS for general polymatrix games on trees and when the number of pure strategies for every player is bounded by a constant they get an expected polynomial-time algorithm (EPTAS).

The work of Ortiz and Irfan [33] applies to a much wider class of games that they call graphical multi-hypermatrix games. They provide a QPTAS for the case where the interaction hypergraph has bounded hypertreewidth. This class includes polymatrix games that have bounded treewdith and logarithmically many actions per player. For the special cases of tree polymatrix games and tree graphical games they go further and provide explicit dynamic programming algorithms that work directly on the game, and avoid the need to solve a CSP.

Gilboa and Zemel [27] showed that it is NP-complete to decide whether there exist Nash equilibria in bimatrix games with certain properties, such as high social welfare. Conitzer and Sandholm [13] extended the list of NP-complete problems of [27]. Bilò and Mavronicolas [5] extended these results to win-lose bimatrix games. Bonifaci et al. [9] showed that it is NP-complete to decide whether a win-lose bimatrix game possesses a Nash equilibrium where every player plays a uniform strategy over their support. Recently, Garg et al. [26] and Bilò and Mavronicolas [6,7] extended these results to many-player games and provided analogous ETR-completeness results.

Elkind et al. have given a polynomial time algorithm for finding exact Nash equilibria in two-action path graphical games [23]. They have also extended this to find good constrained exact equilibria in certain two-action tree graphical games [24]. Greco and Scarcello provide further hardness results for constrained equilibria in graphical games [28].

Computing approximate equilibria in bimatrix games has been well studied [10,14,18,19,25,30,36], but there has been less work for polymatrix games [3, 20,22]. Rubinstein [35] has shown that there is a small constant ϵ such that finding an ϵ-NE of a polymatrix game is PPAD-complete. For constrained ϵ-NE, the only positive results were for bimatrix games and gave algorithms for finding ϵ-NE with constraints on payoffs [15,16].

2 Preliminaries

We start by fixing some notation. We use $[k]$ to denote the set of integers $\{1, 2, \ldots, k\}$, and when a universe $[k]$ is clear, we will use $\bar{S} = \{i \in [k] : i \notin S\}$ to denote the complement of $S \subseteq [k]$. For a k-dimensional vector x, we use x_{-S} to denote the elements of x with indices $\bar{S}$, and in the case where $S = \{i\}$ has only one element, we simply write x_{-i} for x_{-S}.

An n-player polymatrix game is defined by an undirected graph $G = (V, E)$ with n vertices, where each vertex is a player. The edges of the graph specify which players interact with each other. For each $i \in [n]$, we use $N(i) = \{j : (i, j) \in E\}$ to denote the neighbors of player i. Each edge $(i, j) \in E$ specifies a bimatrix game to be played between players i and j. Each player $i \in [n]$ has a fixed number of pure strategies m, so the bimatrix game on edge $(i, j) \in E$ is specified by an $m \times m$ matrix A_{ij}, which gives the payoffs for player i, and an $m \times m$ matrix A_{ji}, which gives the payoffs for player j. We allow the individual payoffs in each matrix to be an arbitrary rational number. We make the standard normalisation assumption that the maximum payoff each player can obtain under any strategy profile is 1 and the minimum is zero, unless specified otherwise. This can be achieved for example by using the procedure described in [22]. A *subgame* of a polymatrix game is obtained by ignoring edges that are not contained within a given subgraph of the game's interaction graph G.

A *mixed strategy* for player i is a probability distribution over player i's pure strategies. A *strategy profile* specifies a mixed strategy for every player. Given a strategy profile $\mathbf{s} = (s_1, \ldots, s_n)$, the pure strategy payoffs, or the payoff vector, of player i under $\mathbf{s}$, where only $\mathbf{s}_{-i}$ is relevant, is the sum of the pure strategy payoffs that he obtains in each of the bimatrix games that he plays. Formally, we define: $\mathbf{p}_i(\mathbf{s}) := \sum_{j \in N(i)} A_{ij} s_j$. The *expected* payoff of player i under the strategy profile $\mathbf{s}$ is defined as $s_i \cdot \mathbf{p}_i(\mathbf{s})$. The *regret* of player i under $\mathbf{s}$ the is difference between i's best response payoff against $\mathbf{s}_{-i}$ and between i's payoff under $\mathbf{s}$. If a strategy has regret ϵ, we say that the strategy is an ϵ-best response. A strategy profile $\mathbf{s}$ is an ϵ-Nash equilibrium, or ϵ-NE, if no player can increase his utility more than ϵ by unilaterally switching from $\mathbf{s}$, i.e., if the regret of every player is at most ϵ. Formally, $\mathbf{s}$ is an ϵ-NE if for every player $i \in [n]$ it holds that $s_i \cdot \mathbf{p}_i(\mathbf{s}) \geq \max \mathbf{p}_i(\mathbf{s}) - \epsilon$. A strategy profile $\mathbf{s}$ is an ϵ-well-supported Nash equilibrium, or ϵ-WSNE, if if the regret of every pure strategy played with positive probability is at most ϵ. Formally, $\mathbf{s}$ is an ϵ-WSNE if for every player $i \in [n]$ it holds that for all $j \in \text{supp}(s_i) = \{k \in [m] \mid (s_i)_k > 0\}$ we have $(\mathbf{p}_i(\mathbf{s}))_j \geq \max \mathbf{p}_i(\mathbf{s}) - \epsilon$.

3   Decision Problems for Approximate Equilibria

In this section, we show NP-completeness for nine decision problems related to constrained approximate Nash equilibria in polymatrix games. Table 1 contains the list of the problems that we study[1]. For Problem 1, we show hardness for all $\epsilon \in [0, 1]$. For the remaining problems, we show hardness for all $\epsilon \in (0, 1)$, i.e., for all approximate equilibria except exact equilibria ($\epsilon = 0$), and trivial approximations ($\epsilon = 1$). All of these problems are contained in NP because a "Yes" instance can be witnessed by a suitable approximate equilibrium (or two in the case of Problem 5). The starting point for all of our hardness results is the NP-complete problem Monotone 1-in-3 SAT.

Definition 1 (Monotone 1-in-3 SAT). *Given a monotone boolean CNF formula ϕ with exactly 3 distinct variables per clause, decide if there exists a satisfying assignment in which exactly one variable in each clause is true. We call such an assignment a 1-in-3 satisfying assignment.*

Every formula ϕ, with variables $V = \{x_1, \ldots, x_n\}$ and clauses $C = \{y_1, \ldots, y_m\}$, can be represented as a bipartite graph between V and C, with an edge between x_i and y_j if and only if x_i appears in clause y_j. We assume, without loss of generality, that this graph is connected. We say that ϕ is *planar* if the corresponding graph is planar. Recall that a graph is called *cubic* if the degree of every vertex is exactly three. We use the following result of Moore and Robson [32].

---

[1] Given probability distributions $\mathbf{x}$ and $\mathbf{x}'$, the *TV* distance between them is $\max_i\{|\mathbf{x}_i - \mathbf{x}'_i|\}$. The TV distance between strategy profiles $\mathbf{s} = (s_1, \ldots, s_n)$ and $\mathbf{s}' = (s'_1, \ldots, s'_n)$ is the maximum TV distance of s_i and s'_i over all i.

Table 1. The decision problems that we consider. All problems take as input an n-player polymatrix game with m actions for each player and an $\epsilon \in [0,1]$.

Problem description	Problem definition
Problem 1: Large total payoff $u \in (0,n]$	Is there an ϵ-NE **s** such that $\sum_{i \in [n]} \mathbf{p}_i(\mathbf{s}) \geq u$?
Problem 2: Small total payoff $u \in [0,n)$	Is there an ϵ-WSNE **s** such that $\sum_{i \in [n]} \mathbf{p}_i(\mathbf{s}) \leq u$?
Problem 3: Small payoff $u \in [0,1)$	Is there an ϵ-WSNE **s** such that $\min_i \mathbf{p}_i(\mathbf{s}) \leq u$?
Problem 4: Restricted support $S \subset [n]$	Is there an ϵ-WSNE **s** with $\text{supp}(s_1) \subseteq S$?
Problem 5: Two ϵ-WSNE $d \in (0,1]$ apart in Total Variation (TV) distance	Are there two ϵ-WSNE with TV distance $\geq d$?
Problem 6: Small largest probability $p \in (0,1)$	Is there an ϵ-WSNE **s** with $\max_j s_1(j) \leq p$?
Problem 7: Large total support size $k \in [n \cdot m]$	Is there an ϵ-WSNE **s** such that $\sum_{i \in [n]} |\text{supp}(s_i)| \geq k$?
Problem 8: Large smallest support size $k \in [n]$	Is there an ϵ-WSNE **s** such that $\min_i |\text{supp}(s_i)| \geq k$?
Problem 9: Large support size $k \in [n]$	Is there an ϵ-WSNE **s** such that $|\text{supp}(s_1)| \geq k$?

Theorem 2 (Sect. 3.1 [32]). *Monotone 1-in-3 SAT is* NP-*complete even when the formula corresponds to a planar cubic graph.*

From now on, we assume that ϕ is a monotone planar cubic formula. We say that ϕ is a "Yes" instance if ϕ admits a 1-in-3 satisfying assignment.

Large Total Payoff for ϵ-NEs. We show that Problem 1 is NP-complete for every $\epsilon \in [0,1]$, even when the interaction graph for the polymatrix game is planar, bipartite, cubic, and each player has at most three pure strategies.

Construction. Given a formula ϕ, we construct a polymatrix game G with $m + n$ players as follows. For each variable x_i we create a player v_i and for each clause y_j we create a player c_j. We now use V to denote the set of variable players and C to denote the clause players. The interaction graph is the bipartite graph between V and C described above. Each edge game has the same structure. Every player in V has two pure strategies called True and False, while every player in C has three pure strategies that depend on the three variables in the clause. If clause y_j contains variables x_i, x_k, x_l, then player c_j has pure strategies i, k and l. The game played between v_i and c_j is shown on the left in Fig. 1. The bimatrix games for v_k and v_l are defined analogously.

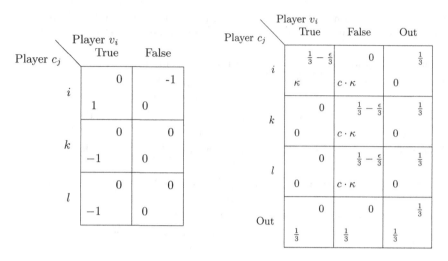

Fig. 1. Left: the game between clause player c_j and variable player v_i for Problem 1. Right: the game between c_j and v_i for Problems 2–9.

Correctness. The constructed game is not normalised. We prove our result for all ϵ, and thus in the normalised game the result will hold for all $\epsilon \in [0, 1]$. We show that for every ϵ, there is an ϵ-NE with social welfare m if and only if ϕ is a "Yes" instance. We begin by showing that if there is a solution for ϕ, then there is an *exact* NE for G with social welfare m, and therefore there is also an ϵ-NE for all ϵ with social welfare m. We start with a simple observation about the maximum and minimum payoffs that players can obtain in G.

Lemma 3. *In G, the total payoff for every variable player is at most 0, and the total payoff for every clause player c_j is at most 1. Moreover, if c_j gets payoff 1, then c_j and the variable players connected to c_j play pure strategies.*

Lemma 4. *If ϕ is a "Yes" instance, there is an NE for G with social welfare m.*

Lemma 5. *If there is a strategy profile for G with social welfare m, then ϕ is a "Yes" instance.*

Together, Lemmas 4 and 5 show that for all possible values of ϵ, it is NP-complete to decide whether there exists an ϵ-NE for G with social welfare m. When we normalise the payoffs in $[0, 1]$, this holds for all $\epsilon \in [0, 1]$.

Theorem 6. *Problem 1 is NP-complete for all $\epsilon \in [0, 1]$, even for degree-3 bipartite planar polymatrix games in which each player has at most 3 pure strategies.*

Hardness of Problems 2–9. To show the hardness Problems 2–9, we modify the game constructed in the previous section. We use G' to denote the new polymatrix game. The interaction graph for G' is exactly the same as for the game G. The bimatrix games are extended by an extra pure strategy for each player, the strategy Out, and the payoffs are adapted. If variable x_i is in clause y_j, then the bimatrix game between clause player c_j and v_i is shown on the right in Fig. 1. To fix the constants, given $\epsilon \in (0,1)$, we choose c to be in the range $(\max(1 - \frac{3\epsilon}{2}, 0), 1)$, and we set $\kappa = \frac{1-\epsilon}{1+2c}$. Observe that $0 < c < 1$, and that $\kappa + 2c \cdot \kappa = 1 - \epsilon$. Furthermore, since $c > 1 - \frac{3\epsilon}{2}$ we have $0 < \kappa < \frac{1}{3}$.

Lemma 7. *If ϕ is a "Yes" instance, then G' possesses an ϵ-WSNE such that no player uses strategy Out.*

Lemma 8. *If ϕ is a "No" instance, then G' possesses a unique ϵ-WSNE where every player plays Out.*

The combination of these two properties allows us to show that Problems 2–5 are NP-complete. For example, for Problem 4, we can ask whether there is an ϵ-WSNE of the game in which player one does not player Out.

Theorem 9. *Problems 2–5 are NP-complete for all $\epsilon \in (0,1)$, even on degree-3 planar bipartite polymatrix games where each player has at most 4 pure strategies.*

Duplicating Strategies. To show hardness for Problems 6–9, we slightly modify the game G' by duplicating every pure strategy except Out for all of the players. Since each player $c_j \in C$ has the pure strategies i, k, l and Out, we give player c_j pure strategies i', k' and l', which each have identical payoffs as the original strategies. Similarly for each player $v_i \in V$ we add the pure strategies True' and False'. Let us denote the game with the duplicated strategies by $\tilde{G}$. Then, if ϕ is a "Yes" instance, we can construct an ϵ-WSNE in which no player plays Out, where each player places at most 0.5 probability on each pure strategy, and where each player uses a support of size 2. These properties are sufficient to show that Problems 6–9 are NP-complete.

Theorem 10. *Problems 6–9 are NP-complete for all $\epsilon \in (0,1)$, even on degree-3 planar bipartite polymatrix games where each player has at most 7 pure strategies.*

4 Constrained Equilibria in Bounded Treewidth Games

In this section, we show that some constrained equilibrium problems can be solved in quasi-polynomial time if the input game has bounded treewidth and at most logarithmically many actions per player. We first present a dynamic programming algorithm for finding approximate Nash equilibria in these games, and then show that it can be modified to find constrained equilibria.

Tree Decompositions. A tree decomposition of a graph $G = (V, E)$ is a pair $(\mathcal{X}, T)$, where $T = (I, F)$ is a tree and $\mathcal{X} = \{X_i | i \in I\}$ is a family of subsets of V such that (1) $\bigcup_{i \in I} X_i = V$ (2) for every edge $(u, v) \in E$ there exists an $i \in I$ such that $\{u, v\} \in X_i$, and (3) for all $i, j, k \in I$ if j is on the path from

i to k in T, then $X_i \cap X_k \subseteq X_j$. The width of a tree decomposition $(\mathcal{X}, T)$ is $\max_i |X_i| - 1$. The treewidth of a graph is the minimum width over all possible tree decompositions of the graph. In general, computing the treewidth of a graph is NP-hard, but there are fixed parameter tractable algorithms for the problem. In particular Bodlaender [8] has given an algorithm that runs in $O(f(w) \cdot n)$ time, where w is the treewidth of the graph, and n is the number of nodes.

4.1   An Algorithm to Find Approximate Nash Equilibria

Let G be a polymatrix game and let $(\mathcal{X}, T)$ be a tree decomposition of G's interaction graph. We assume that an arbitrary node of T has been chosen as the root. Then, given some node v in T, we define $G(X_v)$ to be the subgame that is obtained when we only consider the players in the subtree of v. More formally, this means that we only include players i that are contained in some set X_u where u is in the subtree of v in the tree decomposition. Furthermore, we will use $\widetilde{G}(X_v)$ to denote the players in $G(X_v) \setminus X_v$. For every player $i \in X_v$, we will use $N_i(X_v)$ to denote the neighbours of i in $\widetilde{G}(X_v)$.

k-Uniform Strategies. A strategy s is said to be k-uniform if there exists a multi-set S of k pure strategies such that s plays uniformly over the pure strategies in S. These strategies naturally arise when we sample, with replacement, k pure strategies from a distribution, and play the sampled strategies uniformly. The following is a theorem of [2].

Theorem 11. *Every n-player m-action game has a k-uniform ϵ-NE whenever* $k \geq 8 \cdot \frac{\ln m + \ln n - \ln \epsilon + \ln 8}{\epsilon^2}$.

Candidates and Witnesses. For each node v in the tree decomposition, we compute a set of *witnesses*, where each witness corresponds to an ϵ-NE in $G(X_v)$. Our witnesses have two components: $\mathbf{s}$ provides a k-uniform strategy profile for the players in X_v, while $\mathbf{p}$ contains information about the payoff that the players in X_v obtain from the players in $\widetilde{G}(X_v)$. By summarising the information about the players in $\widetilde{G}(X_v)$, we are able to keep the number of witnesses small.

There is one extra complication, however, which is that the number of possible payoff vectors that can be stored in $\mathbf{p}$ depends on the number of different strategies for the players in $\widetilde{G}(X_v)$, which is exponential, and will cause our dynamic programming table to be too large. To resolve this, we *round* the entries of $\mathbf{p}$ to a suitably small set of rounded payoffs.

Formally, we first define $P = \{x \in [0, 1] : x = \frac{\epsilon}{2n} \cdot k \text{ for some } k \in \mathbb{N}\}$, to be the set of rounded payoffs. Then, given a node v in the tree decomposition, we say that a tuple $(\mathbf{s}, \mathbf{p})$ is a k-*candidate* if:

- $\mathbf{s}$ is a set of strategies of size $|X_v|$, with one strategy for each player in X_v.
- Every strategy in $\mathbf{s}$ is k-uniform.
- $\mathbf{p}$ is a set of payoff vectors of size $|X_v|$. Each element $\mathbf{p}_i \in \mathbf{p}$ is of the form P^m, and assigns a rounded payoff to each pure strategy of player i.

The set of candidates gives the set of possible entries that can appear in our dynamic programming table. Every witness is a candidate, but not every candidate is a witness. The total number of k-candidates for each tree decomposition node v can be derived as follows. Each player has m^k possible k-uniform strategies, and so there are m^{kw} possibilities for $\mathbf{s}$. We have that $|P| = \frac{2n}{\epsilon}$, and that $\mathbf{p}$ contains $m \cdot w$ elements of P, so the total number of possibilities for $\mathbf{p}$ is $(2 \cdot \frac{n}{\epsilon})^{mw}$. Hence, the total number of candidates for v is $m^{kw} \cdot (2 \cdot \frac{n}{\epsilon})^{mw}$.

Next, we define what it means for a candidate to be a witness. We say that a k-candidate is an ϵ, k, r-*witness* if there exists a profile $\mathbf{s}'$ for $G(X_v)$ where

- $\mathbf{s}'$ agrees with $\mathbf{s}$ for the players in X_v.
- Every player in $\widetilde{G}(X_v)$ is ϵ-*happy*, which means that no player in $\widetilde{G}(X_v)$ can increase their payoff by more than ϵ by unilaterally deviating from $\mathbf{s}'$. Note that this does not apply to the players in X_v.
- Each payoff vector $p \in \mathbf{p}$ is within r of the payoff that player i obtains from the players in $\widetilde{G}(X_v)$. More accurately, for every pure strategy l of player i we have that: $|p_l - \sum_{j \in \widetilde{G}(X_v)} (A_{ij} \cdot \mathbf{s}'_j)_l| \leq r$. Note that $\mathbf{p}$ does not capture the payoff obtained from players in X_v, only those in the subtree of v.

The Algorithm. Our algorithm computes a set of witnesses for each tree decomposition node by dynamic programming. At every leaf, the algorithm checks every possible candidate to check whether it is a witness. At internal nodes in the tree decomposition, if a vertex is *forgotten*, that is, if it appears in a child of a node, but not in the node itself, then we use the set of witnesses computed for the child to check whether the forgotten node is ϵ-happy. If this is the case, then we create a corresponding witness for the parent node. The complication here is that, since we use rounded payoff vectors, this check may declare that a player is ϵ-happy erroneously due to rounding errors. So, during the analysis we must be careful to track the total amount of rounding error that can be introduced.

Once a set of witnesses has been computed for every tree decomposition node, a second phase is then used to find an ϵ-NE of the game. This phase picks an arbitrary witness in the root node, and then unrolls it by walking down the tree decomposition and finding the witnesses that were used to generate it. These witnesses collectively assign a k-uniform strategy profile to each player, and this strategy profile will be the ϵ-NE that we are looking for.

Lemma 12. *There is a dynamic programming algorithm that runs in time* $O(n \cdot m^{2kw} \cdot (\frac{n}{\epsilon})^{2mw})$ *that, for each tree decomposition node v, computes a set of candidates $C(v)$ such that: (1) Every candidate $(\mathbf{s}, \mathbf{p}) \in C(v)$ is an ϵ_v, k, r_v-witness for v for some $\epsilon_v \leq 1.5\epsilon$ and $r_v \leq \frac{\epsilon}{4}$. (2) If $\mathbf{s}$ is a k-uniform $\epsilon/4$-NE then $C(v)$ will contain a witness $(\mathbf{s}', \mathbf{p})$ such that $\mathbf{s}'$ agrees with $\mathbf{s}$ for all players in X_v.*

The running time bound arises from the total number of possible candidates for each tree decomposition node. The first property ensures that the algorithm always produces a 1.5ϵ-NE of the game, provided that the root node contains a witness. The second property ensures that the root node will contain a witness provided that game has a k-uniform $\epsilon/4$-NE. Theorem 11 tells us how large k needs to be for this to be the case. These facts yields the following theorem.

Theorem 13. *Let $\epsilon > 0$, G be a polymatrix game with treewidth w, and $k = 128 \cdot \frac{\ln m + \ln n - \ln \epsilon + \ln 8}{\epsilon^2}$. There is an algorithm that finds a 1.5ϵ-NE of G in in $O(n \cdot m^{2kw} + (\frac{n}{\epsilon})^{2mw})$ time.*

Note that if $m \leq \ln n$ (and in particular if m is constant), this is a QPTAS.

Corollary 14. *Let $\epsilon > 0$, and G be a polymatrix game with treewidth w, and $m \leq \ln n$. There is an algorithm that finds a 1.5ϵ-NE of G in $(\frac{n}{\epsilon})^{O(\frac{w \cdot \ln n}{\epsilon^2})}$ time.*

4.2 Constrained Approximate Nash Equilibria

One Variable Decomposable Constraints. We now adapt the algorithm to find a certain class of constrained approximate Nash equilibria. As a motivating example, consider Problem 1, which asks us to find an approximate NE with high social welfare. Formally, this constraint assigns a single rational number (the social welfare) to each strategy profile, and asks us to maximize this number. This constraint also satisfies a *decomposability* property: if a game G consists of two subgames G_1 and G_2, and if there are no edges between these two subgames, then we can maximize social welfare in G by maximizing social welfare in G_1 and G_2 independently. We formalise this by defining a constraint to be *one variable decomposable (OVD)* if the following conditions hold.

- There is a polynomial-time computable function g such that maps every strategy profile in G to a rational number.
- Let **s** be a strategy for game G, and suppose that we want to add vertex v to G. Let s be a strategy choice for v, and **s**$'$ be an extension of **s** that assigns s to v. There is a polynomial-time computable function add such that $g(\mathbf{s}') = \text{add}(G, v, s, g(\mathbf{s}))$.
- Let G_1 and G_2 be two subgames that partition G, and suppose that there are no edges between G_1 and G_2. Let $\mathbf{s}_1$ be a strategy profile in G_1 and $\mathbf{s}_2$ be a strategy profile in G_2. If **s** is the strategy profile for G that corresponds to merging $\mathbf{s}_1$ and $\mathbf{s}_2$, then there is a polynomial-time computable function merge such that $g(\mathbf{s}) = \text{merge}(G_1, G_2, g(\mathbf{s}_1), g(\mathbf{s}_2))$.

Intuitively, the second condition allows us to add a new vertex to a subgame, and the third condition allows us to merge two disconnected subgames. Moreover, observe that the functions add and merge depend only on the value that g assigns to strategies, and not the strategies themselves. This allows our algorithm to only store the value assigned by g, and forget the strategies themselves.

Examples of OVD Constraints. Many of the problems in Table 1 are OVD constraints. Problems 1 and 2 refer to the total payoff of the strategy profile, and so g is defined to be the total payoff of all players, while the functions add and merge simply add the total payoff of the two strategy profiles. Problems 3 and 6 both deal with minimizing a quantity associated with a strategy profile, so for these problems the functions add and merge use the min function to minimize the relevant quantities. Likewise, Problems 7, 8, and 9 seek to maximize a quantity, and so the functions add and merge use the max function. In all cases, proving the required properties for the functions is straightforward.

Finding OVD k-Uniform Constrained Equilibria. We now show that, for every OVD constraint, the algorithm presented in Sect. 4.1 can be modified to find a k-uniform 1.5ϵ-NE that also has a high value with respect to the constraint. More formally, we show that the value assigned by g to the 1.5ϵ-NE is greater than the value assigned to g to all k-uniform $\epsilon/4$-NE in the game.

Given an OVD constraint defined by g, add, and merge, we add an extra element to each candidate to track the variable from the constraint: each candidate has the form $(\mathbf{s}, \mathbf{p}, x)$, where $\mathbf{s}$ and $\mathbf{p}$ are as before, and x is a rational number. The definition of an ϵ, k, r, g-witness is extended by adding the condition:

- Recall that $\mathbf{s}'$ is a strategy profile for $G(X_v)$ whose existence is asserted by the witness. Let $\mathbf{s}''$ be the restriction of $\mathbf{s}'$ to $\widetilde{G}(X_v)$. We have $x = g(\mathbf{s}'')$.

We then modify the algorithm to account for this new element in the witness. At each stage we track the correct value for x. At the leaves, we use g to compute the correct value. At internal nodes, we use add and merge to compute the correct value using the values stored in the witnesses of the children.

If at any point two witnesses are created that agree on $\mathbf{s}$ and $\mathbf{p}$, but disagree on x, then we only keep the witness whose x value is higher. This ensures that we only keep witnesses corresponding to strategy profiles that maximize the constraint. When we reach the root, we choose the strategy profile with maximal value for x to be unrolled in phase 2. The fact that we only keep one witness for each pair $\mathbf{s}$ and $\mathbf{p}$ means that the running time of the algorithm is unchanged.

Theorem 15. *For every $\epsilon > 0$ let $k = 128 \cdot \frac{\ln m + \ln n - \ln \epsilon + \ln 8}{\epsilon^2}$. If G is a polymatrix game with treewidth w, then there is an algorithm that runs in $O(n \cdot m^{2kw} + \left(\frac{n}{\epsilon}\right)^2 mw)$ time and finds a k-uniform 1.5ϵ-NE $\mathbf{s}$ such that $g(\mathbf{s}) \geq g(\mathbf{s}')$ for every strategy profile $\mathbf{s}'$ that is an $\epsilon/4$-NE.*

Results for Non-k-Uniform Strategies. The guarantee given by Theorem 15 is given relative to the best value achievable by a k-uniform $\epsilon/4$-NE. It is also interesting to ask whether we can drop the k-uniform constraint. In the following theorem, we show that if g is defined to be a linear function of the payoffs in the game, then a guarantee can be given relative to *every* $\epsilon/8$-NE of the game. Note that this covers Problems 1, 2, and 3.

Theorem 16. *Suppose that, for a given a OVD constraint, the function g is a linear function of the payoffs. Let $\mathbf{s}$ be the 1.5ϵ-NE found by our algorithm when For every $\epsilon/8$-NE $\mathbf{s}'$ we have that $g(\mathbf{s}) \geq g(\mathbf{s}') - O(\epsilon)$.*

References

1. Austrin, P., Braverman, M., Chlamtac, E.: Inapproximability of NP-complete variants of Nash equilibrium. Theory Comput. **9**, 117–142 (2013)
2. Babichenko, Y., Barman, S., Peretz, R.: Simple approximate equilibria in large games. In: Proceedings of the EC, pp. 753–770 (2014)
3. Barman, S., Ligett, K.: Finding any nontrivial coarse correlated equilibrium is hard. In: Proceedings of EC, pp. 815–816 (2015)
4. Barman, S., Ligett, K., Piliouras, G.: Approximating Nash equilibria in tree polymatrix games. In: Hoefer, M. (ed.) SAGT 2015. LNCS, vol. 9347, pp. 285–296. Springer, Heidelberg (2015). doi:10.1007/978-3-662-48433-3_22
5. Bilò, V., Mavronicolas, M.: The complexity of decision problems about Nash equilibria in win-lose games. In: Serna, M. (ed.) SAGT 2012. LNCS, pp. 37–48. Springer, Heidelberg (2012). doi:10.1007/978-3-642-33996-7_4
6. Bilò, V., Mavronicolas, M.: A catalog of ∃ℝ-complete decision problems about Nash equilibria in multi-player games. In: Proceedings of STACS, pp. 17:1–17:13 (2016)
7. Bilò, V., Mavronicolas, M.: ∃ℝ-complete decision problems about symmetric Nash equilibria in symmetric multi-player games. In: Proceedings of STACS, pp. 13:1–13:14 (2017)
8. Bodlaender, H.L.: A linear-time algorithm for finding tree-decompositions of small treewidth. SIAM J. Comput. **25**(6), 1305–1317 (1996)
9. Bonifaci, V., Di Iorio, U., Laura, L.: The complexity of uniform Nash equilibria and related regular subgraph problems. Theor. Comput. Sci. **401**(1–3), 144–152 (2008)
10. Bosse, H., Byrka, J., Markakis, E.: New algorithms for approximate Nash equilibria in bimatrix games. Theor. Comput. Sci. **411**(1), 164–173 (2010)
11. Braverman, M., Kun-Ko, Y., Weinstein, O.: Approximating the best Nash equilibrium in $n^{o(\log n)}$-time breaks the exponential time hypothesis. In: Proceedings of SODA, pp. 970–982 (2015)
12. Chen, X., Deng, X., Teng, S.-H.: Settling the complexity of computing two-player Nash equilibria. J. ACM **56**(3), 57 (2009). Article No. 14
13. Conitzer, V., Sandholm, T.: New complexity results about Nash equilibria. Games Econ. Behav. **63**(2), 621–641 (2008)
14. Czumaj, A., Deligkas, A., Fasoulakis, M., Fearnley, J., Jurdziński, M., Savani, R.: Distributed methods for computing approximate equilibria. In: Cai, Y., Vetta, A. (eds.) WINE 2016. LNCS, vol. 10123, pp. 15–28. Springer, Heidelberg (2016). doi:10.1007/978-3-662-54110-4_2
15. Czumaj, A., Fasoulakis, M., Jurdziński, M.: Approximate Nash equilibria with near optimal social welfare. In: Proceedings of IJCAI, pp. 504–510 (2015)
16. Czumaj, A., Fasoulakis M., Jurdziński, M.: Approximate plutocratic and egalitarian Nash equilibria. In: Proceedings of AAMAS, pp. 1409–1410 (2016)
17. Daskalakis, C., Goldberg, P.W., Papadimitriou, C.H.: The complexity of computing a Nash equilibrium. SIAM J. Comput. **39**(1), 195–259 (2009)
18. Daskalakis, C., Mehta, A., Papadimitriou, C.H.: Progress in approximate Nash equilibria. In: Proceedings of EC, pp. 355–358 (2007)
19. Daskalakis, C., Mehta, A., Papadimitriou, C.H.: A note on approximate Nash equilibria. Theor. Comput. Sci. **410**(17), 1581–1588 (2009)
20. Deligkas, A., Fearnley, J., Igwe, T.P., Savani, R.: An empirical study on computing equilibria in polymatrix games. In: Proceedings of AAMAS, pp. 186–195 (2016)

21. Deligkas, A., Fearnley, J., Savani, R.: Inapproximability results for approximate Nash equilibria. In: Cai, Y., Vetta, A. (eds.) WINE 2016. LNCS, vol. 10123, pp. 29–43. Springer, Heidelberg (2016). doi:10.1007/978-3-662-54110-4_3
22. Deligkas, A., Fearnley, J., Savani, R., Spirakis, P.G.: Computing approximate Nash equilibria in polymatrix games. Algorithmica 77(2), 487–514 (2017)
23. Elkind, E., Goldberg, L.A., Goldberg, P.W.: Nash equilibria in graphical games on trees revisited. In: Proceedings of EC, pp. 100–109 (2006)
24. Elkind, E., Goldberg, L.A., Goldberg, P.W.: Computing good Nash equilibria in graphical games. In: Proceedings of EC, pp. 162–171 (2007)
25. Fearnley, J., Goldberg, P.W., Savani, R., Sørensen, T.B.: Approximate well-supported Nash equilibria below two-thirds. In: Serna, M. (ed.) SAGT 2012. LNCS, pp. 108–119. Springer, Heidelberg (2012). doi:10.1007/978-3-642-33996-7_10
26. Garg, J., Mehta, R., Vazirani, V.V., Yazdanbod, S.: ETR-completeness for decision versions of multi-player (symmetric) Nash equilibria. In: Halldórsson, M.M., Iwama, K., Kobayashi, N., Speckmann, B. (eds.) ICALP 2015. LNCS, vol. 9134, pp. 554–566. Springer, Heidelberg (2015). doi:10.1007/978-3-662-47672-7_45
27. Gilboa, I., Zemel, E.: Nash and correlated equilibria: some complexity considerations. Games Econ. Behav. 1(1), 80–93 (1989)
28. Greco, G., Scarcello, F.: On the complexity of constrained Nash equilibria in graphical games. Theor. Comput. Sci. 410(38–40), 3901–3924 (2009)
29. Hazan, E., Krauthgamer, R.: How hard is it to approximate the best Nash equilibrium? SIAM J. Comput. 40(1), 79–91 (2011)
30. Kontogiannis, S.C., Spirakis, P.G.: Well supported approximate equilibria in bimatrix games. Algorithmica 57(4), 653–667 (2010)
31. Lipton, R.J., Markakis, E., Mehta, A.: Playing large games using simple strategies. In: Proceedings of EC, pp. 36–41 (2003)
32. Moore, C., Robson, J.M.: Hard tiling problems with simple tiles. Discrete Comput. Geom. 26(4), 573–590 (2001)
33. Ortiz, L.E., Irfan, M.T.: FPTAS for mixed-strategy Nash equilibria in tree graphical games and their generalizations. CoRR, abs/1602.05237 (2016)
34. Ortiz, L.E., Irfan, M.T.: Tractable algorithms for approximate Nash equilibria in generalized graphical games with tree structure. In: Proceedings of AAAI, pp. 635–641 (2017)
35. Rubinstein, A.: Inapproximability of Nash equilibrium. In: Proceedings of STOC, pp. 409–418 (2015)
36. Tsaknakis, H., Spirakis, P.G.: An optimization approach for approximate Nash equilibria. Internet Math. 5(4), 365–382 (2008)

Group Activity Selection on Graphs: Parameterized Analysis

Sushmita Gupta[1], Sanjukta Roy[2(✉)], Saket Saurabh[1,2], and Meirav Zehavi[1]

[1] University of Bergen, Bergen, Norway
{sushmita.gupta,meirav.zehavi}@uib.no
[2] The Institute of Mathematical Sciences, HBNI, Chennai, India
{sanjukta,saket}@imsc.res.in

Abstract. In varied real-life situations, ranging from carpooling to workload delegation, several activities are to be performed, to which end each activity should be assigned to a group of agents. These situations are captured by the Group Activity Selection Problem (GASP). Notably, relevant relations among agents, such as acquaintanceship or physical distance, can often be modeled naturally using graphs. To exploit this modeling ability, Igarashi, Peters and Elkind [AAAI 17] introduced gGASP. Specifically, it is required that each group would correspond to a connected set of the underlying graph. In addition, to enforce the execution of the activities in practice, no individual should desire to desert its group in favor of joining another group. In other words, the assignment should be Nash stable. In this paper, we study gGASP with Nash stability (gNSGA), whose objective is to compute such an assignment. This problem is computationally hard even on such restricted topologies as paths and stars, which naturally led Igarashi, Bredereck, Peters and Elkind [AAAI 17, AAMAS 17] to the study gNSGA in the framework of parameterized complexity. We take this line of investigation forward, significantly advancing the state-of-the-art. First, we show that gNSGA is NP-hard *even when merely one activity is present*. In fact, this special case remains NP-hard when we further restrict the graph to have maximum degree $\Delta = 5$. Consequently, gNSGA is not fixed-parameter tractable (FPT), or even XP, when parameterized by $p + \Delta$, where p is the number of activities. However, we are able to design a parameterized algorithm for gNSGA on *general graphs* with respect to $p + \Delta + t$, where t is the maximum size of a group. Finally, we develop an algorithm that solves gNSGA on graphs of bounded treewidth **tw** in time $4^p \cdot (n + p)^{\mathcal{O}(\mathbf{tw})}$. Here, $\Delta + t$ can be arbitrarily large. Along the way, we resolve several open questions regarding gNSGA.

1 Introduction

Division of labor is required in varied real-world situations. For a task to be accomplished, be it the construction of a building or the development of a product, it is necessary to *assign* agents (such as people or companies) to appropriate activities, and those agents must be willing to contribute towards the common

© Springer International Publishing AG 2017
V. Bilò and M. Flammini (Eds.): SAGT 2017, LNCS 10504, pp. 106–118, 2017.
DOI: 10.1007/978-3-319-66700-3_9

goal. Though workload delegation is perhaps the first example that comes to mind, management of cooperation—or more precisely, formation of groups by agents participating in specific activities—is ubiquitous in almost all aspects of life. Indeed, other examples range from carpooling and seating arrangements to hobbies such as tennis or basketball. All such situations are neatly captured by the Group Activity Selection Problem (GASP) introduced in [4].

An instance of GASP is given by a finite set of agents N, where $|N| = n$, a finite set of activities $A = A^\star \cup \{a_\emptyset\}$, where $A^\star = \{a_1, \ldots, a_p\}$ and $a_\emptyset$ is a *void activity*, and a profile $(\succeq_v)_{v \in N}$ of complete and transitive preference relations over the set of alternatives $X = (A^\star \times \{1, 2, \ldots, n\}) \cup \{(a_\emptyset, 1)\}$.[1] The void activity $a_\emptyset$ is introduced to allow agents to avoid undertaking activities, which also enables agents to be independent. For example, in the case of the development of a product, the set N may consist of employees of some company, each activity in the set A may correspond to the design of a certain component of the product, and the profile $(\succeq_v)_{v \in N}$ may be constructed according to the skills/personal preferences of the employees and their abilities/willingness to function in groups of varied sizes. The void activity would allow to exclude employees from the current project in case no suitable activities can be assigned to them.

The outcome of GASP is defined as an *assignment*, which is a simply function $\pi : N \to A$. Clearly, an arbitrary assignment is extremely undesirable unless the profile $(\succeq_v)_{v \in N}$ is completely meaningless. To take the profile into account, it is first natural to request that π would at least be *individually rational* (IR), which means that for every agent $v \in N$ with $\pi(v) = a \ (\neq a_\emptyset)$, we have $(a, |\pi^a|) \succeq_v (a_\emptyset, 1)$, where $\pi^a = \pi^{-1}(a) = \{v \in N : \pi(v) = a\}$.[2] That is, no agent v would rather "being alone" than being part of a group of size $|\pi_v|$ that performs activity $a = \pi(v)$, where $\pi_v = \pi^a$ (that is, π_v is the set of all those agents that have been assigned the same activity as v). In addition, to enforce the execution of the activities in practice, no individual should desire to act on its own by deserting its group in favor of joining another group. In other words, we would like the assignment to be Nash stable. Formally, an agent $v \in N$ is said to have an *NS-deviation* to an activity $a \in A^\star$ if $a \neq \pi(v)$ and $(a, |\pi^a| + 1) \succ_v (\pi(v), |\pi_v|)$, that is, v prefers to join the activity a, given every one else plays the same activity as before. Accordingly, π is said to be *Nash stable* if it is individually rational and no agent has an NS-deviation.

Let us take a step back and observe that if an assignment π is Nash stable, then the only implication is that no agent has an alternative more preferred than the situation assigned to it by π. However, we do not ensure by any means that the agent would actually be able/willing to cooperate with other members in its group, so that the assignment can actually be executed in a satisfactory manner. Notably, relevant relations among agents, such as acquaintanceship, compatibility or geographical distance, can often be modeled naturally using

---

[1] For the sake of consistency, we follow the notations and definitions of [14].

[2] As we would always work with IR assignments, when specifying preference profiles, we would only explicitly state the alternatives that are preferred more than $a_\emptyset$.

graphs. To exploit this modeling ability, Igarashi et al. [14] introduced gGASP. Specifically, it is required that each group would correspond to a connected set of the underlying graph. For a deeper understanding of the rationale underlying this requirement, let us consider the case where the graph is a social network. Then, by ensuring that each group is a connected set, we ensure that each individual in the group would be acquainted with at least one other person in the group. The desirability of such property is clear when discussing activities such as carpooling, seating arrangements or sports as it is conceivable that people would prefer to share a taxi/sit next to/play with at least one other person whom they know. In the context of workload delegation, apart from a social aspect, it is likely that agents who are familiar with each other would also be able to work more efficiently with each other, with each agent having at least one other agent as a comfortable communication link or a source to "count on".

Formally, an instance of gGASP [14] consists of an instance $(N, (\succeq_v)_{v \in N}, A)$ of GASP and a set of communication links between agents, $L \subseteq \{\{u, v\} \mid u, v \in N, u \neq v\}$. Thus, we assume that we are also given a graph G with vertex set $V(G) = N$ and $E(G) = L$. Here, G is called underlying network (or graph) of gGASP. A set $S \subseteq N$ of agents is said to be a *coalition*, and it is a *feasible coalition* if $G[S]$ is connected where $G[S]$ is the graph induced on the vertices in S. Now, an NS-deviation by an agent v to an activity $a \in A^*$ is called a *feasible NS-deviation* if $\pi^a \cup \{v\}$ is a feasible coalition. Thus, in the context of gGASP, an assignment π is said to be *Nash stable* if it is individually rational and no agent has a feasible NS-deviation. We would be interested in the following question.

Nash Stable gGASP (gNSGA)
Input: An instance $\mathcal{I} = (N, (\succeq_v)_{v \in N}, A, G)$ of gGASP.
Question: Does $\mathcal{I}$ have a feasible Nash stable assignment (fNsa) ?

Igarashi et al. [14] showed that gNSGA is NP-complete even when the underlying graph is a path, a star, or if the size of each connected component is bounded by a constant. In addition, they exhibit FPT algorithms (for the same graph classes) when parameterized by the number of activities. In a more recent work by Igarashi et al. [12], the authors show that when parameterized by the number of players, gNSGA is W[1]-hard on cliques (the classical setting of GASP), but admits an XP-algorithm for the same graph classes. Furthermore, when the underlying graph is a clique, gNSGA is W[1]-hard when parameterized by the number of activities. They also give an FPT algorithm for acyclic graphs, parameterized by the number of activities. Specifically, Igarashi et al. [14] posed the following open question.

For general graphs, the exact parameterized complexity of determining the existence of stable outcomes is unknown… for other networks, including trees, it is not even clear whether our problem is in XP with respect to the number of activities.

Our Contribution. Given that gNSGA is NP-hard even on paths and stars [14], and as this problem inherently encompasses parameters that can be often expected to be small in practice, it is indeed very natural to examine it from the viewpoint of parameterized complexity.[3] In this context, we take the line of investigation initiated by the studies [12,14] several steps forward, significantly advancing the state-of-the-art. In fact, as we explain below, we push some boundaries to their limits, and along the way, we give answer to questions posed by Igarashi et al. [14] that are even stronger than requested.

Hardness: Firstly, we consider $p = |A^*|$, the number of activities, as the parameter. Here, we show that gNSGA is NP-hard *even when merely one activity is present*, that is, $p = 1$. More precisely, we prove the following theorem. Here, Δ is the maximum degree of the graph G.

Theorem 1. *The* gNSGA *problem is* NP-hard *even when* $p = 1$ *and* $\Delta = 5$.

Recall that Igarashi et al. [14] contemplated whether gNSGA is fixed-parameter tractable (FPT) with respect to p. We show that even if $p = 1$, the problem is already NP-hard, and in fact it remains NP-hard even on graphs where the maximum degree Δ is as small as 5! In particular, we derive that gNSGA is para-NP-hard when the parameter is $p + \Delta$. That is, Theorem 1 implies that we do not expect to have an FPT algorithm, or even merely an XP algorithm, on general graphs with respect to both p and Δ *together*. Indeed, the existence of an XP algorithm, that is, an algorithm with running time $n^{f(p,\Delta)}$, where $n = |N|$ and f is any function depending only on p and Δ, would contradict Theorem 1.

FPT Algorithm on General Graphs: In light of Theorem 1, we consider an additional parameter t—the maximum size of any group that can form a feasible coalition. Having this parameter at hand, we are able to design an FPT algorithm that handles *general graphs*. Before we state our theorem formally, note that t is a natural choice for a parameter. Indeed, sport teams/matches usually involve only few players, a taxi or a table have only limited space, and certain tasks are clearly suitable, or best performed, when only few people undertake them. We remark that Δ can also often be expected to be small. For example, when most people in an event (say, a donation evening) do not know each other well, this would indeed be the case when planning a seating arrangement. In addition, when new participants sign-up to organized sport activities, they might only know those friends that are also interested in those exact activities. Moreover, when a company operating across different countries would like to undertake some task, while employees generally know only those other employees with whom they share the same floor, we again arrive at a situation where Δ is small.

Theorem 2. gNSGA *on general graphs is solvable in time* $\mathcal{O}((\Delta p)^{\mathcal{O}(tp)} \cdot n \log n)$.

The proof of Theorem 2 uses the idea of an n-p^*-q^*-*lopsided-universal family*, introduced in [7], to "separate" agents that are assigned non-void activities from

---

[3] For standard definitions concerning parameterized complexity, see [3].

their neighbors. Once this is done, a non-trivial dynamic programming algorithm is developed to test whether there exists an fNsa.

FPT Algorithm for Graphs of Bounded Treewidth: Igarashi [14] designed FPT algorithms for gNSGA on paths, stars and graphs whose connected components are restricted to have constant size. In a more recent article, Igarashi et al. [12] designed an FPT algorithm with running time $\mathcal{O}(p^p \cdot (n + p)^{\mathcal{O}(1)})$ for gNSGA on acyclic graphs (i.e. forests). We generalize this result to a substantially wider class of graphs that includes all the above classes of graphs, namely, graphs of bounded treewidth. This class includes graphs that have an unbounded number of cycles, and in fact it even generalizes the class of all graphs whose feedback vertex set number is small. Formally, we derive the following theorem.

Theorem 3. *The* gNSGA *problem on graphs treewidth* **tw** *is solvable in time* $\mathcal{O}(4^p \cdot (n + p)^{\mathcal{O}(\mathbf{tw}))}$.

Notably, our algorithm solves gNSGA on trees (where **tw** = 1) in time $\mathcal{O}(4^p \cdot (n+p)^{\mathcal{O}(1)})$, that is, significantly faster than the specialized algorithm by Igarashi et al. [12]. In fact, its running time also matches the running time of the even more specialized algorithm by Igarashi et al. [12] for paths.

Related Works. Papers [12,14] that specifically solve gNSGA on restricted classes of graphs were discussed in some details earlier. Here, we give a brief (non-comprehensive) survey of a few results related to problems of flavor similar to that of gGASP, so as to understand the well established roots of this gNSGA.

The literature on graph based cooperative games, to which gGASP is a new addition, can be traced back to Myerson's seminal paper [16] which introduced the graph theoretic model of cooperation, where vertices represent agents participating in the game and edges between pairs of vertices represent cooperative relationship between agents corresponding to the vertices. In cooperative games, there are two basic notions of stability, one based on the individual and the other on the group. The latter notion corresponds to what is known as *core stability*. Hedonic games [1,8,9] form a domain similar to that of GASP where agents have preferences over other agents, but also groups of agents that include themselves. The primary challenges of designing efficient algorithms in hedonic games is that the space requirement for just storing/representing the preference profile is (in general) exponential in the number of agents in the game. Consequently, people have studied the problem in sparse graphs such as trees, or those with a small number of connected components, Igarashi and Elkind's [13] is a recent work in this direction. Papers such as [2,6,13] explore stability in different kinds of cooperative games; but the central findings of these papers are in stark contrast with that of [14] showing that restricted graph classes, such as paths, trees, stars etc., are amenable for algorithms that efficiently compute stable solutions. Building on the work of [14], Igarashi *et al.* [12] studied the parameterized complexity of Nash stability, core stability, as well as individual stability in gGASP with respect to parameters such as the number of activities and the number of players. Finally, we conclude our discussion by pointing the reader to a vast array of literature on *coalition formation games*. Due to lack of space, we refer the reader to [15, pg. 222–223, 330] for an extensive discussion on the topic.

Preliminaries. For standard graph theoretic notation we refer to [5]. For a detailed definition of *(nice) tree decomposition*, see [3, pg. 161].

2   Hardness

In this section, we show that gNSGA is NP-complete *even when there is only one activity* and the maximum degree of the underlying graph is at most 4. Towards this, we give a polynomial time many-to-one reduction from the STEINER TREE* problem (defined below) on graphs of maximum degree at most 4 [11] to gNSGA.

STEINER TREE*
Input: An undirected graph $G^\star$ on n vertices, $K \subseteq V(G^\star)$ (called terminals) and a positive integer ℓ.
Question: Does there exist $H \subseteq V(G^\star)$ such that $G^\star[H]$ is a tree, $K \subseteq H$ and $|H| = \ell$?

STEINER TREE* differs from the usual STEINER TREE problem as follows: here we demand the size of $|H|$ to be *exactly equal* to ℓ rather than at most ℓ. A simple reduction from STEINER TREE (on graphs of maximum degree 4) that maps an instance to itself shows that STEINER TREE* (on graphs of maximum degree 4) is NP-complete as well. We first give our construction.

Construction. We first show how to construct the underlying graph G from $G^\star$. To this end, we take a copy of $G^\star$ and make the following additions to construct G. For every $w \in K$, we construct a path P_w on $n + 1$ *dummy vertices* and add an edge between some end-point of the path and the terminal vertex w. Now, for each vertex $u \in V(G^\star) \setminus K$, we add a new vertex u' and connect u' to u. The vertex u' will act as a *stalker* for the vertex u. A vertex $x \in V(G)$ is called **(i)** a *terminal vertex* if $x \in K$, **(ii)** a *non-terminal vertex* if $x \in V^\star = V(G^\star) \setminus K$, **(iii)** a *dummy vertex* if $x \in \mathsf{Dummy} = \cup_{w \in K} P_w$, and **(iv)** a *stalker vertex* if $x \in \mathsf{Stalker} = \{u' \mid u \in V^\star\}$. This completes the construction of G.

Having constructed G, the instance $\mathcal{I}$ of gNSGA is defined as follows:
- The set of agents is $N = V(G)$ and the set of activities is $A = \{a\}$. That is, we only have one activity (in addition to the void activity $a_\emptyset$).
- Now we define the preference profiles $(\succeq_v)_{v \in N}$ of the agents. Let $|K| = t$ and $\gamma = (n + 2)t + (\ell - t)$. For an agent $v \in N$, the preference profile is

$$\succeq_v := \begin{cases} \langle (a, \gamma), (a_\emptyset, 1) \rangle & \text{if } v \in K \\ \langle (a, \gamma), (a, 1), (a_\emptyset, 1) \rangle & \text{if } v \in V^\star \\ \langle (a, \gamma), (a, \gamma + 1), (a_\emptyset, 1) \rangle & \text{if } v \in \mathsf{Dummy} \\ \langle (a, 2), (a_\emptyset, 1) \rangle & \text{if } v \in \mathsf{Stalker} \end{cases}$$

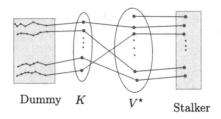

Fig. 1. The construction of the underlying graph G.

This completes the description of the instance of gNSGA.

Correctness. For correctness we show the following equivalence.

Lemma 1. $(G^\star, K, \ell)$ *is a* YES-*instance of* STEINER TREE* *if and only if* $\mathcal{I}$ *is a* YES-*instance of* gNSGA.

Proof. For the forward direction, assume that there exists $H \subseteq V(G^\star)$ such that $G^\star[H]$ is connected, $K \subseteq H$ and $|H| = \ell$. Using the solution graph H, we define an assignment $\pi : N \to A$ as follows. For an agent $v \in N$,

$$\pi(v) := \begin{cases} a & \text{if } v \in (H \cup \text{Dummy}) \\ a_\emptyset & \text{otherwise} \end{cases}$$

Next, we prove that π is an fNsa; with that the forward direction is proved.

Claim. $(\star)^4$ π is an fNsa.

Now we will show the reverse direction of the proof. Let $\pi : N \to A$ be an fNsa. We first derive properties about π.

Property 1: *For every vertex* $v' \in$ Stalker, $\pi(v') = a_\emptyset$. For a contradiction assume that $\pi(v') = a$. Since v' would not join the activity a alone, the vertex v in $V^\star$ joins the activity a. That is, $\pi(v) = a$. But $(a_\emptyset, 1) \succ_v (a, 2)$. Hence, v wants to deviate to the void activity $a_\emptyset$, a contradiction to the stability of π.

Property 2: $\pi^a \neq \emptyset$. For a contradiction assume that no vertex is assigned activity a. Let $v \in V^\star$,[5] then $(a, 1) \succ_v (a_\emptyset, 1)$. That is, v has an NS-deviation to the activity a, again contradicting the stability of π.

Property 3: *If there exists a vertex* $v \in V^\star$ *such that* $v \in \pi^a$, *then* $|\pi^a| > 1$. For a contradiction assume that $|\pi^a| = 1$ (by Property 2 we know that $\pi^a \neq \emptyset$). Then, the stalker vertex v' has an NS-deviation to the activity a as $(a, 2) \succ_{w'} (a_\emptyset, 1)$.

---

[5] Note that, wlog, we can assume $V^\star$ is non empty, otherwise solving STEINER TREE* instance reduces to checking if $G^\star$ is connected.

Property 4: $|\pi^a| \in \{\gamma, \gamma + 1\}$. By Properties 1 and 2, we know that there is a vertex $v \in V^\star \cup K \cup \mathsf{Dummy}$ such that $v \in \pi^a$. Thus, if there exists a vertex $v \in (K \cup \mathsf{Dummy}) \cap \pi^a$, then since $(a, \gamma) \succ_v (a_\emptyset, 1)$ or $(a, \gamma + 1) \succ_v (a_\emptyset, 1)$ (if $v \in \mathsf{Dummy}$) we have that $|\pi^a| \in \{\gamma, \gamma + 1\}$. Furthermore, by Property 3, we know that if $v \in V^\star \cap \pi^a$ then $|\pi^a| > 1$. However, $(a, \gamma) \succ_v (a_\emptyset, 1)$, thus $|\pi^a| \in \{\gamma, \gamma + 1\}$.

Property 5: *For every* $v \in K$, $\pi(v) = a$ *and* $|\pi^a| = \gamma$. For a contradiction assume that there exists a vertex $v \in K$ such that $\pi(v) \neq a$. Then, since $G[\pi^a]$ is connected, we have that all the dummy vertices in the path P_v is not in π^a. This implies that $|\pi^a| \leq (|V^\star| + |K| - 1 + (|K| - 1)(n + 1) < \gamma$. This contradicts Property 5. Finally, since a vertex $v \in K \cap \pi^a$ (in fact every vertex of K is in π^a) and $(a, \gamma) \succ_v (a_\emptyset, 1)$, we have that $|\pi^a| = \gamma$.

Property 6: *For every* $v \in \mathsf{Dummy}$, $\pi(v) = a$. Indeed, otherwise consider a vertex $v \in \mathsf{Dummy}$ such that it has a neighbor in π^a (the fact that $K \subseteq \pi^a$ ensures an existence of such a vertex). Then the fact that $(a, \gamma + 1) \succ_v (a_\emptyset, 1)$ together with Property 5 imply that π is not stable, a contradiction.

Now we are ready to show the reverse direction of the proof. Let $W = \pi^a \cap (V^\star \cup K)$. Since $G[\pi^a]$ is connected, we have that $G[W]$ is connected. The last assertion follows from the fact that if we take a spanning tree L of $G[\pi^a]$, then the paths hanging from the vertices of K can be thought of as "long leaves" and by removing leaves we do not disconnect a tree. Furthermore, since $G[V^\star \cup K]$ is same as $G^\star$, we have that $G^\star[W]$ is connected and contains all the terminals. Finally, for our proof the only thing that remains to show is that $|W| = \ell$. This follows from the fact that $|\pi^a| = \gamma$ and while constructing W we have removed exactly $(n + 1)t$ dummy vertices. This concludes the proof.   ◇

Notice that G has maximum degree bounded by 5. We started with a graph $G^\star$ of maximum degree 4 and we have not added more than one new neighbor to any vertex. So the degree of any vertex in G is at most 5. Thus, our construction and Lemma 1 imply the proof of Theorem 1.

. Our proof of Theorem 1 is robust in the sense that one can start with a family of graphs on which the STEINER TREE* problem is NP-complete and then do the reduction in a way that we remain inside the family of graphs we started with. For example, it is known that STEINER TREE* remains NP-complete on planar graphs of maximum degree 4 [10], and thus our reduction would imply that gNSGA is NP-complete even when there is only one activity and the underlying network is a planar graph of max degree 5.

3   An FPT Algorithm for General Graphs

In the last section we established that gNSGA remains NP-complete even when the number of activities is one and the maximum degree of the underlying network is at most 5. As discussed in the introduction this immediately implies that we can not even have an XP algorithm parameterized by p and the maximum degree Δ of the underlying network. However, with an additional parameter t

– the maximum size of any group that can form in an fNsa—we are able to design an FPT algorithm. We use this notation, let $f : A \rightarrow B$ be some function. Given $A' \subseteq A$, the notation $f(A') = b$ indicates that for all $a \in A'$, it holds that $f(a) = b$.

Let $\mathcal{I} = (N, (\succeq_v)_{v \in N}, A, G)$ be an instance of gNSGA where the maximum degree of G is at most Δ. Our algorithm has two phases: (a) Separation Phase and (b) Validation Phase. We first outline the phases. We start with the Separation Phase.

Let $\pi : N \rightarrow A$ be a *hypothetical* fNsa for the given input $\mathcal{I}$. For our algorithm we would like to have a function f from $N = V(G)$ to $\{1, 2\}$ with the following property:

(P1) Let $N' \subseteq N$ be the set of agents who are assigned a non-void activity (that is an activity in $A^\star$) by π. Then, f labels 1 to every agent in N' and labels 2 to all the agents in $N_G(N')$ (neighbors of N' in G that are not in N').

A function f that satisfies the property (P1) with respect to a fNsa π is called *nice with respect to* π. Furthermore, a function f from $V(G)$ to $\{1, 2\}$ is called *nice* if f satisfies the property (P1) for *some* fNsa π' for $\mathcal{I}$.

In the validation phase, given a nice function f, we construct a fNsa (if it exists), $\pi : N \rightarrow A$, such that all the agents that have been labeled 2 are assigned void activity $a_\emptyset$. In other words, the only agents that get assigned an activity from $A^\star$ are those which are labeled 1 by f. It is possible that π assigns $a_\emptyset$ to some agent that have been labeled 1 by f. To construct an assignment π, if it exists, we employ a dynamic programming procedure. Now we describe both the phases in details.

Separation Phase. We first show the existence of a small sized family of functions such that given an instance of gNSGA on n agents such that it has a fNsa, then there exists a function that is nice with respect to this. Towards this we first introduce the notion of n-$p^\star$-$q^\star$-*lopsided-universal* family. Given a universe U and an integer ℓ by $\binom{U}{\ell}$ we denote all the ℓ-sized subsets of U. We say that a family $\mathcal{F}$ of sets over a universe U of size n is an n-$p^\star$-$q^\star$-*lopsided-universal* family if for every $A \in \binom{U}{p^\star}$ and $B \in \binom{U \setminus A}{q^\star}$ there is an $F \in \mathcal{F}$ such that $A \subseteq F$ and $B \cap F = \emptyset$. An alternative definition that is easily seen to be equivalent is that $\mathcal{F}$ is n-$p^\star$-$q^\star$-lopsided-universal family if for every subset $A \in \binom{U}{p^\star+q^\star}$ and every subset $A' \in \binom{A}{p^\star}$, there is an $F \in \mathcal{F}$ such that $F \cap A = A'$.

Lemma 2 [7]. *There is an algorithm that given n, $p^\star$ and $q^\star$ constructs an n-$p^\star$-$q^\star$-lopsided-universal family $\mathcal{F}$ of cardinality $\binom{p^\star+q^\star}{p^\star} \cdot 2^{o(p^\star+q^\star)} \cdot \log n$ in time $\mathcal{O}(\binom{p^\star+q^\star}{p^\star} \cdot 2^{o(p^\star+q^\star)} \cdot n \log n)$.*

We now show that there exists a family of functions, $\mathcal{H}(\mathcal{I})$, from N to $\{1, 2\}$ such that if $\mathcal{I}$ has a fNsa, π, then there exists a function $f \in \mathcal{H}(\mathcal{I})$ such that f is nice with respect to π. We call the family of functions $\mathcal{H}(\mathcal{I})$, a *nice family*

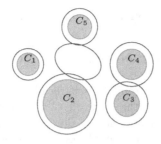

Fig. 2. Depiction of how a nice function f assigns $\{1,2\}$ to the vertices of G. Orange colored parts are assigned 1 by f and the white enclosed parts are assigned 2 by f. $C_1, C_2, \ldots, C_5$ are the components of $G[f^{-1}(1)]$. For $i \in [5]$, the concentric circle outside C_i is $N(C_i)$. It is guaranteed that f assigns 2 to those vertices.

with respect to $\mathcal{I}$. Let the vertex set $N = V(G)$ of the graph G be denoted by $\{v_1, \ldots, v_n\}$. We *identify* the vertex v_i with an integer i and thus we can view the vertex set as $[n]$. To construct the function f, we use Lemma 2. We apply Lemma 2 with universe $U = \{1, 2, \ldots, n\}$, $p^\star = tp$ and $q^\star = \Delta tp = \Delta p^\star$ and obtain a n-$p^\star$-$q^\star$-lopsided-universal family $\mathcal{F}$ of size $\binom{p^\star + q^\star}{p^\star} \cdot 2^{o(p^\star + q^\star)} \cdot \log n$ in time $\mathcal{O}(\binom{p^\star + q^\star}{p^\star} \cdot 2^{o(p^\star + q^\star)} \cdot n \log n)$. Given $\mathcal{F}$, we define $\mathcal{H}(\mathcal{I})$ as follows. For every set $X \in \mathcal{F}$, f_X is defined as follows: $f_X(x) = \begin{cases} 1 & \text{if } x \in X \\ 2 & \text{otherwise.} \end{cases}$

Thus, $\mathcal{H}(\mathcal{I}) := \{f_X \mid X \in \mathcal{F}\}$. Now we show that $\mathcal{H}(\mathcal{I})$ is a nice family. Suppose $\mathcal{I}$ has a fNsa π. Let N' be the set of agents that are assigned non-void activity by π and $W = N_G(N')$. Since the size of any group is upper bounded by t we have that $|N'| \leq tp$ and since the maximum degree of G is upper bounded by Δ we have that $|W| \leq \Delta tp$. Now by the property of n-$p^\star$-$q^\star$-lopsided-universal family $\mathcal{F}$, we know that there exists a set $X \in \mathcal{F}$ such that $N' \subseteq X$ and $W \cap X = \emptyset$. By construction the function f_X is nice with respect to π. This brings us to the following lemma.

Lemma 3. *Let $\mathcal{I}$ be an instance of* gNSGA. *Then, in time* $\mathcal{O}(\binom{p^\star + q^\star}{p^\star} \cdot 2^{o(p^\star + q^\star)} \cdot n \log n)$ *we can construct a nice family* $\mathcal{H}(\mathcal{I})$ *of size* $\binom{p^\star + q^\star}{p^\star} \cdot 2^{o(p^\star + q^\star)} \cdot \log n$. *Here, $p^\star = tp$ and $q^\star = \Delta p^\star$.*

Validation Phase. Now, we give an algorithm that given a function $f : N \to \{1, 2\}$ tests whether f is a nice function. In other words either it *correctly* concludes that f is *not* a nice function or outputs a fNsa, $\pi : N \to A$, such that f is nice with respect to π.

Lemma 4 ($\star$). *Let $\mathcal{I}$ be an instance of* gNSGA *and $f : N \to \{1, 2\}$ be a function. Then in time, $\mathcal{O}(n4^p(p+1)^{tp})$, we can test whether or not f is nice. Moreover, if f is nice, then in the same time we can output an* fNsa, *witness to the property that f is nice.*

Using Lemma 4 we can complete the proof of Theorem 2.

4   FPT Algorithm for Networks of Bounded Treewidth

In this section, we prove Theorem 3. First, recall in time $2^{\mathbf{tw}} \cdot n$ we can obtain a nice tree decomposition of the network G whose width is $\eta = \mathcal{O}(\mathbf{tw})$. In what follows, we let (T, β) denote such a decomposition. For $v \in V(T)$, we say that $\beta(v)$ is the bag of v, and $\gamma(v)$ denotes the union of the bags of v and the descendants of v in T. We apply the method of dynamic programming as described in the following subsections.

The DP Table. Let us begin by describing our DP table, which we denote by M. Each entry of this table is of the form $[x, B, \widehat{B}, f_{act}, f_{tot}, f_{cur}, P]$ for any choice of arguments as follows.

- A node $x \in V(T)$, which indicates that the partial solutions corresponding to the current entry would only assign activities only to those vertices that belong to $\gamma(x)$.
- A subset of activities $B \subseteq A$ such that $a_\emptyset \in B$, which indicates that only activities from B can be assigned to those vertices that belong to $\gamma(x)$. Let us denote $B^\star = B \setminus \{a_\emptyset\}$.
- A subset of activities $\widehat{B} \subseteq B$ such that $a_\emptyset \in \widehat{B}$, which indicates that each activity in $\widehat{B}^\star = \widehat{B} \setminus \{a_\emptyset\}$ has been assigned to at least one vertex in $\gamma(x)$.
- A function $f_{act} : \beta(x) \to \widehat{B}$, which specifies exactly how to assign activities to those vertices that belong to $\beta(x)$.
- A function $f_{tot} : \beta(x) \to [n]$, which specifies exactly, for every vertex v in $\beta(x)$, how many vertices in total (i.e. in $V(G)$) should participate in the same activity as v. Here, the void activity is an exception as for any vertex v assigned to the void activity, we would demand that $f_{tot}(v)$ is set to 1, irrespective of how many vertices participate in this activity.
- A function $f_{cur} : \beta(x) \to [n]$, which specifies exactly, for every vertex v in $\beta(x)$, how many vertices have so far (i.e. in $\gamma(x)$) have been determined to participate in the same activity as v. Here, the void activity is an exception in the same sense as the one specified for f_{tot}.
- A partition P of $\beta(x)$, which is interpreted as follows. For the sake of clarity, we let $f_P : \beta(x) \to 2^{\beta(x)}$ denote the function that, for every vertex v in $\beta(x)$, assigns the set in P to which v belongs. Then, the information captured by P is the specification that for every pair of (distinct) vertices in $\beta(x)$, u and v, satisfying $f_P(u) = f_P(v)$, u and v participate in the same non-void activity, and there exists a path in $G[\gamma(x)]$ between u and v such that all of the vertices of this path participate in the same activity as u and v.

We would say that an entry $\mathsf{M}[x, B, \widehat{B}, f_{act}, f_{tot}, f_{cur}, P]$ is *legal* if the following conditions are satisfied.

1. For all $u, v \in \beta(x)$ such that $f_{act}(u) = f_{act}(v)$, it holds that $f_{tot}(u) = f_{tot}(v)$ and $f_{cur}(u) = f_{cur}(v)$.
2. For all $v \in \beta(x)$, it holds that $f_{cur}(v) \leq f_{tot}(v)$.
3. For all (distinct) $u, v \in \beta(x)$ such that $f_P(u) = f_P(v)$, it holds that $f_{act}(u) = f_{act}(v) \neq a_\emptyset$.
4. For all $u, v \in \beta(x)$ such that $f_{cur}(u) = f_{tot}(u)$ and $f_{act}(u) = f_{act}(v)$, it holds that $f_P(u) = f_P(v)$.
5. For all $v \in \beta(x)$ such that $f_{act}(v) = a_\emptyset$, it holds that $f_{tot}(v) = 1$.
6. For all $v \in \beta(x)$ such that $f_{act}(v) \neq a_\emptyset$, it holds that $(f_{act}(v), f_{tot}(v)) \succ_v (a_\emptyset, 1)$.
7. For all $v \in \beta(x)$, it holds that $(f_{act}(v), f_{tot}(v)) \succ_v (a, 1)$ for every $a \in A \setminus B$.
8. For all $v \in \beta(x)$, there does not exist $u \in N_{G[\beta(x)]}(v)$ such that $f_{act}(v) \neq f_{act}(u)$, $f_{act}(u) \neq a_\emptyset$ and $(f_{act}(u), f_{tot}^\pi(u) + 1) \succ_v (\pi(v), f_{tot}^\pi(v))$.

Formally, we would say that the table M has been *computed correctly* if each of its entries $[x, B, \widehat{B}, f_{act}, f_{tot}, f_{cur}, P]$ stores either 0 or 1, and it stores 1 if and only if this entry is legal and there exists an assignment $\pi : \gamma(x) \to B$ that satisfies the following conditions. Here, for all $v \in \gamma(x)$, we would define $f_{tot}^\pi(v) = |\pi_v|$ if $\pi_v \cap \beta(x) = \emptyset$, and $f_{tot}^\pi(v) = f_{tot}(u)$ for any $u \in \pi_v \cap \beta(x)$ otherwise. Since the entry is legal and in particular satisfies Condition 4, this notation is well defined.

   I. For all $a \in B^\star$, there exists $v \in \gamma(x)$ such that $\pi(v) = a$.
   II. For all $v \in \beta(x)$, it holds that $\pi(v) = f_{act}(v)$.
   III. For all $v \in \beta(x)$, it holds that $|\pi_v| = f_{cur}(v)$.
   IV. For all $v \in \gamma(x)$ such that $\pi(v) \neq a_\emptyset$, it holds that $(\pi(v), f_{tot}^\pi(v)) \succ_v (a_\emptyset, 1)$.
   V. For all $v \in \gamma(x)$, it holds that $(\pi(v), f_{tot}^\pi(v)) \succeq_v (a, 1)$ for every $a \in A \setminus B$.
   VI. For all $v \in \gamma(x)$, there does not exist $u \in N_{G[\gamma(x)]}(v)$ such that $f_{act}(v) \neq f_{act}(u)$, $f_{act}(u) \neq a_\emptyset$ and $(f_{act}(u), f_{tot}^\pi(u) + 1) \succ_v (\pi(v), f_{tot}^\pi(v))$.
   VII. For all $u, v \in \beta(x)$ such that $f_P(u) = f_P(v)$, it holds that there exists a path in $G[\pi_v]$ between v and u.
   VIII. For all $v \in \gamma(x)$ such that $\pi_v \cap \beta(x) = \emptyset$, it holds that π_v is a feasible coalition.

We call an assignment as specified above a *witness* for $[x, B, \widehat{B}, f_{act}, f_{tot}, f_{cur}, P]$. The details about how to fill the table M and its correctness are in the full version.

References

1. Aziz, H., Savani, R., Moulin, H.: Hedonic Games. In: Handbook of Computational Social Choice, pp. 356–376. Cambridge University Press (2016). Chap. 15
2. Chalkiadakis, G., Greco, G., Markakis, E.: Characteristic function games with restricted agent interactions: core-stability and coalition structures. Artif. Intell. **232**, 76–113 (2016)
3. Cygan, M., Fomin, F.V., Kowalik, L., Lokshtanov, D., Marx, D., Pilipczuk, M., Pilipczuk, M., Saurabh, S.: Parameterized Algorithms. Springer, Cham (2015)

4. Darmann, A., Elkind, E., Kurz, S., Lang, J., Schauer, J., Woeginger, G.: Group activity selection problem. In: Goldberg, P.W. (ed.) WINE 2012. LNCS, vol. 7695, pp. 156–169. Springer, Heidelberg (2012). doi:10.1007/978-3-642-35311-6_12
5. Diestel, R.: Graph Theory. Springer, Heidelberg (2000)
6. Elkind, E.: Coalitional games on sparse social networks. In: Liu, T.-Y., Qi, Q., Ye, Y. (eds.) WINE 2014. LNCS, vol. 8877, pp. 308–321. Springer, Cham (2014). doi:10.1007/978-3-319-13129-0_25
7. Fomin, F.V., Lokshtanov, D., Panolan, F., Saurabh, S.: Efficient computation of representative families with applications in parameterized and exact algorithms. J. ACM **63**(4), 29:1–29:60 (2016)
8. Gairing, M., Savani, R.: Computing stable outcomes in hedonic games. In: Kontogiannis, S., Koutsoupias, E., Spirakis, P.G. (eds.) SAGT 2010. LNCS, vol. 6386, pp. 174–185. Springer, Heidelberg (2010). doi:10.1007/978-3-642-16170-4_16
9. Gairing, M., Savani, R.: Computing stable outcomes in hedonic games with voting-based deviations. In: AAMAS 2011, pp. 559–566 (2011)
10. Garey, M.R., Johnson, D.S.: The rectilinear steiner tree problem is NP-complete. SIAM J. Appl. Math. **32**, 826–834 (1977)
11. Garey, M.R., Johnson, D.S.: Computers and Intractability: A Guide to the Theory of NP-Completeness. W.H. Freeman, New York (1979)
12. Igarashi, A., Bredereck, R., Elkind, E.: On parameterized complexity of group activity selection problems on social networks. In: AAMAS 2017 (2017)
13. Igarashi, A., Elkind, E.: Hedonic games with graph-restricted communication. In: AAMAS 2016, pp. 242–250 (2016)
14. Igarashi, A., Elkind, E., Peters, D.: Group activity selection on social network. In: AAAI 2017, pp. 565–571 (2017)
15. Manlove, D.F.: Algorithmics of Matching Under Preferences, vol. 2. WorldScientific, Singapore (2013)
16. Myerson, R.B.: Graphs and cooperation games. Math. Oper. Res. **2**, 225–229 (1977)

The Real Computational Complexity of Minmax Value and Equilibrium Refinements in Multi-player Games

Kristoffer Arnsfelt Hansen$^{(\boxtimes)}$ (iD)

Aarhus University, Aarhus, Denmark
arnsfelt@cs.au.dk

Abstract. We show that for several solution concepts for finite n-player games, where $n \geq 3$, the task of simply verifying its conditions is computationally equivalent to the decision problem of the existential theory of the reals. This holds for trembling hand perfect equilibrium, proper equilibrium, and CURB sets in strategic form games and for (the strategy part of) sequential equilibrium, trembling hand perfect equilibrium, and quasi-perfect equilibrium in extensive form games. For obtaining these results we first show that the decision problem for the minmax value in n-player games, where $n \geq 3$, is also equivalent to the decision problem for the existential theory of the reals.

Our results thus improve previous results of NP-*hardness* as well as SQRT-SUM-*hardness* of the decision problems to *completeness* for $\exists \mathbb{R}$, the complexity class corresponding to the decision problem of the existential theory of the reals. As a byproduct we also obtain a simpler proof of a result by Schaefer and Štefankovič giving $\exists \mathbb{R}$-completeness for the problem of deciding existence of a probability constrained Nash equilibrium.

1 Introduction

From a computational point of view, finite games with three or more players present unique challenges compared to finite 2-player games. This is already indicated by the example due to Nash of a 3-player game with no rational Nash equilibrium [25], but even more strikingly so by constructions of Bubelis [6] and Datta [14]. More precisely, Bubelis constructs a 3-player game with a unique Nash equilibrium giving as equilibrium payoff an arbitrary algebraic number to some player, and Datta shows that any real algebraic variety is isomorphic to the set of fully mixed Nash equilibria in a 3-player game.

The problem of *computing* a Nash equilibrium in finite strategic form games was characterized in seminal work by Daskalakis et al. [13] and Chen and Deng [10] as PPAD-complete for 2-player games and by Etessami and Yannakakis [16] as FIXP-complete for n-player games, when $n \geq 3$.

While a Nash equilibrium is guaranteed to exist, one might be interested in Nash equilibria satisfying certain properties, e.g. having at least a certain social welfare. The corresponding decision problems were characterized as NP-complete

© Springer International Publishing AG 2017
V. Bilò and M. Flammini (Eds.): SAGT 2017, LNCS 10504, pp. 119–130, 2017.
DOI: 10.1007/978-3-319-66700-3_10

for 2-player games by Gilboa and Zemel [19] and by Conitzer and Sandholm [11]. For the analogous problems in games with three or more players a precise characterization was only obtained very recently. Schaefer and Štefankovič [28] obtained the first such result giving ∃ℝ-completeness for deciding the existence of a probability constrained Nash equilibrium. Subsequent work by Garg et al. [17] and Bilò and Mavronicolas [3] extended this to ∃ℝ-completeness for all the analogous decision problems studied for two players. Thus these problems are not only to be considered computationally intractable, as implied by the corresponding results for 2-player games, but are in fact computationally equivalent to each other. Prior to obtaining ∃ℝ-completeness this was an open problem.

In this paper we are interested in several standard *refinements* of Nash equilibrium, all guaranteed to exist. We will not be concerned with the task of actually *computing* these, but rather with the task of *verifying* their conditions. Note that verifying the conditions of a Nash equilibrium is computationally a trivial task. In contrast to this Hansen et al. [22] showed both NP-hardness and SQRT-SUM-hardness for verifying the conditions of the standard refinements we consider here in n-player games, where $n \geq 3$. Here NP-hardness indicates that each of the verification problems are computationally intractable, whereas SQRT-SUM-hardness indicate that the problems might not even be contained in NP. Here we show that the problems are ∃ℝ-complete, thus computationally equivalent to each other and to the decision problem for the existential theory of the reals. The complexity of the corresponding problems for 2-player games is not completely settled, but the results that exist so far give polynomial time algorithms [2, 18, 31].

Like the earlier NP-hardness and SQRT-SUM-hardness results, we prove ∃ℝ-hardness by reduction from the decision problem for computing the minmax value in 3-player games. Thus we first establish ∃ℝ-completeness for this problem. As a byproduct we are able to obtain a simpler and more direct proof of ∃ℝ-completeness of deciding the existence of a probability constrained Nash equilibrium compared to the original proof of Schaefer and Štefankovič.

1.1   The Existential Theory of the Reals

The decision problem for the existential theory of the reals, ETR, is that of deciding validity of sentences of the form $\exists x_1, \ldots, x_n \in \mathbb{R} : \phi(x_1, \ldots, x_n)$, where ϕ is a quantifier free formula with real polynomial inequalities and equalities as atoms. It is easy to see that ETR is NP-hard (cf. [8]); on the other hand, the decision procedure by Canny [9] shows that ETR belongs to PSPACE.

The view of the decision problem of the existential theory of the reals as a complexity class existed only implicitly in the literature, e.g. [8, 30], until being studied more extensively under the name NPR by Bürgisser and Cucker [7] and then under the name ∃ℝ by Schaefer and Štefankovič [26, 28]. In this paper we shall adopt the naming ∃ℝ. We shall also denote co∃ℝ by ∀ℝ.

Bürgisser and Cucker defined the class as the constant-free Boolean part $BP^0(NP_{\mathbb{R}})$ of the analogue class $NP_{\mathbb{R}}$ to NP in the Blum-Shub-Smale model of computation. They showed a large number of problems to be complete for ∃ℝ or $\forall \mathbb{R}$. Interestingly, the corresponding problems with real-valued inputs are not known to be $\mathrm{NP}_{\mathbb{R}}$-complete, but rather complete for classes derived from $\mathrm{NP}_{\mathbb{R}}$ with additional "exotic" quantifiers. Schaefer and Štefankovič simply defined the class $\exists \mathbb{R}$ as the closure of ETR under polynomial time many-one reductions, and proved in particular $\exists \mathbb{R}$-completeness for deciding the existence of a probability constrained Nash equilibrium in 3-player games.

2 $\exists \mathbb{R}$-Completeness of Minmax Value in 3-Player Games

It is well-known that the problem of deciding whether a system of quadratic equations over the reals has a solution is complete for $\exists \mathbb{R}$. By results of Grigor'Ev and Vorobjov [20, 32], whenever a polynomial system has a solution, it has a solution whose coordinates are bounded in magnitude by a doubly-exponential function in the size of the encoding of the system. Combining this with repeated squaring, Schaefer [27] noted that deciding whether a system of quadratic equations has a solution remains $\exists \mathbb{R}$-hard even with a promise that a possible solution can be assumed to be contained in a constant bounded region.

Proposition 1 (Schaefer). *It is $\exists \mathbb{R}$-hard to decide if a system of quadratic equations has a solution under the promise that either the system has no solutions or a solution exists in the unit ball $\mathrm{B}(\mathbf{0}, 1)$.*

Starting from this result we will show $\exists \mathbb{R}$ hardness for computing the minmax value in 3-player games in a few steps. First we will convert the quadratic system into a homogeneous bilinear system over the standard simplex. Then each polynomial of the resulting system is translated into a pair of strategies for Player 1, and a minmax strategy profile for Player 2 and Player 3 obtaining minmax value 0 will correspond to a solution to system of equations. We denote the standard n-simplex $\{x \in \mathbb{R}^{n+1} \mid x \geq 0 \wedge \sum_{i=0}^{n} x_i = 1\}$ by Δ^n and the standard corner n-simplex $\{x \in \mathbb{R}^n \mid x \geq 0 \wedge \sum_{i=1}^{n} x_i \leq 1\}$ by Δ_{c}^n.

Proposition 2. *It is $\exists \mathbb{R}$-complete to decide if a system of homogeneous bilinear equations $Q_k(x, y) = 0$ has a solution $x, y \in \Delta^n$. It remains $\exists \mathbb{R}$-hard under the promise that either the system has no such solution or a solution (x, y) exists such that $x = y$ and where x and y belong to the relative interior of Δ^n.*

Proof. Containment in $\exists \mathbb{R}$ is straightforward. We will show hardness by reduction from the problem of deciding whether a system of quadratic equations has a solution. Let $n > 1$ and let $p_1(x_1, \ldots, x_n), \ldots, p_m(x_1, \ldots, x_n)$ be a system of quadratic equations. By Proposition 1 we will assume that the system either has no solution or has a solution in the unit ball $\mathrm{B}^n(\mathbf{0}, 1)$. Define new quadratic polynomials $q_1(y_1, \ldots, y_n), \ldots, q_m(y_1, \ldots, y_n)$ by $q_k(y_1, \ldots, y_n) = p_k(2ny_1 - 1, \ldots, 2ny_n - 1)$, for $k = 1, \ldots, m$. Suppose $x \in \mathrm{B}(\mathbf{0}, 1)$, and define $y_i = (x_i + 1)/2n$, all i. It follows that $y_i > 0$ for all i and furthermore by the triangle inequality and the Cauchy–Schwarz inequality we have

$$\|y\|_1 \leq \frac{1}{2n}\|x\|_1 + \frac{1}{2} \leq \frac{1}{2n}\sqrt{n}\|x\|_2 + \frac{1}{2} < 1.$$

It follows that y belongs to the interior of the standard corner simplex Δ_c^n, and $q_k(y) = 0$ if and only if $p_k(x) = 0$.

Write $q_k(y_1, \ldots, y_n) = \sum_{i \leq j} a_{ij}^{(k)} y_i y_j + \sum_i b_i^{(k)} y_i + c^{(k)}$ and define the homogeneous bilinear form $Q_k(x_0, x_1, \ldots, x_n, y_0, y_1, \ldots, y_n)$ by

$$Q_k(x, y) = \sum_{1 \leq i \leq j} a_{ij}^{(k)} x_i y_j + \sum_{i=1}^{n} \sum_{j=0}^{n} b_i^{(k)} x_i y_j + \sum_{i=0}^{n} \sum_{j=0}^{n} c^{(k)} x_i y_j.$$

Additionally, for $j = 1, \ldots, n$, define $Q_{m+j}(x, y) = \sum_{i=0}^{n} x_i y_j - x_j y_i$. Essentially we just have introduced the slack variable $y_0 = 1 - \sum_{i=1}^{n} y_i$, replaced quadratic terms $y_i y_j$ by bilinear quadratic terms $x_i y_j$, homogenized using the equality $\sum_{i=0}^{n} y_i = 1$, and finally ensured that $x = y$. If $(y_1, \ldots, y_n)$ belongs to the interior of Δ_c^n, then extending with the slack variable y_0 we have that $(y_0, \ldots, y_n)$ belongs to the relative interior of Δ^n, and furthermore $Q_k(y_0, \ldots, y_n, y_0, \ldots, y_n) = 0$ whenever $q_k(y_1, \ldots, y_n) = 0$ for all $k = 1, \ldots, m+n$. Conversely, if $x, y \in \Delta^n$ and $Q_k(x, y) = 0$ for all $k = 1, \ldots, m + n$ we must have $x_j = y_j$ using $Q_{m+j}(x, y) = 0$ for $j = 1, \ldots, n$. Then removing the slack variable from y again we have $q_k(y_1, \ldots, y_n) = 0$ for all $k = 1, \ldots, m$, using $Q_k(y, y) = 0$ for $k = 1, \ldots, m$.   □

Theorem 3. *It is $\exists\mathbb{R}$-complete to decide for a given 3-player game in strategic form if the minmax value for Player 1 is at most 0. It remains $\exists\mathbb{R}$-hard under the promise that either the minmax value for Player 1 is strictly greater than 0 or the minmax value for Player 1 is equal to 0 and Player 2 and Player 3 can enforce this using a fully mixed strategy profile (x, y), where $x = y$, and against which all strategies of Player 1 yield payoff 0.*

Proof. Containment in $\exists\mathbb{R}$ is straightforward. Namely, the question to decide is existence of $x \in \Delta^{n_2-1}$ and $y \in \Delta^{n_3-1}$ such that $\sum_{i,j} u_1(k, i, j) x_i y_j \leq 0$ for all k, where u_1 is the payoff function of Player 1. We will show hardness by reduction from the promise problem given by Proposition 2. Let $Q_1(x, y), \ldots, Q_m(x, y)$ be homogeneous bilinear polynomials in variables $x = (x_0, \ldots, x_n)$ and $y = (y_0, \ldots, y_n)$. We form a 3-player game where Player 2 and Player 3 each have $n + 1$ strategies given by the set $\{0, \ldots, n\}$ and Player 1 has $2m$ strategies given by the set $\{1, \ldots, 2m\}$. Let $Q_k(x, y) = \sum_{i=0}^{n} \sum_{j=0}^{n} a_{ij}^{(k)} x_i y_j$ for $k = 1, \ldots, m$. The payoff to Player 1 is simply given by $u_1(2k - 1, i, j) = -u_1(2k, i, j) = a_{ij}^{(k)}$. It follows that if (x, y) is a strategy profile for Player 2 and Player 3, or equivalently $x, y \in \Delta^n$, then

$$\mathop{E}_{i \sim x, j \sim y}[u_1(2k - 1, i, j)] = - \mathop{E}_{i \sim x, j \sim y}[u_1(2k, i, j,)] = Q_k(x, y).$$

If (x, y) is a fully mixed strategy profile such that $\max_k \mathrm{E}_{i \sim x, j \sim y}[u_1(k, i, j)] \leq 0$ and $x = y$ this means that $Q_k(x, y) \leq 0$ and $-Q_k(x, y) \leq 0$ for all k, implying $Q_k(x, y) = 0$ for all k, and x and y belong to the relative interior of Δ^n, as well as $\max_k \mathrm{E}_{i \sim x, j \sim y}[u_1(k, i, j)] = 0$. Conversely, if $Q_k(x, y) = 0$ for all k with x and y being in the relative interior of Δ^n and $x = y$, we have that (x, y) is a fully mixed strategy profile and $\mathrm{E}_{i \sim x, j \sim y}[u_1(k, i, j)] = 0$ for all k, which in particular means that $\max_k \mathrm{E}_{i \sim x, j \sim y}[u_1(k, i, j)] = 0$.   □

2.1 ∃ℝ-Completeness of Constrained Nash Equilibrium

The decision problem NASHINABALL is the following: Given a game G in strategic form and a rational r, decide whether G has a Nash equilibrium strategy profile in which all pure strategies of all players are played with probability at most r.

Schaefer and Štefankovič [28] showed that NASHINABALL is ∃ℝ-complete for 3-player games. Their hardness proof goes via Brouwer functions and uses the rather involved transformation of Etessami and Yannakakis [16] of Brouwer functions into 3-player games using an intermediate construction of 10-player games. Here we give a simpler and more direct proof of ∃ℝ-hardness of NASHINABALL as a consequence of Theorem 3.

Theorem 4 (Schaefer and Štefankovič). NASHINABALL *∃ℝ-complete.*

Proof. We show ∃ℝ-hardness by reduction from the minmax problem in 3-player games. Let G be the 3-player game given by Theorem 3 where all strategies of Player 2 and Player 3 are duplicated. This duplication has the effect that in case the minmax value of Player 1 is 0, then the minmax strategy profile of Player 2 and Player 3 can be assumed to use only probabilities of magnitude at most $\frac{1}{2}$. We next give each player an additional strategy $\bot$. The payoffs are as follows. The payoff to all players is 0 when at least one player plays $\bot$. For the strategy combinations where no player plays $\bot$, the payoff to Player 1 is the same as it would have been in G, and the payoffs to Player 2 and Player 3 is the *negative* of the payoff to Player 1.

Suppose first that the minmax value of Player 1 in G is 0. Let (τ_2, τ_3) be a minmax strategy profile for Player 2 and Player 3 in which each strategy is played with probability at most $\frac{1}{2}$, and against which all strategies of Player 1 yield payoff 0. Let τ_1 be the uniform strategy profile for Player 1 in G. We claim that $\tau = (\tau_1, \tau_2, \tau_3)$ is a Nash equilibrium. Indeed, since Player 1 plays uniformly, all players receive payoff 0 regardless of the strategies of Player 2 and Player 3, since in G we have $u_1(2k-1, i, j) = -u_1(2k, i, j,)$ for all $k = 1, \ldots, m$. Furthermore, since Player 2 and Player 3 play the minmax strategy profile (τ_2, τ_3), all players receive payoff 0 regardless of the strategy of Player 1. In other words, no player can gain from deviating which means that τ is a Nash equilibrium.

Suppose now that $\tau = (\tau_1, \tau_2, \tau_3)$ is a Nash equilibrium profile in which all strategies of all players is played with probability at most $\frac{1}{2}$. In particular each player plays $\bot$ with probability at most $\frac{1}{2}$. Let $\tau' = (\tau_1', \tau_2', \tau_3')$ be the strategy profile where τ_j' is the conditional distribution obtained from τ_j given that Player j does not play the strategy $\bot$. We claim that (τ_2', τ_3') is a minmax strategy profile in G ensuring minmax value 0 for Player 1. Indeed, since for a strategy combination where no player plays $\bot$, Player 1 receives the negative payoff of Player 2 and Player 3, the strategy profile τ', which is played with nonzero probability, must give all players payoff 0, since otherwise either Player 1 or Player 2 and Player 3 would gain from playing $\bot$ with probability 1. Also, Player 1 does not have a best reply in G to (τ_2', τ_3') ensuring payoff strictly greater than 0, which proves the claim.   □

3   ∃ℝ-Completeness of Equilibrium Refinements

In this section we prove ∃ℝ-completeness for decision problems associated with several equilibrium refinements. These results directly improve NP-hardness and SQRT-SUM-hardness results obtained by Hansen et al. [22], and thereby settles their complexity exactly. The previous NP-hardness and SQRT-SUM-hardness results were proved by reduction from the minmax value problem we considered in Sect. 2. These reductions do not work directly for our purposes however, since they actually reduced from a *gap* version of the minmax value problem, deciding whether the minmax value is either *strictly less* or *strictly greater* than a given number r. A gap version of the problem is NP-hard even with any inverse polynomial gap [5,21]. SQRT-SUM-hardness of the gap problem follows directly from the fact that a sum of square-roots can be compared for equality to a rational in polynomial time as shown by Blömer [4][1]. For ∃ℝ-hardness it is not possible to ensure such a gap; fortunately we are able to modify the reductions to circumvent this, instead relying on our ability to directly assume fully mixed minmax strategy profiles of Player 2 and Player 3.

We very briefly give the definitions of the solution concepts we consider before each theorem, and refer to Hansen et al. [22] and references therein for more details.

In contrast to the hardness results of Hansen, Miltersen and Sørensen we obtain *completeness* results. Showing that the problems that arise from Nash equilibrium refinements are contained in ∃ℝ is *not* straightforward and hinges on the fact that one may construct a virtual infinitesimal by means of repeated squaring. As shown by Bürgisser and Cucker [7, Theorem 9.2] and Schaefer [27, Lemma 3.12] (cf. [28, Lemma 4.1]) this extends the power of the existential theory of the reals with universal quantification over an arbitrarily small $\varepsilon > 0$.

Lemma 5. *The problem of deciding validity of the sentence*

$$\exists \epsilon_0 > 0 \; \forall \varepsilon \in (0, \varepsilon_0) \; \exists x : \varphi(\varepsilon, x)$$

when given a quantifier-free formula over the reals $\varphi(\varepsilon, x)$ belongs to ∃ℝ.

The Nash equilibrium refinements concepts we consider can all be expressed as being limit points x of a sequence $x^{(\varepsilon)}$ of points that can be expressed by an existentially quantified formula, which in many of the cases is straightforward to derive. Then containment in ∃ℝ follows from Lemma 5.

3.1   Trembling Hand Perfect and Proper Equilibrium of Games in Strategic Form

We first consider the concept of trembling hand perfect equilibrium, introduced by Selten [29]. We state an equivalent definition due to Myerson [24].

---

[1] This crucial point of the reduction by Hansen et al. was unfortunately omitted in the paper [22].

Definition 6 (Trembling hand perfect equilibrium). *Let G be a m-player game in strategic form. A strategy profile σ for G is an ε-perfect equilibrium if and only if it is fully mixed and only pure strategies that are best replies get probability more than ε. A strategy profile σ for G is a trembling hand perfect equilibrium if and only if it is the limit point of a sequence of ε-perfect equilibria with $\varepsilon \to 0^+$.*

Theorem 7. *For any $m \geq 3$ it is $\exists\mathbb{R}$-complete to decide if a given pure strategy Nash equilibrium of a given m-player game in strategic form is trembling hand perfect.*

Proof. It is not hard to express that a given strategy profile in a m-player game is a trembling hand perfect equilibrium by a first-order formula over the reals. We reproduce below such a formulation by Etessami et al. [15] for completeness and illustration. Suppose that player i has n_i strategies. A strategy profile is thus a tuple $x \in \Delta^{n_1-1} \times \cdots \times \Delta^{n_m-1} \subseteq \mathbb{R}^n$, $n = n_1 + \cdots + n_m$. Define

$$R_i(x \setminus k) = \sum_{a_{-i}} u_i(k; a_{-i}) \prod_{j \neq i} x_{j,a_j}.$$

Let EPS-PE(x', ε) be the quantifier-free formula defined by the conjunction of the following formulas, together expressing that x' is an ε-perfect equilibrium.

$$x'_{i,k} > 0, \text{ for } i = 1, \ldots, m \text{ and } k = 1, \ldots, n_i$$
$$x'_{i,1} + \cdots + x'_{i,n_i} = 1, \text{ for } i = 1, \ldots, m$$
$$(R_i(x' \setminus k) \geq R_i(x' \setminus \ell)) \vee (x'_{i,k} \leq \varepsilon), \text{ for } i = 1, \ldots, m \text{ and } k, \ell = 1, \ldots, n_i$$

We can now express that x is a trembling hand perfect equilibrium by

$$\forall \varepsilon > 0\ \exists x' : \text{EPS-PE}(x', \varepsilon) \wedge \|x - x'\|_2^2 \leq \varepsilon.$$

Thus deciding whether a given strategy profile x is a trembling hand perfect equilibrium is contained in $\exists\mathbb{R}$ by Lemma 5.

In order to show hardness for $\exists\mathbb{R}$ we reduce from the promise minmax problem given by Theorem 3. Let G be the given 3-player game in strategic form. We define a new game G' from G where each player is given an additional pure strategy $\perp$. The payoffs to Player 2 and Player 3 are 0 for all strategy combinations. The payoff to Player 1 is 0 when at least one player plays $\perp$. For the strategy combinations where no player plays $\perp$ the payoff to Player 1 is the same as it would have been in G.

Clearly, in G' the strategy profile $\mu = (\perp, \perp, \perp)$ is a Nash equilibrium. We claim that the minmax value for Player 1 in G is 0 precisely when μ is a trembling hand perfect equilibrium in G'.

Suppose first that the minmax value for Player 1 in G is 0 and let (τ_2, τ_3) be a minmax strategy profile of Player 2 and Player 3, which we by assumption can assume is fully mixed and against which all strategies of Player 1 yield payoff 0.

Let τ_1 be the uniform strategy for Player 1 in G, and let $\tau = (\tau_1, \tau_2, \tau_3)$. Define for $k \geq 1$

$$\sigma^{(k)} = (1 - \frac{1}{k})\mu + \frac{1}{k}\tau.$$

By assumption and construction $\sigma^{(k)}$ is a fully mixed strategy profile of G' converging to μ as k increases. Note that all strategies of all players are best replies to $\sigma^{(k)}$. Thus μ is trembling hand perfect.

Suppose next that the minmax value for Player 1 in G is strictly greater than 0. Suppose $\sigma^{(k)} = (\sigma_1^{(k)}, \sigma_2^{(k)}, \sigma_3^{(k)})$ is a sequence of fully mixed strategy profiles converging to μ. Since $\sigma_2^{(k)}$ and $\sigma_3^{(k)}$ do not place probability 1 on $\perp$, Player 1 has a reply to $(\sigma_2^{(k)}, \sigma_3^{(k)})$ with an expected payoff strictly larger than 0. Thus $\perp$ is not a best reply of Player 1 to $(\sigma_2^{(k)}, \sigma_3^{(k)})$. Since this holds for all sequences $\sigma^{(k)}$ it follows that μ is not trembling hand perfect. $\qquad \square$

With minimal changes we obtain the same result for the concept of proper equilibrium of Myerson [24] that further refines that of a trembling hand perfect equilibrium.

Definition 8 (Proper equilibrium). *Let G be a m-player game in strategic form. A strategy profile σ for G is an ε-proper equilibrium if and only if is fully mixed, and the following condition holds: Given two pure strategies, p_k and p_ℓ, of the same player, if p_k is a worse reply against σ than p_ℓ, then σ must assign a probability to p_k that is at most ε times the probability it assign to p_ℓ. A strategy profile σ for G is a proper equilibrium if and only if is the limit point of a sequence of ε-proper equilibria with $\varepsilon \to 0^+$.*

Theorem 9. *For any $m \geq 3$ it is $\exists \mathbb{R}$-complete to decide if a given pure strategy Nash equilibrium of a given m-player game in strategic form is proper.*

Proof. We can show containment in $\exists \mathbb{R}$ exactly in the same way as the proof of Theorem 7 by replacing the last set of formulas making up EPS-PE by

$$(R_i(x \setminus k) \geq R_i(x \setminus \ell)) \vee (x_{i,k} \leq \varepsilon x_{i,\ell}), \text{ for } i = 1, \ldots, m \text{ and } k, \ell = 1, \ldots, n_i.$$

For establishing $\exists \mathbb{R}$-hardness we can simply observe that this follows already from the proof of Theorem 7. When the minmax value in G is 0, the sequence $\sigma^{(k)}$ establishes that μ is even a proper equilibrium, since all strategies of all players are best replies to $\sigma^{(k)}$. When the minmax value in G is strictly greater than 0, since μ is not trembling hand perfect μ is not proper as well. $\qquad \square$

3.2 Sequential Equilibrium of Games in Extensive Form

For extensive form games we first consider the concept of sequential equilibrium, introduced by Kreps and Wilson [23]. Kreps and Wilson defined a sequential equilibrium to be a pair (b, μ), which is called an assessment, consisting of a behavior strategy profile b together with a belief profile μ. The belief profile

μ specifies for each information set I a probability distribution on the nodes contained in I. Here we will only be concerned with the strategy part b of a sequential equilibrium. For this reason we will adopt an alternative definition of (the strategy part of) a sequential equilibrium, also given by Kreps and Wilson [23, Proposition 6]. Kreps and Wilson used the terminology *weakly perfect equilibrium* for this.

Definition 10 (Sequential equilibrium). *Let G be a m-player game in extensive form with imperfect information but having perfect recall and let u be the tuple of utilities of G. A behavior strategy b is (the strategy part of) a sequential equilibrium if there is a sequence $(b^{(k)}, u^{(k)})$, where $b^{(k)}$ is a fully mixed behavior strategy profile and $u^{(k)}$ is a tuple of utilities with $\lim_{k \to \infty}(b^{(k)}, u^{(k)}) = (b, u)$ and for every k and j, $b_j^{(k)}$ is a best reply to $b_{-j}^{(k)}$ with respect to utilities $u_j^{(k)}$.*

Theorem 11. *For any $m \geq 3$ it is $\exists \mathbb{R}$-complete to decide if a given pure strategy profile of a given m-player game in extensive form is part of a sequential equilibrium.*

Proof. We first sketch the proof for containment in $\exists \mathbb{R}$. Let u be the tuple of utilities of G. First we express by a quantifier free formula $\varphi(b', u')$ that b' is a fully mixed behavior strategy profile and that b_j' is a best response to b_{-j}' with respect to u_j' for all j. We can now express that b is (the strategy part of) a sequential equilibrium by

$$\forall \varepsilon > 0 \; \exists b', u' : \varphi(b', u') \wedge \|b - b'\|_2^2 \leq \varepsilon \wedge \|u - u'\|_2^2 \leq \varepsilon.$$

Thus by Lemma 5 it follows that deciding whether a given behavior strategy profile b is (the strategy part of) a sequential equilibrium is contained in $\exists \mathbb{R}$.

For establishing $\exists \mathbb{R}$ hardness we again reduce from the promise minmax problem given by Theorem 3. Let G be the given 3-player game in strategic form. We define an extensive form game G' from G as follows. The players each get to choose an action of G in turn, without revealing the choice to the other players. Player 2 chooses first, then Player 3, and finally Player 1. In G' each player has a single information set thus making G' strategically equivalent to G. The payoffs to Player 2 and Player 3 are 0 for all combinations of actions. The payoff to Player 1 is inherited from G. We now give each player an additional pure action $\perp$ in their respective information set. Choosing this action results in the game ending immediately with all players receiving payoff 0.

Clearly, in G' the behavior strategy profile $b = (\perp, \perp, \perp)$ is a Nash equilibrium. We claim that the minmax value for Player 1 in G is 0 precisely when b is (the strategy part of) a sequential equilibrium.

Suppose first that the minmax value for Player 1 in G is 0 and let (τ_2, τ_3) be a minmax strategy profile of Player 2 and Player 3, which we by assumption can assume is fully mixed and against which all strategies of Player 1 yield payoff 0. Let τ_1 be the uniform strategy for Player 1 in G and let $\tau = (\tau_1, \tau_2, \tau_3)$. We now define for $k \geq 1$

$$b^{(k)} = (1 - \frac{1}{k})b + \frac{1}{k}\tau.$$

By assumption and construction $b^{(k)}$ is a fully mixed behavior strategy profile converging to b as k increases. For the sequence of utilities we simply let $u^{(k)} = u$ for all k. Note that in $b^{(k)}$ all players are playing a best reply, and it follows that b is (the strategy part of) a sequential equilibrium.

Suppose next that the minmax value for Player 1 in G is strictly greater then 0. Thus let $\varepsilon > 0$ be such that the minmax value for Player 1 in G is at least ε. Let $(b^{(k)}, u^{(k)})$ be a sequence of of pairs of fully mixed behavior strategy profiles $b^{(k)}$ and tuples of utilities $u^{(k)}$ with $\lim_{k\to\infty}(b^{(k)}, u^{(k)}) = (b, u)$. Let k be such that $\|u - u^{(k)}\|_\infty < \varepsilon/2$ and let I denote the information set of Player 1. Since $b^{(k)}$ is fully mixed, I is reached with non-zero probability. Conditioned on reaching I, the action $\perp$ gives Player 1 payoff 0, whereas the best reply gives payoff at least ε. Since $\|u - u^{(k)}\|_\infty < \varepsilon/2$ the payoff to Player 1 conditioned on reaching I with respect to $u^{(k)}$ differs by less than $\varepsilon/2$ to the payoff with respect to u. Thus, conditioned on reaching I, the action $\perp$ gives Player 1 payoff strictly less than $\varepsilon/2$, whereas the best reply gives payoff at least $\varepsilon/2$. But $b^{(k)}$ is fully mixed and therefore is Player 1 not playing a best reply. Since this holds for all sequences $(b^{(k)}, u^{(k)})$ is follows that b is not (the strategy part of) a sequential equilibrium.   □

We also obtain analogous $\exists\mathbb{R}$-completeness results for trembling hand perfect equilibrium for extensive form games as defined by Selten [29] and for quasi-perfect equilibrium of van Damme [12]. Here the main work consists in showing containment in $\exists\mathbb{R}$, since $\exists\mathbb{R}$-hardness essentially follows from the proof of Theorem 11. These results are omitted in this version of the paper due to space constraints.

3.3 CURB Sets in Games in Strategic Form

Here we consider the set valued solution concept of Closed Under Rational Behavior (CURB) strategy sets by Basu and Weibull [1].

Definition 12 (CURB set). *In a m-player game, a family of sets of pure strategies, $S_1, S_2, \ldots, S_m$ with S_i being a subset of the strategy set of player i, is closed under rational behavior (CURB) if and only if for all pure strategies k of Player i so that k is a best reply to a product distribution x on $S_1 \times S_2 \times \cdots \times S_{i-1} \times S_{i+1} \times \cdots \times S_m$, we have that $k \in S_i$.*

Theorem 13. *For any $m \geq 3$ it is $\forall\mathbb{R}$-complete to decide whether a family of sets of pure strategies $S_1, S_2, \ldots, S_m$ is CURB in a given m-player game in strategic form.*

Proof. We show that the problem of deciding whether a family of sets of pure strategies $S_1, S_2, \ldots, S_m$ is *not* CURB is complete for $\exists\mathbb{R}$. Containment in $\exists\mathbb{R}$ is straightforward. We existentially quantify over a mixed strategy profile x and have a disjunction for all i and all $k \notin S_i$ over inequalities expressing that k is a best reply to x.

In order to show hardness for $\exists\mathbb{R}$ we reduce from the promise minmax problem given by Theorem 3. Let G be the given 3-player game in strategic form. We define a new game G' from G where Player 1 is given an additional strategy $\perp$. The payoffs to Player 2 and Player 3 are 0 for all strategy combinations. The payoff to Player 1 is 0 if he plays $\perp$ and otherwise is the same as it would have been in G.

By construction we now have that the minmax value of Player 1 in G is 0 if and only if the set of all pure strategies except $\perp$ is not CURB in G'. Namely, if the minmax value of G is 0, then $\perp$ is a best reply for Player 1 to the minmax strategy profile of Player 2 and Player 3, and the set of all pure strategies except $\perp$ is not CURB. On the other hand, if the minmax value of G is strictly greater than 0, then $\perp$ is never a best reply in G', and the set of all pure strategies except $\perp$ is therefore CURB. $\qquad\square$

References

1. Basu, K., Weibull, J.W.: Strategy subsets closed under rational behavior. Econ. Lett. **36**(2), 141–146 (1991)
2. Benisch, M., Davis, G.B., Sandholm, T.: Algorithms for rationalizability and CURB sets. In: Proceedings of the Twenty-First National Conference on Artificial Intelligence, pp. 598–604. AAAI Press (2006)
3. Bilò, V., Mavronicolas, M.: A catalog of $\exists\mathbb{R}$-complete decision problems about Nash equilibria in multi-player games. In: Ollinger, N., Vollmer, H. (eds.) STACS 2016. LIPIcs, vol. 47, p. 17:1–17:13. Schloss Dagstuhl - Leibniz-Zentrum fuer Informatik (2016)
4. Blömer, J.: Computing sums of radicals in polynomial time. In: 32nd Annual Symposium on Foundations of Computer Science (FOCS 1991), pp. 670–677. IEEE Computer Society Press (1991)
5. Borgs, C., Chayes, J., Immorlica, N., Kalai, A.T., Mirrokni, V., Papadimitriou, C.: The myth of the folk theorem. Games Econ. Behav. **70**(1), 34–43 (2010)
6. Bubelis, V.: On equilibria in finite games. Int. J. Game Theor. **8**(2), 65–79 (1979)
7. Bürgisser, P., Cucker, F.: Exotic quantifiers, complexity classes, and complete problems. Found. Comput. Math. **9**(2), 135–170 (2009)
8. Buss, J.F., Frandsen, G.S., Shallit, J.O.: The computational complexity of some problems of linear algebra. J. Comput. Syst. Sci. **58**(3), 572–596 (1999)
9. Canny, J.F.: Some algebraic and geometric computations in PSPACE. In: Simon, J. (ed.) Proceedings of the 20th Annual ACM Symposium on Theory of Computing (STOC 1988), pp. 460–467. ACM (1988)
10. Chen, X., Deng, X.: Settling the complexity of two-player Nash equilibrium. In: 47th Annual IEEE Symposium on Foundations of Computer Science (FOCS 2006), pp. 261–272. IEEE Computer Society Press (2006)
11. Conitzer, V., Sandholm, T.: New complexity results about Nash equilibria. Games Econ. Behav. **63**(2), 621–641 (2008)
12. van Damme, E.: A relation between perfect equilibria in extensive form games and proper equilibria in normal form games. Int. J. Game Theor. **13**, 1–13 (1984)
13. Daskalakis, C., Goldberg, P.W., Papadimitriou, C.H.: The complexity of computing a Nash equilibrium. SIAM J. Comput. **39**(1), 195–259 (2009)

14. Datta, R.S.: Universality of Nash equilibria. Math. Oper. Res. **28**(3), 424–432 (2003)
15. Etessami, K., Hansen, K.A., Miltersen, P.B., Sørensen, T.B.: The complexity of approximating a trembling hand perfect equilibrium of a multi-player game in strategic form. In: Lavi, R. (ed.) SAGT 2014. LNCS, vol. 8768, pp. 231–243. Springer, Heidelberg (2014). doi:10.1007/978-3-662-44803-8_20
16. Etessami, K., Yannakakis, M.: On the complexity of Nash equilibria and other fixed points. SIAM J. Comput. **39**(6), 2531–2597 (2010)
17. Garg, J., Mehta, R., Vazirani, V.V., Yazdanbod, S.: ETR-completeness for decision versions of multi-player (symmetric) Nash equilibria. In: Halldórsson, M.M., Iwama, K., Kobayashi, N., Speckmann, B. (eds.) ICALP 2015. LNCS, vol. 9134, pp. 554–566. Springer, Heidelberg (2015). doi:10.1007/978-3-662-47672-7_45
18. Gatti, N., Panozzo, F.: New results on the verification of Nash refinements for extensive-form games. In: Proceedings of the 11th International Conference on Autonomous Agents and Multiagent Systems (AAMAS 2012), pp. 813–820. International Foundation for Autonomous Agents and Multiagent Systems (2012)
19. Gilboa, I., Zemel, E.: Nash and correlated equilibria: some complexity considerations. Games Econ. Behav. **1**(1), 80–93 (1989)
20. Grigor'Ev, D.Y., Vorobjov, N.: Solving systems of polynomial inequalities in subexponential time. J. Symb. Comput. **5**(1–2), 37–64 (1988)
21. Hansen, K.A., Hansen, T.D., Miltersen, P.B., Sørensen, T.B.: Approximability and parameterized complexity of minmax values. In: Papadimitriou, C., Zhang, S. (eds.) WINE 2008. LNCS, vol. 5385, pp. 684–695. Springer, Heidelberg (2008). doi:10.1007/978-3-540-92185-1_74
22. Hansen, K.A., Miltersen, P.B., Sørensen, T.B.: The computational complexity of trembling hand perfection and other equilibrium refinements. In: Kontogiannis, S., Koutsoupias, E., Spirakis, P.G. (eds.) SAGT 2010. LNCS, vol. 6386, pp. 198–209. Springer, Heidelberg (2010). doi:10.1007/978-3-642-16170-4_18
23. Kreps, D.M., Wilson, R.: Sequential equilibria. Econometrica **50**(4), 863–894 (1982)
24. Myerson, R.B.: Refinements of the Nash equilibrium concept. Int. J. Game Theor. **15**, 133–154 (1978)
25. Nash, J.: Non-cooperative games. Ann. Math. **2**(54), 286–295 (1951)
26. Schaefer, M.: Complexity of some geometric and topological problems. In: Eppstein, D., Gansner, E.R. (eds.) GD 2009. LNCS, vol. 5849, pp. 334–344. Springer, Heidelberg (2010). doi:10.1007/978-3-642-11805-0_32
27. Schaefer, M.: Realizability of graphs and linkages. In: Pach, J. (ed.) Thirty Essays on Geometric Graph Theory, pp. 461–482. Springer, Heidelberg (2013). doi:10.1007/978-1-4614-0110-0_24
28. Schaefer, M., Štefankovič, D.: Fixed points, Nash equilibria, and the existential theory of the reals. Theor. Comput. Syst. **60**(2), 172–193 (2017)
29. Selten, R.: A reexamination of the perfectness concept for equilibrium points in extensive games. Int. J. Game Theor. **4**, 25–55 (1975)
30. Shor, P.W.: Stretchability of pseudolines is NP-hard. In: Gritzmann, P., Sturmfels, B. (eds.) Applied Geometry And Discrete Mathematics. DIMACS Series in Discrete Mathematics and Theoretical Computer Science, vol. 4, pp. 531–554. DIMACS/AMS (1990)
31. van Damme, E.: Stability and Perfection of Nash Equilibria, 2nd edn. Springer, Heidelberg (1991)
32. Vorob'ev, N.N.: Estimates of real roots of a system of algebraic equations. J. Sov. Math. **34**(4), 1754–1762 (1986)

Conditional Value-at-Risk: Structure and Complexity of Equilibria

Marios Mavronicolas[1(✉)] and Burkhard Monien[2]

[1] Department of Computer Science, University of Cyprus, 1678 Nicosia, Cyprus
mavronic@ucy.ac.cy
[2] Faculty of Electrical Engineering, Computer Science and Mathematics,
University of Paderborn, 33102 Paderborn, Germany
bm@upb.de

Abstract. *Conditional Value-at-Risk,* denoted as CVaR_α, is becoming the prevailing measure of risk over two paramount economic domains: the insurance domain and the financial domain; $\alpha \in (0, 1)$ is the *confidence level.* In this work, we study the strategic equilibria for an economic system modeled as a game, where risk-averse players seek to minimize the Conditional Value-at-Risk of their costs. Concretely, in a CVaR_α-*equilibrium,* the mixed strategy of each player is a *best-response.* We establish two significant properties of CVaR_α at equilibrium: (1) The *Optimal-Value* property: For any best-response of a player, each mixed strategy in the support gives the same cost to the player. This follows directly from the concavity of CVaR_α in the involved probabilities, which we establish. (2) The *Crawford* property: For every α, there is a 2-player game with no CVaR_α-equilibrium. The property is established using the *Optimal-Value* property and a new functional property of CVaR_α, called *Weak-Equilibrium-for-*VaR_α, we establish. On top of these properties, we show, as one of our two main results, that deciding the existence of a CVaR_α-equilibrium is strongly $\mathcal{NP}$-hard even for 2-player games. As our other main result, we show the strong $\mathcal{NP}$-hardness of deciding the existence of a V-equilibrium, over 2-player games, for any valuation V with the *Optimal-Value* and the *Crawford* properties. This result has a rich potential since we prove that the very significant and broad class of *strictly quasiconcave* valuations has the *Optimal-Value* property.

1 Introduction

1.1 Conditional Value-at-Risk and its Equilibria

Risk management is a principal component of any volatile economy; it amounts to defining and analyzing risk metrics in order to evaluate and minimize its adverse impacts. Starting with the Markowitz's *Mean-Variance* model [13], where

---

Partially supported by the German Research Foundation (DFG) within the Collaborative Research Centre "On-the-Fly-Computing" (SFB 901), and by funds for the promotion of research at University of Cyprus.

V. Bilò and M. Flammini (Eds.): SAGT 2017, LNCS 10504, pp. 131–143, 2017.
DOI: 10.1007/978-3-319-66700-3_11

risk is identified with the *Variance* of loss, there is an extensive literature on formulating and optimizing risk valuations—see, e.g., [12,26].

We shall focus on a prominent risk measure: *Conditional Value-at-Risk*, denoted as CVaR_α; $\alpha \in (0,1)$ is the *confidence level*, and values of α close to 1, such as 90%, 95% or 99%, are of interest. CVaR_α is closely related to *Value-at-Risk*, denoted as VaR_α. Stated simply, VaR_α is an upper percentile of a given loss distribution: it is the lowest amount which, with probability at least α, the loss will not exceed. VaR_α attracted the interest of practitioners as a conceptually simple and intuitive quantification of risk, and prevailed as a benchmark. In spite of its success, VaR_α lacks an important property: by not looking beyond the α-percentile, VaR_α does not account for losses exceeding it; hence, it provides an inadequate picture of risks by failing to capture "tail risks".

In the discrete setting, CVaR_α is a weighted average of VaR_α and the expected losses strictly above VaR_α. CVaR_α is computationally attractive as it can be optimized using Linear Programming and it is expressible as a minimization formula [20,21]; it is continuous in α and convex in the variables (cf. [29]). CVaR_α is recognized as a suitable model of risk in volatile economic circumstances; it is widely used for modeling and optimizing hedge and credit risks. There were recently two quite notable developments in the insurance domain and the financial domain, namely the issue of the *Solvency II Directive* in the issurance domain, and of *Basel III Rules* in the financial domain, which enforce a tremendous increase to the role and applicability of CVaR_α.

To model the strategic interactions incurred in the management of risk, we adopt the game-theoretic framework from [15, Sect. 2] with risk-averse players seeking to minimize their costs. Each player is using a *mixed strategy*: a probability distribution over her *strategies*; her cost is evaluated by CVaR_α, which is a particular case of a *valuation* V, mapping a tuple of mixed strategies to reals. We are interested in the strategic equilibria incurred by CVaR_α, call them CVaR_α-*equilibria, where no player could unilaterally decrease her* CVaR_α *cost*.

The first counterxample game with no equilibrium for a certain valuation was given by Crawford [5, Sect. 4]; more counterexamples appeared in [14]. Naturally, the properties of a valuation bear a significant impact on the existence of an equilbrium. The most prominent valuation considered in Game Theory is *Expectation*, denoted as E; although inadequate to accomodate risk, it guarantees the existence of a *Nash equilibrium* [16,17]. Stated for *maximization* games, equilibrium existence is still guaranteed for concave valuations [6,8].

Concave valuations like *Variance* (Var) are well-suited to model risk; so, for a concave V, a V-equilibrium is the proper generalization of Nash equilibrium to risk-aversion. The seminal Markowitz's *Mean-Variance* approach [13] to portfolio maximization is built around Var. Deciding the existence of an (E + Var)-equilibrium is strongly $\mathcal{NP}$-hard [15, Theorem 5.12] for 2-player games. This was the first concrete instance of the strong $\mathcal{NP}$-hardness of ∃V-EQUILIBRIUM: deciding, given a game, the existence of a V-equilibrium, for 2-player games; it assumed that V has the *Weak-Equilibrium-for-Expectation* property [15, Definition 3.1] and fulfills a few technical conditions [15, Theorem 5.1]. *Weak-Equilibrium-for-Expectation* means that all strategies supported in a player's best-response mixed strategy induce the same expectation. Towards proving hardness results for ∃CVaR$_\alpha$-EQUILIBRIUM, we shall explore corresponding properties for CVaR$_\alpha$.

1.2   Results and Significance

We shall focus on (E+R)-*valuations* [15], combining expected cost and a measure of risk, where R has the *Risk-Positivity* property: its value is always non-negative, and 0 only if no risk is incurred. So, V being an (E+R)-valuation means that V can be written as V = E+R, for some R with the *Risk-Positivity* property. An example of an (E + R)-valuation where R appears explicitly is E + Var, which was first explicitly proposed in the seminal paper of Markowitz [13]; Var is known to have the *Risk-Positivity* property [14]. There are examples of (E+R)-valuations where R appears implicitly, such as CVar$_\alpha$ and *Extended Sharpe Ratio* ESR (Theorems 1 and 10, respectively). An (E + R)-valuation is an example of an approach to combine two different optimization goals, namely expectation and riskiness, into a single formula; the *Sharpe Ratio* [24, 25] is another successful example.

   We show five interesting properties of CVaR$_\alpha$, falling into two classes. The first class addresses CVaR$_\alpha$ as a mathematical function.

- CVaR$_\alpha$ is an (E + R)-valuation, where R has the *Risk-Positivity* property (Theorem 1).
- CVaR$_\alpha$ is concave in the probabilities (Theorem 2). The concavity of CVaR$_\alpha$ is of independent interest and may find applications in Convex Optimization.

The second class includes properties of CVaR$_\alpha$ in reference to equilibria of games:

- A significant consequence of the concavity of CVaR$_\alpha$, following from [15, Proposition 3.1], is that it has the *Optimal-Value* property: For every game, every convex combination of strategies in the support of a player's best-response mixed strategy yields the same minimum cost. The *Optimal-Value* property gives necessary conditions on best-response mixed strategies.
- We show, using a counterexample, that CVaR$_\alpha$ lacks the *Weak-Equilibrium-for-Expectation* property (Theorem 5). Since CVaR$_\alpha$ has the *Optimal-Value* property, this implies a separation between the two properties. Nevertheless, we identify a new functional property of CVaR$_\alpha$, termed *Weak-Equilibrium-for*-VaR$_\alpha$: all strategies supported in a player's CVaR$_\alpha$-best response mixed strategy induce the same VaR$_\alpha$ unless some strong condition holds.
- We show that CVaR$_\alpha$ has the *Crawford* property: there is a so called *Crawford game* with no CVaR$_\alpha$-equilibrium (Theorem 6). The Crawford property for a valuation V implies that the problem ∃V-EQUILIBRIUM is non-total. We adopt the Crawford game G$_C(\delta)$ from [15], which involves a parameter δ. We use *Weak-Equilibrium-for*-VaR$_\alpha$ to prove the Crawford property.

As one of our two main results, we show the generic result (Theorem 7) that ∃V-EQUILIBRIUM is strongly $\mathcal{NP}$-hard when V is an (E + R)-valuation with the *Optimal-Value* property, provided that there is a parameter δ, with $0 < \delta \leq \frac{1}{4}$, such that three additional conditions hold; δ has to be a rational number since it explicitly enters the reduction. The *Optimal-Value* property is very crucial for the reduction. The first two such conditions ((2/a) and (2/b)) stipulate two natural analytical properties of V. We give an existential proof that Condition (2/a) holds under some very mild continuity assumptions on the risk valuation R (Lemma 4). The third condition (2/c) requires that for the chosen value of δ, the *Crawford game* $G_C(\delta)$; has no V-equilibrium. So, Theorem 7 addresses a very broad class of (E + R)-valuations with three natural properties: *Risk-Positivity* of R, *Optimal-Value* and *Crawford*; hence, Theorem 7 can be used as a recipe for proving $\mathcal{NP}$-hardness results for other valuations. Indeed, we explicitly determine a parameter δ for which CVaR_α fulfills both additional conditions (2/a) and (2/b) in Theorem 7. Hence our other main result: $\exists\text{CVaR}_\alpha\text{-EQUILIBRIUM}$ is strongly $\mathcal{NP}$-hard for 2-player games (Theorem 8).

Recall that the *Optimal-Value* property of CVaR_α followed directly from its concavity. Since the proof of the $\mathcal{NP}$-hardness of $\exists\text{CVaR}_\alpha\text{-EQUILIBRIUM}$ used the *Optimal-Value* property, it is natural to ask how crucial concavity is to the $\mathcal{NP}$-hardness of $\exists\text{V-EQUILIBRIUM}$ in general. Does $\mathcal{NP}$-hardness extend beyond concave valuations? Towards this end, we consider *quasiconcave* valuations [7, 9, 18], a broad generalization of concave valuations, and their proper subclass of *strictly quasiconcave* valuations [19]. As a nice historical survey in [11] shows, the notion of quasiconcavity emerged from the works of three authors, namely von Neumann [18], DeFinetti [7] and Fenchel [9], in Economics and Mathematics, and started the research field of *generalized convexity*. Quasiconcave functions make a very rich class with many applications in Mathematical Economics; for example, in Microeconomics, the convexity of customers' preference relations is a very desirable feature, implying the quasiconcavity of their utilities [22].

A strictly quasiconcave valuation incurs to a player mixing two particular mixed strategies a cost larger than the minimum of the individual costs incurred by the two mixed strategies when the two costs are not equal. A significant property of strictly quasiconcave valuations, making them interesting to consider in the context of equilibrium computation, is that they do not exclude the existence of mixed equilibria, whereas strictly concave valuations do (cf. [15, Sect. 1.5]); this is because the notion of strictness is weaker for strictly quasiconcave valuations than for strictly concave valuations. We show that every strictly quasiconcave valuation has the *Optimal-Value* property (Theorem 9).

To test the applicability of the *Optimal-Value* property of strictly quasiconcave valuations to deriving concrete $\mathcal{NP}$-hardness, we consider the *Extended Sharpe Ratio* ESR := $\dfrac{M \cdot \text{E}}{\text{M} - \text{Var}}$; M is a constant. ESR is inspired from the famous *Sharpe Ratio* [24, 25], applying to revenue maximization, which measures how well an asset's return compensates the investor for the risk; ESR is an adjustment of Sharpe Ratio to minimization games. We prove that ESR is an (E + R)-valuation. The strict quasiconcavity of ESR (Theorem 11) follows from the new result that the ratio of a concave over a convex function is strictly quasiconcave (Lemma 6). Further we prove that ESR has the *Weak-Equilibrium-for-Expectation* property; it is used for proving that $G_C(\delta)$ has no ESR-equilibrium (Theorem 12). Hence, the strong $\mathcal{NP}$-hardness of deciding the existence of an ESR-equilibrium follows (Theorem 13), settling an open problem from [15, Sect. 6]. This witnesses that Theorem 7 is applicable beyond concave valuations.

1.3 Related Work and Comparison

This work falls into the research field of computing equilibria for games where players are not expectation-maximizers, started by Fiat and Papadimitriou in [10].

Theorem 7 extends the strong $\mathcal{NP}$-hardness from valuations with the *Weak-Equilibrium-for-Expectation* property in [15, Theorem 5.1] to ones with *Optimal-Value* property, encompassing, by Theorem 9, the much broader class of strictly quasiconcave valuations. The extra conditions for Theorem 7 and [15, Theorem 5.1] are quite similar; *Risk-Positivity* is used for both. Like the reduction in [15, Theorem 5.1], the reduction in Theorem 7 uses ideas from the reduction in [4], used for the $\mathcal{NP}$-hardness proof of decision problems about Nash equilibria.

Concrete strong $\mathcal{NP}$-hardness results for 2-player games followed from [15, Theorem 5.1], for concave valuations like $\mathsf{E} + \mathsf{Var}$ [15, Theorem 5.12]; other concrete results for 2-strategies games, and for many significant choices of V, are given in [15, Theorem 4.6]. Further complexity results had appeared in [14].

To the best of our knowledge, there are in the literature only three hardness results about VaR_α, which, however, are not related to equilibrium computation: (1) Maximizing the expectation of portfolio return of an investment, for a given α and a given threshold on VaR_α, is $\mathcal{NP}$-complete [2]. (2) In the discrete setting, maximizing the expectation of expected return over all *scenaria* is $\mathcal{NP}$-complete [1]. (A scenario is a set of options with potential investment losses.) (3) For a given α, minimizing VaR, subject to the constraints that *(i)* the "scenario-weighted" portfolio price is no more than the initial endowment, and *(ii)* the loss exceeds VaR with probability no more than α, is $\mathcal{NP}$-hard [30].

2   (E + R)-Valuations

A *valuation* is a special case of a function, which maps probability vectors to reals. In this work, we shall focus on an $(\mathsf{E} + \mathsf{R})$-*valuation* [15, Definition 2.1], which is a valuation of the form $\mathsf{V} = \mathsf{E} + \mathsf{R}$, where E is the *expectation valuation* with $\mathsf{E}(\mathbf{p}) = \sum_{k \in [\ell]} p_k a_k$, and R is the *risk valuation,* a continuous valuation with the *Risk-Positivity* property: For each probability vector $\mathbf{p}$, (C.1) $\mathsf{R}(\mathbf{p}) \geq 0$, and (C.2) $\mathsf{R}(\mathbf{p}) = 0$ if and only if the corresponding random variable P is constant.

A function $\mathsf{f} : \mathsf{D} \to \mathbb{R}$ on a convex set $\mathsf{D} \subseteq \mathbb{R}^n$ is *concave* if for every pair of points $x, y \in \mathsf{D}$, for all $\lambda \in [0, 1]$, $\mathsf{f}(\lambda y + (1 - \lambda) x) \geq \lambda \mathsf{f}(y) + (1 - \lambda) \mathsf{f}(x)$; f is *strictly concave* if the inequality is strict for all $\lambda \in (0, 1)$. A function $\mathsf{f} : \mathsf{D} \to \mathbb{R}$ on a convex set D is (strictly) *convex* if $-\mathsf{f}$ is (strictly) concave.

Quasiconcavity generalizes concavity: A function $\mathsf{f} : \mathsf{D} \to \mathbb{R}$ is *quasiconcave* if for every pair of points $x, y \in \mathsf{D}$, for all $\lambda \in (0, 1)$, $\mathsf{f}(\lambda y + (1 - \lambda)x) \geq \min\{\mathsf{f}(y), \mathsf{f}(x)\}$. A function $\mathsf{f} : \mathsf{D} \to \mathbb{R}$ on a convex set $\mathsf{D} \subset \mathbb{R}^n$ is *quasiconvex* if

$-f$ is quasiconcave. A quasiconcave function need not be concave. Thus, these functions miss many significant properties of concave functions. For example, a quasiconcave function is not necessarily continuous, while a concave function on an open interval is also continuous on it.

A quasiconcave function $f : D \rightarrow \mathbb{R}$ is *strictly quasiconcave* [19] if for every pair of points $x, y \in D$, $f(x) \neq f(y)$ implies that for all $\lambda \in (0, 1)$, $f(\lambda y + (1-\lambda)x) > \min\{f(y), f(x)\}$; f is *strictly quasiconvex* if $-f$ is strictly quasiconcave. The notion of strictness for strictly quasiconcave functions is weaker than for strictly concave functions: for the former, strictness allows that $f(\lambda x + (1-\lambda)y) = \min\{f(x), f(y)\}$, with $\lambda \in (0, 1)$, in the case $f(x) = f(y)$, whereas the equality is excluded for the latter. A notable property of a strictly quasiconcave function is a property of concave functions: a local maximum is also a global maximum [19, Theorem 2].

3   Conditional Value-at-Risk (CVaR_α)

Our presentation draws from [20, 21]. Fix throughout a confidence level $\alpha \in [0, 1)$. Some of the tools to study the properties of CVaR_α come from *Value-at-Risk*, denoted as VaR_α; they are both examples of a *valuation*: a function from probability distributions to reals. We consider a random variable P, representing a distribution of loss; the cumulative distribution function of P may be continuous or discontinuous. A *discrete* random variable P has a discrete distribution function, making a special case of a discontinuous cumulative distribution function: it takes on values $0 \leq a_1 < a_2 < \ldots < a_\ell$ with probabilities $p_j := \mathbb{P}(\mathsf{P} = a_j)$, $1 \leq j \leq \ell$; the values $a_1, a_2, \ldots, a_\ell$ are called *scenaria* in the risk literature. In this case, we shall identify P with the probability vector $\mathbf{p}$. Note that in defining $\mathsf{VaR}_\alpha(\mathbf{p})$ and $\mathsf{CVaR}_\alpha(\mathbf{p})$, the values $a_1, \ldots, a_\ell$ have to be fixed. Denote the cumulative distribution function of $\mathbf{p}$ as $\Pi(\mathbf{p}, z) = \sum_{k \in [\ell] | a_k \leq z} p_k$.

3.1   Value-at-Risk (VaR_α)

The *Value-at-Risk* of P at *confidence level* $\alpha \in [0, 1)$, denoted as $\mathsf{VaR}_\alpha(\mathsf{P})$, is defined as $\mathsf{VaR}_\alpha(\mathsf{P}) = \inf\{\zeta \mid \mathbb{P}[\mathsf{P} \leq \zeta] \geq \alpha\}$; this definition applies to both continuous and discontinuous distributions. So, $\mathbb{P}[\mathsf{P} \leq \mathsf{VaR}_\alpha(\mathsf{P})] \geq \alpha$, with equality holding when the cumulative distribution function is continuous at $\mathsf{VaR}_\alpha(\mathsf{P})$— put in other words, when there is no *probability atom* at $\mathsf{VaR}_\alpha(\mathsf{P})$. When P is a discrete random variable, $\mathsf{VaR}_\alpha(\mathbf{p}) = \min\{a_k \mid \sum_{j=1}^k p_j \geq \alpha\}$; note that VaR_α is a discontinuous function of α. In this case, denote as $\kappa_\alpha = \kappa_\alpha(\mathbf{p})$ the unique index such that $\sum_{k \in [\kappa_\alpha]} p_k \geq \alpha > \sum_{k \in [\kappa_\alpha - 1]} p_k$; so, $a_{\kappa_\alpha} = \mathsf{VaR}_\alpha(\mathbf{p})$. In the full version of the paper, we give an example establishing that VaR_α is not an $(\mathsf{E} + \mathsf{R})$-valuation. We shall later use an interesting property of VaR_α (cf. [3, Exercise 3.24 (f)]), for which we provide a new proof.

Lemma 1. *With $\alpha \in (0, 1)$, $\mathsf{VaR}_\alpha(\mathbf{p})$ is quasiconcave and quasiconvex in the probabilities, but neither strictly quasiconcave nor strictly quasiconvex.*

3.2   Conditional Value-at-Risk (CVaR_α)

CVaR_α accounts for losses no smaller than VaR_α. For random variables with a continuous cumulative distribution function, the *Conditional Value-at-Risk* of P, denoted as $\mathsf{CVaR}_\alpha(\mathsf{P})$, is the conditional expectation of P subject to $\mathsf{P} \geq \mathsf{VaR}_\alpha(\mathsf{P})$ [23, Definition 2]; so, $\mathsf{CVaR}_\alpha(\mathsf{P}) = \mathsf{E}(\mathsf{P} \mid \mathsf{P} \geq \mathsf{VaR}_\alpha(\mathsf{P}))$. Note that $\alpha = 0$ is a degenerate case with $\mathsf{CVaR}_0(\mathsf{P}) = \mathsf{E}(\mathsf{P})$. The general definition of CVaR_α for random variables with a possibly discontinuous cumulative distribution function is obtained as follows (cf. [21, Proposition 6]): Denote as $\mathsf{CVaR}_\alpha^+(\mathsf{P})$ the conditional expectation of P subject to $\mathsf{P} > \mathsf{VaR}_\alpha(\mathsf{P})$; $\mathsf{CVaR}_\alpha^+(\mathsf{P})$ is also called the *Condtional Tail Expectation* or *Upper* CVaR [23]. Set $\lambda_\alpha(\mathsf{P}) := \dfrac{\mathbb{P}\left[\mathsf{P} \leq \mathsf{VaR}_\alpha(\mathsf{P})\right] - \alpha}{1 - \alpha}$. Note that $\lambda_\alpha(\mathsf{P})$ provides for the possibility that there is a probability atom (that is, the cumulative distribution function is discontinuous) at $\mathsf{VaR}_\alpha(\mathsf{P})$. The numerator of $\lambda_\alpha(\mathsf{P})$ is the "excess" probability mass at $\mathsf{VaR}_\alpha(\mathsf{P})$, i.e., the probability mass beyond α; the denominator $1 - \alpha$ is the support size of $\mathbb{P}_\alpha$, the α-*tail distribution of* P, denoted as $\mathbb{P}_\alpha$ and defined as follows: $\mathbb{P}_\alpha\left[\mathsf{P} \leq \zeta\right] = 0$ if $\zeta < \mathsf{VaR}_\alpha(\mathsf{P})$, and $\mathbb{P}_\alpha\left[\mathsf{P} \leq \zeta\right] = \dfrac{\mathbb{P}\left[\mathsf{P} \leq \zeta\right] - \alpha}{1 - \alpha}$ if $\zeta \geq \mathsf{VaR}_\alpha(\mathsf{P})$; thus, $\lambda_\alpha(\mathsf{P}) = \mathbb{P}_\alpha\left[\mathsf{P} \leq \mathsf{VaR}_\alpha(\mathsf{P})\right]$. Hence, we obtain the *weighted average formula* for $\mathsf{CVaR}_\alpha(\mathsf{P})$ [21, Proposition 6]:

$$\mathsf{CVaR}_\alpha(\mathsf{P}) := \lambda_\alpha(\mathsf{P}) \cdot \mathsf{VaR}_\alpha(\mathsf{P}) + (1 - \lambda_\alpha(\mathsf{P})) \cdot \mathsf{CVaR}_\alpha^+(\mathsf{P}) \,;$$

so, CVaR_α is the weighted average of VaR_α and the expectation of losses strictly exceeding VaR_α. Note that $\mathsf{CVaR}_\alpha(\mathsf{P}) = \mathsf{VaR}_\alpha(\mathsf{P})$ when $\lambda_\alpha(\mathsf{P}) = 1$; note also that for continuous cumulative distributions, $\lambda_\alpha(\mathsf{P}) = 0$ and $\mathsf{CVaR}_\alpha(\mathsf{P}) = \mathsf{CVaR}_\alpha^+(\mathsf{P})$, the expected loss strictly exceeding VaR_α. For a discrete random variable P, identified with the probability vector $\mathbf{p}$, the weighted average formula reduces to the *discrete weighted average formula* (cf. [21, Proposition 8]):

$$\mathsf{CVaR}_\alpha(\mathbf{p}) = \frac{1}{1 - \alpha}\left(\left(\sum_{k \in [\kappa_\alpha]} p_k - \alpha\right) a_{\kappa_\alpha} + \sum_{k \geq \kappa_\alpha + 1} p_k a_k\right).$$

Recall the *minimization function* $\mathsf{F}_\alpha(\mathbf{p}, z) := z + \dfrac{1}{1 - \alpha}\sum_{k \in [n]\mid a_k > z}(a_k - z)p_k$, for a probability vector $\mathbf{p}$ and a number $z \in \mathbb{R}$, from [20, Sect. 2]. Note that F_α is linear in its first argument. The continuity of F_α in z follows from its convexity (in z) [21, Theorem 10]. We provide a direct and simple proof, bypassing convexity, for the discrete setting.

Lemma 2. *The function $\mathsf{F}_\alpha(\mathbf{p}, z)$ is continuous in z.*

We now prove some monotoniciy properties of $\mathsf{F}_\alpha(\mathbf{p}, z)$ in z, drawing on the continuity of F_α (Lemma 2); they offer refinements of properties from [21, Proposition 5], which will be needed later. In order to take into account the possibility that some probabilities may be 0, denote, for an index $j \in [\ell]$, $\mathrm{next}(\mathbf{p}, j) := \min\{k > j \mid p_k \neq 0\}$; $\mathrm{next}(\mathbf{p}, j) := \ell + 1$ when $p_k = 0$ for all $j < k \leq \ell$.

Lemma 3. *The following implications hold:*

(1) *If* $\Pi\left(\mathbf{p}, \mathsf{VaR}_\alpha(\mathbf{p})\right) > \alpha$, *then:*
   (1/a) $\mathsf{F}_\alpha\left(\mathbf{p}, z\right)$ *increases strictly monotone in z for $z \geq \mathsf{VaR}_\alpha(\mathbf{p})$.*
   (1/b) $\mathsf{F}_\alpha\left(\mathbf{p}, z\right)$ *decreases strictly monotone in z for $z \leq \mathsf{VaR}_\alpha(\mathbf{p})$.*
(2) *If* $\Pi\left(\mathbf{p}, \mathsf{VaR}_\alpha(\mathbf{p})\right) = \alpha$, *then:*
   (2/a) $\mathsf{F}_\alpha\left(\mathbf{p}, z\right)$ *increases strictly monotone for $z \geq a_{\mathsf{next}(\mathbf{p}, \kappa_\alpha(\mathbf{p}))}$.*
   (2/b) $\mathsf{F}_\alpha\left(\mathbf{p}, z\right)$ *is constant for $\mathsf{VaR}_\alpha(\mathbf{p}) \leq z \leq a_{\mathsf{next}(\mathbf{p}, \kappa_\alpha(\mathbf{p}))}$.*
   (2/c) $\mathsf{F}_\alpha\left(\mathbf{p}, z\right)$ *decreases strictly monotone for $z \leq \mathsf{VaR}_\alpha(\mathbf{p})$.*

By Lemmas 2, 3 immediately implies a result from [21, Theorem 10]:

Corollary 1. $\mathsf{CVaR}_\alpha(\mathbf{p}) = \min_z \mathsf{F}_\alpha(\mathbf{p}, z) = \mathsf{F}_\alpha(\mathbf{p}, \mathsf{VaR}_\alpha(\mathbf{p}))$.

We continue to show two significant properties:

Theorem 1. *With $\alpha \in (0,1)$, CVaR_α is an $(\mathsf{E} + \mathsf{R})$-valuation.*

Theorem 2. *With $\alpha \in (0,1)$, CVaR_α is concave in the probabilities.*

We continue with a characterization of the cases where CVaR_α is constant over all convex combinations of probability vectors $\mathbf{p}$ and $\mathbf{q}$; interestingly, one of the two cases requires that $\mathsf{VaR}_\alpha(\mathbf{p}) = \mathsf{VaR}_\alpha(\mathbf{q})$. We use the quasiconcavity and quasiconvexity of VaR_α in the probabilities (Lemma 1) to prove:

Theorem 3. *Fix a confidence level $\alpha \in (0,1)$ and consider probability vectors $\mathbf{p}$ and $\mathbf{q}$ with $\mathsf{VaR}_\alpha(\mathbf{p}) \leq \mathsf{VaR}_\alpha(\mathbf{q})$ and $\mathsf{CVaR}_\alpha(\mathbf{p}) = \mathsf{CVaR}_\alpha(\mathbf{q})$. Then, $\mathsf{CVaR}_\alpha\left(\lambda \cdot \mathbf{p} + (1 - \lambda) \cdot \mathbf{q}\right)$ is constant for $\lambda \in [0,1]$ if and only if:*

(1) $\mathsf{VaR}_\alpha(\mathbf{p}) = \mathsf{VaR}_\alpha(\mathbf{q})$, *or:*
(2) $\alpha = \Pi\left(\mathbf{p}, \mathsf{VaR}_\alpha(\mathbf{p})\right)$ *and* $\kappa_\alpha(\mathbf{q}) \leq \mathsf{next}(\mathbf{p}, \kappa_\alpha(\mathbf{p}))$.

4   Properties of CVaR_α at Equilibrium

Our definitions and notation for the game-theoretic framework are standard; they are patterned after those in [15, Sects. 2 and 3].

For a player $i \in [n]$, a *valuation function*, or *valuation* for short, V_i is a real-valued function, yielding a value $\mathsf{V}_i(\mathbf{p})$ to each mixed profile $\mathbf{p}$, so that in the special case when $\mathbf{p}$ is a profile $\mathbf{s}$, $\mathsf{V}_i(\mathbf{s}) = \mu_i(\mathbf{s})$. The *expectation valuation* $\mathsf{E}_i(\mathbf{p}) = \sum_{\mathbf{s} \in S} \mathbf{p}(\mathbf{s}) \mu_i(\mathbf{s})$, with $i \in [n]$. A *valuation* $\mathsf{V} = \langle \mathsf{V}_1, \ldots, \mathsf{V}_n \rangle$ is a tuple of valuations, one per player; G^V denotes G together with V. We shall view each valuation V_i, with $i \in [n]$, as a function of the mixed strategy p_i, for a fixed partial mixed profile $\mathbf{p}_{-i}$; this view incurs corresponding definitions in the game-theoretic framework for *Risk-Positivity* and (strict) concavity and quasiconcavity, which we now provide. We shall assume that V is an $(\mathsf{E} + \mathsf{R})$-valuation. By Condition (C.2) in the *Risk-Positivity* property, it follows that for each player $i \in [n]$ and mixed profile $\mathbf{p}$, $\mathsf{R}_i(\mathbf{p}) = 0$ if and only if for each profile $\mathbf{s} \in \mathcal{S}$ with $\mathbf{p}(\mathbf{s}) > 0$, $\mu_i(\mathbf{s})$ is constant over all choices of strategies by the other players; in such case, $\mathsf{V}_i(\mathbf{p}) = \mathsf{E}_i(\mathbf{p}) = \mu_i(\mathbf{s})$ for any profile $\mathbf{s} \in \mathcal{S}$ with $\mathbf{p}(\mathbf{s}) > 0$. The valuation V is *concave* (resp., *(strictly) quasiconcave*) if the following holds for every game G: For each player $i \in [n]$, the valuation $V_i(p_i, \mathbf{p}_{-i})$ is concave (resp., (strictly) quasiconcave) in p_i, for a fixed $\mathbf{p}_{-i}$.

A mixed strategy p_i of player i is a V_i-*best response* to $\mathbf{p}_{-i}$ if p_i minimizes $V_i(., \mathbf{p}_{-i})$ over Δ_i; so, $V_i(p_i, \mathbf{p}_{-i}) = \min \{V_i(p'_i, \mathbf{p}_{-i}) \mid p'_i \in \Delta(S_i)\}$. The mixed profile $\mathbf{p}$ is a V-*equilibrium* if for each player $i \in [n]$, the mixed strategy p_i is a V_i-best response to $\mathbf{p}_{-i}$; so, no player could unilaterally deviate to another mixed strategy to reduce her cost. Denote as $\exists V$-EQUILIBRIUM the problem of deciding, given a game G, the existence of a V-equilibrium for G^V.

The valuation V has the *Optimal-Value* property if the following condition holds for every game G: For each player $i \in [n]$, if $\widehat{p}_i$ is a V_i-best response to a partial mixed profile $\mathbf{p}_{-i}$, then, for any mixed strategy q_i with $\sigma(q_i) \subseteq \sigma(\widehat{p}_i)$, $V_i(q_i, \mathbf{p}_{-i}) = V_i(\widehat{p}_i, \mathbf{p}_{-i})$. By a sufficient condition from [15, Proposition 3.1], every concave valuation has the *Optimal-Value* property; hence, by the concavity of CVaR_α (Theorem 2), CVaR_α has the *Optimal-Value* property. Here is an immediate implication of the *Optimal-Value* property for CVaR_α-best responses: all strategies supported in a player's CVaR_α-best response mixed strategy induce the same (conditional) VaR_α (unless some other, strong condition holds).

Theorem 4 (Weak-Equilibrium-for-VaR_α Property). *Fix a confidence level $\alpha \in (0,1)$. Take a CVaR_α-best response p_i of player $i \in [n]$ to the partial mixed profile $\mathbf{p}_{-i}$, where $p^1, p^2 \in \sigma(p_i)$ with $\mathsf{VaR}_\alpha(\langle p^1, \mathbf{p}_{-i}\rangle) \leq \mathsf{VaR}_\alpha(\langle p^2, \mathbf{p}_{-i}\rangle)$. Then:*

(1) $\mathsf{VaR}_\alpha(\langle p^1, \mathbf{p}_{-i}\rangle) = \mathsf{VaR}_\alpha(\langle p^2, \mathbf{p}_{-i}\rangle)$, *or:*
(2) $\alpha = \Pi\left(\langle p^1, \mathbf{p}_{-i}\rangle, \mathsf{VaR}_\alpha(\langle p^1, \mathbf{p}_{-i}\rangle)\right)$
and $\kappa_\alpha(\langle p^2, \mathbf{p}_{-i}\rangle) \in \left[\kappa_\alpha(\langle p^1, \mathbf{p}_{-i}\rangle), \mathsf{next}(\mathbf{p}, \kappa_\alpha(\langle p^1, \mathbf{p}_{-i}\rangle))\right]$.

The $(E + R)$-valuation V has the *Weak-Equilibrium-for-Expectation* property [15, Definition 3.1] if the following condition holds for every game G: For each player $i \in [n]$, the *Weak-Equilibrium-for Expectation* property for player i holds: If p_i is a V_i-best response to a partial mixed profile $\mathbf{p}_{-i}$, then for each pair of strategies $j, k \in \sigma(p_i)$, $E_i\left(p_i^j, \mathbf{p}_{-i}\right) = E_i\left(p_i^k, \mathbf{p}_{-i}\right)$. We show:

Theorem 5. *For a fixed confidence level $\alpha \in (0,1)$, CVaR_α does not have the Weak-Equilibrium-for-Expectation property.*

For $0 < \delta < 1$, the *Crawford game* $G_C(\delta)$ is the 2-player game with bimatrix $\begin{pmatrix} \langle 1+\delta, 1+\delta\rangle & \langle 1, 1+2\delta\rangle \\ \langle 1, 1+2\delta\rangle & \langle 1+2\delta, 1\rangle \end{pmatrix}$ from [15]; it is a generalization of a game from [5, Sect. 4]. $G_C(\delta)$ has no pure equilibrium [15, Lemma 5.11]. We show:

Theorem 6. *With $\alpha \in \left(\frac{1}{3}, 1\right)$ and $\delta > 0$, $G_C(\delta)$ has no mixed CVaR_α-equilibrium.*

5 $\mathcal{NP}$-Hardness from Optimal-Value Property

We first recall some notation from [15, Sect. 2.2]. For an $(E + R)$-valuation V, we shall deal with the so called *bivalued* cases where for a player $i \in [n]$ and a mixed profile $\mathbf{p}$, $\{\mu_i(\mathbf{s}) \mid \mathbf{p}(\mathbf{s}) > 0\} = \{a, b\}$ with $a < b$, so that $\mathsf{R}_i(\mathbf{p})$ is a function of the three parameters a, b and q with $q = \sum_{\mathbf{s} \in \mathcal{S} \mid \mu_i(\mathbf{s}) = b} \mathbf{p}(\mathbf{s})$. Then, denote $\widehat{\mathsf{R}}_i(a, b, q) := \mathsf{R}_i(\mathbf{p})$ and $\widehat{\mathsf{V}}_i(a, b, q) := a + q(b - a) + \widehat{\mathsf{R}}_i(a, b, q)$. We show:

Theorem 7. *Fix an* $(E + R)$*-valuation* V *such that:*

(1) V *has the Optimal-Value property.*
(2) *There is a rational number* δ*, with* $0 < \delta \le \frac{1}{4}$*, such that:*
    (2/a) $\widehat{\mathsf{R}}(1, 1 + \varrho, q) < \frac{1}{2}$ *for each probability* $q \in [0, 1]$*, with* $0 < \varrho \le 2\delta$*.*
    (2/b) $\widehat{\mathsf{V}}(1, 1 + \varrho, r) < \widehat{\mathsf{V}}(1, 2, q)$ *for each pair of probabilities* $0 \le r \le q \le 1$*,*
    with $0 < \varrho \le 2\delta$*.*
    (2/c) *The Crawford game* $\mathsf{G}_C(\delta)$ *has no* V*-equilibrium.*

Then, $\exists\mathsf{V}$-EQUILIBRIUM *is strongly* $\mathcal{NP}$*-hard for 2-player games.*

We present a proof with a general reduction involving the parameter δ from Condition (2), which is a rational number. The reduction uses the *Crawford game* $\mathsf{G}_C(\delta)$, with $0 < \delta \le \frac{1}{4}$, as a "gadget"; so, δ enters the reduction through $\mathsf{G}_C(\delta)$. We employ a reduction from SAT. An instance of SAT is a propositional formula ϕ in the form of a conjunction of *clauses* $\mathsf{C} = \{c_1, \ldots, c_k\}$ over a set of *variables* $\mathcal{V} = \{v_1, \ldots, v_m\}$. Denote as $\mathsf{L} = \{\ell_1, \bar{\ell}_1, \ldots, \ell_m, \bar{\ell}_m\}$ the set of *literals* corresponding to the variables in $\mathcal{V}$. We shall use lower-case letters $c, c_1, c_2, \ldots$, $v, v_1, v_2, \ldots$, and $\ell, \ell_1, \ell_2, \ldots$ to denote clauses from C, variables from $\mathcal{V}$ and literals from L, respectively. Denote $\Lambda := \mathsf{C} \cup \mathcal{V} \cup \mathsf{L}$. We shall use the *Crawford set* $\mathcal{F} = \{\mathsf{f}_1, \mathsf{f}_2\}$ with the two strategies f_1 and f_2 from $\mathsf{G}_C(\delta)$; f denotes f_1 or f_2.

Proof. From an instance ϕ of SAT, form $\mathsf{G} = \mathsf{G}(\phi) = \langle [2], \{S_i\}_{i \in [2]}, \{\mu_i\}_{i \in [2]} \rangle$ with: For $i \in [2]$, $S_i := \Lambda \cup \mathcal{F}$; see the cost functions $\{\mu_i\}_{i \in [2]}$ are given in Fig. 1. For a player $i \in [2]$, denote $p_i(\mathcal{F}) := \sum_{\mathsf{f} \in \mathcal{F}} p_i(\mathsf{f})$, $p_i(\mathsf{L}) := \sum_{\ell \in \mathsf{L}} p_i(\ell)$ and $p_i(\Lambda) := \sum_{\lambda \in \Lambda} p_i(\lambda)$; note that $p_i(\mathcal{F}) + p_i(\Lambda) = 1$. We first prove that in a V-equilibrium $\langle p_1, p_2 \rangle$ for G, $p_1(\Lambda) \cdot p_2(\Lambda) > 0$; further, we prove that $p_1(\Lambda) = p_2(\Lambda) = 1$. We finally establish some stronger properties for a V-equilibrium from which strong $\mathcal{NP}$-hardness follows. $\square$

Condition (2/a) in Theorem 7 is fulfilled under some very mild conditions on $\widehat{\mathsf{R}}$:

Lemma 4. *Consider a function* $\widehat{\mathsf{R}}(1, 1 + \varrho, q)$ *with the following properties:* (C.1) $\widehat{\mathsf{R}}(1, 1, q) = 0$ *for* $q \in [0, 1]$*.* (C.2) $\widehat{\mathsf{R}}$ *is continuous in* ϱ *and in* q*. Then, there is a rational number* δ*,* $0 < \delta \le \frac{1}{4}$*, such that* $\widehat{\mathsf{R}}(1, 1 + \varrho, q)$ *fulfills Condition (2/a).*

We now use Theorem 7 to derive a concrete strong $\mathcal{NP}$-hardness result for CVaR_α, with $\alpha \in \left(\frac{1}{3}, 1\right)$. Condition (1) from Theorem 7 is fulfilled since CVaR_α has the *Optimal-Value* property. We continue to prove:

Lemma 5. *For* $0 < \delta < \frac{1}{4} \cdot (1 - \alpha)$*,* CVaR_α *fulfills (2/a) and (2/b) from Theorem 7.*

Profile s	Condition on s	$\langle \mu_1(\mathbf{s}), \mu_2(\mathbf{s}) \rangle$
$\langle \ell_1, \ell_2 \rangle$	$\ell_1 \neq \overline{\ell}_2$	$\langle 1, 1 \rangle$
$\langle \ell_1, \ell_2 \rangle$	$\ell_1 = \overline{\ell}_2$	$\langle 2, 2 \rangle$
$\langle \ell, v \rangle$	ℓ is a literal for v	$\langle 2, m \rangle$
$\langle \ell, v \rangle$	ℓ is not a literal for v	$\langle 2, 0 \rangle$
$\langle \ell, c \rangle$	$\ell \notin c$	$\langle 2, 0 \rangle$
$\langle \ell, c \rangle$	$\ell \in c$	$\langle 2, m \rangle$
$\langle \ell, f \rangle$	—	$\langle 2, 1 \rangle$
$\langle v_1, v_2 \rangle$ or $\langle c_1, c_2 \rangle$ or $\langle v, c \rangle$	—	$\langle 2, 2 \rangle$
$\langle v, f \rangle$ or $\langle c, f \rangle$	—	$\langle 2, 1 \rangle$
$\langle f_1, f_1 \rangle$	—	$\langle 1 + \delta, 1 + \delta \rangle$
$\langle f_1, f_2 \rangle$ or $\langle f_2, f_1 \rangle$	—	$\langle 1, 1 + 2\delta \rangle$
$\langle f_2, f_2 \rangle$	—	$\langle 1 + 2\delta, 1 \rangle$

Fig. 1. The cost functions for the game $\mathsf{G}(\phi)$. For a profile $\langle s_1, s_2 \rangle$ not in the table, set $\mu_i(s_1, s_2) := \mu_{\overline{i}}(s_2, s_1)$, with $i \in [2]$ and $\overline{i} \neq i$; so, $\overline{i}$ is the opponent player.

By Theorem 6, CVaR_α with $\alpha \in \left(\frac{1}{3}, 1 \right)$ fulfills Condition (2/c) from Theorem 7. Thus, we obtain:

Theorem 8. *With $\alpha \in \left(\frac{1}{3}, 1 \right)$, $\exists \mathsf{CVaR}_\alpha$-EQUILIBRIUM is strongly $\mathcal{NP}$-hard for 2-player games.*

In the full version of the paper, we improve Theorem 8 to cover all $\alpha \in (0, 1)$.

6   Strict Quasiconcavity and Extended Sharpe Ratio

We extend the *Optimal-Value* property to any strictly quasiconcave valuation:

Theorem 9. *A strictly quasiconcave valuation has the Optimal-Value property.*

So Theorem 9 extends Theorem 7 by replacing the *Optimal-Value* property in Theorem 7 (Condition (2/a)) with the strict quasiconcavity property.

Given a mixed profile $\mathbf{p}$ and a player $i \in [n]$, the *Extended Sharpe Ratio* is $\mathsf{ESR}_i(\mathbf{p}) = \dfrac{M \cdot \mathsf{E}_i(\mathbf{p})}{M - \mathsf{Var}_i(\mathbf{p})}$, where M is a constant independent of $\mathbf{p}$, with $M > \left(\max_{i \in [n], \mathbf{s} \in \mathcal{S}} \mu_i(\mathbf{s}) \right)^2$. We prove:

Theorem 10. *ESR is an (E + R)-valuation.*

It was known that the ratio of a concave function and a convex function is strictly quasiconcave (cf. [3, Example 3.38]). We present a new, simple proof to get a stronger result with quasiconcavity replaced by strict quasiconcavity.

Lemma 6. *Consider a pair of a concave function $\mathsf{g} : \mathsf{D} \to \mathbb{R}_+$ and a convex function $\mathsf{h} : \mathsf{D} \to \mathbb{R}_+$. Then, the function $\mathsf{f} := \frac{\mathsf{g}}{\mathsf{h}}$ is strictly quasiconcave.*

Since $M \cdot \mathsf{E}$ is concave and $M - \mathsf{Var}$ is convex, Lemma 6 implies:

Theorem 11. ESR *is strictly quasiconcave.*

By Theorems 9 and 11 sets the *Optimal-Value* property for ESR. We prove:

Lemma 7. *For δ with $0 < \delta \leq \frac{1}{4}$,* ESR *fulfills* (2/a) *and* (2/b) *from Theorem 7.*

We now show the Crawford property for ESR:

Theorem 12. *For $\delta > 0$, the game* $\mathsf{G}_C(\delta)$ *has no mixed* ESR-*equilibrium.*

By the *Optimal-Value* property, Lemma 7 and Theorems 7, 12 implies:

Theorem 13. ∃ESR-EQUILIBRIUM *is strongly $\mathcal{NP}$-hard for 2-player games.*

7   Conclusion and Open Problems

This work presents the *first* complexity results for the equilibrium computation problem about CVaR_α; these are shown using significant mathematical properties for CVaR_α, such as concavity in the probabilities, we establish. The strong $\mathcal{NP}$-hardness of ∃CVaR_α-EQUILIBRIUM (Theorem 8) follows from two significant properties of CVaR_α: *(i)* the *Optimal-Value* property and *(ii)* the *Crawford* property (Theorem 6), shown using its *Weak-Equilibrium-for-*VaR_α property. Using a few additional technical conditions, we prove that for any 2-player game with a valuation V with the *Optimal-Value* and the *Crawford* properties, it is strongly $\mathcal{NP}$-hard to decide if the game has a V-equilibrium (Theorem 7).

Theorems 7 and 9 show together the $\mathcal{NP}$-hardness of deciding the existence of equilibria for valuations with the strict quasiconcavity property from [19], fulfilling some extra conditions. Since a wide spectrum of functions from Mathematical Economics are quasiconcave, this is a very general result. *So it is rather likely that we have to cope with $\mathcal{NP}$-hardness when we want to model risk, and this is the most significant insight following from our $\mathcal{NP}$-hardness results.* The field of *Operations Research* has done this successfully in the last 40 years. In *Algorithmic Game Theory*, we now have to seek for positive results in the directions of *(i)* special classes of games, *(ii)* approximate equilibria and *(iii)* other solution concepts, such as *correlated equilibria*. A spectrum of concrete open questions remain. Since CVaR_α is becoming the prevailing measure of risk in contemporary financial engineering, it would be interesting to design algorithms to compute approximate CVaR_α-equilibria, or to prove inapproximability results.

References

1. Benati, S.: The computation of the worst conditional expectation. Eur. J. Oper. Res. **155**, 414–425 (2004)
2. Benati, S., Rizzi, R.: A mixed integer linear programming formulation of the optimal mean/value-at-risk portfolio problem. Eur. J. Oper. Res. **176**, 423–434 (2007)
3. Boyd, S., Vandenberghe, L.: Convex Optimization. Cambridge University Press, Cambridge (2004)

4. Conitzer, V., Sandholm, T.: New complexity results about Nash equilibria. Games Econ. Behav. **63**(2), 621–641 (2008)
5. Crawford, V.P.: Equilibrium without independence. J. Econ. Theory **50**(1), 127–154 (1990)
6. Debreu, G.: A social equilibrium existence theorem. Proc. Nat. Acad. Sci. U.S.A. **38**, 886–893 (1952)
7. DeFinetti, B.: Sulle stratificazioni convesse. Annal. Mat. **30**, 173–183 (1949)
8. Fan, K.: Fixed point and minimax theorems in locally convex topological linear spaces. Proc. Nat. Acad. Sci. **38**, 121–126 (1952)
9. Fenchel, W.: Convex Cones, Sets and Functions. Lecture Notes, Department of Mathematics, Princeton University (1953)
10. Fiat, A., Papadimitriou, C.H.: When the players are not expectation maximizers. In: Kontogiannis, S., Koutsoupias, E., Spirakis, P.G. (eds.) SAGT 2010. LNCS, vol. 6386, pp. 1–14. Springer, Heidelberg (2010). doi:10.1007/978-3-642-16170-4_1
11. Guerraggio, A., Molho, E.: The origins of quasi-concavity: a development between mathematics and economics. Hist. Math. **31**, 62–75 (2004)
12. Krokhmal, P., Zabarankin, M., Uryasev, S.: Modeling and optimization of risk. Surv. Oper. Res. Manag. Sci. **16**, 49–66 (2011)
13. Markowitz, H.: Portfolio selection. J. Finan. **7**, 77–91 (1952)
14. Mavronicolas, M., Monien, B.: Minimizing expectation plus variance. Theory Comput. Syst. **57**, 617–654 (2015)
15. Mavronicolas, M., Monien, B.: The complexity of equilibria for risk-modeling valuations. Theor. Comput. Sci. **634**, 67–96 (2016)
16. Nash, J.F.: Equilibrium points in n-person games. Proc. Nat. Acad. Sci. U.S.A. **36**, 48–49 (1950)
17. Nash, J.F.: Non-cooperative games. Ann. Math. **54**, 286–295 (1951)
18. von Neumann, J.: Zur theorie der gesellschaftsspiele. Math. Ann. **100**, 295–320 (1928)
19. Ponstein, J.: Seven kinds of convexity. SIAM Rev. **9**, 115–119 (1967)
20. Rockafellar, R.T., Uryasev, S.: Optimization of conditional value-at-risk. J. Risk **2**, 21–42 (2000)
21. Rockafellar, R.T., Uryasev, S.: Conditional value-at-risk for general loss distributions. J. Bank. Finan. **26**, 1443–1471 (2002)
22. Rubinstein, A.: Lecture Notes in Microeconomic Theory. Princeton University Press, Princeton (2006)
23. Sarykalin, S., Serraino, G., Uryasev, S.: Value-at-risk vs. conditional value-at-risk in risk management and optimization. In: Tutorials in Operations Research, Chap. 13 (2008)
24. Sharpe, W.F.: A simplified model for portfolio analysis. Manag. Sci. **9**, 277–293 (1963)
25. Sharpe, W.F.: Mutual fund performance. J. Bus. **39**, 119–138 (1966)
26. Steinbach, M.C.: Markowitz revisited: mean-variance models in financial portfolios. SIAM Rev. **43**, 31–85 (2001)
27. Stoyanov, S.V., Rachev, S.T., Fabozzi, F.: Optimal financial portfolios. Appl. Math. Fin. **14**, 401–436 (2007)
28. Uryasev, S.: Conditional value-at-risk: optimization algorithms and applications. In: Financial Engineering News, no. 14, February 2000
29. Uryasev, S.: Optimization Using CV@R – Algorithms and Applications. Lecture Notes, Notes 7, Stochastic Optimization ESI 6912, University of Florida
30. Yang, X., Tao, S., Liu, R., Cai, M.: Complexity of scenario-based portfolio optimization problem with VaR objective. Int. J. Found. Comput. Sci. **13**, 671–679 (2002)

Congestion Games, Network and
Opinion Formation Games

Reconciling Selfish Routing with Social Good

Soumya Basu$^{(\boxtimes)}$, Ger Yang, Thanasis Lianeas, Evdokia Nikolova,
and Yitao Chen

The University of Texas at Austin, Austin, USA
basusoumya@utexas.edu

Abstract. Selfish routing is a central problem in algorithmic game theory, with one of the principal applications being that of routing in road networks. Inspired by the emergence of routing technologies and autonomous driving, we revisit selfish routing and consider three possible outcomes of it: (i) θ-Positive Nash Equilibrium flow, where every path that has non-zero flow on all of its edges has cost no greater than θ times the cost of any other path, (ii) θ-Used Nash Equilibrium flow, where every used path that appears in the path flow decomposition has cost no greater than θ times the cost of any other path, and (iii) θ-Envy Free flow, where every path that appears in the path flow decomposition has cost no greater than θ times the cost of any other path in the path flow decomposition. We first examine the relations of these outcomes among each other and then measure their possible impact on the network's performance. Right after, we examine the computational complexity of finding such flows of minimum social cost and give a range for θ for which this task is easy and a range of θ for which this task is NP-hard for the concepts of θ-Used Nash Equilibrium flow and θ-Envy Free flow. Finally, we propose strategies which, in a worst-case approach, can be used by a central planner in order to provide good θ-flows.

1 Introduction

Two Sides of the Coin: Social Welfare vs Selfishness. A fundamental problem arising in the management of road-traffic and communication networks is routing traffic to optimize network performance. In the setting of road-traffic networks the average delay incurred by a unit of flow quantifies the cost of a routing assignment. From a collective perspective minimizing the average cost translates to maximizing the welfare obtained by society. Starting from the seminal works of Wardrop [24] and Beckman et al. [2], the literature on network games has differentiated between (1) the objective of a central planner to minimize average cost and thus find a socially optimal (SO) flow, and (2) the selfish objectives of users minimizing their respective costs. In the latter case, the network users acting in their own interest are assumed to converge to a Nash Equilibrium (NE) flow as further rerouting fails to improve their own objective.

The tension between the central planner and individual users has been an object of intense study and solutions such as toll placement or Stackelberg routing (e.g., [2,15,19]) have been proposed in the past, each facing criticism in terms

© Springer International Publishing AG 2017
V. Bilò and M. Flammini (Eds.): SAGT 2017, LNCS 10504, pp. 147–159, 2017.
DOI: 10.1007/978-3-319-66700-3_12

of implementation and fairness towards various users. To mitigate this tension in a way that is more fair to the users, we set out to explore the properties of alternative solution concepts where users under some reasonable incentive condition adopt a more "socially desirable" routing of traffic in between the Nash equilibrium (which has high social cost) and the social optimum (which may be undesirable/unfair to users on the longer paths) [20]. The advent of routing applications and the growing dependence of users on these applications places us at an epoch when such new ideas in mechanism design may be more relevant and also more readily integrated to practice.

Consider the scenario where some routing application presents the uninformed users with routes alongside the guarantees of "relative fairness" and "reasonable delay" and the users adopt the paths. In this scenario, one may naturally bring forth the questions of whether there exist solutions (flows) where good social welfare is achieved under an appropriate incentive condition for the users and if such solutions can be efficiently computed. An example of such a solution could be enforcing a θ-approximate Nash equilibrium of low social cost, where users are guaranteed to get assigned a path of cost no greater than θ times the cost of the shortest path and as such, the solution is "relatively fair".[1] Yet, other solution concepts seem to arise naturally and are introduced below.

Selfishness and Envy. To achieve the coveted middle ground between the social optimum and Nash equilibrium, by combining good social welfare with satisfied users, we consider equilibria notions related to: (1) selfishness and (2) envy. First, we consider selfishness where users tend to selfishly improve their own cost. Here we make the distinction between *positive* paths, i.e. paths that have positive flow in all of their edges (note, this is independent of the path flow decomposition), and *used* paths, i.e. paths that appear in the path flow decomposition with positive flow. With these definitions we define a θ-Positive Nash Equilibrium (θ-PNE)[2] to be a flow in which the length of any *positive* path in the network is less than or equal to θ times the length of any other path, and a θ-Used Nash Equilibrium (θ-UNE) to be a flow in which the length of any *used* path in the network is less than or equal to θ times the length of any other path. As we shall see, the set of θ-PNE flows is in general a strict subset of the set of θ-UNE flows, though for $\theta = 1$ these sets coincide. The definition of θ-approximate Nash equilibrium [9] corresponds to that of θ-UNE.

Next, we consider the notion of envy where for the same source and destination a user experiences envy against another user if the latter incurs smaller delay compared to the former under a given path flow. Similarly to the approximate Nash equilibrium flow we can consider a notion of approximately envy free flows where in a θ-Envy Free (θ-EF) flow, the ratio of any two *used* paths in

---

[1] The concept of fairness in selfish routing has been considered in the past, with the two main approaches defining fairness as: (1) the ratio of the maximum path delay in a given flow to the average delay under Nash equilibrium [20] and (2) the ratio of the maximum path delay to the minimum path delay in a given flow [11].

[2] In the literature, PNE is typically used for abbreviating Pure Nash Equilibrium, but we always use it to denote Positive Nash Equilibrium, as defined here.

the network is upper bounded by θ, for some $\theta \geq 1$. Note, the difference from the θ-UNE definition is that a used path's cost is compared only to other used paths' costs. Envy free flows arise naturally once we consider the routing applications setup where users only collect information about the routes provided by the application. On the one hand, the possible costs for the current users in some sense compare to the costs of the users that have already used the network. On the other hand, routes for which there is no (sufficient) information, i.e., routes that have not been chosen in the past (sufficiently many times), potentially may never appear as an option. Another motivation for a θ-EF flow arises from the literature on imitation games, e.g. [14], where users imitate other users with lower delay and jointly reach a fixed point which is a 1-EF flow.

An example of how the concepts of θ-PNE, θ-UNE, and θ-EF may differ from each other is illustrated in Fig. 1, with the details discussed in Sect. 3, where these notions are formally introduced.

Related Work. Starting from the seminal work of Koutsoupias and Papadmitriou [18], quantifying the worst case inefficiency of various non-cooperative games, including routing games, quickly became an intense area of research. In a routing game with arbitrary latency functions the ratio between the cost of a Nash equilibrium (NE) flow to the cost of a socially optimal (SO) flow may grow unbounded, as shown by Roughgarden and Tardos [22]. A series of papers have focused on developing techniques for bounding the inefficiency of the NE flow (e.g., [12,16,22]).

Considering the generalization to approximate NE flows, Caragiannis et al. [5–7] provided existential and computational results regarding approximate equilibria in weighted and unweighted atomic congestion games. Feldmann et al. [13] also considered computational issues for approximate equilibria and applied the method of randomized rounding to analyze which approximation guarantees can be achieved for atomic congestion games with latency functions in specific classes. Chen and Sinclair [8], focusing again on atomic congestion games, studied questions related to convergence times to approximate Nash equilibria. Christodoulou et al. [9] studied the performance of approximate Nash equilibria for atomic and non-atomic congestion games with polynomial latency functions by considering how much the price of anarchy worsens and how much the price of stability improves as a function of the approximation factor θ.

In a related thread of research, Jahn et al. [17] formalized the notion of constrained system optimal, where additional constraints were added along with the flow feasibility constraints. The additional constraints were introduced to reduce the unfairness of the resulting flow. Further, useful insights were obtained by Schulz and Stier-Moses [23] about the social welfare and fairness of these constrained system optimal flows. Recently, there have been efforts [3,4] in quantifying the inefficiency needed to guarantee fairness among users. The authors there defined the 'price of fairness' as the proportional decrease of utility under fair resource allocation. As mentioned earlier, in routing games the fairness of socially optimal flows under different but related definitions has been studied by Roughgarden [20] and Correa et al. [10]. Further, Correa et al. [11] considered the fairness and efficiency of min-max flows, where the objective is to minimize the maximum length of any used path in the network and noted how different path flows affect the fairness in the network even when the induced edge flows are identical.

Contribution. When users ask their routing devices for good origin to destination paths, they care about the end-to-end delay on their paths without (directly) caring about local (subpath) optimality conditions. This highlights that the path flows may play a key role in achieving the full potential for route planning mechanisms. On the conceptual side, through the study of the proposed solution concepts, i.e., the θ-PNE, θ-UNE and θ-EF, we clearly differentiate path flows from edge flows and study their varied effects in the balance between fairness and social cost.

On the technical side, we start by observing that the 1-UNE and the 1-PNE are indeed identical, which explains why the 'used' and the 'positive' paths have not been explicitly differentiated before this work. Beyond the case of $\theta = 1$, we notice that θ-PNE, θ-UNE and θ-EF flows are progressively larger sets, each containing the previous one, with promise of better tradeoff between the social welfare and fairness. In order to grasp the large separation between these concepts note that for some networks the θ-UNE is not contained in $\Omega(n\theta)$-PNE, where n is the number of nodes in the network (Lemma 1).

Motivated from the classical study of the price of anarchy (PoA) of equilibrium flows we investigate the PoA of θ-UNE and θ-EF. In general we expect that as we move from θ-PNE to θ-EF flows, from a worst case perspective, we will encounter flows with larger social cost. As a worst case example we show that the PoA can be unbounded for 1-EF flows. However, the PoA upper bounds for both θ-PNE and θ-UNE turn out to be identical (Lemma 3). Our PoA bound generalizes the PoA bound of 1-PNE from [16]. Through a similar reasoning we show that the price of stability is non increasing from θ-PNE to θ-EF flows.

We next focus on computing a θ-PNE, a θ-UNE or a θ-EF flow with low social cost. The convex optimization approach for computing a socially optimal flow fails due to the non-convexity of the sets of θ-PNE, θ-UNE and θ-EF flows for $\theta > 1$ (Proposition 1). Formally, we prove (Theorem 1) that obtaining the best θ-UNE or the best θ-EF flow is NP-hard. Indeed given a socially optimal flow it is NP-hard to decide whether it admits a path flow decomposition which is θ-UNE (θ-EF) for arbitrarily large θ (assuming arbitrary latency functions). However, we leave open the complexity of finding the best θ-PNE flow ($\theta > 1$).

In the positive direction, we provide an approximation algorithm, based on a modified potential function, for designing a θ-PNE flow—which generates θ-EF and θ-UNE flows—with social cost guarantees. We explicitly derive the approximation ratio upper bound for solving minimization of social cost under the solution concepts for $\theta \geq 1$ for two classes of latency functions which are used in congestion networks, namely (1) polynomials with positive coefficients, (2) M/M/1 delay function (Theorem 3). This modified function approach was used by Christodoulou et al. [9] to derive upper bound for PoS(θ-UNE) with polynomial latency functions.

2 Preliminaries

Network and Flows. Consider a directed graph $G = (V, E)$ and a set of commodities $\mathcal{K}$ with $K = |\mathcal{K}|$. Each commodity $k \in \mathcal{K}$ is associated with a source s_k, a sink t_k, and a demand $d_k > 0$. Each edge $e \in E$ is given a *latency function* $\ell_e(x)$, assumed to be standard, i.e. nonnegative, differentiable, and nondecreasing. We consider the standard *nonatomic network congestion game*, where each user routes an infinitesimal amount of flow. Let $\mathcal{P}^k$ be the set of simple paths from s_k to t_k, and denote $\mathcal{P} = \cup_{k \in \mathcal{K}} \mathcal{P}^k$. A *feasible* flow can be represented as a *path flow vector* $\boldsymbol{f} = (f_\pi)_{\pi \in \mathcal{P}}$ that satisfies all demands, i.e. $\sum_{\pi \in \mathcal{P}^k} f_\pi = d_k$ for all $k \in \mathcal{K}$. The set of feasible path flow vectors is denoted by $\mathcal{D}_p$. A feasible path flow vector $\mathbf{f}$, induces a feasible *edge flow vector* in the network given as $\boldsymbol{x} = (x_e^k = \sum_{\pi \in \mathcal{P}^k : e \in \pi} f_\pi)_{e \in E, k \in \mathcal{K}}$. The congestion through edge e is the aggregate flow $x_e = \sum_{k \in \mathcal{K}} x_e^k$, for all $e \in E$. There may exist multiple feasible path flows, denoted as the set $\mathcal{D}_p(\boldsymbol{x})$, that give the same edge flow $\boldsymbol{x}$. We denote the set of feasible edge flows by $\mathcal{D}_E$.

Used and Positive Paths. Given a commodity $k \in \mathcal{K}$ and an edge flow vector $\boldsymbol{x}$, path $\pi \in \mathcal{P}^k$ is *positive* for commodity k if for all edges $e \in \pi$, $x_e^k > 0$. (Edge flow x_e is insufficient for this definition.) Given a commodity $k \in \mathcal{K}$ and a path flow vector $\boldsymbol{f}$, path $\pi \in \mathcal{P}^k$ is *used* by commodity k if $f_\pi > 0$ and *unused* otherwise. For each commodity $k \in \mathcal{K}$, we define the set of *positive* paths under edge flow $\boldsymbol{x}$ as $\mathcal{P}_+^k(\boldsymbol{x})$ and the set of *used* paths under path flow $\boldsymbol{f}$ as $\mathcal{P}_u^k(\boldsymbol{f})$. Note that a used path is always positive but a positive path may be unused.

Costs and Socially Optimal Flow. Under a path flow $\boldsymbol{f} \in \mathcal{D}_p$, the cost (latency) of a path π is defined to be the sum of latencies of edges along the path: $\ell_\pi(\boldsymbol{f}) = \ell_\pi(\boldsymbol{x}) = \sum_{e \in \pi} \ell_e(x_e)$ for $\boldsymbol{f} \in \mathcal{D}_p(\boldsymbol{x})$. The *social cost* (SC) of an edge flow $\boldsymbol{x} \in \mathcal{D}_E$ is $SC(\boldsymbol{x}) = \sum_{e \in E} x_e \ell_e(x_e)$. The social cost of a path flow is the social cost of its corresponding edge flow. The socially optimal edge/path flow is the flow that minimizes social cost among all feasible edge/path flows. The set of socially optimal edge/path flows is denoted by $SO_E = \{\boldsymbol{x} \in \arg\min SC(\boldsymbol{x})\}$ and $SO_p = \{\boldsymbol{f} \in \arg\min SC(\boldsymbol{f})\}$, respectively.

Nash Equilibrium and Efficiency. A path flow $\boldsymbol{f}$ is a *Nash Equilibrium* if for any commodity $k \in \mathcal{K}$ and any used path $p \in \mathcal{P}_u^k(\boldsymbol{f})$ we have $\ell_p(\boldsymbol{f}) \leq \ell_q(\boldsymbol{f})$, for all paths $q \in \mathcal{P}^k$. The efficiency of an equilibrium is often measured via the *price of anarchy* and the *price of stability*. Here we generalize them for arbitrary set of flows as in the following definitions. Given a set of path flows $\mathcal{F}$ and socially optimal edge flow $\boldsymbol{x}^* \in SO_E$, the price of anarchy (PoA) and the price of stability (PoS) are defined as

$$PoA(\mathcal{F}) = \max\left\{ \frac{SC(\boldsymbol{f})}{SC(\boldsymbol{x}^*)} : \boldsymbol{f} \in \mathcal{F} \right\}, \quad PoS(\mathcal{F}) = \min\left\{ \frac{SC(\boldsymbol{f})}{SC(\boldsymbol{x}^*)} : \boldsymbol{f} \in \mathcal{F} \right\}.$$

3 Solution Concepts

Here we give the formal definition of the solution concepts we introduced in Sect. 1. We also provide an example to illustrate their differences, and prove that each solution concept may correspond to a non-convex set of flows. We begin with the definitions of the solution concepts:

Definition 1 (θ-PNE, θ-UNE, and θ-EF).

1. *An edge flow $\boldsymbol{x}$ is a θ-Positive Nash Equilibrium (θ-PNE) flow if for any commodity $k \in \mathcal{K}$, any positive path $p \in \mathcal{P}_+^k(\boldsymbol{x})$, and all paths $q \in \mathcal{P}^k$, we have $\ell_p(\boldsymbol{x}) \leq \theta \ell_q(\boldsymbol{x})$.*
2. *A path flow $\boldsymbol{f}$ is a θ-Used Nash Equilibrium (θ-UNE) flow if for any commodity $k \in \mathcal{K}$, any used path $p \in \mathcal{P}_u^k(\boldsymbol{f})$, and all paths $q \in \mathcal{P}^k$, we have $\ell_p(\boldsymbol{f}) \leq \theta \ell_q(\boldsymbol{f})$.*
3. *A path flow $\boldsymbol{f}$ is θ-Envy Free (θ-EF) if for any commodity $k \in \mathcal{K}$, any used path $p \in \mathcal{P}_u^k(\boldsymbol{f})$, and all used paths $q \in \mathcal{P}_u^k(\boldsymbol{f})$, we have $\ell_p(\boldsymbol{f}) \leq \theta \ell_q(\boldsymbol{f})$.*

For simplicity, we use θ-PNE, θ-UNE or θ-EF to describe the set of θ-PNE, θ-UNE or θ-EF flows, respectively. Also, we refer to them as θ-fair flows. To see how these concepts may differ from each other, we give an example in Fig. 1.

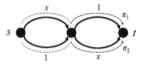

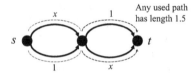

(a) Paths π_1 and π_2 have 1/2 unit of flow. This path flow assignment is a social optimum.

(b) The path flow assignment in Figure 1(a) is 1-EF but not 1-UNE.

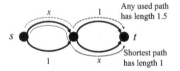

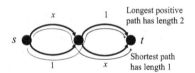

(c) The path flow assignment in Figure 1(a) is 1.5-UNE but not 1.5-PNE.

(d) The path flow assignment in Figure 1(a) is 2-PNE.

Fig. 1. Example illustrating the three solution concepts θ-UNE, θ-PNE and θ-EF.

Our goal is to examine the properties of θ-fair flows and provide ways to obtain such flows with good social cost. Regarding the second direction, in general, the sets of θ-PNE, θ-UNE, and θ-EF flows may not be convex and may contain multiple path flows, which raises the level of difficulty for computing good or optimal such flows. Next, we present an example that demonstrates the non-convexity of these sets (Fig. 2).

Proposition 1 (Non-convexity of θ flows). *There exists an instance such that the sets θ-PNE, θ-UNE, and θ-EF are not convex, for some $\theta > 1$.*

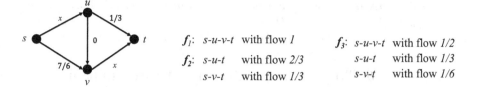

Fig. 2. Non-convexity of θ-flows. Both f_1 and f_2 are 3/2-PNE/UNE/EF, but their convex combination (with even weights) f_3 is not.

The following two lemmas establish the hierarchy among the proposed solution concepts by showing a crisp containment of various flows. Due to space constraints, the proofs are presented in the full version [1], Sect. 4.

Lemma 1 (Hierarchy of θ-flows). *Given a multi-commodity network, for $\theta' > \theta \geq 1$, $\mathcal{F} \in \{PNE, UNE, EF\}$ satisfies θ-$\mathcal{F} \subseteq \theta'$-$\mathcal{F}$. Further, for any $\theta \geq 1$, θ-PNE $\subseteq \theta$-UNE $\subseteq \theta$-EF holds. On the other hand, for any $\theta \geq 1$, there exists a network such that 1-EF $\not\subseteq \theta$-UNE.*

In the following lemma, we further demonstrate the relationship between the θ-UNE and θ-PNE. We can see that 1-UNE and 1-PNE both coincide with the familiar Nash equilibrium.

Lemma 2 (θ-UNE and θ-PNE). *Given a multi-commodity network with n nodes, for any path flow f and its induced edge flow x, $f \in 1$-UNE if and only if $x \in 1$-PNE. Further, for any $\theta > 1$, θ-UNE $\subset ((n-1)\theta)$-PNE holds. On the other hand, for any $\theta \geq 1.5$, there exists a network such that θ-UNE $\not\subseteq ((n-3)\theta/3)$-PNE.*

We next analyze the cost of the θ-flows. Note that the θ flows are not unique for $\theta > 1$ and this implies that potentially under each solution concept we can have a range of attainable costs. From the containment relations of the θ flows (Lemma 1, Part 2), it follows that for any $\theta \geq 1$,

$$\text{PoA}(\theta\text{-EF}) \geq \text{PoA}(\theta\text{-UNE}) \geq \text{PoA}(\theta\text{-PNE}),$$
$$\text{PoS}(\theta\text{-EF}) \leq \text{PoS}(\theta\text{-UNE}) \leq \text{PoS}(\theta\text{-PNE}).$$

Further, we present upper bounds on the PoA for θ-UNE and θ-PNE flows, and show that the PoA for 1-EF flow is unbounded. In that effort, we generalize techniques presented in [16] which was used for bounding PoA(1-PNE). We need the following definitions in order to bound PoA:

$$\omega(\mathcal{L}, \lambda) = \sup_{\ell \in \mathcal{L}} \sup_{x, x' \geq 0} \frac{(\ell(x) - \lambda \ell(x')) x'}{x\ell(x)}, \quad \Lambda(\theta) = \{\lambda \in \mathbb{R}^+ : \omega(\mathcal{L}, \lambda) \leq 1/\theta\}.$$

The following lemma summarizes the PoA results for the solution hierarchies. For the proof refer to the full version [1], Sect. 5.

Lemma 3 (PoA of θ-flows). *For latency functions in class $\mathcal{L}$, $PoA(\theta\text{-}UNE)$ $\leq \inf_{\lambda \in \Lambda(\theta)} \theta\lambda(1 - \theta\omega(\mathcal{L}, \lambda))^{-1}$. On the other hand, there exists a network with linear latency functions for which the $PoA(1\text{-}EF)$ is unbounded.*

4   Optimal θ-Flows: Complexity and Approximation

In this section, we first discuss the possibility of designing a flow that balances the fairness and the social cost in the network under the new solution concepts. The standard convex optimization approaches fail to find socially optimal θ flow as the sets of θ-PNE, θ-UNE and θ-EF flows are all non-convex. We formally prove that finding socially optimal θ-UNE or θ-EF flows is NP hard. Then, using a modified potential function technique we provide approximation guarantees for two common classes of latency functions used in congestion network modeling, namely (1) polynomial and (2) M/M/1.

Consider the instance in Fig. 3. The next proposition states that though the socially optimal flow—uniquely determined by the edge flow—is unfair in the worst case, there exist path flows which are fair or almost fair under the concepts of UNE and EF flows. The proof is in the full version [1], Sect. 7.1.

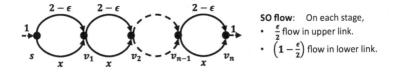

SO flow: On each stage,
- $\frac{\epsilon}{2}$ flow in upper link.
- $\left(1 - \frac{\epsilon}{2}\right)$ flow in lower link.

Fig. 3. Improved balance: example.

Proposition 2 (Balanced path flows). *For the n-stage instance depicted in Fig. 3 with $n\epsilon = 2$, the socially optimal edge flow is a 2-PNE. Moreover, the socially optimal flow admits path flows which are 1-EF or $(1 + 1/n)$-UNE.*

4.1   Existence and Complexity

The previous motivating example naturally leads to the following computational problems given a $\theta \geq 1$.

- **(P1)** Find a θ-EF path flow with the minimal social cost.
- **(P2)** Find a θ-UNE path flow with the minimal social cost.
- **(P3)** Find a θ-PNE edge flow with the minimal social cost.

Existence of Polynomial-Size Solutions. An observation to Problem (P1) and (P2) is that the outputs of these two problems are path flow vectors, which are potentially of exponential size relative to the problem instances. The following lemma proves the existence of polynomial sized path flows, in absence of which there is no hope to find a polynomial time algorithm for these problems.

Lemma 4 (Existence of polynomial-size solutions). *Given a θ-EF (or a θ-UNE) path flow $\boldsymbol{f}$, there exists a θ-EF (resp., a θ-UNE) path flow $\boldsymbol{f}'$ that uses at most $|E|$ paths for each source-sink pair and has the same edge flow as $\boldsymbol{f}$.*

Computational Complexity. We show that for large θ, the socially optimal flow is guaranteed to be contained in those θ-flows, and hence the optimal θ-flows can be computed efficiently. However, for small θ, we will show that solving Problem (P1) and Problem (P2) is NP-hard, while it remains open whether Problem (P3) can be computed efficiently. More precisely, for a latency class $\mathcal{L}$, this particular threshold is $\gamma(\mathcal{L}) = \min\{\gamma : \ell^*(x) \le \gamma\ell(x), \forall \ell \in \mathcal{L}, \forall x \ge 0\}$, where $\ell^*(x) = \ell(x) + x\ell'(x)$. The main result of this section is:

Theorem 1 (Computational Complexity of (P1)–(P3)). *For any multi commodity instance with latency functions in any class $\mathcal{L}$, there is a polynomial time algorithm for solving Problem (P1)–(P3) for $\theta \ge \gamma(\mathcal{L})$. On the other hand, it is NP-hard to solve Problem (P1) for $\theta \in [1, \gamma(\mathcal{L}))$ and Problem (P2) for $\theta \in (1, \gamma(\mathcal{L}))$, for arbitrary single commodity instances with latency functions in an arbitrary class $\mathcal{L}$.*

The first part of Theorem 1 follows easily from the following lemma in [11].

Lemma 5 (P1)–(P3) for $\theta \ge \gamma(\mathcal{L})$ are easy [11]). *For an instance with latency functions in $\mathcal{L}$, any socially optimal path flow $\boldsymbol{f} \in SO_p$ is $\gamma(\mathcal{L})$-PNE.*

For the proof of the second part of Theorem 1, we consider the class of polynomial functions of degree at most p, denoted by $\mathcal{L}_p$. We note that $\gamma(\mathcal{L}_p) = p+1$. We show that when the latency functions are in $\mathcal{L}_p$, then the related decision problems we state in Theorem 2, stated below, have polynomial-time reductions from the NP-complete problem PARTITION.

Theorem 2 (NP-hardness of (P1) and (P2)). *For an arbitrary single commodity instance with latency functions in class $\mathcal{L}_p$ for $p \ge 1$, it is NP-hard to*

1. *decide whether a socially optimal flow has a θ-UNE path flow decomposition for $\theta \in (1, p+1)$.*
2. *decide whether a socially optimal flow has a θ'-EF path flow decomposition for $\theta' \in [1, p+1)$.*

Proof. (Proof Sketch) For lack of space we present a proof sketch mentioning the flow of key ideas behind the proof. The proof is divided into two parts. For the first part, we show the NP-hardness for 1.5-UNE and 1-EF path flow decompositions under the social optimum in Lemma 6, based on the construction in Theorem 3.3 in Correa et al. [11]. Then, in the second part, we propose a novel way to generalize the construction to the entire range of θ and θ' specified in Theorem 2.

Construction. Let $G(q)$ be the two-link parallel network with the top link e_u having latency $\ell_u(x) = q$ and the bottom link e_b having latency $\ell_b(x) = qx$. Given an instance of the PARTITION problem, $q_1, \ldots, q_n$, $\sum_{i=1}^{n} q_i = 2B$, we now construct a single commodity network as the two-link n-stage network G, shown in Fig. 4. In stage i we connect $G(q_{i-1})$ to $G(q_i)$ to the right for $i = 2$ to n. A unit demand has to be routed from the source in $G(q_1)$ to the destination in $G(q_n)$. Finally, we augment to the right of G a two-link parallel network G'. For G' the top link latency is $\ell_{u,(n+1)}(x) = ax^p + b$ and bottom link latency is $\ell_{d,(n+1)}(x) = cx^p$. We set $a = \frac{\alpha B}{(1-3/8B)^p}$, $b = \beta B(p+1)$, and $c = \frac{(\alpha+\beta)B}{(3/8B)^p}$, where $\alpha, \beta > 0$ are appropriate parameters (specified in the proof of Theorem 1 in the full version [1]). We call the entire network H.

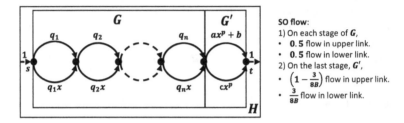

Fig. 4. An instance of congestion game constructed from a given instance of PARTI-TION.

Lemma 6 states that it is NP-hard to find 1.5-UNE and 1-EF path flow decompositions under the social optimum.

Lemma 6 (Hardness Result in G). *For single commodity instances with linear latency functions it is NP-hard to decide whether a social optimum flow has a 1.5-UNE flow decomposition or a 1-EF flow decomposition.*

Amplification of Hardness. The next step is to amplify the hardness result to all $\theta \in (1, p+1)$ for UNE and to all $\theta' \in [1, p+1)$ for EF flows. The key observation that facilitates this amplification is the following claim.

Claim 1. If the answer to PARTITION is NO then in the sub-network G any path decomposition of the socially optimal flow o routes at least $\frac{1}{2B}$ amount of flow through paths of length strictly greater than $\frac{3}{2}B$.

Through careful combination of the paths it is shown that the socially optimal flow is a c_1-UNE and c_2-EF flow if and only if the given PARTITION instance has a YES answer. Here c_1 and c_2 are constants given by

$$c_1 = \frac{\alpha + \beta + \beta p + \frac{3}{2}}{1 + \alpha + \beta} = 1 + \frac{\frac{1}{2} + \beta p}{1 + \alpha + \beta}, \quad c_2 = \frac{\alpha + \beta + \beta p + \frac{3}{2}}{2 + \alpha + \beta} = 1 + \frac{-\frac{1}{2} + \beta p}{2 + \alpha + \beta}.$$

4.2   Approximation Using Modified Potential Functions

In this section, we provide a modified potential function based approximation algorithm to problems (P1), (P2) and (P3). The idea of modified potential functions was introduced for bounding the PoS of approximate Nash equilibria in [9]. For a given θ this approach produces a θ-PNE and, due to containment, any feasible path flow corresponding to the edge flow will be in θ-UNE and θ-EF.

Algorithm 1. Modified Potential Algorithm

Input: Multi-commodity network $\mathcal{G}$, θ.
Output: Edge flow $\mathbf{x}_A \in \theta$-PNE and path flow $\mathbf{f}_A \in \theta$-UNE $\cap \theta$-EF .
 1: For all $e \in E$ choose $\phi_e(x) \in [\ell_e(s)/\theta, \ell_e(x)]$ (Specified later.)
 2: Compute $\mathbf{x}_A = \operatorname{argmin}_{\mathbf{x} \in \mathcal{F}} \sum_{e \in E} \int_{x=0}^{x_e} \phi_e(x)dx$.
 3: Compute any path decomposition $\mathbf{f}_A$ of $\mathbf{x}_A$.

Theorem 3 characterizes the performance of Algorithm 1 for two important classes of latency functions which are used for modeling congestion networks—(1) Polynomial latency with positive coefficients, and (2) M/M/1 latency.

Theorem 3 (Performance of Algorithm 1). *Given a multi-commodity network $\mathcal{G}$ and $\theta \geq 1$ the algorithm produces an edge flow that is θ-PNE and a path flow that is both θ-UNE and θ-EF.*

*(1) **Polynomial Latency:** Additionally, let the latency function $\ell_e(x) = \sum_{k=0}^{p} a_{e,k} x^k$, $a_{e,k} \geq 0$, for all $e \in E$ and some finite p. Algorithm 1 with $\phi_e(x) = \sum_{k=0}^{p} \zeta_k a_{e,k} x^k$, $\zeta_k = (1 + \min\{k, \theta - 1\})/\theta$ for all $e \in E, k \leq p$, is a $\left(\theta \left(1 - \frac{p}{1+p} \left(\frac{\theta}{1+p} \right)^{1/p} \right) \right)^{-1}$-approximation algorithm for the problems (P1), (P2) and (P3).*

*(2) **M/M/1 Latency:** Additionally, let the latency function $\ell_e(x) = 1/(u_e - x)$ with $u_e > 0, \rho_e = d_{tot}/u_e$ and $\rho_{\max} = \max_{e \in E} \rho_e < 1$, for all $e \in E$. Algorithm 1 with $\phi_e(x) = 1/(a_e u_e - x)$, $a_e = \{0, 1 - \theta(1 - \rho)\}/\rho$ for all $e \in E$, is a $\frac{1}{2} \left(1 + \frac{1}{\sqrt{1 - \rho_{\max}(\theta)}} \right)$-approximation algorithm for the problems (P1), (P2) and (P3), where $\rho_{\max}(\theta) = \max\{0, 1 - \theta(1 - \rho_{\max})\}$.*

The first part of Theorem 3 (i.e., the output being θ flows) follows from the idea, that a 1-PNE flow under the modified potential is a θ-PNE flow under the original latency functions. To prove the second part, we bound the inefficiency of the flow $\mathbf{x}_A$ as the PoA(1-PNE) under the modified potential functions, which in turn serves as an approximation ratio for the minimization of social cost under θ-PNE (P3), θ-UNE (P2) and θ-EF (P1). Using proper functions $\phi_e(\cdot)$ along with the λ-μ smoothness framework [21] we strictly improve the approximation ratio from PoA(1-PNE). Recall the trivial solution—1-PNE flow which can be computed efficiently gives an approximation ratio of PoA(1-PNE). The choice of $\phi_e(\cdot)$ and the subsequent bounds for polynomial latencies were presented in [9], but for upper bounding PoS(θ-UNE). The detailed proofs are presented in the full version [1], Sect. 7.2.

Acknowledgements. This work was supported in part by NSF grant numbers CCF-1216103, CCF-1350823 and CCF-1331863. Part of the research was performed while a subset of the authors were at the Simons Institute in Berkeley, CA in Fall 2015.

References

1. Basu, S., Yang, G., Lianeas, T., Nikolova, E., Chen, Y.: Reconciling selfish routing with social good. arXiv:1707.00208 (2017)
2. Beckmann, M., McGuire, C., Winsten, C.B.: Studies in the economics of transportation. Technical report (1956)
3. Bertsimas, D., Farias, V.F., Trichakis, N.: The price of fairness. Oper. Res. **59**(1), 17–31 (2011)
4. Bertsimas, D., Farias, V.F., Trichakis, N.: On the efficiency-fairness trade-off. Manag. Sci. **58**(12), 2234–2250 (2012)
5. Caragiannis, I., Fanelli, A., Gravin, N., Skopalik, A.: Computing approximate pure nash equilibria in congestion games. ACM SIGecom Exch. **11**(1), 26–29 (2012)
6. Caragiannis, I., Fanelli, A., Gravin, N., Skopalik, A.: Approximate pure nash equilibria in weighted congestion games: existence, efficient computation, and structure. ACM Trans. Econ. Comput. **3**(1), 2 (2015)
7. Caragiannis, I., Flammini, M., Kaklamanis, C., Kanellopoulos, P., Moscardelli, L.: Tight bounds for selfish and greedy load balancing. In: Bugliesi, M., Preneel, B., Sassone, V., Wegener, I. (eds.) ICALP 2006. LNCS, vol. 4051, pp. 311–322. Springer, Heidelberg (2006). doi:10.1007/11786986_28
8. Chien, S., Sinclair, A.: Convergence to approximate nash equilibria in congestion games. In: SODA (2007)
9. Christodoulou, G., Koutsoupias, E., Spirakis, P.G.: On the performance of approximate equilibria in congestion games. Algorithmica **61**(1), 116–140 (2011)
10. Correa, J.R., Schulz, A.S., Stier Moses, N.E.: Computational complexity, fairness, and the price of anarchy of the maximum latency problem. In: Bienstock, D., Nemhauser, G. (eds.) IPCO 2004. LNCS, vol. 3064, pp. 59–73. Springer, Heidelberg (2004). doi:10.1007/978-3-540-25960-2_5
11. Correa, J.R., Schulz, A.S., Stier-Moses, N.E.: Fast, fair, and efficient flows in networks. Oper. Res. **55**(2), 215–225 (2007)
12. Correa, J.R., Schulz, A.S., Stier-Moses, N.E.: A geometric approach to the price of anarchy in nonatomic congestion games. Games Econ. Behav. **64**(2), 457–469 (2008)
13. Feldmann, A.E., Röglin, H., Vöcking, B.: Computing approximate nash equilibria in network congestion games. In: Shvartsman, A.A., Felber, P. (eds.) SIROCCO 2008. LNCS, vol. 5058, pp. 209–220. Springer, Heidelberg (2008). doi:10.1007/978-3-540-69355-0_18
14. Fischer, S., Vöcking, B.: On the evolution of selfish routing. In: Albers, S., Radzik, T. (eds.) ESA 2004. LNCS, vol. 3221, pp. 323–334. Springer, Heidelberg (2004). doi:10.1007/978-3-540-30140-0_30
15. Fleischer, L., Jain, K., Mahdian, M.: Tolls for heterogeneous selfish users in multicommodity networks and generalized congestion games. In: FOCS (2004)

16. Harks, T.: On the price of anarchy of network games with nonatomic and atomic players. Technical report, available at Optimization Online (2007)
17. Jahn, O., Möhring, R.H., Schulz, A.S., Stier-Moses, N.E.: System-optimal routing of traffic flows with user constraints in networks with congestion. Oper. Res. **53**(4), 600–616 (2005)
18. Koutsoupias, E., Papadimitriou, C.: Worst-case equilibria. In: Meinel, C., Tison, S. (eds.) STACS 1999. LNCS, vol. 1563, pp. 404–413. Springer, Heidelberg (1999). doi:10.1007/3-540-49116-3_38
19. Roughgarden, T.: Stackelberg scheduling strategies. In: STOC (2001)
20. Roughgarden, T.: How unfair is optimal routing? In: SODA (2002)
21. Roughgarden, T.: Intrinsic robustness of the price of anarchy. J. ACM (JACM) **62**(5), 32 (2015)
22. Roughgarden, T., Tardos, É.: How bad is selfish routing? J. ACM (JACM) **49**(2), 236–259 (2002)
23. Schulz, A.S., Stier-Moses, N.E.: Efficiency and fairness of system-optimal routing with user constraints. Networks **48**(4), 223–234 (2006)
24. Wardrop, J.G.: Some Theoretical Aspects of Road Traffic Research (1952)

Selfish Network Creation with Non-uniform Edge Cost

Ankit Chauhan, Pascal Lenzner[✉], Anna Melnichenko, and Louise Molitor

Algorithm Engineering Group, Hasso Plattner Institute, Potsdam, Germany
{ankit.chauhan,pascal.lenzner,anna.melnichenko,louise.molitor}@hpi.de

Abstract. Network creation games investigate complex networks from a game-theoretic point of view. Based on the original model by Fabrikant et al. [PODC'03] many variants have been introduced. However, almost all versions have the drawback that edges are treated uniformly, i.e. every edge has the same cost and that this common parameter heavily influences the outcomes and the analysis of these games.

We propose and analyze simple and natural parameter-free network creation games with non-uniform edge cost. Our models are inspired by social networks where the cost of forming a link is proportional to the popularity of the targeted node. Besides results on the complexity of computing a best response and on various properties of the sequential versions, we show that the most general version of our model has constant Price of Anarchy. To the best of our knowledge, this is the first proof of a constant Price of Anarchy for any network creation game.

1 Introduction

Complex networks from the Internet to various (online) social networks have a huge impact on our lives and it is thus an important research challenge to understand these networks and the forces that shape them. The emergence of the Internet was one of the driving forces behind the rise of Algorithmic Game Theory [26] and it has also kindled the interdisciplinary field of Network Science [6], which is devoted to analyzing and understanding real-world networks. Game-theoretic models for network creation lie in the intersection of both research directions and yield interesting insights into the structure and evolution of complex networks. In these models, agents are associated to nodes of a network and choose their neighbors selfishly to minimize their cost. Many such models have been proposed, most prominently the models of Jackson and Wolinsky [16], Bala and Goyal [5] and Fabrikant et al. [14], but almost all of them treat edges equally, that is, they assume a fixed price for establishing any edge which is considered as a parameter of these games. This yields very simple models but has severe influence on the obtained equilibria and their properties.

We take a radical departure from this assumption by proposing and analyzing a variant of the Network Creation Game [14] in which the edges have non-uniform cost which solely depends on the structure of the network. In particular, the cost

© Springer International Publishing AG 2017
V. Bilò and M. Flammini (Eds.): SAGT 2017, LNCS 10504, pp. 160–172, 2017.
DOI: 10.1007/978-3-319-66700-3_13

of an edge between agent u and v which is bought by agent u is proportional to v's degree in the network, i.e. edge costs are proportional to the degree of the other endpoint involved in the edge. Thus, we introduce individual prices for edges and at the same time we obtain a simple model which is parameter-free.

Our model is inspired by social networks in which the nodes usually have very different levels of popularity which is proportional to their degree. In such networks connecting to a celebrity usually is expensive. Hence, we assume that establishing a link to a popular high degree node has higher cost than connecting to an unimportant low degree node. Moreover, in social networks links are formed mostly locally, e.g. between agents with a common neighbor, and it rarely happens that links are removed, on the contrary, such networks tend to get denser over time [20]. This motivates two other extensions of our model which consider locality and edge additions only.

1.1 Model and Notation

Throughout the paper we will consider unweighted undirected networks $G = (V, E)$, where V is the set of nodes and E is the set of edges of G. Since edges are unweighted, the distance $d_G(u, v)$ between two nodes u, v in G is the number of edges on a shortest path between u and v. For a given node u in a network G let $N_k(u)$ be the set of nodes which are at distance at most k from node u in G and let $B_k(u)$ be the set of nodes which are at exactly distance k from node u (the distance-k ball around u). We denote the diameter of a network G by $D(G)$, the degree of node u in G, which is the number of edges incident to u, by $deg_G(u)$. We will omit the reference to G whenever it is clear from the context.

We investigate a natural variant of the well-known Network Creation Game (NCG) by Fabrikant et al. [14] which we call the *degree price network creation game (degNCG)*. In a NCG the selfish agents correspond to nodes in a network and the strategies of all agents determine which edges are present. In particular, the strategy S_u of an agent u is any subset of V, where $v \in S_u$ corresponds to agent u owning the undirected edge $\{u, v\}$. For $v \in S_u$ we will say that agent u buys the edge $\{u, v\}$. Any strategy vector $\mathbf{s}$ which specifies a strategy for each agent then induces the network $G(\mathbf{s})$, where

$$G(\mathbf{s}) = \left(V, \bigcup_{u \in V} \bigcup_{v \in S_u} \{u, v\} \right).$$

Here we assume that $G(\mathbf{s})$ does not contain multi-edges, which implies that every edge has exactly one owner. Since edge-ownership is costly (see below) this assumption trivially holds in any equilibrium network. Moreover, any network G together with an ownership function, which assigns a unique owner for every edge, determines the corresponding strategy vector. Hence, we use strategy vectors and networks interchangeably and we assume that the owner of every edge is known. In our illustrations we indicate edge ownership by directing the edges away from their owner. We will draw undirected edges if the ownership does not matter.

The cost function of agent u in network $G(\mathbf{s})$ consists of the sum of edge costs for all edges owned by agent u and the distance cost, which is defined as the sum of the distances to all other nodes in the network if it is connected and ∞ otherwise. The main novel feature which distinguishes the degNCG from the NCG is that each edge has an individual price which is proportional to the degree of the endpoint which is not the owner. That is, if agent u buys the edge $\{u, v\}$ then u's cost for this edge is proportional to node v's degree. For simplicity we will mostly consider the case where the price of edge $\{u, v\}$ for agent u is exactly v's degree without counting edge $\{u, v\}$. Thus the cost of agent u in network $G(\mathbf{s})$ is

$$cost_u(G(\mathbf{s})) = \sum_{v \in S_u} (\deg_{G(\mathbf{s})}(v) - 1) + dist_{G(\mathbf{s})}(u).$$

Note that in contrast to the NCG our variant of the model does not depend on any parameter and the rather unrealistic assumption that all edges have the same price is replaced with the assumption that buying edges to well-connected nodes is more expensive than connecting to low degree nodes.

Given any network $G(\mathbf{s})$, where agent u has chosen strategy S_u. We say that S_u is a best response strategy of agent u, if S_u minimizes agent u's cost, given that the strategies of all other agents are fixed. We say that a network $G(\mathbf{s})$ is in pure Nash equilibrium (NE), if no agent can strictly decrease her cost by unilaterally replacing her current strategy with some other strategy. That is, a network $G(\mathbf{s})$ is in NE if all agents play a best response strategy.

Observe that in the degNCG we assume that agents can buy edges to every node in the network. Especially in modeling large social networks, this assumption seems unrealistic. To address this, we also consider a restricted version of the model which includes locality, i.e. where only edges to nodes in distance at most k, for some fixed $k \geq 2$, may be bought. We call this version the *k-local degNCG* (degkNCG) and its pure Nash equilibria are called k-local NE (kNE). We will mostly consider the case strongest version where $k = 2$.

We measure the quality of a network $G(\mathbf{s})$ by its *social cost*, which is simply the sum over all agents' costs, i.e. $cost(G(\mathbf{s})) = \sum_{u \in V} cost_u(G(\mathbf{s}))$. Let $worst_n$ and $best_n$ denote the social cost of a (k)NE network on n nodes which has the highest and lowest social cost, respectively. Moreover, let opt_n be the minimum social cost of any network on n nodes. We measure the deterioration due to selfishness by the Price of Anarchy (PoA) which is the maximum over all n of the ratio $\frac{worst_n}{opt_n}$. Moreover, the more optimistic Price of Stability (PoS) is the maximum over all n of the ratio $\frac{best_n}{opt_n}$.

The use case of modeling social networks indicates another interesting version of the degNCG, which we call the *degree price add-only game (degAOG)* and its k-local version degkAOG. In these games, agents can only add edges to the network whereas removing edges is impossible. This mirrors social networks where an edge means that both agents know each other.

1.2   Related Work

Our model is a variant of the well-known Network Creation Game (NCG) proposed by Fabrikant et al. [14]. The main difference to our model is that in [14] it is assumed that every edge has price $\alpha > 0$, where α is some fixed parameter of the game. This parameter heavily influences the game, e.g. the structure of the equilibrium networks changes from a clique for very low α to trees for high α. For different regimes of α different proof techniques yield a constant PoA [1,13–15,21,23] but it is still open whether the PoA is constant for all α. In particular, constant upper bounds on the PoA are known for $\alpha < n^{1-\varepsilon}$, for any fixed $\varepsilon > \frac{1}{\log n}$ [13], and if $\alpha > 65n$ [21]. The best general upper bound is $2^{\mathcal{O}\sqrt{\log n}}$ [13]. The dynamics of the NCG have been studied in [17] where it was shown that there cannot exist a generalized ordinal potential function for the NCG. Also the complexity of computing a best response has been studied and its NP-hardness was shown in [14]. If agents resort to simple strategy changes then computing a best response can trivially be done in polynomial time and the obtained equilibria approximate Nash equilibria well [19].

Removing the parameter α by restricting the agents to edge swaps was proposed and analyzed in [2,24]. The obtained results are similar, e.g. the best known upper bound on the PoA is $2^{\mathcal{O}\sqrt{\log n}}$, there cannot exist a potential function [18] and computing a best response is NP-hard. However, allowing only swaps leads to the unnatural effects that the number of edges cannot change and that the sequential version heavily depends on the initial network.

Several versions for augmenting the NCG with locality have been proposed and analyzed recently. It was shown that the PoA may deteriorate heavily if agents only know their local neighborhood or only a shortest path tree of the network [7,8]. In contrast, a global view with a restriction to only local edge-purchases yields only a moderate increase of the PoA [11].

The idea of having nodes with different popularity was also discussed in the so called celebrity games [3,4]. There, nodes have a given popularity and agents buy fixed-price edges to get highly popular nodes within some given distance bound. Hence, this model differs heavily from our model.

To the best of our knowledge, there are only two related papers which analyze a variant of the NCG with non-uniform edge price. In [12] agents can buy edges of different quality which corresponds to their length and the edge price depends on the edge quality. Distances are then measured in the induced weighted network. Closer to our model is [22] where heterogeneous agents, important and unimportant ones, are considered and both classes of agents have different edge costs. Here, links are formed with bilateral agreement [10,16] and important nodes have a higher weight, which increases their attractiveness.

1.3   Our Contribution

We introduce and analyze the first parameter-free variants of Network Creation Games [14] which incorporate non-uniform edge cost. In almost all known versions the outcomes of the games and their analysis heavily depend on the edge cost parameter α. We depart from this by assuming that the cost of an edge solely depends on structural properties of the network, in particular, on the degree of the endpoint to which the edge is bought. Essentially, our models incorporate that the cost of an edge is proportional to the popularity of the node to which it connects. This appears to be a realistic feature, e.g. for modeling social networks.

On the first glance, introducing non-uniform edge cost seems to be detrimental to the analysis of the model. However, in contrast to this, we give a simple proof that the PoA of the degNCG is actually constant. To the best of our knowledge, our model is the first version of the NCG for which this strong statement could be established. A constant PoA is widely conjectured for the original NCG [14] but its proof is despite serious efforts of the community still an open question. Besides this strongest possible bound on the PoA, which we also generalize to arbitrary linear functions of a node's degree and to the 4-local version, we prove a PoA upper bound of $\mathcal{O}(\sqrt{n})$ for the deg2NCG, where agents are restricted to act within their 2-neighborhood and we show for this version that computing a best response strategy is NP-hard. Moreover, we investigate the dynamic properties of the deg(2)NCG and prove that improving response dynamics may not converge to an equilibrium, that is, there cannot exist a generalized ordinal potential function.

We contrast these negative convergence results by analyzing a version where agents can only add edges, i.e. the deg(2)AOG, where convergence of the sequential version is trivially guaranteed, and by analyzing the speed of convergence for different agent activation schemes. The restriction to only edge additions has severe impact on the PoA, yielding a $\Theta(n)$ bound, but we show that the impact on the social cost is low, if round-robin dynamics starting from a path are considered, where agents buy their best possible single edge in each step.

Due to space constraints, all omitted details can be found in [9].

2   Hardness

In this section we investigate the computational hardness of computing a cost minimizing strategy, i.e. a best response, in the deg2NCG and in the deg2AOG.

Theorem 1. *Computing the best response in the deg2NCG and the deg2AOG is* NP-*hard.*

3   Analysis of Equilibria

We start with the most fundamental statement about equilibria which is their existence. We use the center sponsored spanning star S_n, see Fig. 1(a), for the proof and provide some other examples of NE and 2NE networks in Fig. 1.

Theorem 2. *The star S_n is a (k)NE for the deg(k)NCG and the deg(k)AOG for any k.*

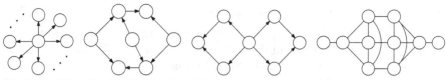

(a) The NE network S_n

(b) A NE in the degAOG/degNCG with $D = 3$.

(c) A 2NE in the deg2NCG with $D = 4$.

(d) A 2NE in the deg2AOG with $D = 5$.

Fig. 1. Examples of NE and 2NE networks

3.1 Bounding the Diameter of Equilibrium Networks

We investigate the diameter of (2)NE networks. Analogously to the original NCG [14], bounding the diameter plays an important role in bounding the PoA.

Theorem 3. *Consider variants of the degAOG and the degNCG where the price of any edge $\{u, v\}$ bought by agent u is any linear function of v's degree in G, that is, $price_u(\{u, v\}) = \beta \cdot deg_{G(s)}(v) + \gamma$, where $\beta, \gamma \in \mathbb{R}$. Then the diameter of any NE network in the degAOG and the degNCG is constant.*

Proof. We consider a NE network $G = (V, E)$ and assume that the diameter D of G is at least 4. Then there exist nodes $a, b \in V$, such that $d_G(a, b) = D$. Therefore, the distance cost of a agent a in G is at least $D + |B_1(b)|(D - 1) + |N_2(a)|$. Thus, if agent a buys the edge $\{a, b\}$ then this improves agent a's distance cost by at least $D - 1 + |B_1(b)|(D - 3)$. Since the network G is in NE, the distance cost improvement must be less than agent u's cost for buying the edge $\{a, b\}$:

$$D - 1 + |B_1(b)|(D - 3) \leq \beta \cdot deg_G(b) + \gamma$$
$$\iff D - 1 + (D - 3) \cdot deg_G(b) \leq \beta \cdot deg_G(b) + \gamma.$$

Solving for D under the assumption $deg_G(b) \geq 1$ yields

$$D \leq \frac{(\beta + 3)deg_G(b) + \gamma + 1}{deg_G(b) + 1} < \beta + 3 + \frac{\gamma + 1}{deg_G(b) + 1} \in \mathcal{O}(1).$$

□

Using $\beta = 1$ and $\gamma = -1$ yields the edge price for our version of the degNCG and the degAOG. This, and the NE example in Fig. 1(b) yields the following:

Corollary 1. *The diameter of any NE network in the degAOG and the degNCG is at most 3 and this upper bound is tight.*

Since in the proof of Theorem 3 in the case of $\beta = 1$ and $\gamma = -1$ buying an edge to a node in distance 4 suffices, we get the following statement.

Corollary 2. *Any 4NE network has diameter at most 3.*

Note that the examples in Fig. 1(c) and (d) show that the diameter in the 2-local version, i.e. in the deg2NCG and the deg2AOG, can exceed 3. We prove a higher upper bound on the diameter for the 2-local versions.

Theorem 4. *The diameter of any 2NE network is in $\mathcal{O}(\sqrt{n})$.*

Proof. Consider a 2NE network $G = (V, E)$ with $|V| = n$ and let D denote its diameter. Consider two nodes $a, b \in V$ such that $d_G(a, b) = D$ and a shortest-path tree $T_a = (V, E_a)$ which is rooted at node a (see Fig. 2).

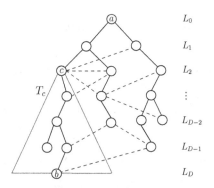

Fig. 2. The shortest-path tree T_a. Dashed lines denote edges of G which are not in the tree, i.e. the non-tree edges.

The height of T_a is D and there must be a subtree T_c which contains node b and which has node c as root, where c is chosen such that $d_G(a, c) = 2$ and c belongs to the path from a to b in T_a. Since the height of T_c is $D - 2$ it follows that the number of nodes in T_c must be at least $D - 1$. Let $|T_x|$ denote the number of nodes in the subtree of T_a rooted at node x. Hence, we have $|T_c| \geq D - 1$.

Note that if agent a buys any edge $\{a, x\}$ in network G then this improves a's distance cost by at least $|T_x|$. Since G is in 2NE, we know that buying the edge $\{a, c\}$ is not an improving move for agent a which implies that $|T_c|$ is at most the cost of the edge $\{a, c\}$ which is equal to $deg_G(c)$. Since $|T_c| \geq D - 1$ it follows that $deg_G(c) \geq D - 1$.

Let L_i denote the set of nodes which are in distance i from the root a in the tree T_a. For example $L_0 = \{a\}$, $c \in L_2$ and $b \in L_D$. Thus, we have $D - 1 \leq deg_G(c) \leq |L_1| + (|L_2| - 1) + |L_3|$.

Analogously, since G is in 2NE, we have that no agent v_i in layer L_i on the $c - b$ path in T_a can decrease her cost by buying an edge to a node in layer L_{i+2} which is a neighbor of a neighbor in T_a. With analogue reasoning as above we get $D - (i - 1) \leq deg_G(v_i) \leq |L_{i-1}| + (|L_i| - 1) + |L_{i+1}|$.

Note that not only agents from lower layers cannot improve by buying edges towards nodes in upper layers but also agents from upper layers cannot improve by buying edges towards nodes in lower layers. Thus we have

$$D - (i - 1) \le deg_G(v_i) \le |L_{i-1}| + (|L_i| - 1) + |L_{i+1}|$$

and

$$D - (i - 1) \le deg_G(v_{D-i}) \le |L_{D-i-1}| + (|L_{D-i}| - 1) + |L_{D-i+1}|$$

for any $2 \le i \le \lfloor \frac{D}{2} \rfloor - 1$. Summing up all inequalities yields:

$$2 \sum_{i=2}^{\lfloor \frac{D}{2} \rfloor - 1} (D - (i - 1)) \le 3 \left(\sum_{i=1}^{D} |L_i| - (D - 1) \right).$$

For the left side we have

$$\frac{3D^2}{4} - 4D - 3 < \left(\left\lfloor \frac{D}{2} \right\rfloor - 2 \right) \left(2D + 1 - \left\lfloor \frac{D}{2} \right\rfloor \right) = 2 \sum_{i=2}^{\lfloor \frac{D}{2} \rfloor - 1} (D - (i - 1))$$

and the right side gives $3 \left(\sum_{i=1}^{D} |L_i| - (D - 1) \right) \le 3n - 3D + 3$, which yields

$$\frac{3D^2}{4} - 4D - 3 < 3n - 3D + 3 \Rightarrow D < \frac{2}{3} \left(1 + \sqrt{9n + 19} \right) \in \mathcal{O}(\sqrt{n}). \qquad \square$$

3.2 Price of Stability

For analyzing the Price of Stability, we have to investigate the network which has the minimum possible social cost.

Lemma 1. *The center sponsored spanning star S_n is an optimal solution of the deg(k)NCG and the deg(k)AOG for any k.*

We have shown in the proof of Theorem 2 that the center sponsored spanning star S_n is in (k)NE for any k. With Lemma 1 this yields the following for $k \ge 2$.

Corollary 3. *The Price of Stability of the deg(k)NCG and the deg(k)AOG is 1.*

3.3 Price of Anarchy

For investigating the quality of the equilibria of our games, we first adapt an important lemma by Fabrikant et al. [14] to our setting.

Lemma 2. *If a (k)NE network G in the deg(k)NCG has diameter D, then its social cost is at most $\mathcal{O}(D)$ times the minimum possible social cost.*

From Corollaries 1 and 2 we know that the diameter of any NE in the degNCG and any 4NE in the deg4NCG is at most 3. Also, from the Lemma 2 we know that the social cost of any NE network G is at most $\mathcal{O}(D(G))$ times the minimum possible social cost. This implies the following statement.

Theorem 5. *The Price of Anarchy of the degNCG and the deg4NCG is in $\mathcal{O}(1)$.*

A straightforward adaptation of Lemma 2 together with Theorem 3 yields:

Corollary 4. *The Price of Anarchy of variants of the degNCG where the price of any edge $\{u, v\}$ bought by agent u is linear in v's degree in G, is constant.*

Using Theorem 4 and Lemma 2 yields the following statement.

Corollary 5. *The Price of Anarchy of the deg2NCG is in $\mathcal{O}(\sqrt{n})$.*

We conclude with analyzing the PoA in the $\deg(k)$AOG. The upper bound is trivially in $\mathcal{O}(n)$, the matching lower bound holds, since a clique is in (k)NE for the $\deg(k)$AOG for any k.

Observation 6. *The Price of Anarchy of deg(k)AOG is in $\Theta(n)$ for any k.*

4   Dynamics

In this section we consider the dynamic properties of the sequential version of the $\deg(k)$NCG and the $\deg(k)$AOG. The sequential version corresponds to an iterative process, called *improving response dynamics (IRD)*, which starts with some initial strategy vector **s** and its corresponding initial network $G(\mathbf{s})$ and then agents are activated one at a time according to some activation scheme, e.g. a random or adversarially chosen move order or round-robin activation, and active agents are allowed to myopically update their current strategy. They will do so only if the new strategy yields strictly less cost than their current strategy. For the $\deg(2)$AOG we will also consider the *best single edge dynamics*, which is a special case of the improving response dynamics, in which active agents buy the best possible single edge, if this strictly decreases their current cost.

The most important dynamic property of a game is the finite improvement property (FIP) [25], which states that any sequence of improving moves must be finite. The seminal paper [25] established that having the FIP is equivalent to being a generalized ordinal potential game. Thus, games having the FIP are guaranteed to converge to an equilibrium under improving move dynamics.

4.1   Dynamics in the deg(k)NCG

We investigate the convergence properties of the $\deg(k)$NCG and prove that the $\deg(k)$NCG may not converge under improving move dynamics.

Theorem 7. *The deg(k)NCG does not have the FIP for any k, which implies that these games cannot have a generalized ordinal potential function.*

Proof (Sketch). See Fig. 3 for an improving response cycle.

Remark 1. The presented improving response cycle in Fig. 3 is not a best response cycle for the $\deg(k)$NCG since in network G_3 agent j has a strictly better local move: Buying the edge to agent h and swapping her edge from i to e.

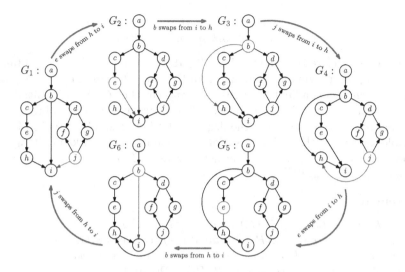

Fig. 3. Example of an improving response cycle for the deg(k)NCG.

4.2   Dynamics in the Add-Only Model

We consider dynamics in the deg(k)AOG. First of all, since agents can only add edges, the deg(k)AOG trivially has the FIP, i.e. it is an ordinal potential game with the number of bought edges serving as a generalized ordinal potential function.

Since convergence is guaranteed, we focus on investigating the speed of convergence and the quality of the obtained networks. For the latter, Observation 6 yields a devastating result. However, we contrast this for the deg2AOG by proving that if round-robin best single edge dynamics starting on a path as initial network are used, then the social cost is actually close to the best possible achievable social cost.

Theorem 8. *Let $P_n = \{v_1 \cdots v_n\}$ be the path of length n, with v_1 and v_n as leaf nodes, as a initial graph for the deg(k)AOG:*

1. *If in any step the active agent is chosen uniformly at random then IRD in the deg(k)AOG converge in $\mathcal{O}(n^3)$ steps in expectation.*
2. *If in any step the active agent and her improving response is chosen adversarially then IRD in the deg(k)AOG converge in $\Theta(n^2)$ steps.*
3. *If round-robin best single edge dynamics are used in the deg2AOG, the process converges in at most $\mathcal{O}(n \log n)$ steps to a network with diameter $\mathcal{O}(1)$.*

We contrast the upper bounds by showing that convergence in $\mathcal{O}(n)$ many improving responses is possible.

Theorem 9. *Let P_n be the initial network then there exists a sequence of improving responses which takes*

1. $n - 2 + \frac{n-7}{3}$ steps to obtain a NE network in the degAOG;
2. $n - 1$ steps to obtain a 2NE network in the deg2AOG.

Finally, we investigate the quality of the (2)NE networks which can be obtained by improving move dynamics starting from the path P_n. For this, we introduce a measure which is similar to the Price of Anarchy. Let G_0 be any initial connected network and let $Z(G_0)$ be the set of networks which can be obtained via improving response dynamics in the deg(2)AOG. Let $Best(G_0) \in Z(G_0)$ be the reachable network with minimum social cost among all networks in $Z(G_0)$. We can now measure the quality of any network $G \in Z(G_0)$ by investigating the ratio $\rho(G, G_0) = \frac{cost(G)}{cost(Best(G_0))}$.

Theorem 10.

1. Let G be any network in $Z(G_0)$ then $\rho(G, G_0) \in \mathcal{O}(n)$.
2. There is a network $G \in Z(P_n)$ for the deg(2)AOG with $\rho(G, P_n) \in \Theta(n)$.
3. Let G^* be the network obtained by the round-robin best single edge dynamics in the deg2AOG, then we have $\rho(G^*, P_n) \in \mathcal{O}(\log n)$.

5   Conclusion

We have introduced natural variants of the well-known NCG by Fabrikant et al. [14], which have the distinctive features that they are parameter-free and at the same time incorporate non-uniform edge costs. Besides proving that computing a best response is NP-hard and that improving response dynamics may never converge to an equilibrium, we have also established that the degNCG has a constant Price of Anarchy. This strong statement holds whenever the edge price is any linear function of the degree of the non-owner endpoint of the edge or if agents are allowed to buy edges to nodes in their 4-neighborhood. For the version which includes stronger locality, i.e. the deg2NCG, we have shown that the PoA is in $\mathcal{O}(\sqrt{n})$ and, as a contrast, for the add-only version the PoA is in $\Theta(n)$. We also demonstrate how to circumvent the latter negative result by using suitable activation schemes on a sparse initial network.

Studying the bilateral version of our model, where both endpoints of the edge have to agree and pay proportionally to the degree of the other endpoint for establishing an edge, is an obvious future research direction. For this version, we have already established that most of our proofs can be easily adapted, which implies that our results, with minor modifications, still hold. Another interesting extension would be to consider an edge price function which depends on the degree of *both* involved nodes. This could be set up such that edges between nodes of similar degree are cheap and edges become expensive when the degree of both nodes differs greatly.

References

1. Albers, S., Eilts, S., Even-Dar, E., Mansour, Y., Roditty, L.: On Nash equilibria for a network creation game. ACM TEAC **2**(1), 2 (2014)

2. Alon, N., Demaine, E.D., Hajiaghayi, M.T., Leighton, T.: Basic network creation games. SIAM J. Discret. Math. **27**(2), 656–668 (2013)
3. Álvarez, C., Blesa, M.J., Duch, A., Messegué, A., Serna, M.: Celebrity games. Theor. Comput. Sci. **648**, 56–71 (2016)
4. Álvarez, C., Messeguè, A.: Max celebrity games. In: Bonato, A., Graham, F.C., Prałat, P. (eds.) WAW 2016. LNCS, vol. 10088, pp. 88–99. Springer, Cham (2016). doi:10.1007/978-3-319-49787-7_8
5. Bala, V., Goyal, S.: A noncooperative model of network formation. Econometrica **68**(5), 1181–1229 (2000)
6. Barabási, A.-L.: Network Science. Cambridge University Press, Cambridge (2016)
7. Bilò, D., Gualà, L., Leucci, S., Proietti, G.: Locality-based network creation games. In: SPAA 2014, pp. 277–286. ACM, New York (2014)
8. Bilò, D., Gualà, L., Leucci, S., Proietti, G.: Network creation games with traceroute-based strategies. In: Halldórsson, M.M. (ed.) SIROCCO 2014. LNCS, vol. 8576, pp. 210–223. Springer, Cham (2014). doi:10.1007/978-3-319-09620-9_17
9. Chauhan, A., Lenzner, P., Melnichenko, A., Molitor, L.: Selfish network creation with non-uniform edge cost (2017). arXiv preprint arXiv:1706.10200
10. Corbo, J., Parkes, D.: The price of selfish behavior in bilateral network formation. In: PODC 2005, pp. 99–107. ACM, New York (2005)
11. Cord-Landwehr, A., Lenzner, P.: Network creation games: think global – act local. In: Italiano, G.F., Pighizzini, G., Sannella, D.T. (eds.) MFCS 2015. LNCS, vol. 9235, pp. 248–260. Springer, Heidelberg (2015). doi:10.1007/978-3-662-48054-0_21
12. Cord-Landwehr, A., Mäcker, A., auf der Heide, F.M.: Quality of service in network creation games. In: Liu, T.-Y., Qi, Q., Ye, Y. (eds.) WINE 2014. LNCS, vol. 8877, pp. 423–428. Springer, Cham (2014). doi:10.1007/978-3-319-13129-0_34
13. Demaine, E.D., Hajiaghayi, M.T., Mahini, H., Zadimoghaddam, M.: The price of anarchy in network creation games. ACM Trans. Algorithms **8**(2), 13 (2012)
14. Fabrikant, A., Luthra, A., Maneva, E., Papadimitriou, C.H., Shenker, S.: On a network creation game. In: PODC 2003, pp. 347–351. ACM, New York (2003)
15. Graham, R., Hamilton, L., Levavi, A., Loh, P.-S.: Anarchy is free in network creation. ACM Trans. Algorithms (TALG) **12**(2), 15 (2016)
16. Jackson, M.O., Wolinsky, A.: A strategic model of social and economic networks. J. Econ. Theory **71**(1), 44–74 (1996)
17. Kawald, B., Lenzner, P.: On dynamics in selfish network creation. In: SPAA 2013, pp. 83–92. ACM (2013)
18. Lenzner, P.: On dynamics in basic network creation games. In: Persiano, G. (ed.) SAGT 2011. LNCS, vol. 6982, pp. 254–265. Springer, Heidelberg (2011). doi:10. 1007/978-3-642-24829-0_23
19. Lenzner, P.: Greedy selfish network creation. In: Goldberg, P.W. (ed.) WINE 2012. LNCS, vol. 7695, pp. 142–155. Springer, Heidelberg (2012). doi:10.1007/ 978-3-642-35311-6_11
20. Leskovec, J., Kleinberg, J., Faloutsos, C.: Graphs over time: densification laws, shrinking diameters and possible explanations. In: SIGKDD 2005, pp. 177–187. ACM (2005)
21. Mamageishvili, A., Mihalák, M., Müller, D.: Tree Nash equilibria in the network creation game. In: Bonato, A., Mitzenmacher, M., Prałat, P. (eds.) WAW 2013. LNCS, vol. 8305, pp. 118–129. Springer, Cham (2013). doi:10.1007/ 978-3-319-03536-9_10
22. Meirom, E.A., Mannor, S., Orda, A.: Network formation games with heterogeneous players and the internet structure. In: EC 2014, pp. 735–752. ACM (2014)

23. Mihalák, M., Schlegel, J.C.: The price of anarchy in network creation games is (mostly) constant. In: Kontogiannis, S., Koutsoupias, E., Spirakis, P.G. (eds.) SAGT 2010. LNCS, vol. 6386, pp. 276–287. Springer, Heidelberg (2010). doi:10.1007/978-3-642-16170-4_24
24. Mihalák, M., Schlegel, J.C.: Asymmetric swap-equilibrium: a unifying equilibrium concept for network creation games. In: Rovan, B., Sassone, V., Widmayer, P. (eds.) MFCS 2012. LNCS, vol. 7464, pp. 693–704. Springer, Heidelberg (2012). doi:10.1007/978-3-642-32589-2_60
25. Monderer, D., Shapley, L.S.: Potential games. Games Econ. Behav. **14**(1), 124–143 (1996)
26. Papadimitriou, C.H.: Algorithms, games, and the internet. In: Proceedings on 33rd Annual ACM Symposium on Theory of Computing, pp. 749–753 (2001)

Opinion Formation Games with Aggregation and Negative Influence

Markos Epitropou[1], Dimitris Fotakis[2($\boxtimes$)], Martin Hoefer[3],
and Stratis Skoulakis[2]

[1] Department of Electrical and Systems Engineering, University of Pennsylvania,
Philadelphia, USA
mep@seas.upenn.edu
[2] School of Electrical and Computer Engineering, NTU Athens, Zografou, Greece
fotakis@cs.ntua.gr, sskoul@corelab.ntua.gr
[3] Institut Für Informatik, Goethe-Universität Frankfurt/Main, Frankfurt, Germany
mhoefer@cs.uni-frankfurt.de

Abstract. We study continuous opinion formation games with aggregation aspects. In many domains, expressed opinions of people are not only affected by local interaction and personal beliefs, but also by influences that stem from global properties of the opinions in the society. To capture the interplay of such global and local effects, we propose a model of opinion formation games with aggregation, where we concentrate on the *average public opinion* as a natural way to represent a global trend in the society. While the average alone does not have good strategic properties as an aggregation rule, we show that with a reasonable influence of the average public opinion, the good properties of opinion formation models are preserved. More formally, we prove that a unique equilibrium exists in average-oriented opinion formation games. Simultaneous best-response dynamics converge to within distance ε of equilibrium in $O(n^2 \ln(n/\varepsilon))$ rounds, even in a model with *outdated information* on the average public opinion. For the Price of Anarchy, we show a small bound of $9/8 + o(1)$, almost matching the tight bound for games without aggregation. Moreover, some of the results apply to a general class of opinion formation games with negative influences, and we extend our results to the case where expressed opinions come from a restricted domain.

1 Introduction

The formation and dynamics of opinions are an important aspect in society and have been studied extensively for decades (see e.g., [16]). Opinion formation is based on information exchange, which is often *local* in the sense that socially connected people (e.g., family, friends, colleagues) interact more often and affect each other's opinion more strongly. Moreover, opinion formation is often *dynamic* in the sense that discussions and interactions lead to changes in

This work was supported by DFG Cluster of Excellence MMCI at Saarland University. Stratis Skoulakis is supported by a scholarship from the Onassis Foundation.

V. Bilò and M. Flammini (Eds.): SAGT 2017, LNCS 10504, pp. 173–185, 2017.
DOI: 10.1007/978-3-319-66700-3_14

the expressed opinions. With the advent of the internet and social media, local and dynamic aspects of opinion formation have become ever more dominant. To capture opinion formation on a formal level, several models have been proposed (see e.g., [4,6,9,12,13,15] for continuous opinions and [5,10,20] for discrete ones).

Motivation and Opinion Formation Model. We build on the influential model of Friedkin and Johnsen (FJ) [12] for continuous opinion formation, following the game-theoretic viewpoint of [6]. Each agent i holds an *intrinsic belief* $s_i \in [0,1]$, which is private and invariant over time, and a *public opinion* $z_i \in [0,1]$. Agent i selects her opinion so as to minimize the total (weighted) disagreement of z_i to her belief and to the opinions in her social neighborhood. In a dynamic setting, the agents start with their beliefs and in each round $t \geq 1$, update their opinion $z_i(t)$ to the minimizer of their disagreement cost, given the opinions of the others in round $t - 1$. The FJ model is extensively studied and has nice algorithmic properties. It admits a unique equilibrium [6,12], which is approached quickly by the simultaneous best-response dynamics [13]. The Price of Anarchy (PoA) is 9/8 for undirected social networks and $\Omega(n)$ for general directed networks [6]. Moreover, tight PoA bounds can be obtained by an elegant local smoothness argument both for undirected [4] and for directed [8] networks.

Despite these favorable properties, the FJ model disregards influences from global properties of the opinions, and also the nature of the dynamics of consensus formation. In many domains, public opinions are not only affected by local interaction and personal beliefs, as in e.g., [4,6,7,9,12,13], but also by influences that stem from global properties of the opinions in the society. People are getting exposed to global trends, societal norms, results from voting and polling, etc., which are usually interpreted as the consensus view of the society and may affect opinion formation. Furthermore, groups of people (or networks of agents) often need to agree on a common action, even if their beliefs and/or their expressed opinions are totally different. This happens, e.g., when networked devices need to implement a common action, when people vote over a set of alternatives, or when a wisdom-of-the-crowd opinion is formed in a social network. In similar situations, an *aggregation rule* maps the public opinions to a *global* opinion that represents the consensus view on the issue at hand. E.g., in the FJ model, the global opinion might be the average or the median of the equilibrium opinions.

In presence of aggregation, an agent can also anticipate the impact of its chosen public opinion on the global one and might incorporate it in its choice. Hence, the disagreement cost should also reflect the distance of an agent's intrinsic belief to the global opinion. To address these issues, we consider a variant of the opinion formation game of [6,12,13] with opinion aggregation. Each agent i selects her opinion z_i so as to minimize:

$$C_i(\mathbf{z}) = w_i(s_i - z_i)^2 + \sum_{j \neq i} w_{ij}(z_j - z_i)^2 + \alpha_i(\mathrm{aggr}(\mathbf{z}) - s_i)^2. \tag{1}$$

In (1), $\mathbf{z} = (z_1, \ldots, z_n) \in \mathbb{R}^n$ is the public opinion vector, $s_i \in [0,1]$ is the belief of agent i, and $\mathrm{aggr}(\mathbf{z})$ maps $\mathbf{z}$ to a global opinion $\mathrm{aggr}(\mathbf{z})$. The weights $w_{ij} \geq 0$ quantify how much the public opinion of agent j influences i, $w_i > 0$ quantifies i's self-confidence, and $\alpha_i > 0$ quantifies the appeal of $\mathrm{aggr}(\mathbf{z})$ to i.

Motivated by previous work on the wisdom of the crowd (see e.g., [16, Sect. 8.3], [14]), we concentrate on *average-oriented* opinion formation games, where the aggregation rule $\mathrm{aggr}(\mathbf{z})$ maps $\mathbf{z}$ to its average $\mathrm{avg}(\mathbf{z}) = \sum_{j=1}^n z_j/n$. Then, the best response of each agent i to a public opinion vector $\mathbf{z}$ is:

$$z_i = \frac{\left(w_i + \frac{\alpha_i}{n}\right) s_i + \sum_{j \neq i}\left(w_{ij} - \frac{\alpha_i}{n^2}\right) z_j}{w_i + \frac{\alpha_i}{n^2} + \sum_{j \neq i} w_{ij}}. \tag{2}$$

Contribution. The aggregation rule in (1) might significantly affect the dynamics and the equilibrium of opinion formation. This becomes evident in (2), where i's influence from some opinions z_j can be negative. Negative influence models agent competition for dragging the average public opinion close to their intrinsic beliefs. An important side-effect is that the best-response (and equilibrium) opinions may become polarized and be pushed towards opposite directions, far away from the agent intrinsic beliefs. This is a significant departure from the FJ model, where the equilibrium opinions lie between the minimum and maximum intrinsic beliefs of the agents. Interestingly, we prove that the nice algorithmic properties of the FJ model are not affected – neither by negative influence nor by outdated information on the average opinion.

We show (Lemma 1) that average-oriented games admit a unique equilibrium, and simultaneous best-response dynamics converges to it within distance $\varepsilon > 0$ in $O(n^2 \ln(n/\varepsilon))$ rounds. For this result, all agents have access to the average public opinion in each round. Since the average is global information and thus difficult to monitor in large networks, we consider average-oriented opinion dynamics with outdated information. Here the average public opinion is announced to all the agents simultaneously every few rounds (e.g., a polling agency publishes this information every now and then). We prove (Theorem 1) that opinion dynamics with outdated information about the average converges to the unique equilibrium within distance $\varepsilon > 0$ after $O(n^2 \ln(n/\varepsilon))$ updates on the average. Both these results are proven for a more general setting with negative influence between the agents and with partially outdated information about the agent public opinions. The main point here is that negative influence and outdated information do not introduce undesirable oscillating phenomena to opinion dynamics.

In Sect. 4, we bound the PoA of average-oriented opinion formation games. We consider symmetric games, where $w_{ij} = w_{ji} \geq 0$ for all agent pairs $i \neq j$, all agents have the same self-confidence w and the same influence α from the average (for non-symmetric games the PoA is $\Omega(n)$, even without aggregation, see [6, Fig. 2]). We show (Theorem 2) that the PoA is at most $9/8 + O(\alpha/(wn^2))$. In general, this bound cannot be improved since for $\alpha = 0$, $9/8$ is a tight bound for the PoA under the FJ model [6]. While the proof builds on [4], local smoothness cannot be directly applied to symmetric average-oriented games, because the function $(\text{avg}(\boldsymbol{z}) - s_i)^2$ is not locally smooth. To overcome this difficulty, we combine local smoothness with the fact that the average opinion at equilibrium is equal to the average belief, a consequence of symmetry (Proposition 1).

A frequent assumption on continuous opinion formation is that agent beliefs and opinions take values in a finite interval of non-negative real numbers. By scaling, one can then treat beliefs and opinions as numbers in $[0, 1]$. Here, we also assume that agent beliefs $s_i \in [0, 1]$. However, due to negative influence, the equilibrium opinions in our model may become polarized and end up far away from $[0, 1]$. We believe that such opinion polarization is natural and should be allowed under negative influence. Therefore, in Sects. 3 and 4, we assume that public opinions can take arbitrary real values. Then, in Sect. 5, we also consider *restricted* average-oriented games with public opinions restricted to $[0, 1]$, and study how convergence properties and price of anarchy are affected.

Existence and uniqueness of equilibrium for restricted games follow from [18]. We prove (Theorem 3) that the convergence rate of opinion dynamics with negative influence and with outdated information is not affected by restriction of public opinions to $[0, 1]$. As for the PoA of restricted symmetric games, we consider the special case where $w_i = \alpha_i = 1$, for all agents i, and show that the PoA does not exceed $(\sqrt{2} + 2)^2/2 + O(\frac{1}{n})$ (Theorem 4). A technical challenge is that partial derivatives of the agent cost functions in the local smoothness inequality do not need to be 0 at equilibrium, due to the opinion restriction to $[0, 1]$. So, we first show that if $w_{ij} = 0$ for all $i \neq j$, the PoA is at most $1 + 1/n^2$. Then we combine the PoA of this simpler game with the local smoothness inequality of [4] and obtain an upper bound on the PoA of the restricted game. Due to lack of space, some proofs are omitted from this extended abstract.

Clearly, there are many alternative ways to model aggregation, which offer interesting directions for future research. For example, a possible aggregation is the *median* instead of the average. The median aggregation rule is prominent in Social Choice (see e.g., [3,17]). However, it turns out that the FJ model with median aggregation has significantly less favorable properties. There are examples where median-oriented games lack exact equilibria (and, hence, convergence of best-response dynamics), but they can be shown to have approximate equilibria. A study of the median rule is beyond the scope of this paper.

Further Related Work. To the best of our knowledge, this is the first work to analyze the convergence of simultaneous best-response dynamics of the FJ model with negative influence and outdated information, or the price of anarchy of the FJ model with average opinion aggregation. However, there is some recent work on properties of opinion formation either with global information, or with negative influence, or where consensus is sought. We concentrate here on related previous work most relevant to ours. Discrete opinion formation is considered in [11] in the binary voter model, where each agent i has a certain probability of adopting the opinion of an agent outside i's local neighborhood (this is conceptually equivalent to estimating the average opinion with random sampling). The authors analyze the convergence time and the probability that consensus is reached. Necessary and sufficient conditions under which local interaction in social networks with positive and negative influence reaches consensus are derived in [1]. Recently, a model of discrete opinion formation was introduced in [2] with generalized social relations, which include positive and negative influence. The authors show that generalized discrete opinion formation games admit a potential function, and thus, best-response dynamics converge to a Nash equilibrium.

2 Model and Preliminaries

Notation and Conventions. We define $[n] \equiv \{1, \dots, n\}$. For a vector z, z_{-i} is z without its i-th coordinate and (z, z_{-i}) is the vector obtained from z if we replace z_i with z. Let $\mathbf{0} \equiv (0, \dots, 0)$ and $\mathbb{I}$ be the $n \times n$ identity matrix. We use capital letters for matrices and lowercase letters for their elements such that, e.g., a_{ij} is the (i, j) element of a matrix A.

$\|A\|$ and $\|z\|$ denote the infinity norms of matrix A and vector z, resp. We repeatedly use the standard properties of matrix norms without explicitly referring to them, i.e., (i) for any matrices A and B, $\|AB\| \le \|A\| \|B\|$ and $\|A + B\| \le \|A\| + \|B\|$; (ii) for any matrix A and any $\lambda \in \mathbb{R}$, $\|\lambda A\| \le |\lambda| \|A\|$; and (iii) for any matrix A and any integer ℓ, $\|A^\ell\| \le \|A\|^\ell$. Moreover, we use that for any $n \times n$ real matrix A with $\|A\| < 1$, $\sum_{\ell=0}^{\infty} A^\ell = (\mathbb{I} - A)^{-1}$.

Average-Oriented Opinion Formation. We consider average-oriented opinion formation games with n agents as introduced in Sect. 1. Wlog., we assume that agent beliefs $s \in [0, 1]^n$. For the public opinions z, we initially assume values in $\mathbb{R}$. In Sect. 5, we explain what changes if we restrict them to $[0, 1]$. An average-oriented game $\mathcal{G}$ is *symmetric* if $w_{ij} = w_{ji}$ for all $i \ne j$, and $w_i = w$ and $\alpha_i = \alpha$ for all agents i. $\mathcal{G}$ is *nonsymmetric* otherwise. If $\mathcal{G}$ is symmetric, we let $w = 1$, by scaling other weights accordingly. Our convergence results hold for nonsymmetric games, our PoA bounds hold only for symmetric ones.

A vector z^* is an *equilibrium* of an opinion formation game $\mathcal{G}$ if for any agent i and any opinion z, $C_i(z^*) \le C_i(z, z^*_{-i})$, i.e., the agents cannot improve their individual cost at z^* by unilaterally changing their opinions. The *social cost* $C(z)$ of $\mathcal{G}$ is $C(z) = \sum_{i \in N} C_i(z)$. An opinion vector o is *optimal* if for any z, $C(o) \le C(z)$. An optimal vector exists because the social cost function is proper. The *Price of Anarchy* of $\mathcal{G}$ (PoA($\mathcal{G}$)) is $C(z^*)/C(o)$, where z^* is the unique equilibrium and o an optimal vector.

It will be convenient to write (2) in matrix form. Let $S_i = w_i + \frac{\alpha_i}{n^2} + \sum_{j \ne i} w_{ij}$. We always assume that $S_i > 0$ so that $C_i(z)$ is strictly convex in z_i. We define two $n \times n$ matrices A and B. Matrix A has $a_{ii} = 0$, for all $i \in N$, and $a_{ij} = (w_{ij} - \frac{\alpha_i}{n^2})/S_i$, for all $j \ne i$. Matrix B is diagonal and has $b_{ii} = (w_i + \frac{\alpha_i}{n})/S_i$, for all $i \in N$, and $b_{ij} = 0$, for all $j \ne i$.

We assume that $\alpha_i \le S_i \le n w_i$, for all agents i (i.e., the agents neither are overwhelmed by the average opinion nor have extremely low self-confidence). This implies $\|A\| \le 1 - \frac{2}{n^2}$, which is crucial for the convergence of best response dynamics. We term a matrix similar to A (i.e., with infinity norm less than 1 and 0s in its diagonal) *influence matrix*, and a matrix similar to B (i.e., diagonal one with positive elements) *self-confidence matrix*.

The simultaneous best-response dynamics of an average-oriented game $\mathcal{G}$ starts with $z(0) = s$ and proceeds in rounds. In each round $t \geq 1$, the public opinion vector $z(t)$ is:

$$z(t) = Az(t-1) + Bs \tag{3}$$

We refer to (3) and similar equations as *opinion formation processes*. An opinion formation process $\{z(t)\}_{t \in \mathbb{N}}$ *converges* to a stable state z^* if for all $\varepsilon > 0$, there is a $t^*(\varepsilon)$, such that for all $t \geq t^*(\varepsilon)$, $\|z(t) - z^*\| \leq \varepsilon$. Iterating (3) over t (see also [13, Sect. 2]) implies that for all rounds $t \geq 1$,

$$z(t) = Az(t-1) + Bs = \cdots = A^t s + \sum_{\ell=0}^{t-1} A^\ell Bs. \tag{4}$$

Outdated Information of the Average Opinion. We study opinion formation when the agents have outdated information about the average public opinion. There is an infinite increasing sequence of rounds $0 = \tau_0 < \tau_1 < \tau_2 < \cdots$ that describes an *update schedule* for the average opinion. At the end of round τ_p, the average $\mathrm{avg}(z(\tau_p))$ is announced to the agents. We refer to the rounds between two updates as an *epoch*, where rounds $\tau_p + 1, \ldots, \tau_{p+1}$ comprise epoch p. The length of each epoch p, denoted by $k_p = \tau_{p+1} - \tau_p \geq 1$, is assumed to be finite. The update schedule is the same for all agents, but the agents might not be aware of it. They are only assumed to be aware of the most recent value of the average public opinion provided to them.

In this case, we need to distinguish in (2) and (3) between the influence from social neighbors, for which the most recent opinions $z(t-1)$ are used, and the influence from the average public opinion, where possibly outdated information is used. As such, we now rely on three different $n \times n$ matrices D, E and B. Self-confidence matrix B is defined as before. Influence matrix D has $d_{ii} = 0$, for all $i \in [n]$, and $d_{ij} = w_{ij}/S_i$, for all $j \neq i$, and accounts for the influence from social neighbors. Influence matrix E has $e_{ii} = 0$, for all $i \in [n]$, and $e_{ij} = -\alpha_i/(n^2 S_i)$, for all $j \neq i$, and accounts for the influence from the average public opinion. By definition, $A = D + E$. Moreover, $\|D\| \leq 1 - 1/n$ and that $\|E\| \leq (n-1)/n^2$.

At the beginning of the opinion formation process, $z(0) = s$. For each round t in epoch p, $\tau_p + 1 \leq t \leq \tau_{p+1}$, the agent opinions are updated according to:

$$z(t) = Dz(t-1) + Ez(\tau_p) + Bs \tag{5}$$

Note that at the beginning of each epoch p, every agent i can subtract $z_i(\tau_p)$ from $n\,\mathrm{avg}(z(\tau_p))$ and compute the term $Ez(\tau_p)$ as $-\frac{\alpha_i}{n^2 S_i}(n\,\mathrm{avg}(z(\tau_p)) - z_i(\tau_p))$.

Opinion Formation with Negative Influence. An interesting aspect of average-oriented games is that the influence matrix A may contain negative elements. Motivated by this observation, we prove our convergence results for a general domain of opinion formation games that may have negative weights w_{ij}. Similarly to [6,13], the individual cost function of each agent i is $C_i(\mathbf{z}) = w_i(z_i - s_i)^2 + \sum_{j \neq i} w_{ij}(z_i - z_j)^2$, and i's best response to $\mathbf{z}_{-i}$ is

$$z_i = \frac{w_i s_i + \sum_{j \neq i} w_{ij} z_j}{w_i + \sum_{j \neq i} w_{ij}}. \tag{6}$$

The important difference is that now some w_{ij} may be negative. We require that for each agent i, $w_i > 0$ and $S_i = w_i + \sum_{j \neq i} w_{ij} > 0$ (and thus, $C_i(\mathbf{z})$ is strictly convex in z_i). The matrices A and B are defined as before. Namely, $a_{ij} = w_{ij}/S_i$, for all $i \neq j$, and B has $b_{ii} = w_i/S_i$ for all i. We always require that $\|A\| < 1 - \beta$, for some $\beta > 0$ (β may depend on n). Simultaneous best-response dynamics is again defined by (3).

3   Convergence of Average-Oriented Opinion Formation

For any nonnegative influence matrix A with largest eigenvalue at most $1 - \beta$, [13, Lemma 5] shows that (3) converges to $\mathbf{z}^* = (\mathbb{I} - A)^{-1}B\mathbf{s}$ within distance ε in $O(\ln(\frac{\|B\|}{\varepsilon\beta})/\beta)$ rounds. We generalize [13, Lemma 5] to average-oriented games, where A may contain negative elements. Thus, we show the following lemma for generalized opinion formation games with negative influence between the agents.

Lemma 1. *Let A be any influence matrix, possibly with negative elements, with $\|A\| \leq 1 - \beta$, for some $\beta > 0$. Then, for any self-confidence matrix B, any $\mathbf{s} \in [0,1]^n$ and any $\varepsilon > 0$, the opinion formation process $\mathbf{z}(t) = A\mathbf{z}(t-1) + B\mathbf{s}$ converges to $\mathbf{z}^* = (\mathbb{I} - A)^{-1}B\mathbf{s}$ within distance ε in $O(\ln(\frac{\|B\|}{\varepsilon\beta})/\beta)$ rounds.*

Since $\mathbb{I} - A$ is nonsingular, $\mathbf{z}^*$ is the unique vector that satisfies $\mathbf{z}^* = A\mathbf{z}^* + B\mathbf{s}$. Thus, $\mathbf{z}^*$ is the unique equilibrium of the corresponding opinion formation game. Moreover, since for average-oriented games $\|A\| \leq 1 - 2/n^2$, Lemma 1 implies that any average-oriented game admits a unique equilibrium $\mathbf{z}^* = (\mathbb{I} - A)^{-1}B\mathbf{s}$, and for any $\varepsilon > 0$, (3) converges to $\mathbf{z}^*$ within distance ε in $O(n^2 \ln(n/\varepsilon))$ rounds.

We next extend Lemma 1 to the case where the agents use possibly outdated information about the average public opinion in each round. In fact, we establish convergence for a general domain with negative influence between the agents, which includes average-oriented opinion formation processes as a special case.

Theorem 1. *Let D and E be influence matrices, possibly with negative elements, such that $\|D\| \leq 1 - \beta_1$, $\|E\| \leq 1 - \beta_2$, for some $\beta_1, \beta_2 \in (0,1)$ with $\beta_1 + \beta_2 > 1$. Then, for any self-confidence matrix B, any $\mathbf{s} \in [0,1]^n$, any update schedule $0 = \tau_0 < \tau_1 < \tau_2 < \cdots$ and any $\varepsilon > 0$, the opinion formation process (5) converges to $\mathbf{z}^* = (\mathbb{I} - (D + E))^{-1}B\mathbf{s}$ within distance ε in $O(\ln(\frac{\|B\|}{\varepsilon\beta})/\beta)$ epochs, where $\beta = \beta_1 + \beta_2 - 1 > 0$.*

Proof. We observe that $z^* = (\mathbb{I} - (D+E))^{-1}Bs$ is the unique solution of $z^* = Dz^* + Ez^* + Bs$ (as in Lemma 1, since $\|E+D\| \leq 1-\beta$, with $\beta > 0$, the matrix $\mathbb{I} - (D+E)$ is non-singular). Hence, if (5) converges, it converges to z^*. To show convergence, we bound the distance of $z(t)$ to z^* by a decreasing function of t and show an upper bound on $t^*(\varepsilon) = \min\{t : e(t) \leq \varepsilon\}$.

As in the proof of Lemma 1, for each round $t \geq 1$, we define $e(t) = \|z(t) - z^*\|$ as the distance of the opinions at time t to z^*. For convenience, we also define

$$f(\beta_1, \beta_2, k) = (1-\beta_1)^k + (1-\beta_2)\frac{1 - (1-\beta_1)^k}{\beta_1}.$$

For any fixed value of $\beta_1, \beta_2 \in (0,1)$ with $\beta_1 + \beta_2 > 1$, $f(\beta_1, \beta_2, k)$ is a decreasing function of k. Indeed, the derivative of f with respect to k is equal to $\ln(1-\beta_1)(1-\beta_1)^k(1 - \frac{1-\beta_2}{\beta_1})$, which is negative, because $1 > (1-\beta_2)/\beta_1$, since $\beta_1 + \beta_2 > 1$.

We next show that (i) for any epoch $p \geq 0$ and any round k, $0 \leq k \leq k_p$, in epoch p, $e(\tau_p + k) \leq f(\beta_1, \beta_2, k)e(\tau_p)$; and (ii) that in the last round $\tau_{p+1} = \tau_p + k_p$ of each epoch $p \geq 0$, $e(\tau_{p+1}) \leq (1-\beta)e(\tau_p)$. The first claim shows that the distance to equilibrium decreases from each round to the next within each epoch, while the second claim shows that the distance to equilibrium decreases geometrically from the last round of each epoch to the last round of the next epoch. Combining the two claims, we obtain that for any epoch $p \geq 0$ and any round k, $0 \leq k \leq k_p$, in epoch p, $e(\tau_p + k) \leq f(\beta_1, \beta_2, k)(1-\beta)^p e(0)$. Therefore, for any update schedule $\tau_0 < \tau_1 < \tau_2 < \cdots$, the opinion formation process (5) converges to $(\mathbb{I} - (D+E))^{-1}Bs$ in $O(\ln(e(0)/\varepsilon)/\beta)$ epochs.

To prove (i), we fix any epoch $p \geq 0$ and apply induction on k. The basis, where $k = 0$, holds because $f(\beta_1, \beta_2, 0) = 1$. For any round k, with $1 \leq k \leq k_p$, in p, we have that:

$$\begin{aligned}
e(\tau_p + k) &= \|Dz(\tau_p + k - 1) + Ez(\tau_p) + Bs - (Dz^* + Ez^* + Bs)\| \\
&\leq \|D\|\,\|z(\tau_p + k - 1) - z^*\| + \|E\|\,\|z(\tau_p) - z^*\| \\
&\leq (1-\beta_1)e(\tau_p + k - 1) + (1-\beta_2)e(\tau_p) \\
&\leq (1-\beta_1)f(\beta_1, \beta_2, k-1)e(\tau_p) + (1-\beta_2)e(\tau_p) = f(\beta_1, \beta_2, k)e(\tau_p).
\end{aligned}$$

The first inequality follows from the properties of matrix norms. The second inequality holds because $\|D\| \leq 1-\beta_1$ and $\|E\| \leq 1-\beta_2$. The third inequality follows from the induction hypothesis. Finally, we use that for any $k \geq 1$, $(1-\beta_1)f(\beta_1, \beta_2, k-1) + 1 - \beta_2 = f(\beta_1, \beta_2, k)$.

To prove (ii), we fix any epoch $p \geq 0$ and apply claim (i) to the last round $\tau_{p+1} = \tau_p + k_p$, with $k_p \geq 1$, of epoch p. Hence, $e(\tau_{p+1}) = \|z(\tau_p + k_p) - z^*\| \leq f(\beta_1, \beta_2, k_p)e(\tau_p)$.

We next show that $f(\beta_1, \beta_2, k_p) \leq 2 - (\beta_1 + \beta_2) = 1 - \beta$, which concludes the proof of the claim. The inequality holds because for any integer $k \geq 1$, $f(\beta_1, \beta_2, k)$ is a convex function of β_1. For a formal proof, we fix any $k \geq 1$ and any $\beta_2 \in (0,1)$, and consider the functions $g(x) = (1-x)^k + \frac{1-(1-x)^k}{x}(1-\beta_2)$ and $h(x) = 2 - \beta_2 - x$, where $x \in [1-\beta_2, 1]$ (since we assume that $\beta_1 \in (0,1)$ and that $\beta_1 > 1 - \beta_2$). For any fixed value of $\beta_2 \in (0,1)$, $h(x)$ is a linear function of x with

$h(1-\beta_2) = 1$ and $h(1) = 1-\beta_2$. For any fixed value of $k \geq 1$ and $\beta_2 \in (0,1)$, $g(x)$ is a convex function of x with $g(1-\beta_2) = 1 = h(1-\beta_2)$ and $g(1) = 1-\beta_2 = h(1)$. Therefore, for any $\beta_1 \in [1 - \beta_2, 1]$, $g(\beta_1) \leq h(\beta_1) = 2 - (\beta_1 + \beta_2)$.

To obtain an upper bound on $e(0) = \|s - z^*\|$, we work as in the proof of Lemma 1, using the fact that $\|D + E\| \leq 1 - \beta$, and show first that $\|(\mathbb{I} - (D + E))^{-1}\| \leq 1/\beta$ and then that $\|z^*\| \leq \|B\|/\beta$. Since $z(0) = s$, we have that $e(0) = \|s - z^*\| \leq 1 + \|B\|/\beta$. Using the fact that for each epoch $p \geq 0$ and for every round k, $0 \leq k \leq k_p$, in p, $e(\tau_p + k) \leq f(\beta_1, \beta_2, k)(1 - \beta)^p e(0)$, we obtain that $t^*(\varepsilon) = O(\ln(\frac{\|B\|}{\varepsilon\beta})/\beta)$ epochs.   □

For average-oriented games, $D+E = A$, $\|D\| \leq 1-1/n$ and $\|E\| \leq (n-1)/n^2$. Hence, applying Theorem 1 with $\beta \geq 1/n^2$, we conclude that for any $\varepsilon > 0$, the opinion formation process (5) with outdated information about $\mathrm{avg}(z(t))$ converges to $z^* = (\mathbb{I} - A)^{-1}Bs$ within distance ε in $O(n^2 \ln(n/\varepsilon))$ epochs.

4   The PoA of Symmetric Average-Oriented Games

We proceed to bound the PoA of average-oriented opinion formation games. We now concentrate on the most interesting case of symmetric games, since nonsymmetric opinion formation games can have a PoA of $\Omega(n)$, even if $\alpha = 0$ (see e.g., [6, Fig. 2]). We recall that for symmetric games, $w_{ij} = w_{ji}$ for all agent pairs i, j, and $w_i = 1$ and $\alpha_i = \alpha$, for all agents i.

Our analysis generalizes a local smoothness argument put forward in [4, Sect. 3.1]. A function $C(z)$ is (λ, μ)-locally smooth [19] if there exist $\lambda > 0$ and $\mu \in (0,1)$, such that for all $z, x \in \mathbb{R}^n$,

$$C(z) + (x - z)^T C'(z) \leq \lambda C(x) + \mu C(z), \tag{7}$$

where $C'(z) = (\frac{\vartheta C_1(z)}{\vartheta z_1}, \frac{\vartheta C_2(z)}{\vartheta z_2}, \cdots, \frac{\vartheta C_n(z)}{\vartheta z_n})$ is the vector with the partial derivative of $C_i(z)$ with respect to z_i, for each agent i. At the equilibrium z^*, $C'(z^*) = \mathbf{0}$. Hence, applying (7) for the equilibrium z^* and for the optimal solution o, we obtain that PoA $\leq \lambda/(1 - \mu)$. For symmetric games without aggregation, [4, Sect. 3.1] shows that for any $s \in [0,1]^n$, the cost function $\sum_{i=1}^n (z_i - s_i)^2 + \sum_{i \in N} \sum_{j \neq i} w_{ij}(z_i - z_j)^2$ is (λ, μ)-locally smooth for any $\lambda \geq \max\{1/(4\mu), 1/(\mu+1)\}$. Using $\lambda = 3/4$ and $\mu = 1/3$, we obtain that the PoA of symmetric opinion formation games without aggregation is at most $9/8$ [4], which is tight [6, Fig. 1].

This elegant approach cannot be directly generalized to symmetric average-oriented games, because the function $\sum_{i \in N}(\mathrm{avg}(z) - s_i)^2$ is not (λ, μ)-locally smooth for any $\mu < 1$. So, instead of trying to find λ, μ so that (7) holds for all $z \in \mathbb{R}^n$, we identify values of λ, μ such that (7) holds for all opinion vectors z with $\mathrm{avg}(z) = \mathrm{avg}(s)$. This suffices for bounding the PoA, since we need to apply (7) only for the optimal opinion vector o and the equilibrium opinion vector z^*. Moreover, the following proposition shows that for the equilibrium vector z^*, we have that $\mathrm{avg}(z^*) = \mathrm{avg}(s)$.

Proposition 1. *Let z^* be the equilibrium and s the agent belief vector of any symmetric average-oriented opinion formation game. Then, $\mathrm{avg}(z^*) = \mathrm{avg}(s)$.*

Based on Proposition 1, we show that the PoA of symmetric average-oriented games tends to $9/8$, which is the PoA of symmetric opinion formation games without aggregation.

Theorem 2. *Let $\mathcal{G}$ be any symmetric average-oriented opinion formation game with n agents and influence $\alpha \geq 0$ from the average public opinion. Then, $\mathrm{PoA}(\mathcal{G}) \leq \frac{9}{8} + O(\frac{\alpha}{n^2})$.*

Proof. We find appropriate parameters $\lambda > 0$ and $\mu \in (0, 1)$ such that (7) holds for any $x \in \mathbb{R}^n$ and any $z \in \mathbb{R}^n$ with $\mathrm{avg}(z) = \mathrm{avg}(s)$. Since the equilibrium z^* of any symmetric average-oriented game $\mathcal{G}$ has $\mathrm{avg}(z^*) = \mathrm{avg}(s)$, by Proposition 1, $\mathrm{PoA}(\mathcal{G}) \leq \lambda/(1-\mu)$.

We divide agent's i personal cost $C_i(z)$ into three parts $C_i(x) = F_i(z) + I_i(z) + A_i(z)$, where $F_i(z) = \sum_{j \neq i} w_{ij}(z_i - z_j)^2$, $I_i(z) = (x_i - s_i)^2$ and $A_i(z) = (\mathrm{avg}(z) - s_i)^2$. Following the previous notation:

$$F(z) = \sum_{i \in N} F_i(z) = \sum_{i \in N} \sum_{j \neq i} w_{ij}(z_i - z_j)^2 = 2 \sum_{i,j:i \neq j} w_{ij}(z_i - z_j)^2$$

$$I(z) = \sum_{i \in N} I_i(z) = \sum_{i \in N}(z_i - s_i)^2 = (z - s)^T(z - s)$$

$$A(z) = \sum_{i \in N} A_i(z) = \alpha \sum_{i \in N}(\mathrm{avg}(z) - s_i)^2 = \alpha(\mathrm{avg}(z) - s)^T(\mathrm{avg}(z) - s).$$

Hence, the social cost is $C(z) = F(z) + I(z) + A(z)$. We let $F'(z) = (\frac{\vartheta F_1(z)}{\vartheta z_1}, \cdots, \frac{\vartheta F_n(z)}{\vartheta z_n})$, $I'(z) = (\frac{\vartheta I_1(z)}{\vartheta z_1}, \cdots, \frac{\vartheta I_n(z)}{\vartheta z_n})$ and $A'(z) = (\frac{\vartheta A_1(z)}{\vartheta z_1}, \cdots, \frac{\vartheta A_n(z)}{\vartheta z_n})$ be the vectors with the partial derivatives of $F_i(z)$, $I_i(z)$ and $A_i(z)$, respectively, with respect to z_i, for each agent i. Note that $A'(z) = (2\alpha/n)(\mathrm{avg}(z) - s)$. The following two propositions are proven in [4, Sect. 3.1].

Proposition 2 [4]. *For any symmetric matrix $W = (w_{ij})$, any $z, x \in \mathbb{R}^n$, and any $\lambda > 0$ and $\mu \in (0, 1)$ with $\lambda \geq 1/(4\mu)$,*

$$F(z) + (x - z)^T F'(z) \leq \lambda F(x) + \mu F(z).$$

Proposition 3 [4]. *For any $z, x, s \in \mathbb{R}^n$, $\lambda > 0$ and $\mu \in (0, 1)$ with $\lambda \geq 1/(\mu + 1)$, it holds that $I(z) + (x - z)^T I'(z) \leq \lambda I(x) + \mu I(z)$.*

Using Proposition 1 and increasing the right-hand side by a small fraction of $I(x)$ and $I(z)$, we can prove an upper bound on $A(z) + (x - z)^T A'(z)$.

Proposition 4. *For any $\alpha > 0$, any $z, x, s \in \mathbb{R}^n$ with $\mathrm{avg}(z) = \mathrm{avg}(s)$, any $\delta \geq 0$, and any $\lambda > 0$ and $\mu \in (0, 1)$ such that $\lambda\mu \geq \alpha/n^2$,*

$$A(z) + (x - z)^T A'(z) \leq \delta A(x) + \mu I(x) + (1 - \delta + 2\lambda)A(z) + \mu I(z). \qquad (8)$$

Applying Propositions 2 and 3 with $\lambda = 3/4$ and $\mu = 1/3$, and Proposition 4, and summing up the corresponding inequalities, we obtain that for any $\delta \geq 0$, and any $\lambda > 0$ and $\mu \in (0, 1)$ with $\lambda\mu \geq \alpha/n^2$,

$$\text{PoA}(\mathcal{G}) \leq \frac{\max\{3/4, \delta\} + \mu}{1 - \max\{1/3, 1 - \delta + 2\lambda\} - \mu} \tag{9}$$

If α/n^2 is small enough, e.g., if $\alpha/n^2 \leq 1/2400$, we use $\delta = 3/4$, $\lambda = 1/24$ and $\mu = 24\alpha/n^2$ in (9) and obtain that $\text{PoA}(\mathcal{G}) \leq 9/8 + O(\frac{\alpha}{n^2})$. Otherwise, we use $\mu = 1/3$, $\lambda = 3\alpha/n^2$ and $\delta = 6\alpha/n^2 + 2/3$, and obtain that $\text{PoA}(\mathcal{G}) = O(\frac{\alpha}{n^2})$. $\square$

5 Average-Oriented Games with Restricted Opinions

A frequent assumption in the literature on opinion formation is that agent beliefs come from a finite interval of nonnegative real numbers. Then, by scaling we can assume beliefs $s_i \in [0, 1]$. If the influence matrix A is nonnegative, then since $b_{ii} + \sum_{j=1}^{n} a_{ij} = 1$ for all $i \in [n]$, we have that the equilibrium opinions are $\boldsymbol{z}^* = (\mathbb{I} - A)^{-1}B\boldsymbol{s} \in [0, 1]^n$. In contrast, for the more general domain we treat here, an important side-effect of negative influence is that the best-response (and equilibrium) opinions may not belong to $[0, 1]$. Motivated by this observation, we consider a *restricted* variant of opinion formation games, where the (best-response and equilibrium) opinions are restricted to $[0, 1]$.

To distinguish restricted opinion formation processes from their unrestricted counterparts, we use $\boldsymbol{y}(t)$ to denote the opinion vectors restricted to $[0, 1]^n$. For restricted average-oriented games and restricted games with negative influence, the best-response opinion y_i of each agent i to $\boldsymbol{y}_{-i}$ is computed by (2) and (6), respectively. But now, if the resulting value is $y_i < 0$, we increase it to $y_i = 0$, while if $y_i > 1$, we decrease it to $y_i = 1$. Since the individual cost $C_i(\boldsymbol{y})$ is a strictly convex function of y_i, the restriction of y_i to $[0, 1]$ results in a minimizer $y^* \in [0, 1]$ of $C_i(y, \boldsymbol{y}_{-i})$. So, the restricted opinion formation process is

$$\boldsymbol{y}(t) = [A\boldsymbol{y}(t - 1) + B\boldsymbol{s}]_{[0,1]}, \tag{10}$$

where $[\cdot]_{[0,1]}$ is the restriction of opinions $\boldsymbol{y}(t)$ to $[0, 1]^n$. The influence matrix A (and the influence matrices D and E for processes with outdated information) and the self-confidence matrix B are defines as in unrestricted opinion formation.

We show a general result for restricted opinion formation processes that is equivalent to Theorem 1. As in Sect. 3, we prove our result for the more general setting of negative influence. Using Theorem 3, we can immediately bound the convergence time for restricted average-oriented processes. The proof of the following is similar to the proof of Theorem 1.

Theorem 3. *Let D and E be influence matrices, possibly with negative elements, such that $\|D\| \leq 1 - \beta_1$, $\|E\| \leq 1 - \beta_2$, for some $\beta_1, \beta_2 \in (0, 1)$ with $\beta_1 + \beta_2 > 1$. Then, for any self-confidence matrix B, any $\boldsymbol{s} \in [0, 1]^n$, any update schedule $0 = \tau_0 < \tau_1 < \tau_2 < \cdots$, the restricted opinion formation process*

$y(t) = [Dy(t-1) + Ey(\tau_p) + Bs]_{[0,1]}$ converges to the unique equilibrium point y^* of $y'(t) = [(D+E)y'(t-1) + Bs]_{[0,1]}$. For any $\varepsilon > 0$, $y(t)$ is within distance ε to y^* after $O(\ln(\frac{1}{\varepsilon})/\beta)$ epochs, where $\beta = \beta_1 + \beta_2 - 1$.

We also bound the PoA of restricted symmetric average-oriented games. Due to opinion restriction to $[0,1]$, the average opinion at equilibrium may be far from avg(s). Therefore, we cannot rely on Proposition 4 anymore. Moreover, the PoA of restricted games increases fast with α (e.g., if $s = (0,\ldots,0,1/n)$, $w_{ij} = 0$ for all $i \neq j$, and $\alpha = n^2$, PoA $= \Omega(n)$). Therefore, we restrict our attention to the case where $\alpha = w = 1$ and show that the PoA of restricted symmetric average-oriented games remains constant. An interesting intermediate result of our analysis is that if all agents only value the distance of their opinion to their belief and to the average, i.e., if $w_{ij} = 0$ for all $i \neq j$, the PoA of such games is at most $1 + 1/n^2$.

Theorem 4. *Let $\mathcal{G}$ be any symmetric average-oriented opinion formation game with $n \geq 2$ agents, $w = \alpha = 1$, and opinions restricted to $[0,1]$. Then, PoA$(\mathcal{G}) \leq (\sqrt{2}+2)^2/2 + O(\frac{1}{n})$, where $(2+\sqrt{2})^2/2 < 5.8285$.*

References

1. Altafini, C.: Consensus problems on networks with antagonistic interactions. IEEE Trans. Autom. Control **58**(4), 935–946 (2013)
2. Auletta, V., Caragiannis, I., Ferraioli, D., Galdi, C., Persiano, G.: Generalized discrete preference games. In: Proceedings of 25th International Joint Conference on Artificial Intelligence (IJCAI 2016), pp. 53–59 (2016)
3. Barberà, S.: An introduction to strategy proof social choice functions. Soc. Choice Welf. **18**, 619–653 (2001)
4. Bhawalkar, K., Gollapudi, S., Munagala, K.: Coevolutionary opinion formation games. In: Proceedings of 45th ACM Symposium on Theory of Computing (STOC 2013), pp. 41–50 (2013)
5. Bilò, V., Fanelli, A., Moscardelli, L.: Opinion formation games with dynamic social influences. In: Cai, Y., Vetta, A. (eds.) WINE 2016. LNCS, vol. 10123, pp. 444–458. Springer, Heidelberg (2016). doi:10.1007/978-3-662-54110-4_31
6. Bindel, D., Kleinberg, J.M., Oren, S.: How bad is forming your own opinion? In: Proceeding of 52nd IEEE Symposium on Foundations of Computer Science (FOCS 2011), pp. 57–66 (2011)
7. Chazelle, B., Wang, C.: Inertial Hegselmann-Krause systems. IEEE Trans. Autom. Control (2017, to appear)
8. Chen, P.-A., Chen, Y.-L., Lu, C.-J.: Bounds on the price of anarchy for a more general class of directed graphs in opinion formation games. Oper. Res. Lett. **44**(6), 808–811 (2016)
9. DeGroot, M.H.: Reaching a consensus. J. Am. Stat. Assoc. **69**, 118–121 (1974)
10. Ferraioli, D., Goldberg, P.W., Ventre, C.: Decentralized dynamics for finite opinion games. In: Serna, M. (ed.) SAGT 2012. LNCS, pp. 144–155. Springer, Heidelberg (2012). doi:10.1007/978-3-642-33996-7_13
11. Fotouhi, B., Rabbat, M.G.: The effect of exogenous inputs and defiant agents on opinion dynamics with local and global interactions. IEEE J. Sel. Top. Sig. Process. **7**(2), 347–357 (2013)

12. Friedkin, N.E., Johnsen, E.C.: Social influence and opinions. J. Math. Sociol. **15**(3–4), 193–205 (1990)
13. Ghaderi, J., Srikant, R.: Opinion dynamics in social networks with stubborn agents: equilibrium and convergence rate. Automatica **50**, 3209–3215 (2014)
14. Golub, B., Jackson, M.O.: Naïve learning in social networks and the wisdom of crowds. Am. Econ. J.: Microecon. **2**(1), 112–149 (2010)
15. Hegselmann, R., Krause, U.: Opinion dynamics and bounded confidence models, analysis, and simulation. J. Artif. Soc. Soc. Simul. **5**, 2 (2002)
16. Jackson, M.O.: Social and Economic Networks. Princeton University Press, Princeton (2008)
17. Moulin, H.: On strategy-proofness and single-peakedness. Publ. Choice **35**, 437–455 (1980)
18. Rosen, J.B.: Existence and uniqueness of equilibrium points in concave n-person games. Econometrica **33**, 520–534 (1965)
19. Roughgarden, T., Schoppmann, F.: Local smoothness and the price of anarchy in splittable congestion games. J. Econ. Theory **156**, 317–342 (2015)
20. Yildiz, E., Ozdaglar, A., Acemoglu, D., Saberi, A., Scaglione, A.: Binary opinion dynamics with stubborn agents. ACM Trans. Econ. Comput. **1**(4), 19:1–19:30 (2013)

The Efficiency of Best-Response Dynamics

Michal Feldman[1], Yuval Snappir[1(✉)], and Tami Tamir[2]

[1] School of Computer Science, Tel-Aviv University, Tel Aviv, Israel
yuval.snappir@gmail.com
[2] School of Computer Science, The Interdisciplinary Center, Herzliya, Israel

Abstract. Best response (BR) dynamics is a natural method by which players proceed toward a pure Nash equilibrium via a local search method. The quality of the equilibrium reached may depend heavily on the order by which players are chosen to perform their best response moves. A *deviator rule* S is a method for selecting the next deviating player. We provide a measure for quantifying the performance of different deviator rules. The *inefficiency* of a deviator rule S with respect to an initial strategy profile p is the ratio between the social cost of the worst equilibrium reachable by S from p and the social cost of the best equilibrium reachable from p. The inefficiency of S is the maximum such ratio over all possible initial profiles. This inefficiency always lies between 1 and the *price of anarchy*.

We study the inefficiency of various deviator rules in network formation games and job scheduling games (both are congestion games, where BR dynamics always converges to a pure NE). For some classes of games, we compute optimal deviator rules. Furthermore, we define and study a new class of deviator rules, called *local* deviator rules. Such rules choose the next deviator as a function of a restricted set of parameters, and satisfy a natural independence condition called *independence of irrelevant players*. We present upper bounds on the inefficiency of some local deviator rules, and also show that for some classes of games, no local deviator rule can guarantee inefficiency lower than the price of anarchy.

Keywords: Congestion games · Best-response dynamics · Deviator rules · Price of anarchy

1 Introduction

Nash equilibrium (NE) is perhaps the most popular solution concept in games. It is a strategy profile from which no individual player can benefit by a unilateral deviation. However, a Nash equilibrium is a declarative notion, not an algorithmic one. To justify equilibrium analysis, we have to come up with a natural

This work was partially supported by the European Research Council under the European Union's Seventh Framework Programme (FP7/2007-2013)/ERC grant agreement number 337122.

V. Bilò and M. Flammini (Eds.): SAGT 2017, LNCS 10504, pp. 186–198, 2017.
DOI: 10.1007/978-3-319-66700-3_15

behavior model that leads the players of a game to a Nash equilibrium. Otherwise, the prediction that players play an equilibrium is highly questionable. Best response (BR) dynamics is a simple and natural method by which players proceed toward a NE via the following local search method: as long as the strategy profile is not a NE, an arbitrary player is chosen to improve her utility by deviating to her best strategy given the profile of others.

Work on BR dynamics advanced in two main avenues: The first studies whether BR dynamics converges to a NE, if one exists [17,21]. The second explores how fast it takes until BR dynamics converges to a NE [11,13,18,25]. It is well known that BR dynamics does not always converge to a NE, even if one exists. However, for the class of finite *potential games* [22,24], a pure NE (PNE) always exists, and BR dynamics is guaranteed to converge to one of the equilibria of the game. A potential game is one that admits a *potential function*—a function that assigns a real value to every strategy profile, and has the miraculous property that for any unilateral deviation, the change in the utility of the deviating player is mirrored accurately in the potential function. This mirroring, combined with the fact that the game is finite, guarantees that any BR sequence must terminate and this happens at some (local) minimum of the potential function, which is a NE by definition. While BRD is guaranteed to converge, convergence may take an exponential number of iterations, even in a potential game [3].

Our focus in this work is different than the directions mentioned above. The description of BR dynamics leaves the choice of the deviating player unspecified. Thus, BR dynamics is essentially a large family of dynamics, differing from one another in the choice of who would be the next player to perform her best response move. In this paper, we study how the choice of the deviating player (henceforth a *deviator rule*) affects the efficiency of the equilibrium reached via BR dynamics. Our contribution is the following: (i) We introduce a new measure for quantifying the inefficiency of deviator rules, (ii) we introduce a natural class of simple and local deviator rules, and (iii) we analyze the inefficiency of deviator rules in network formation games and job scheduling games. Our results distinguish between games where local deviator rules can lead to good outcomes and games for which any local deviator rule performs poorly.

1.1 Model and Problem Statement

A game G has a set N of n players. Each player i has a strategy space P_i, and the player chooses a strategy $p_i \in P_i$. A strategy profile is a vector of strategies for each player, $p = (p_1, \ldots, p_n)$. The strategy profile of all players except player i is denoted by p_{-i}, and it is convenient to denote a strategy profile p as $p = (p_i, p_{-i})$. Similarly, for a set of players I, we denote by p_I and p_{-I} the strategy profile of players in I and in $N \setminus I$, respectively, and we write $p = (p_I, p_{-I})$. Each player has a cost function $c_i : P \to \mathbb{R}^{\geq 0}$, where $c_i(p)$ denotes player i's cost in the strategy profile p. Every player wishes to minimize her cost. There is also a social objective function, mapping each strategy profile to a social cost.

Given a strategy profile p, the best response of player i is the set of strategies that minimize player i's cost, fixing the strategies of all other players, formally $BR_i(p) = \arg \min_{p'_i \in P_i} c_i(p'_i, p_{-i})$. Player i is said to be *suboptimal* in p if the player can reduce her cost by a unilateral deviation, i.e., if $p_i \notin BR_i(p)$. If no player is suboptimal in p, then p is a *Nash equilibrium* (NE) (in this paper we restrict attention to pure NE; i.e., an equilibrium in pure strategies).

Given an initial strategy profile p^0, a best response (BR) sequence from p^0 is a sequence $\langle p^0, p^1, \ldots \rangle$ in which for every $T = 0, 1, \ldots$ there exists a player $i \in N$ s.t $p^{T+1} = (BR_i(p^T_{-i}), p^T_{-i})$. In this paper we restrict attention to games in which every BR sequence is guaranteed to converge to a NE.

Deviator Rules and Their Inefficiency. A *deviator rule* is a function $S : P \to N$ that given a profile p, chooses a deviator among all suboptimal players in p. The chosen player then performs a best response move (breaking ties arbitrarily). Given an initial strategy profile p^0 and a deviator rule S we denote by $NE_S(p^0)$ the set of NE that can be obtained as the final profile of a BR sequence $\langle p^0, p^1, \ldots \rangle$, where for every $T \geq 0$, p^{T+1} is a profile resulting from a deviation of $S(p^T)$ (recall that players break ties arbitrarily, thus this is a set of possible Nash equilibria).

Given an initial profile p^0, let $NE(p^0)$ be the set of Nash equilibria reachable from p^0 via a BR sequence, and let $p^\star(p^0)$ be the best NE reachable from p^0 via a BR sequence, that is, $p^\star(p^0) = \arg \min_{p \in NE(p^0)} SC(p)$, where $SC : P \to \mathbb{R}$ is some social cost function.

The *inefficiency* of a deviator rule S in a game G, denoted α_S^G, is defined as the worst ratio, among all initial profiles p^0, and all NE in $NE_S(p^0)$, between the social cost of the worst NE reachable by S (from p^0) and the social cost of the best NE reachable from p^0. I.e., $\alpha_S^G = \sup_{p^0} \max_{p \in NE_S(p^0)} \frac{SC(p)}{SC(p^\star(p^0))}$. For a class of games $\mathcal{G}$, the inefficiency of a deviator rule S with respect to $\mathcal{G}$ is defined as the worst case inefficiency over all games in $\mathcal{G}$: $\alpha_S^{\mathcal{G}} = \sup_{G \in \mathcal{G}} \{\alpha_S^G\}$. A deviator rule with inefficiency 1 is said to be *optimal*, i.e., an optimal deviator rule is one that for every initial profile reaches a best equilibrium reachable from that initial profile.

The following observation shows that the inefficiency of every deviator rule is bounded from above by the *price of anarchy* (PoA) [20,23]. Recall that the PoA is the ratio between the cost of the worst NE and the cost of the social optimum, and is used to quantify the loss incurred due to selfish behavior.

Observation 1. *For every game G and for every deviator rule S it holds that the inefficiency of S is at least 1 and bounded from above by the PoA.*

Local Deviator Rules. We define and study a class of simple deviator rules, called *local* deviator rules. Local deviator rules are defined with respect to state vectors, that represent the state of the players in a particular profile. Given a profile p, every player i is associated with a state vector v_i, consisting of several parameters that describe her state in p and in the strategy profile obtained by her best response. The specific parameters may vary from one application to another.

A vector profile is a vector $\boldsymbol{v} = (v_1, \ldots, v_n)$, consisting of the state vectors of all players. A deviator rule is said to be local if it satisfies the independence of irrelevant players condition, defined below.

Definition 1. *A deviator rule S satisfies independence of irrelevant players (IIP) if for every two state vectors v_{i_1}, v_{i_2}, and every two vector profiles $\boldsymbol{v}, \boldsymbol{v}'$ such that $\boldsymbol{v} = (v_{i_1}, v_{i_2}, v_{-\{i_1,i_2\}})$, and $\boldsymbol{v}' = (v_{i_1}, v_{i_2}, v'_{-\{i_1,i_2\}})^1$, if $S(\boldsymbol{v}) = i_1$, then $S(\boldsymbol{v}') \neq i_2$.*

The IIP condition means that if the deviator rule chooses a state vector v_i over a state vector v_j in one profile, then, whenever these two state vectors exist, the deviator rule would not choose vector v_j over v_i. Note that this condition should hold even across different game instances and even when the number of players is different. Many natural deviator rules satisfy the IIP condition. For example, suppose that the state vector of a player contains her cost in the current profile and her cost in the profile obtained by her best response; then, both (i) max-cost, which chooses the player with the maximum current cost, and (ii) max-improvement, which chooses the player with the maximum improvement, are local deviator rules.

Congestion Games. A congestion game has a set E of m resources, and the strategy space of every player i is a collection of sets of resources; i.e., $P_i \subseteq 2^E$. Every resource $e \in E$ has a cost function $f_e : \mathbb{N} \to \mathbb{R}$, where $f_e(\ell)$ is the cost of resource e if ℓ players use resource e. The cost of player i in a strategy profile p is $c_i(p) = \sum_{e \in p_i} f_e(\ell_e(p))$, where $\ell_e(p)$ is the number of players that use resource e in the profile p. Every congestion game is a potential game [22], thus admits a pure NE, and moreover, every BR sequence converges to a pure NE. In this paper we study the efficiency of deviator rules in the following congestion games:

Network Formation Games [3]: There is an underlying graph, and every player is associated with a pair of source and target nodes s_i, t_i. The strategy space of every player i is the set of paths from s_i to t_i. The resources are the edges of the graph, every edge e is associated with some fixed cost c_e, which is evenly distributed by the players using it. That is, the cost of an edge e in a profile p is $f_e(p) = c_e/\ell(p)$. In network formation games the cost of a resource decreases in the number of players using it. We also consider a weighted version of network formation games on parallel edge networks, where players have weights and the cost of an edge is shared proportionally by its users. The social cost function here is the sum of the players' costs; that is $SC(p) = \sum_{i \in N} c_i(p)$.

The state vector of a player in a network formation game, in a profile p, consists of: (1) player i's cost in p: $c_i(p)$, (2) the cost of player i's path: $\sum_{e \in p_i} c_e$, (3) player i's cost in the profile obtained from a best response of i: $c_i(p'(i))$ (where $p'(i) = (p_{-i}, BR_i(p_{-i}))$ is the profile obtained from a best response of i), and (4) the cost of player i's path in the profile $p'(i)$: $\sum_{e \in BR_i(p_{-i})} c_e$. In weighted instances, the state vector includes player i's weight as well.

---

[1] Note that the vectors $\boldsymbol{v}$ and $\boldsymbol{v}'$ may correspond to different sets of players.

Job Scheduling Games [26]: The resources are machines, and players are jobs that need to be processed on one of the machines. Each job has some length, and the strategy space of every player is the set of the machines. The load on a machine in a strategy profile p is the total length of the jobs assigned to it. The cost of a job is the load on its chosen machine. We also consider games with conflicting congestion effect [7,14], where jobs have unit length and in addition to the cost associated with the load, every machine has an activation cost B, shared by the jobs assigned to it. The social cost function here is the *makespan*, that is $SC(p) = \max_{i \in N} c_i(p)$. The state vector of a job (player) in a job scheduling game, in a profile p, consists of the job's length, the job's current machine and the loads on the machines.

1.2   Our Results

In Sect. 2 we present our results for network formation games. We first study symmetric games, where all the players share the same source and target nodes. We observe that the local *Min-Path* deviator rule, which chooses the player with the cheapest best response path, is optimal. In contrast, the local *Max-Cost* deviator rule has the worst possible inefficiency, n (which matches the PoA for this game). We then consider asymmetric network formation games. Unfortunately, the optimality of Min-Path does not carry over to asymmetric network formation games, even when played on series of parallel paths (SPP) networks. In particular, the inefficiency of Min-Path in single-source multi-target instances is $\theta(|V|)$, and for multi-source multi-target instances, it further grows to $\theta(2^{|V|})$. On the positive side, we show $poly(n, |V|)$ dynamic-programming algorithms for finding an optimal BR sequence for network formation games played on SPP networks, for single-source multi-target instances, and for multi-source multi-target instances with proper intervals (i.e., where no player's strategy is a subset of another player's strategy). The specification of these algorithms is deferred to the full version due to space constraints. For network formation games played on extension-parallel networks we show that every local deviator rule has an inefficiency of $\Omega(n)$.

In Sect. 2.4 we study network formation games with weighted players. It turns out that weighted players lead to quite negative results. We show that even in the simplest case of parallel-edge networks, it is NP-hard to find an optimal BR-sequence, and no local deviator rule can ensure a constant inefficiency. Moreover, the Min-Path deviator rule has inefficiency $\Omega(n)$, even in symmetric games on series-parallel graphs, and even if the ratio between the maximal and minimal weights approaches 1.

The analysis of job scheduling games is deferred to the dull version. A job's (= player's) state vector in job scheduling games includes the job's lengths and the machines' loads. Local deviator rules capture many natural rules, such as Longest-Job, Max-Cost, Max-Improvement, and more. We show that in an instance with m identical machines, no local deviator rule can guarantee inefficiency better than the *PoA*, which is $\frac{2m}{m+1}$. In contrast, for job scheduling games with conflicting congestion effects [14], we present an optimal local deviator rule.

Positive results on local deviator rules imply that a centralized authority that can control the order of deviations can lead the population to a good outcome, by considering merely local information captured in the close neighborhood of the current state. In contrast, negative results for local deviator rules imply that even if a centralized authority can control the order of deviations, in order to converge to a good outcome, it cannot rely only on local information; rather, it must be able to perform complex calculations and to consider a large search space.

1.3   Related Work

Congestion games have been widely studied from a game theoretic perspective. The questions that are most commonly analyzed are the existence of a pure NE, the convergence of BRD to a NE, and the loss incurred due to selfish behavior – commonly quantified by the *price of anarchy* [20,23] and *price of stability* [3].

Our work addresses congestion games and some variants thereof. It is well known that every congestion game is a potential game [22,24] and therefore admits a PNE and possesses the finite improvement property (FIP). In particular, every BRD converges to a PNE. However, the convergence time may, in general, be exponentially long. It is shown in [3,13] that finding a PNE in network formation games is PLS-complete. Examples for exponential convergence of job scheduling games are presented in [10]. It has been shown in [9] that in random potential games with n players in which every player has at most a strategies, the worst case convergence time is $n \cdot a^{n-1}$.

The observation that the convergence of BRD can be exponentially long has led to a large amount of work aiming to identify special classes of congestion games for which BRD converges to a PNE in polynomial time (or even linear time). Examples include [3] for games with positive congestion effects, and [10,16] for games with negative congestion effects. For resource selection games (i.e., where feasible strategies are composed of singletons), polynomial convergence has been proven in [18].

Variants of congestion games such as weighted network formation games and resource selection games with player-specific cost functions have been also considered in the literature [8,17,21]. Some classes of cost functions that always admit PNE or the FIP were identified in [17]. It was also shown that singleton weighted congestion games that always admit a PNE do not always have the FIP. [4,12] present variants of network formation games in which computing a player's best response is NP-hard. A cost-sharing game on unrelated machines has been studied in [5], where it was shown that a PNE exists only for instances with unit-cost machines. Moreover, even when a PNE exists and BRD is guaranteed to converge, the implementation of BRD can be computationally hard.

BRD has been studied also in games that do not converge to a PNE. The notion of *dynamic inefficiency* was defined in [6] as the average social cost in a BR infinite sequence (for games that do not possess the finite improvement property), and different deviator rules are analyzed with respect to the dynamic inefficiency measure.

The effect of the deviator rule on the convergence time of job scheduling games was studied in [10]. This paper considered the convergence time under the Max-Weight-Job, Min-Weight-Job, FIFO and random deviator rules. The Max-Cost deviator rule was considered also in [15] for conflicting congestion games [7,14], and in [19] for swap-games [2]. In both cases Max-Cost significantly improves convergence time to $\mathcal{O}(n)$.

2   Network Formation Games

In this section we study network formation games. We consider two natural local deviator rules, namely *Max-Cost* and *Min-Path*. The Max-Cost deviator rule chooses a suboptimal player that currently incurs the highest cost, i.e., $Max - Cost(p) \in \arg\max_{\{i \in N | p_i \notin BR_i(p)\}} c_i(p)$. The Min-Path deviator rule chooses a suboptimal player whose path in the profile obtained from a best response move is cheapest, i.e., $Min - Path(p) \in \arg\min_{\{i \in N | p_i \notin BR_i(p)\}} \sum_{e \in BR_i(p)} c_e$.

2.1   Warmup: Symmetric Network Formation Games

A network-formation game is *symmetric* if all the players have the same source and target nodes. Recall that the inefficiency of any deviator rule is upper bounded by the price of anarchy (PoA) of the game. It is well known that the PoA of network formation games is n.

We first show that the Max-Cost rule may perform as poorly as the PoA, even in symmetric games on parallel-edge networks.

Observation 2. *The inefficiency of Max-Cost in symmetric network formation games on parallel-edge networks is* n.

On the other hand, we show that Min-Path is an optimal deviator rule, i.e., it always reaches the best NE reachable from any initial profile. Our analysis of Min-Path is based on the following Lemma:

Lemma 1. *In symmetric network formation games, the path chosen by the first deviator is the unique path that will be chosen by all subsequent players, regardless of the order in which they deviate.*

Lemma 1 directly implies the optimality of Min-Path:

Theorem 1. *Min-Path is an optimal deviator rule for symmetric network formation games.*

Proof. By Lemma 1 the first deviation dictates the NE to be reached. Thus, the set of reachable NE is the set of BR paths in p^0 (i.e, $\{BR_i(p^0) | i$ is suboptimal in $N\}$). Clearly choosing the cheapest one among them is optimal.   □

2.2 Series of Parallel Paths (SPP) Networks

In this section we study NFGs played on SPP networks. An SPP network consists of m *segments*, where each segment is a parallel-edge network. Let $\{u_0, \ldots, u_m\}$ denote the vertex set, and for every $j \leq m$, let E_j denote the set of edges in segment j (i.e., the parallel edges connecting u_{j-1} and u_j). For a player i, let $E(i) = \cup_{s_i < k \leq t_i} E_k$, denote the set of edges player i may choose.

Note that in an SPP network, a player's choice of an edge in E_j is independent of any other segment in her path. This implies that a NFG on an SPP network consists of a sequence of symmetric games, where the set of players participating in each game varies. Combining this observation with Lemma 1 implies:

Lemma 2. *In every network formation game played on an SPP network with m segments, for every $1 \leq j \leq m$, and every BR sequence, let i be the first player in the sequence such that $E_j \in E(i)$, and let e be the edge in E_j chosen by player i. Then e is the unique edge in E_j players deviate to.*

Based on the above lemma, it is possible to develop polynomial-time algorithms, based on dynamic programming, for finding optimal BR sequences for SPP networks, for both single-source multi-targets games and multi-source multi-target games with proper intervals. Due to space constraints, the algorithms are omitted.

The Performance of Min-Path in SPP Networks. Recall that Min-Path was shown to be optimal for symmetric NFGs. We now analyze its inefficiency for SPP networks. Given an SPP network and a BR-sequence, we say that a segment is *unresolved* if there are at least two players whose intervals include the segment, and each of them will select a different edge in the segment if chosen to perform a BR next. The other segments are denoted *resolved*. By Lemma 2, after player i performs her best response, all the segments in her interval are resolved. Thus, no player migrates more than once. The reachable NEs have the same edges in the resolved segments and can only differ in unresolved segments. Therefore, the migrations of players who use only resolved segments does not influence the reachable NEs and in the following analyses we ignore them. Thus, all the deviations we consider resolve at least one segment. We denote by R_i the resolved segments after i such deviations. Let OPT denote the minimal cost of a NE reachable from p^0 by some BR sequence. Formally, $OPT = SC(p^\star(p_0))$.

Lemma 3. *For any BR sequence of an SPP network instance, as long as there are unresolved segments, there exists a suboptimal player whose interval includes unresolved segments, and if this player is chosen next, then the cost of the unresolved segments she would set is at most OPT.*

Proof. Let p be an intermediate strategy profile in the BR sequence. Consider the players according to the order they deviate in some optimal BR sequence. Let i' be the first player in this order who is suboptimal in p. Since no player prior to i' in the optimal sequence is suboptimal, the segments that i' would resolve by a deviation from p are a subset of the segments she resolves in the optimal sequence. In the optimal sequence she obviously resolves these segments such that the selected edges are of total cost at most OPT, and therefore this is an upper bound on the total cost of unresolved segments she would set by deviating from p.   □

Using the above lemma, we provide tight analysis on the performance of Min-Path for SPPs with multi-targets and single or multiple sources. Note that in a single-source instance, every player resolves the prefix of the network corresponding to her interval.

Theorem 2. *The inefficiency of Min-Path in SPP NFGs with single-source and multi-targets is $\theta(m)$.*

Proof. We show that the total cost determined for the segments resolved in every iteration is at most OPT. Since at least one segment is resolved in each iteration, the whole network's cost is bounded by $m \cdot OPT$. Let i be the i-th player chosen to deviate by Min-Path and assume i has unresolved segments. Let i' be the player guaranteed by Lemma 3. It may be that $i = i'$. Both players have the same BR path in the resolved segments and therefore differ only in their unresolved segments. Since Min-Path chose i, the cost of her unresolved segments is at most the cost of i''s unresolved segments, which is at most OPT by Lemma 3.

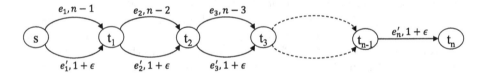

Fig. 1. A network on which Min-Path has inefficiency $\Omega(m)$

We show that the analysis is tight: Consider the network depicted in Fig. 1. There are $n = m$ players, where t_i is the target of player i. In the initial strategy profile for every $1 \leq i \leq n$, $p_i^0 = \langle e_1, e_2, \ldots, e_{i-1}, e_i' \rangle$. Note that in every segment, E_i, connecting t_{i-1} and t_i, the upper edge costs $n - i$ and is used by the $n - i$ players $i+1, \ldots, n$ and the lower edge costs $1 + \epsilon$ and is used only by player i.

In every segment, the players using the upper edge will benefit from deviating to the lower one and the player using the lower edge will benefit from deviating to the upper one. By Lemma 2, the first deviation will determine the edge that will be used in the NE reached. Therefore, a first deviation of player n to $\langle e_1', \ldots, e_n' \rangle$ will result in the NE corresponding to that path and has a social cost $n \cdot (1 + \epsilon)$.

Player i's BR path's cost is $\displaystyle\sum_{1 \leq t \leq i-1} (1+\epsilon) + (n-i) = (i-1) \cdot (1+\epsilon) + (n-i) = (n-1) + (i-1) \cdot \epsilon$. Therefore, Min-Path chooses Player 1 to deviate first. After her deviation all the rest of the players would use e_1 in their BR and treating t_1 as source shows that the next Player to deviate will be Player 2 and then Player 3 etc. Min-Path's BR sequence's NE will consist of $\langle e_1, \ldots e_{n-1}, e'_n \rangle$ and therefore its inefficiency for this strategy profile is $\frac{(n-1)\frac{n}{2}+1+\epsilon}{n(1+\epsilon)} \xrightarrow{\epsilon \to 0} \frac{(n-1)\frac{n}{2}+1}{n} \approx \frac{n}{2}$. Since $n = m$, we conclude that the inefficiency of Min-Path in SPP networks with single-source and multi-targets is $\theta(m)$. □

We next show that Min-Path performs poorly on more general instances.

Theorem 3. *The inefficiency of Min-Path in SPP network formation games with multi-sources and multi-targets is $\theta(2^m)$.*

Proof. Let $c(R_i)$ denote the total cost of resolved segments after i deviations of players whose deviation resolved at least one segment. Since the initially resolved segments has to be included in the NE reached, it holds that $c(R_0) \leq OPT$. We prove that $c(R_i) - c(R_{i-1}) \leq c(R_{i-1}) + OPT$ for every i; i.e., $c(R_i) \leq 2c(R_{i-1}) + OPT$. This implies that the total network's cost is $c(R_m) \leq 2^{m+1} \cdot OPT$.

Let i be the ith player chosen to deviate by Min-Path that has some unresolved segments. The total cost of the unresolved segments that i resolves is $c(R_i) - c(R_{i-1})$. Let i' be a player guaranteed by Lemma 3. Since i was chosen by Min-Path, the cost of i's BR path is lower than the cost of i''s BR path. But the cost of i's BR path is at least the cost of the unresolved segments in i's BR path. On the other hand, the cost of i''s BR path equals the sum of the cost of her resolved segments, which is bounded by $c(R_{i-1})$ and the cost she would set to her unresolved segments, bounded by OPT (by Lemma 3). Putting it all together, we get $c(R_i) - c(R_{i-1}) \leq c(R_{i-1}) + OPT$, as required. In the full version, we present a matching lower bound. □

2.3   Local Rules for Extension Parallel (EP) Graphs

We now show that the inefficiency of any local deviator rule is $\Omega(n)$, even in the restricted class of EP networks. Recall that the state vector of a player consists of the player's cost in her current profile and in the profile obtained by a deviation of the player, and the total cost of the path used by the player in the two profiles.

Theorem 4. *For the class of single-source network formation games played on extension-parallel networks, the inefficiency of every local deviator rule is $\Omega(n)$.*

Proof. Consider the network depicted in Fig. 2(a). There are n players, all sharing the source s, and the targets are as depicted in the figure. Consider the following profile: (i) Player 1 uses the path $\langle e_1 \rangle$. (ii) Player 2 uses the path $\langle e_1, e_2 \rangle$. (iii) Players 3, 4 use the path $\langle e_3 \rangle$. (iv) Players 5 to n use the path $\langle e_5 \rangle$.

Recall that the state vector of player i consists of her current cost, the total cost of her current path, her post-deviation cost, and the total cost of her post-deviation path. Consider player 2, who uses the path $\langle e_1, e_2 \rangle$. Her current cost is 22 (she shares the cost of edge e_1 with player 1 and pays fully for edge e_2), the total cost of her path is 34, her post-deviation cost is 10

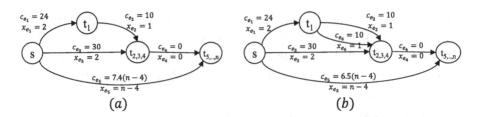

Fig. 2. A local deviator rule fails (a) if v_2 is preferred, and (b) if v_3 is preferred. Every edge is labelled by the edge cost and the number of players using it in the initial strategy profile. E.g., the edge e_1 costs 24 and is used by 2 players in the initial strategy profile.

(obtained by deviating to e_3, and sharing this cost with players $3, 4$), and the total cost of her post-deviation path is 30. Thus, the state vector of player 2 is $v_2 = (22, 34, 10, 30)$. Similarly, one can verify that the state vector of player 3 (or layer 4) is $v_3 = (15, 30, 13, 34)$ (obtained by deviating to the path $\langle e_1, e_2 \rangle$). The suboptimal players in this profile are players 2 and 3 (or 4). If the deviator rule chooses the state vector v_2 over v_3, then player 2 will deviate to e_3, reaching a NE whose social cost is $54 + 7.4(n - 4)$. On the other hand, if the deviator rule chooses the state vector v_3 over v_2, then player 3 will deviate to $\langle e_1, e_2 \rangle$, and from this point on all players will deviate to $\langle e_1, e_2 \rangle$, reaching a NE whose social cost is 34. We conclude that a deviator rule that prefers v_2 over v_3 reaches an inefficiency of $\frac{54 + 7.4(n - 4)}{34} = \Omega(n)$.

Consider next the network depicted in Fig. 2(b). There are n players, all sharing the source s, with targets as depicted in the figure. Consider the following profile: (i) Player 1 uses the path $\langle e_1, e_2 \rangle$. (ii) Player 2 uses the path $\langle e_1, e_6 \rangle$. (iii) Players $3, 4$ use the path $\langle e_3 \rangle$. (iv) Players 5 to n use the path $\langle e_5 \rangle$.

One can verify in a similar analysis to the one showed for the game in Fig. 2(a) that if a deviator rule prefer v_3 over v_2 then it has inefficiency of $\frac{34 + 6.5(n - 4)}{30} = \Omega(n)$. We conclude that any local deviator rule has inefficiency of $\Omega(n)$. □

2.4   Weighted Symmetric Network Formation Games

In this section we consider network formation games with weighted players [1,8,17], where every player is associated with a cost w_i. If an edge of cost c_e is shared by k players with weights $w_1, w_2, \ldots, w_k$, then player i pays $\frac{w_i}{\sum_{j=1}^{k} w_j} \cdot c_e$.

For weighted NFGs on parallel-edge graphs, we prove the following.

Theorem 5. *In weighted network formation games on parallel edge networks: (a) it is NP-hard to calculate the social cost of an optimal reachable NE from a given profile; (b) finding a reachable NE that approximates the social optimum by factor $\frac{3}{2}$ is NP-hard; (c) any local deviator rule has inefficiency $\Omega(\sqrt{n})$.*

A direct corollary of the last theorem (part (a)) is that the problem of finding an optimal BR sequence is NP-hard.

Weighted symmetric network formation games with strategies consisting of two resources (i.e., a 2-segment SPP) are potential games [3]. Theorem 1 shows that in unweighted symmetric games, the Min-Path deviator rule ensures convergence to the optimal reachable NE. Here we show that in weighted games the efficiency of Min-Path can be as poor as the PoA, even if the weights are arbitrarily close to each other, and the strategies are sets of two resources.

Theorem 6. *In weighted network formation games on a 2-segment SPP, the Min-Path deviator rule has inefficiency $\Omega(n)$. This bound is valid even if the ratio $\max_i w_i / \min_i w_i$ is arbitrarily close to 1.*

References

1. Ackermann, H., Röglin, H., Vöcking, B.: Pure Nash equilibria in player-specific and weighted congestion games. Theor. Comput. Sci. **410**(17), 1552–1563 (2009)
2. Alon, N., Demaine, E.D., Hajiaghayi, M., Leighton, T.: Basic network creation games. In: Proceedings of the 22nd ACM Symposium on Parallelism in Algorithms and Architectures, pp. 106–113 (2010)
3. Anshelevich, E., Dasgupta, A., Kleinberg, J., Tardos, E., Wexler, T., Roughgarden, T.: The price of stability for network design with fair cost allocation. SIAM J. Comput. **38**(4), 1602–1623 (2008)
4. Avni, G., Kupferman, O., Tamir, T.: Network-formation games with regular objectives. J. Inf. Comput. **251**, 165–178 (2016)
5. Avni, G., Tamir, T.: Cost-sharing scheduling games on restricted unrelated machines. Theor. Comput. Sci. **646**, 26–39 (2016)
6. Berger, N., Feldman, M., Neiman, O., Rosenthal, M.: Dynamic inefficiency: anarchy without stability. In: Persiano, G. (ed.) SAGT 2011. LNCS, vol. 6982, pp. 57–68. Springer, Heidelberg (2011). doi:10.1007/978-3-642-24829-0_7
7. Chen, B., Gürel, S.: Efficiency analysis of load balancing games with and without activation costs. J. Sched. **15**(2), 157–164 (2011)
8. Chen, H., Roughgarden, T.: Network design with weighted players. Theory Comput. Syst. **45**(2), 302–324 (2009)
9. Durand, S., Gaujal, B.: Complexity and optimality of the best response algorithm in random potential games. In: Gairing, M., Savani, R. (eds.) SAGT 2016. LNCS, vol. 9928, pp. 40–51. Springer, Heidelberg (2016). doi:10.1007/978-3-662-53354-3_4
10. Even-Dar, E., Kesselman, A., Mansour, Y.: Convergence time to Nash equilibria. In: Baeten, J.C.M., Lenstra, J.K., Parrow, J., Woeginger, G.J. (eds.) ICALP 2003. LNCS, vol. 2719, pp. 502–513. Springer, Heidelberg (2003). doi:10.1007/3-540-45061-0_41
11. Even-Dar, E., Mansour, Y.: Fast convergence of selfish rerouting. In: Proceedings of SODA, pp. 772–781 (2005)
12. Fabrikant, A., Luthra, A., Maneva, E., Papadimitriou, C., Shenker, S.: On a network creation game. In: Proceedings of PODC, pp. 347–351 (2003)
13. Fabrikant, A., Papadimitriou, C., Talwar, K.: The complexity of pure Nash equilibria. In: Proceedings of STOC, pp. 604–612 (2004)
14. Feldman, M., Tamir, T.: Conflicting congestion effects in resource allocation games. J. Oper. Res. **60**(3), 529–540 (2012)
15. Feldman, M., Tamir, T.: Convergence of best-response dynamics in games with conflicting congestion effects. Inf. Process. Lett. **115**(2), 112–118 (2015)

16. Fotakis, D.: Congestion games with linearly independent paths: convergence time and price of anarchy. Theory Comput. Syst. **47**(1), 113–136 (2010)
17. Harks, T., Klimm, M.: On the existence of pure Nash equilibria in weighted congestion games. Math. Oper. Res. **37**(3), 419–436 (2012)
18. Ieong, S., Mcgrew, R., Nudelman, E., Shoham, Y., Sun, Q., Fast, C.: A simple class of congestion games. In: Proceedinhgs of AAAI, pp. 489–494 (2005)
19. Kawald, B., Lenzner, P.: On dynamics in selfish network creation. In: Proceedings of SPAA, pp. 83–92 (2013)
20. Koutsoupias, E., Papadimitriou, C.: Worst-case equilibria. Comput. Sci. Rev. **3**(2), 65–69 (1999)
21. Milchtaich, I.: Congestion games with player specific payoff functions. Games Econ. Behav. **13**, 111–124 (1996)
22. Monderer, D., Shapley, L.S.: Potential games. Games Econ. Behav. **14**, 124–143 (1996)
23. Papadimitriou, C.H.: Algorithms, games, and the internet. In: Proceedings of 33rd STOC, pp. 749–753 (2001)
24. Rosenthal, R.W.: A class of games possessing pure-strategy Nash equilibria. Int. J. Game Theory **2**, 65–67 (1973)
25. Syrgkanis, V.: The complexity of equilibria in cost sharing games. In: Saberi, A. (ed.) WINE 2010. LNCS, vol. 6484, pp. 366–377. Springer, Heidelberg (2010). doi:10.1007/978-3-642-17572-5_30
26. Vöcking, B.: Selfish load balancing (Chap. 20). In: Algorithmic Game Theory. Cambridge University Press, Cambridge (2007)

Efficient Best Response Computation for Strategic Network Formation Under Attack

Tobias Friedrich, Sven Ihde, Christoph Keßler, Pascal Lenzner[⊠],
Stefan Neubert, and David Schumann

Algorithm Engineering Group, Hasso Plattner Institute, Potsdam, Germany
pascal.lenzner@hpi.de

Abstract. Inspired by real world examples, e.g. the Internet, researchers have introduced an abundance of strategic games to study natural
phenomena in networks. Unfortunately, almost all of these games have
the conceptual drawback of being computationally intractable, i.e. computing a best response strategy or checking if an equilibrium is reached is
NP-hard. Thus, a main challenge in the field is to find tractable realistic
network formation models. We address this challenge by investigating
a very recently introduced model by Goyal et al. [14] which focuses on
robust networks in the presence of a strong adversary who attacks (and
kills) nodes in the network and lets this attack spread virus-like through
the network via neighboring nodes.

Our main result is to establish that this natural model is one of the
few exceptions which are both realistic and computationally tractable.
In particular, we answer an open question of Goyal et al. by providing an efficient algorithm for computing a best response strategy, which
implies that deciding whether the game has reached a Nash equilibrium
can be done efficiently as well. Our algorithm essentially solves the problem of computing a minimal connection to a network which maximizes
the reachability while hedging against severe attacks on the network
infrastructure and may thus be of independent interest.

1 Introduction

Many of today's important networks, most prominently the Internet, are essentially the outcome of an unsupervised decentralized network formation process
among many selfish entities [22]. In the case of the Internet these selfish entities
are Autonomous Systems (AS) which interconnect via peering agreements and
thereby create a connected network of networks. Each AS can be understood
as a selfish player who strategically chooses a subset of other ASs to directly
connect with. Each inter-AS-connection is costly and yields a benefit and a risk.
The benefit is a reliable direct link towards the other AS. However, such a connection may be used by malicious software and thus harbors the risk of collateral
damage if a neighboring AS is attacked.

© Springer International Publishing AG 2017
V. Bilò and M. Flammini (Eds.): SAGT 2017, LNCS 10504, pp. 199–211, 2017.
DOI: 10.1007/978-3-319-66700-3_16

The field of strategic network formation, started by the seminal works of Jackson and Wolinsky [15], Bala and Goyal [2] and Fabrikant et al. [11], studies the global structure and properties of networks formed by individual players making decentralized local strategic choices. In all considered models there are players trying to optimize their own benefit, while minimizing their individual cost. It is far from obvious why a collection of individual selfish strategies eventually results in useful and reliable network topologies like the Internet. Studying the properties of such models aims for revealing insights about properties of existing naturally grown networks and inspiring methods to improve them.

Required features of any Internet-like communication network are reachability and robustness. Such networks have to ensure that even in case of cascading edge or node failures caused by technical defects or malicious attacks, e.g. DDoS-attacks or viruses, most participating nodes can still communicate. This important focus on network robustness has long been neglected and is now a very recent endeavor in the strategic network formation community, see e.g. [6,14,17,20]. We contribute to this endeavor by proving that the very recently introduced natural model by Goyal et al. [13,14] is one of the few exceptions of a tractable network formation model. In particular, we provide an efficient algorithm for computing a utility maximizing strategy for their elegant model, which can be used to efficiently decide whether a network is in Nash equilibrium. Thus, our algorithm allows the model of Goyal et al. to be used to predict real world phenomena in large scale simulations and to analyze real world networks.

Related Work: We focus on the model for strategic network formation with attack and immunization recently proposed by Goyal et al. [13,14]. This model essentially augments the well-known reachability model by Bala and Goyal [2] with robustness considerations. In particular, different types of adversaries are introduced which attack (and destroy) a node of the network. This attack then spreads virus-like to neighboring nodes and destroys them as well. Besides deciding which links to form, players also decide whether they want to buy immunization against eventual attacks. The model is the first model which incorporates network formation and immunization decisions at the same time.

The authors of [13,14] provide beautiful structural results for their model. For example, showing that equilibrium networks are much more diverse than in the non-robust version, that the amount of edge overbuilding due to robustness concerns is small and that equilibrium networks generally achieve very high social welfare. Besides this, the authors raise the intriguing open problem of settling the complexity of computing a best response strategy in their model[1].

Computing a best response in network formation games can be done in polynomial time for the non-robust reachability model [2] and if the allowed strategy changes are very simple [16,19]. However, these examples are exceptions. The existence of an efficient best response algorithm for a network formation game is in general a rare gem. For almost all related network formation models, e.g. [4–6,8,10,11,21], where players strive for a central position in the network,

---

[1] This question was raised in [13] for the maximum carnage adversary and is replaced in [14] with a reference to our preprint [12] of the present paper.

it has been shown that the problem is indeed NP-hard. The model by Goyal et al. [13,14] seems on the first glance computationally easier than the above mentioned centrality models since players only strive for reaching all other players. However, the presence of a strong adversary and the possibility of immunization renders finding a best possible strategy a non-trivial problem.

To the best of our knowledge, besides the model by Goyal et al. [13,14] there are only a few other models which combine selfish network formation with robustness considerations and all of them consider a much weaker adversary which can only destroy a single edge. The earliest are models by Bala and Goyal [3] and Kliemann [17], both essentially augment the model by Bala and Goyal [2] with single edge failures. Other related models are by Meirom et al. [20] and Chauhan et al. [6]. Both latter models consider players who try to be as central as possible in the created networks but at the same time want to protect themselves against single edge failures. In [20] heterogeneous players are considered whereas in [6] all players are homogeneous. The complexity of computing a best response was only settled for the model by Chauhan et al. [6] where it was proven to be NP-hard.

Apart from network formation games, also vaccination games, e.g. [1,7,18,23], are related. There the network is fixed and the selfish nodes only have to decide if the want to immunize or not. Computing a best response in these models is trivial (there are only two strategies) but pure Nash equilibria may not exist.

Our Contribution: We establish that the natural model by Goyal et al. [13,14] is one of the few examples of a tractable realistic model for strategic network formation and thereby answer an open question by these authors. In particular, we provide an efficient algorithm for computing a best response strategy for their main model, i.e. the "maximum carnage" adversary which tries to kill as many nodes as possible, and for the natural variant which employs the even stronger random attack adversary.

Due to space constraints, we refer to [12] for all omitted details.

2 Model

We consider the model proposed by Goyal et al. [13,14] and mostly use their notation. In this model the n nodes of a network $G = (V, E)$ correspond to individual players $v_1, \ldots, v_n$. We will thus use the terms node, vertex and player interchangeably. The edge set E is determined by the players' strategic behavior as follows. Each player $v_i \in V$ can decide to buy undirected edges to a subset of other players, paying $\alpha > 0$ per edge, where α is some fixed parameter.

If player v_i decides to buy the edge to node v_j, then we say that the edge $\{v_i, v_j\}$ is owned and paid for by player v_i. Buying an undirected edge entails connectivity benefits and risks for both participating endpoints. In order to cope with these risks, each player can also decide to buy immunization against attacks at a cost of $\beta > 0$, which is also a fixed parameter of the model. We call a player *immunized* if this player decides to buy immunization, and *vulnerable* otherwise.

The strategy $s_i = (x_i, y_i)$ of player v_i consists of the set $x_i \subseteq V \setminus \{v_i\}$ of the nodes to buy an edge to, and the immunization choice $y_i \in \{0, 1\}$, where $y_i = 1$ if and only if player v_i decides to immunize. The strategy profile $\mathbf{s} = (s_1, \ldots, s_n)$ of all players then induces an undirected graph $G(\mathbf{s}) = \left(V, \bigcup_{v_i \in V} \bigcup_{v_j \in x_i} \{v_i, v_j\}\right)$. The immunization choices $y_1, \ldots, y_n$ in $\mathbf{s}$ partition V into the set of immunized players $\mathcal{I} \subseteq V$ and vulnerable players $\mathcal{U} = V \setminus \mathcal{I}$. The components in the induced subgraph $G[\mathcal{U}]$ are called *vulnerable regions* and the set of those regions is $\mathcal{R}_\mathcal{U}$. The vulnerable region of any player $v_i \in \mathcal{U}$ is $\mathcal{R}_\mathcal{U}(v_i)$. Immunized regions $\mathcal{R}_\mathcal{I}$ are defined analogously as the components of the induced subgraph $G[\mathcal{I}]$.

After the network $G(\mathbf{s})$ is built, we assume that an adversary attacks one vulnerable player according to a strategy known to the players. We consider mostly the maximum carnage adversary [13,14] which tries to destroy as many nodes of the network as possible. To achieve this, the adversary chooses a vulnerable region of maximum size and attacks some player in that region. If there is more than one such region with maximum size, then one of them is chosen uniformly at random. If a player $v_i \in \mathcal{U}$ is attacked, then v_i will be destroyed and the attack spreads to all vulnerable neighbors of v_i, eventually destroying all players in $\mathcal{R}_\mathcal{U}(v_i)$. Let $t_{max} = \max_{R \in \mathcal{R}_\mathcal{U}} \{|R|\}$ be the number of nodes in the vulnerable region of maximum size and $\mathcal{T} = \{v_i \in \mathcal{U} \mid |\mathcal{R}_\mathcal{U}(v_i)| = t_{max}\}$ be the corresponding set of nodes which may be targeted. The set of targeted regions is $\mathcal{R}_\mathcal{T} = \{R \in \mathcal{R}_\mathcal{U} \mid |R| = t_{max}\}$, and $\mathcal{R}_\mathcal{T}(v_i)$ is the targeted region of a player $v_i \in \mathcal{T}$. Thus, if $v_i \in \mathcal{T}$ is attacked, then all players in $\mathcal{R}_\mathcal{T}(v_i)$ will be destroyed.

The *utility* of a player v_i in network $G(\mathbf{s})$ is defined as the expected number of nodes reachable by v_i after the adversarial attack on network $G(\mathbf{s})$ (zero in case v_i was destroyed) less v_i's expenditures for buying edges and immunization. More formally, let $CC_i(t)$ be the connected component of v_i after an attack to node $v_t \in \mathcal{T}$ and let $|CC_i(t)|$ denote its number of nodes. Then the utility (or profit) $u_i(\mathbf{s})$ of v_i in the strategy profile $\mathbf{s}$ is

$$u_i(\mathbf{s}) = \frac{1}{|\mathcal{T}|} \left(\sum_{v_t \in \mathcal{T}} |CC_i(t)| \right) - |x_i| \cdot \alpha - y_i \cdot \beta.$$

Fixing the strategies of all other players, the *best response* of a player v_i is a strategy $s_i^* = (x_i^*, y_i^*)$ which maximizes v_i's utility $u_i((s_1, \ldots, s_{i-1}, s_i^*, s_{i+1}, \ldots, s_n))$. We will call the strategy change to s_i^* a best response for player v_i in the network $G(\mathbf{s})$, if changing from strategy $s_i \in \mathbf{s}$ to strategy s_i^* is the best possible strategy for player v_i if no other player changes her strategy.

Consider what happens if we remove node v_i from the network $G(\mathbf{s}) = (V, E)$ and we call the obtained network $G(\mathbf{s}) \setminus v_i$. In this case, $G(\mathbf{s}) \setminus v_i$ consists of connected components $C_1, \ldots, C_\ell$. The edge-set x_i^* can thus be partitioned into ℓ subsets $x_i^*(C_1), \ldots, x_i^*(C_\ell)$, where $x_i^*(C_z)$ denotes the set of nodes in C_z to which v_i buys an edge under best response strategy s_i^*. We will say that $x_i^*(C_z)$ is an *optimal partner set* for component C_z. Therefore, x_i^* is the union of optimal partner sets for all connected components in $G(\mathbf{s}) \setminus v_i$.

A best response is calculated for one arbitrary but fixed player v_a, which we call the *active player*. Furthermore let $\mathcal{C}$ be the set of connected components which exist in $G(\mathbf{s}) \setminus v_a$. Let $\mathcal{C}_{\mathcal{U}} = \{C \in \mathcal{C} \mid C \cap \mathcal{I} = \emptyset\}$, $\mathcal{C}_{\mathcal{I}} = \mathcal{C} \setminus \mathcal{C}_{\mathcal{U}}$ and $\mathcal{C}_{inc} = \{C \in \mathcal{C} \mid \exists u \in C : \{u,v\} \in E\}$, where $\mathcal{C}_{\mathcal{U}}$ is the set of components in which all vertices are vulnerable, $\mathcal{C}_{\mathcal{I}}$ is the set of components which contain at least one immunized vertex and $\mathcal{C}_{inc}$ is the set of components to which player v_a is connected through incoming edges bought by some other player.

3   The Best Response Algorithm

A naive approach to calculate the best response for player v_a would consider all 2^n possible strategies and select one that yields the best utility. This is clearly infeasible for a larger number of players.

3.1   Key Observations

Our algorithm exploits three observations to reduce the complexity from exponential to polynomial:

Observation 1: The network $G(\mathbf{s}) \setminus v_a$ may consist of ℓ connected components that can be dealt with independently for most decisions. As long as the set of possible targets of the adversary does not change, the best response of v_a can be constructed by first choosing components to which a connection is profitable and then choosing for each of those components an optimal set of nodes within the respective component to build edges to.

Observation 2: Homogeneous components in $G(\mathbf{s}) \setminus v_a$, which consist of only vulnerable or only immunized nodes, provide the same benefit no matter whether v_a connects to them with one or with more than one edge. Thus the connection decision is a binary decision for those components.

Observation 3: Mixed components in $G(\mathbf{s}) \setminus v_a$, which contain both immunized and vulnerable nodes, consist of homogeneous regions that again have the property that at most one edge per homogeneous region can be profitable. Merging those regions into block nodes forms an auxilliary tree, called Meta Tree, which we use in an efficient dynamic programming algorithm to compute the most profitable subset of regions to connect with.

3.2   Main Algorithm

Our algorithm, called BESTRESPONSECOMPUTATION, is described in Algorithm 1 and a schematic overview can be found in Fig. 1.

Algorithm 1: BESTRESPONSE-COMPUTATION

Input: Strategies $\mathbf{s} = (s_1, \ldots, s_n)$, Player $v_a, 1 \leq a \leq n$

Output: Best response strategy of player v_a denoted by $s_a = (x_a, y_a)$

1   $s_\emptyset = (\emptyset, 0)$;

2   Let $G(\mathbf{s}')$ be the induced game state with $\mathbf{s}' = (s_1, \ldots, s_{a-1}, s_\emptyset, s_{a+1}, \ldots, s_n)$;

3   Let $\mathcal{A}_t, \mathcal{A}_v$ be the solutions of SUBSETSELECT on $\mathcal{C}_{\mathcal{U}}$;

4   Let $\mathcal{A}_g$ be the solution of GREEDYSELECT on $\mathcal{C}_{\mathcal{U}}$;

5   $s_t = \text{POSSIBLESTRATEGY}(\mathcal{A}_t, 0)$;

6   $s_v = \text{POSSIBLESTRATEGY}(\mathcal{A}_v, 0)$;

7   $s_g = \text{POSSIBLESTRATEGY}(\mathcal{A}_g, 1)$;

8   $S = \{s_\emptyset, s_t, s_v, s_g\}$;

9   **return** strategy $s \in S$ which maximizes v_a's utility;

Algorithm 2: POSSIBLESTRATEGY

Input: Set of components $\mathcal{A}$, immunization choice y_a

Output: Best strategy with single edges to components in $\mathcal{A}$, given immunization y_a

1   $M := \emptyset$;

2   **foreach** $C \in \mathcal{A}$ **do**

3     $M = M \cup v_i$ for an arbitrary node $v_i \in C$;

4   Locally, add edges to nodes in M, update $\mathcal{R}_{\mathcal{I}}, \mathcal{R}_{\mathcal{U}}, \mathcal{R}_{\mathcal{T}}$ according to y_a and M;

5   $B \leftarrow \emptyset$;

6   **foreach** *component* $C \in \mathcal{C}_{\mathcal{I}}$ **do**

7     $B \leftarrow B \cup \text{PARTNERSETSELECT}(C)$

8   **return** $(M \cup B, y_a)$;

Our algorithm solves the problem of finding a best response strategy by considering both options of buying or not buying immunization and computing for both cases the best possible set of edges to buy. Thus, the first step of BESTRESPONSECOMPUTATION is to drop the current strategy of the active player v_a and to replace it with the empty strategy $s_\emptyset = (\emptyset, 0)$ in which player v_a does not buy any edge and does not buy immunization. Then the resulting strategy profile $\mathbf{s}' = (s_1, \ldots, s_{a-1}, s_\emptyset, s_{a+1}, \ldots, s_n)$ and the set of connected components $\mathcal{C}_{\mathcal{U}}$ and $\mathcal{C}_{\mathcal{I}}$ with respect to network $G(\mathbf{s}') \setminus v_a$ is considered.

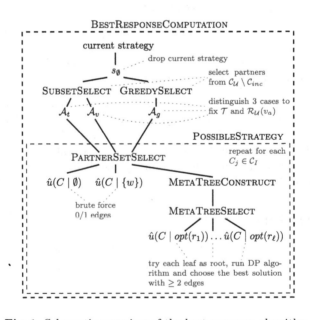

Fig. 1. Schematic overview of the best response algorithm.

The subroutine SUBSETSELECT determines the optimal sets of components of $\mathcal{C}_\mathcal{U}$ to connect to if v_a does not immunize. This is done by solving an adjusted Knapsack problem which involes only small numbers. Two such sets of components, called $\mathcal{A}_t$ and $\mathcal{A}_v$, are computed depending on player v_a becoming targeted or not by connecting to these components. Additionally the subroutine GREEDYSELECT greedily computes a best possible subset of components of $\mathcal{C}_\mathcal{U}$ to connect with in case v_a buys immunization.

The challenging part of the problem is to cope with the connected components in $\mathcal{C}_\mathcal{I}$ which also contain immunized nodes. For such components our algorithm detects and merges equivalent nodes and thereby simplifies these components to an auxiliary tree structure, which we call the Meta Tree. This tree is then used in a dynamic programming fashion to efficiently compute the best possible set of edges to buy towards nodes within the respective component. Thus, our approach for handling components containing immunized nodes can be understood as first performing a data-reduction similar to many approaches for kernelization in the realm of Parameterized Algorithmics [9] and then solving the reduced problem via dynamic programming.

The subroutine POSSIBLESTRATEGY, see Algorithm 2, obtains the best set of nodes in components in $\mathcal{C}_\mathcal{I}$. As this set depends on the number of targeted regions, it has to be determined for several cases independently. These cases are v_a not being immunized and not being targeted, v_a not being immunized but being targeted, and v_a being immunized. The correctness of this is guaranteed by the following lemma.

Lemma 1. *Player v_a can deal with distinct components from $\mathcal{C}_\mathcal{I}$ independently, if $\mathcal{T}$ and $\mathcal{R}_\mathcal{U}(v_a)$ do not change.*

For each case, POSSIBLESTRATEGY first chooses an arbitrary single edge to buy into the previously selected components from $\mathcal{C}_\mathcal{U}$. This is correct since we have:

Lemma 2. *Buying at most one edge into any component $C \in \mathcal{C}_\mathcal{U}$ yields maximum profit for player v_a.*

Then the best set of edges to buy into components in $\mathcal{C}_\mathcal{I}$ is computed independently for each component $C \in \mathcal{C}_\mathcal{I}$ via the subroutines PARTNERSETSELECT, METATREECONSTRUCT and METATREESELECT. The union of the obtained sets is then returned. Finally, the algorithm compares the empty strategy and the individually obtained best possible strategies for the above mentioned cases and selects the one which maximizes player v_a's utility. All in all we get:

Theorem 1. *The algorithm BESTRESPONSECOMPUTATION is correct and runs in polynomial time.*

The run time of our best response algorithm heavily depends on the size of the largest obtained Meta Tree and we achieve a worst-case run time of $\mathcal{O}(n^4 + k^5)$ for the maximum carnage adversary and $\mathcal{O}(n^4 + nk^5)$ for the random attack adversary, where n is the number of nodes in the network and k is the number of blocks in the largest Meta Tree. In the worst case, this yields a run time of $\mathcal{O}(n^5)$ and $\mathcal{O}(n^6)$, respectively. To contrast this worst-case bound, we also provide in [12] empirical results showing that k is usually much smaller than n, which emphasizes the effectiveness of our data-reduction and thereby shows that our algorithm is expected to be much faster than the worst-case upper bound.

3.3   Partner Selection for Components in $\mathcal{C}_\mathcal{I}$

Let $C_1, \ldots, C_c \in \mathcal{C}_\mathcal{I}$ be the components v_a might buy edges into. By definition, each of those components contains at least one immunized node. The next statement ensures that we only need to consider buying edges to such nodes.

Lemma 3. *Player v_a has an optimal partner set for $C \in \mathcal{C}_\mathcal{I}$ which only buys edges to immunized players.*

For computing an optimal partner set for a component $C \in \mathcal{C}_\mathcal{I}$, we consider the expected contribution of C to v_a's profit given that v_a buys edges to all nodes in a set Δ, and denote this profit by $\hat{u}_{v_a}(C \mid \Delta)$.

PartnerSetSelect. For each component $C \in \mathcal{C}_\mathcal{I}$ we compute three candidate sets of players to buy edges to and finally select the candidate set that yields the highest profit contribution for the considered component C for player v_a. The three candidate sets for component C are obtained as follows:

Case 1: The player considers buying no additional edges into C. In this case the resulting player set is empty.

Case 2: The player considers buying one additional edge into C. The resulting player set contains the immunized partner that maximizes the profit for C.

Case 3: The player considers buying at least two edges. An optimal set of at least two immunized partners is obtained via the algorithm META TREE SELECT.

As all possible cases are covered, the most profitable set of those three candidate solutions must be the optimal partner set for component C. This optimal partner set is returned. We refer to this subroutine as PARTNERSETSELECT.

The first two cases, buying either no or exactly one edge into component C are easily solved: if no edge is purchased by v_a, then the expected profit contribution is $\hat{u}_{v_a}(C \mid \emptyset)$. If exactly one edge is bought then the expected profit contribution is $\hat{u}_{v_a}(C \mid \{w\})$, where w is the vertex in C which maximizes v_a's expected profit for component C.

Case 3 is much more difficult to handle. It is the main point where we need to employ algorithmic techniques to avoid a combinatorial explosion. To ease the strategy selection, for each component $C \in \mathcal{C}_\mathcal{I}$ we create an auxiliary graph to identify sets of nodes which offer equivalent benefits with respect to connection. This graph is a bipartite tree which we call the Meta Tree of C. Figure 2 shows a conversion of a graph component into its Meta Tree by merging adjacent nodes of the same type into regions and collapsing regions into blocks. So called Bridge Blocks (orange) of the Meta Tree represent targeted regions of C that would, if destroyed, decompose C into at least two components. If the adversary however chooses to attack a player in a so-called Candidate Block (blue or violet), C would remain connected. Details of the Meta Tree are discussed in [12]. An important property is guaranteed by the following lemma.

Lemma 4. *All leaves of the Meta Tree are Candidate Blocks.*

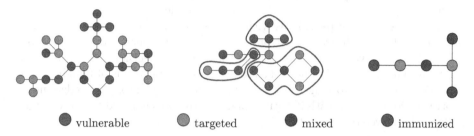

● vulnerable   ● targeted   ● mixed   ● immunized

Fig. 2. A graph component (left), the corresponding Meta Graph (middle), which is an intermediate step in the construction, and the obtained Meta Tree (right). (Color figure online)

We now use the Meta Tree for maximizing the expected component profit for v_a.

Solving Case 3 of PARTNERSETSELECT. In the following let M be the Meta Tree of component C. Moreover, we assume that M has at least two Candidate Blocks, since otherwise, by Lemma 3 buying at most one edge suffices.

Idea of the METATREESELECT *Algorithm.* The following two lemmas imply that we only have to consider to buy single edges into leaves of the Meta Tree which are Candidate Blocks. Thus, we only have to find the optimal combination of leaves of M to which to buy an edge.

Lemma 5. *Buying more than 1 edge to a Candidate Block is never beneficial.*

Lemma 6. *Let M be the Meta Tree of component C. If player v_a has an optimal partner set for C which contains buying at least two edges, then v_a also has an optimal partner set for C which contains only leaves of M.*

Probing all possible combinations of leaves of M yields exponential runtime. We use the following two observations to compute the best possible combination of leaves efficiently. Both observations are based on the assumption that player v_a buys an edge to some leaf r of M and we consider the tree M rooted at r. Later we ensure this assumption by rooting M at each possible leaf. Let w be any vertex of M.

Observation 1: If player v_a has an edge to w, then it can be decided efficiently whether it is beneficial to buy exactly one or no edge into a subtree of w, as the influence of any additional edges into M does not propagate over w. Hence decisions are independent for subtrees.

Observation 2: Let the children of w in M be $x_1, \ldots, x_\ell$. Consider that v_a has an edge to w and it has already been decided for each subtree rooted at $x_1, \ldots, x_\ell$ whether or not to buy an edge into that subtree. If there exists at least one edge between v_a and any of those subtrees, then it cannot be beneficial to buy additional edges into the subtree rooted at w. Either w is destroyed, and the previous edge-buy decisions apply to the disconnected subtrees, or w survives, and v_a is connected to all subtrees via node w.

These observations provide us with the foundation for a dynamic programming algorithm which decides bottom-up whether it is beneficial to buy at most one edge into a given subtree by reusing the edge buy decisions of its subtrees.

Note that the algorithm never has to compare combinations of bought edges, as the only decision to make is, whether or not to buy exactly one edge into a subtree in combination with iteratively shifting the presumed edge to the parent node of the leaves to the root r.

The METATREESELECT *Algorithm:* The METATREESELECT algorithm can be found in Algorithm 3. It roots M at every leaf and assumes buying an edge towards some immunized node within the root Candidate Block. Then the subroutine ROOTEDMETATREESELECT, see Algorithm 4, gets the rooted Meta Tree $M(r)$ and some vertex r_T (which initially is the only child of the currently considered root leaf) as input and recursively computes the expected profit contribution of one additional edge from v_a to a block in the subtree T rooted at r_T under the assumption that v_a is already connected to the parent block $p(r_T)$ of r_T in $M(r)$. Let $|T|$ denote the number of players represented by the union of all blocks in T and $opt(r_T)$ will be the set of blocks in T the algorithm decided to buy an edge to.

Algorithm 3: METATREESELECT

Input: Meta Tree M for component C

Output: Player v_a's optimal partner set for C consisting of at least two partners

1 **foreach** *leaf r of M* **do**
2      $M(r) \leftarrow$ root M at vertex r;
3      $w \leftarrow r$'s only child in $M(r)$;
4      $opt(r) \leftarrow$ {some immunized node in r} $\cup$ ROOTEDMETATREESELECT$(M(r), w)$;

5 $best \leftarrow opt(r)$ which maximizes $\hat{u}_{v_a}(C \mid opt(r))$;
6 **if** $|best| \geq 2$ **then**
7   **return** $best$;

8 **else**
9   **return** $\emptyset$;

Algorithm 4: ROOTEDMETATREESELECT

Input: rooted Meta Tree $M(r)$, vertex r_T of $M(r)$

Output: Set of nodes from T to buy an edge to

1 $opt(r_T) \leftarrow \emptyset$;
2 **foreach** *child w of r_T* **do**
3      $opt(r_T) \leftarrow opt(r_T) \cup$ ROOTEDMETATREESELECT$(M(r), w)$;

4 **if** r_T *is a Bridge Block or* $opt(r_T) \neq \emptyset$ *or a player in T bought an edge to v_a* **then**
5   **return** $opt(r_T)$;

6 **foreach** *leaf l of T* **do**
7      $profit(l) \leftarrow$ additional profit of v_a with edge to l;

8 $best \leftarrow l$ which maximizes $profit(l)$;
9 **if** $profit(best) > \alpha$ **then**
10     $opt(r_T) \leftarrow opt(r_T) \cup$ {some immunized node in $best$};

11 **return** $opt(r_T)$;

After processing all subtrees of r_T (Algorithm 4 lines 2–3), the algorithm distinguishes three cases (Algorithm 4 line 4):

Case 1: r_T is a Bridge Block. Then, as M is bipartite, $p(r_T)$ must be a Candidate Block. As the algorithm assumes the existence of an edge from v_a to $p(r_T)$, there also exists a path from v_a to r_T via $p(r_T)$ in all attack scenarios. Thus, no additional edge is needed (Algorithm 4 line 5).

Case 2: There exists an edge between v_a and some node x in T, either through an edge v_a buys according to the results of the recursive invocations, or through a preexisting edge bought by player x. Then, depending on the attack target, there either exists a path from v_a to r_T via x or via $p(r_T)$. Hence, no additional edge is needed (Algorithm 4 line 5).

Case 3: Player v_a can get disconnected from r_T by an attack on $p(r_T)$. Then the algorithm considers each leaf l of T as possible partner (Algorithm 4 line 6), computes the profit contribution of an edge to l (Algorithm 4 line 7) and selects a leaf that maximizes this profit contribution (Algorithm 4 line 8).

The additional profit of an edge to l is computed as follows: An edge to l only yields profit, if a Bridge Block t is attacked which either belongs to T or $t = p(r_T)$, and l is located in a subtree of t. In this case, the profit contribution equals the size of this subtree. Therefore let profit$(l \mid t)$ be the additional profit an edge to l contributes to the utility of v_a in case t is attacked and let $\mathcal{B}$ be the set of all Bridge Blocks in T. Thus profit$(l) = \frac{|p(r_T)|}{|T|}|T| + \sum_{t \in \mathcal{B}} \frac{|t|}{|T|}$profit$(l \mid t)$, with

$$\text{profit}(l \mid t) = \begin{cases} 0, & \text{if } l \text{ is not in any subtree of } t \\ |Y|, & \text{if } Y \text{ is a subtree of } t \text{ and } l \text{ is in } Y. \end{cases}$$

Finally, If the additional profit of the best possible leaf exceeds the edge costs, l is added to the set of partners of v_a (Algorithm 4 line 10).

The correctness of METATREESELECT is based in the following statement:

Lemma 7. *If v_a has an edge to $p(r_T)$ and $opt(r_T)$ is returned by* ROOTEDME- TATREESELECT*$(M(r), r_T)$, then there exists an optimal partner set for component C which contains r^* and $opt(r_T)$.*

Theorem 2. *If there is an optimal partner set with at least two nodes for component C, then* METATREESELECT *algorithm outputs such a set.*

Proof. Assume that there exists an optimal partner set with at least two nodes for component C and assume that the Meta Tree M of component C is rooted at some leaf r. Since the algorithm compares all possibilities to root M at a leaf and by Lemma 6, at least one of those leaves must be contained in an optimal partner set. Assume that r is indeed such a leaf.

Thus, by buying r we satisfy the assumption needed for ROOTEDMETA- TREESELECT. By Lemma 7, ROOTEDMETATREESELECT returns a set of nodes, which together with r^* yields an optimal partner set for C. Hence, the algorithm METATREESELECT is correct.   □

4   Conclusion

For most models of strategic network formation computing a utility maximizing strategy is known to be NP-hard. In this paper, we have proven that the model by Goyal et al. [13,14] is a notable exception to this rule. The presented efficient algorithm for computing a best response for a player circumvents a combinatorial explosion essentially by simplifying the given network and thereby making it amenable to a dynamic programming approach. An efficient best response computation is the key ingredient for using the model in large scale simulations and for analyzing real world networks. Moreover, our algorithm can be adapted to a significantly stronger adversary and we are confident that further modifications for coping with other variants of the model are possible.

Future Work: Settling the complexity of computing a best response strategy with respect to the maximum disruption adversary is left as an open problem. Besides this, it seems worthwhile to consider a variant with directed edges, originally introduced by Bala and Goyal [2]. Directed edges would more accurately model the differences in risk and benefit which depend on the flow direction. Using the analogy of the WWW, a user who downloads information benefits from it, but also risks getting infected. In contrast, the user providing the information is exposed to little or no risk.

References

1. Aspnes, J., Chang, K., Yampolskiy, A.: Inoculation strategies for victims of viruses and the sum-of-squares partition problem. J. Comput. Syst. Sci. **72**(6), 1077–1093 (2006)
2. Bala, V., Goyal, S.: A noncooperative model of network formation. Econometrica **68**(5), 1181–1229 (2000)
3. Bala, V., Goyal, S.: A strategic analysis of network reliability. Rev. Econ. Des. **5**(3), 205–228 (2000). doi:10.1007/s100580000019. ISSN 1434-4750
4. Bilò, D., Gualà, L., Leucci, S., Proietti, G.: Locality-based network creation games. In: SPAA 2014, pp. 277–286 (2014)
5. Bilò, D., Gualà, L., Proietti, G.: Bounded-distance network creation games. ACM TEAC **3**(3), 16:1–16:20 (2015)
6. Chauhan, A., Lenzner, P., Melnichenko, A., Münn, M.: On selfish creation of robust networks. In: Gairing, M., Savani, R. (eds.) SAGT 2016. LNCS, vol. 9928, pp. 141–152. Springer, Heidelberg (2016). doi:10.1007/978-3-662-53354-3_12
7. Chen, P.-A., David, M., Kempe, D.: Better vaccination strategies for better people. In: EC 2010, pp. 179–188. ACM (2010)
8. Cord-Landwehr, A., Lenzner, P.: Network creation games: think global - act local. In: MFCS 2015, pp. 248–260 (2015)
9. Downey, R.G., Fellows, M.R.: Fundamentals of Parameterized Complexity. Texts in Computer Science. Springer, Heidelberg (2013)
10. Ehsani, S., Fadaee, S.S., Fazli, M., Mehrabian, A., Sadeghabad, S.S., Safari, M.A., Saghafian, M.: A bounded budget network creation game. ACM Trans. Algorithms **11**(4), 34 (2015)

11. Fabrikant, A., Luthra, A., Maneva, E.N., Papadimitriou, C.H., Shenker, S.: On a network creation game. In: PODC 2003, pp. 347–351 (2003)
12. Friedrich, T., Ihde, S., Keßler, C., Lenzner, P., Neubert, S., Schumann, D.: Efficient best-response computation for strategic network formation under attack. CoRR, abs/1610.01861 (2016)
13. Goyal, S., Jabbari, S., Kearns, M., Khanna, S., Morgenstern, J.: Strategic Network Formation with Attack and Immunization. arXiv preprint arXiv:1511.05196 (2015)
14. Goyal, S., Jabbari, S., Kearns, M., Khanna, S., Morgenstern, J.: Strategic network formation with attack and immunization. In: Cai, Y., Vetta, A. (eds.) WINE 2016. LNCS, vol. 10123, pp. 429–443. Springer, Heidelberg (2016). doi:10.1007/978-3-662-54110-4_30
15. Jackson, M.O., Wolinsky, A.: A strategic model of social and economic networks. J. Econ. Theory **71**(1), 44–74 (1996)
16. Kawald, B., Lenzner, P.: On dynamics in selfish network creation. In: SPAA 2013, pp. 83–92. ACM (2013)
17. Kliemann, L.: The price of anarchy for network formation in an adversary model. Games **2**(3), 302–332 (2011)
18. Kumar, V.A., Rajaraman, R., Sun, Z., Sundaram, R.: Existence theorems and approximation algorithms for generalized network security games. In: ICDCS 2010, pp. 348–357. IEEE (2010)
19. Lenzner, P.: Greedy selfish network creation. In: WINE 2012, pp. 142–155 (2012)
20. Meirom, E.A., Mannor, S., Orda, A.: Formation games of reliable networks. In: INFOCOM 2015, pp. 1760–1768 (2015)
21. Mihalák, M., Schlegel, J.C.: The price of anarchy in network creation games is (mostly) constant. In: Kontogiannis, S., Koutsoupias, E., Spirakis, P.G. (eds.) SAGT 2010. LNCS, vol. 6386, pp. 276–287. Springer, Heidelberg (2010). doi:10.1007/978-3-642-16170-4_24
22. Papadimitriou, C.H.: Algorithms, games, and the internet. In: STOC 2001, pp. 749–753 (2001)
23. Saha, S., Adiga, A., Vullikanti, A.K.S.: Equilibria in epidemic containment games. In: AAAI, pp. 777–783 (2014)

Path Deviations Outperform Approximate Stability in Heterogeneous Congestion Games

Pieter Kleer[1] and Guido Schäfer[1,2(✉)]

[1] Centrum Wiskunde & Informatica (CWI), Networks and Optimization Group,
Amsterdam, The Netherlands
{kleer,schaefer}@cwi.nl
[2] Department of Econometrics and Operations Research,
Vrije Universiteit Amsterdam, Amsterdam, The Netherlands

Abstract. We consider non-atomic network congestion games with heterogeneous players where the latencies of the paths are subject to some bounded deviations. This model encompasses several well-studied extensions of the classical Wardrop model which incorporate, for example, risk-aversion, altruism or travel time delays. Our main goal is to analyze the worst-case deterioration in social cost of a *deviated Nash flow* (i.e., for the perturbed latencies) with respect to an original Nash flow.

We show that for homogeneous players deviated Nash flows coincide with approximate Nash flows and derive tight bounds on their inefficiency. In contrast, we show that for heterogeneous populations this equivalence does not hold. We derive tight bounds on the inefficiency of both deviated and approximate Nash flows for *arbitrary* player sensitivity distributions. Intuitively, our results suggest that the negative impact of path deviations (e.g., caused by risk-averse behavior or latency perturbations) is less severe than approximate stability (e.g., caused by limited responsiveness or bounded rationality).

We also obtain a tight bound on the inefficiency of deviated Nash flows for matroid congestion games and homogeneous populations if the path deviations can be decomposed into edge deviations. In particular, this provides a tight bound on the Price of Risk-Aversion for matroid congestion games.

1 Introduction

In 1952, Wardrop [17] introduced a simple model, also known as the *Wardrop model*, to study outcomes of selfish route choices in traffic networks which are affected by congestion. In this model, there is a continuum of non-atomic players, each controlling an infinitesimally small amount of flow, whose goal is to choose paths in a given network to minimize their own travel times. The latency (or delay) of each edge is prescribed by a non-negative, non-decreasing latency function which depends on the total flow on that edge. Ever since its introduction, the Wardrop model has been used extensively, both in operations research and traffic engineering studies, to investigate various aspects of selfish routing in networks.

© Springer International Publishing AG 2017
V. Bilò and M. Flammini (Eds.): SAGT 2017, LNCS 10504, pp. 212–224, 2017.
DOI: 10.1007/978-3-319-66700-3_17

More recently, the classical Wardrop model has been extended in various ways to capture more complex player behaviors. Examples include the incorporation of uncertainty attitudes (e.g., risk-aversion, risk-seeking), cost alterations (e.g., latency perturbations, road pricing), other-regarding dispositions (e.g., altruism, spite) and player biases (e.g., responsiveness, bounded rationality).

Several of these extensions can be viewed as defining some modified cost for each path which combines the original latency with some 'deviation' (or perturbation) along that path. Such deviations are said to be β-*bounded* if the total deviation along each path is at most β times the latency of that path. The player objective then becomes to minimize the combined cost of latency and deviation along a path (possibly using different norms). An equilibrium outcome corresponds to a β-*deviated Nash flow*, i.e., a Nash flow with respect to the combined cost. The deviations might be given explicitly (e.g., as in the altruism model of Chen et al. [1]) or be defined implicitly (e.g., as in the risk-aversion model of Nikolova and Stier-Moses [13]). Further, different fractions of players might perceive these deviations differently, i.e., players might be heterogeneous with respect to the deviations.

Another extension, which is closely related to the one above, is to incorporate different degrees of 'responsiveness' of the players. For example, each player might be willing to deviate to an alternative route only if her latency decreases by at least a certain fraction. In this context, an equilibrium outcome corresponds to an ϵ-*approximate Nash flow* for some $\epsilon \geq 0$, i.e., for each player the latency is at most $(1 + \epsilon)$ times the latency of any other path. Here, ϵ is a parameter which reflects the responsiveness of the players. An analogue definition can be given for populations with heterogeneous responsiveness parameters.

To illustrate the relation between deviated and approximate Nash flows, suppose we are given a β-deviated Nash flow f for some $\beta \geq 0$, where the latency $\ell_P(f)$ of each path P is perturbed by an arbitrary β-bounded deviation $\delta_P(f)$ satisfying $0 \leq \delta_P(f) \leq \beta \ell_P(f)$. Intuitively, the deviations inflate the latency on each path by at most a factor of $(1 + \beta)$. Further, assume that the population is homogeneous. From the Nash flow conditions (see Sect. 2 for formal definitions), it follows trivially that f is also an ϵ-approximate Nash flow with $\epsilon = \beta$. But does the converse also hold? That is, can every ϵ-approximate Nash flow be induced by a set of bounded path deviations? More generally, what about the relation between deviated and approximate Nash flows for heterogenous populations? Can we bound the inefficiency of these flows?

In this paper, we answer these questions by investigating the relation between the two equilibrium notions. Our main goal is to quantify the inefficiency of deviated and approximate Nash flows, both for homogeneous and heterogeneous populations. To this aim, we study the (relative) worst-case deterioration in social cost of a β-deviated Nash flow with respect to an original (unaltered) Nash flow; we use the term β-*deviation ratio* to refer to this ratio. This ratio has recently been studied in the context of risk aversion [9,13] and in the more general context of bounded path deviations [6]. Similarly, for approximate Nash flows we are interested in bounding the ϵ-*stability ratio*, i.e., the worst-case deterioration in social cost of an ϵ-approximate Nash flow with respect to an original Nash flow.

Note that these notions differ from the classical *price of anarchy* notion [8], which refers to the worst-case deterioration in social cost of a β-deviated (respectively, ε-approximate) Nash flow with respect to an *optimal* flow. While the price of anarchy typically depends on the class of latency functions (see, e.g., [1,2,6,13] for results in this context), the deviation ratio is independent of the latency functions but depends on the topology of the network (see [6,13]).

Our Contributions. The main contributions of this paper are as follows:

1. We show that for homogeneous populations the set of β-deviated Nash flows coincides with the set of ϵ-approximate Nash flows for $\beta = \epsilon$. Further, we derive an upper bound on the ϵ-stability ratio (and thus also on the ϵ-deviation ratio) which is at most $(1 + \epsilon)/(1 - \epsilon n)$, where n is the number of nodes, for single-commodity networks. We also prove that the upper bound we obtain is tight for *generalized Braess graphs*. These results are presented in Sect. 4.

2. We prove that for heterogenous populations the above equivalence does not hold. We derive tight bounds for both the β-deviation ratio and the ϵ-stability ratio for single-commodity instances on series-parallel graphs and arbitrary sensitivity distributions of the players. To the best of our knowledge, these are the first inefficiency results in the context of heterogenous populations which are tight for *arbitrary* sensitivity distributions. Our bounds show that both ratios depend on the demands and sensitivity distribution γ of the heterogenous players (besides the respective parameters β and ϵ). Further, it turns out that the β-deviation ratio is always at most the ϵ-stability ratio for $\epsilon = \beta\gamma$. These results are given in Sect. 3.

3. We also derive a tight bound on the β-deviation ratio for single-commodity matroid congestion games and homogeneous populations if the path deviations can be decomposed into edge deviations. To the best of our knowledge, this is the first result in this context which goes beyond network congestion games. In particular, this gives a tight bound on the Price of Risk-Aversion [13] for matroid congestion games. This result is of independent interest and presented in Sect. 4.

In a nutshell, our results reveal that for homogeneous populations there is no quantitative difference between the inefficiency of deviated and approximate Nash flows in the worst case. In contrast, for heterogenous populations the β-deviation ratio is always at least as good as the ϵ-stability ratio with $\epsilon = \beta\gamma$. Intuitively, our results suggest that the negative impact of path deviations (e.g., caused by risk-averse behavior or latency perturbations) is less severe than approximate stability (e.g., caused by limited responsiveness or bounded rationality).

Related Work. We give a brief overview of the works which are most related to our results. Christodoulou et al. [2] study the inefficiency of approximate equilibria in terms of the price of anarchy and price of stability (for homogeneous populations). Generalized Braess graphs were introduced by Roughgarden [14] and are used in many other lower bound constructions (see, e.g., [3,6,14]). Chen et al. [1] study an altruistic extension of the Wardrop model and, in particular, also consider heterogeneous altruistic populations. They obtain an upper bound on the ratio between an altruistic Nash flow and a social optimum for parallel graphs, which is tight for two sensitivity classes. It is mentioned that this bound is most likely not tight in general. Meir and Parkes [11] study player-specific cost functions in a smoothness framework [15]. Some of their inefficiency results are tight, although none of their bounds seems to be tight for arbitrary sensitivity distributions. Matroids have also received some attention in the Wardrop model. In particular, Fujishige et al. [5] show that matroid congestion games are immune against the Braess paradox (and their analysis is tight in a certain sense). We refer the reader to [6] for additional references and relations of other models to the bounded path deviation model considered here.

2   Preliminaries

Let $\mathcal{I} = (E, (l_e)_{e \in E}, (\mathcal{S}_i)_{i \in [k]}, (r_i)_{i \in [k]})$ be an instance of a non-atomic congestion game. Here, E is the set of resources (or edges, or arcs) that are equipped with a non-negative, non-decreasing, continuous latency function $l_e : \mathbb{R}_{\geq 0} \to \mathbb{R}_{\geq 0}$. Each commodity $i \in [k]$ has a strategy set $\mathcal{S}_i \subseteq 2^E$ and demand $r_i \in \mathbb{R}_{>0}$. Note that in general the strategy set $\mathcal{S}_i$ of player i is defined by arbitrary resource subsets. If each strategy $P \in \mathcal{S}_i$ corresponds to an s_i, t_i-path in a given directed graph, then the corresponding game is called a *network* congestion game.[1] We slightly abuse terminology and use the term *path* also to refer to a strategy $P \in \mathcal{S}_i$ of player i (which does not necessarily correspond to a path in a graph); no confusion shall arise. We denote by $\mathcal{S} = \cup_i \mathcal{S}_i$ the set of all paths.

An outcome of the game is a (feasible) flow $f^i : \mathcal{S}_i \to \mathbb{R}_{\geq 0}$ satisfying $\sum_{P \in \mathcal{S}_i} f_P^i = r_i$ for every $i \in [k]$. We use $\mathcal{F}(\mathcal{S})$ to denote the set of all feasible flows $f = (f^1, \ldots, f^k)$. Given a flow $f = (f^i)_{i \in [k]} \in \mathcal{F}(\mathcal{S})$, we use f_e^i to denote the total flow on resource $e \in E$ of commodity $i \in [k]$, i.e., $f_e^i = \sum_{P \in \mathcal{S}_i : e \in P} f_P^i$. The total flow on edge $e \in E$ is defined as $f_e = \sum_{i \in [k]} f_e^i$.

The latency of a path $P \in \mathcal{S}$ with respect to f is defined as $l_P(f) := \sum_{e \in P} l_e(f_e)$. The cost of commodity i with respect to f is $C_i(f) = \sum_{P \in \mathcal{S}_i} f_P l_P(f)$. The *social cost* $C(f)$ of a flow f is given by its total average latency, i.e., $C(f) = \sum_{i \in [k]} C_i(f) = \sum_{e \in E} f_e l_e(f_e)$. A flow that minimizes $C(\cdot)$ is called *(socially) optimal*.

If the population is heterogenous, then each commodity $i \in [k]$ is further partitioned in h_i *sensitivity classes*, where class $j \in [h_i]$ has demand r_{ij} such that $r_i = \sum_{j \in [h_i]} r_{ij}$. Given a path $P \in \mathcal{S}_i$, we use $f_{P,j}$ to refer to the amount of flow on path P of sensitivity class j (so that $\sum_{j \in [h_i]} f_{P,j} = f_P$).

---

[1] If a network congestion game with a single commodity is considered (i.e., $k = 1$), we omit the commodity index for ease of notation.

Deviated Nash flows. We consider a *bounded deviation model* similar to the one introduced in [6].[2] We use $\delta = (\delta_P)_{P \in \mathcal{S}}$ to denote some arbitrary path deviations, where $\delta_P : \mathcal{F}(\mathcal{S}) \to \mathbb{R}_{\geq 0}$ for all $P \in \mathcal{S}$. Let $\beta \geq 0$ be fixed. Define the set of *β-bounded path deviations* as $\Delta(\beta) = \{(\delta_P)_{P \in \mathcal{S}} \mid 0 \leq \delta_P(f) \leq \beta l_P(f) \text{ for all } f \in \mathcal{F}(\mathcal{S})\}$.

Every commodity $i \in [k]$ and sensitivity class $j \in [h_i]$ has a non-negative sensitivity γ_{ij} with respect to the path deviations. The population is *homogeneous* if $\gamma_{ij} = \gamma$ for all $i \in [k]$, $j \in [h_i]$ and some $\gamma \geq 0$; otherwise, it is *heterogeneous*. Define the *deviated latency* of a path $P \in \mathcal{S}_i$ for sensitivity class $j \in [h_i]$ as $q_P^j(f) = l_P(f) + \gamma_{ij} \delta_P(f)$.

We say that a flow f is a *β-deviated Nash flow* if there exist some β-bounded path deviations $\delta \in \Delta(\beta)$ such that

$$\forall i \in [k], \forall j \in [h_i], \forall P \in \mathcal{S}_i, f_{P,j} > 0: \qquad q_P^j(f) \leq q_{P'}^j(f) \ \forall P' \in \mathcal{S}_i. \quad (1)$$

We define the *β-deviation ratio* β-DR$(\mathcal{I})$ as the maximum ratio $C(f^\beta)/C(f^0)$ of an β-deviated Nash flow f^β and an original Nash flow f^0. Intuitively, the deviation ratio measures the worst-case deterioration in social cost as a result of (bounded) deviations in the path latencies. Note that here the comparison is done with respect to an *unaltered* Nash flow to measure the impact of these deviations.

The set $\Delta(\beta)$ can also be restricted to path deviations which are defined as a function of edge deviations along that path. Suppose every edge $e \in E$ has a deviation $\delta_e : \mathbb{R}_{\geq 0} \to \mathbb{R}_{\geq 0}$ satisfying $0 \leq \delta_e(x) \leq \beta l_e(x)$ for all $x \geq 0$. For example, feasible path deviations can then be defined by the L_1-norm objective $\delta_P(f) = \sum_{e \in P} \delta_e(x)$ (as in [6,13]) or the L_2-norm objective $\delta_P(f) = \sqrt{\sum_{e \in P} \delta_e(x)^2}$ (as in [9,13]). The *Price of Risk-Aversion* introduced by Nikolova and Stier-Moses [13] is technically the same ratio as the deviation ratio for the L_1- and L_2-norm (see [6] for details).

Approximate Nash Flows. We introduce the notion of an approximate Nash flow. Also here, each commodity $i \in [k]$ and sensitivity class $j \in [h_i]$ has a non-negative sensitivity ϵ_{ij}. We say that the population is *homogeneous* if $\epsilon_{ij} = \epsilon$ for all $i \in [k]$, $j \in [h_i]$ and some $\epsilon \geq 0$; otherwise, it is *heterogeneous*.

A flow f is an *ϵ-approximate Nash flow* with respect to sensitivities $\epsilon = (\epsilon_{ij})_{i \in [k], j \in [h_i]}$ if

$$\forall i \in [k], \ \forall j \in [h_i], \ \forall P \in \mathcal{S}_i, f_{P,j} > 0: \qquad l_P(f) \leq (1 + \epsilon_{ij}) l_{P'}(f) \ \forall P' \in \mathcal{S}_i \quad (2)$$

Note that a 0-approximate Nash flow is simply a Nash flow. We define the *ϵ-stability ratio* ϵ-SR$(\mathcal{I})$ as the maximum ratio $C(f^\epsilon)/C(f^0)$ of an ϵ-approximate Nash flow f^ϵ and an original Nash flow f^0.

Some of the proofs are missing in the main text below and can be found in [7].

---

[2] In fact, in [6] more general path deviations are introduced; the path deviations considered here correspond to $(0, \beta)$-*path deviations* in [6].

3   Heterogeneous Populations

We first elaborate on the relation between deviated and approximate Nash flows for general congestion games with heterogeneous populations.

Proposition 1. *Let $\mathcal{I}$ be a congestion game with heterogeneous players. If f is a β-deviated Nash flow for $\mathcal{I}$, then f is an ϵ-approximate Nash flow for $\mathcal{I}$ with $\epsilon_{ij} = \beta\gamma_{ij}$ for all $i \in [k]$ and $j \in [h_i]$ (for the same demand distribution r).*

Discrete Sensitivity Distributions. Subsequently, we show that the reverse of Proposition 1 does not hold. We do this by providing tight bounds on the β-deviation ratio and the ϵ-stability ratio for instances on (single-commodity) series-parallel graphs and arbitrary discrete sensitivity distributions.

Theorem 1. *Let $\mathcal{I}$ be a single-commodity network congestion game on a series-parallel graph with heterogeneous players, demand distribution $r = (r_i)_{i\in[h]}$ normalized to 1, i.e., $\sum_{j\in[h]} r_i = 1$, and sensitivity distribution $\gamma = (\gamma_i)_{i\in[h]}$, with $\gamma_1 < \gamma_2 < \cdots < \gamma_h$. Let $\beta \geq 0$ be fixed and define $\epsilon = (\beta\gamma_i)_{i\in[h]}$. Then the ϵ-stability ratio and the β-deviation ratio are bounded by:*

$$\epsilon\text{-}SR(\mathcal{I}) \leq 1 + \beta \sum_{j=1}^{h} r_j\gamma_j \quad and \quad \beta\text{-}DR(\mathcal{I}) \leq 1 + \beta \cdot \max_{j\in[h]} \left\{ \gamma_j \left(\sum_{p=j}^{h} r_p \right) \right\}. \quad (3)$$

Further, both bounds are tight for all distributions r and γ.

It is not hard to see that the bound on the β-deviation ratio is always smaller than the bound on the ϵ-stability ratio.[3] Our bound on the β-deviation ratio also yields tight bounds on the *Price of Risk-Aversion* [13] for series-parallel graphs and arbitrary heterogeneous risk-averse populations, both for the L_1-norm and L_2-norm objective.[4]

   We need the following technical lemma for the proof of the β-deviation ratio.

Lemma 1. *Let $0 \leq \tau_{k-1} \leq \cdots \leq \tau_1 \leq \tau_0$ and $c_i \geq 0$ for $i = 1, \ldots, k$ be given. We have $c_1\tau_0 + \sum_{i=1}^{k-1}(c_{i+1} - c_i)\tau_i \leq \tau_0 \cdot \max_{i=1,\ldots,k}\{c_i\}$.*

Proof (Theorem 1, β-deviation ratio). Let $x = f^\beta$ be a β-deviated Nash flow with path deviations $(\delta_P)_{P\in\mathcal{S}} \in \Delta(\beta)$ and let $z = f^0$ be an original Nash flow. Let $X = \{a \in A : x_a > z_a\}$ and $Z = \{a \in A : z_a \geq x_a \text{ and } z_a > 0\}$ (arcs with $x_a = z_a = 0$ may be removed without loss of generality).

   In order to analyze the ratio $C(x)/C(z)$ we first argue that we can assume without loss of generality that the latency function $l_a(y)$ is constant for values $y \geq x_a$ for all arcs $a \in Z$. To see this, note that we can replace the function $l_a(\cdot)$ with the function $\hat{l}_a$ defined by $\hat{l}_a(y) = l_a(x_a)$ for all $y \geq x_a$ and $\hat{l}_a(y) = l_a(y)$

---

[3] This follows from Markov's inequality: for a random variable Y, $P(Y \geq t) \leq E(Y)/t$.
[4] Observe that we show tightness of the bound on parallel arcs, in which case these objectives coincide.

for $y \leq x_a$. In particular, this implies that the flow x is still a β-deviated Nash flow for the same path deviations as before. This holds since for any path P the latency $l_P(x)$ remains unchanged if we replace the function l_a by $\hat{l}_a$.

By definition of arcs in Z, we have $x_a \leq z_a$ and therefore $\hat{l}_a(z_a) = l_a(x_a) \leq l_a(z_a)$. Let z' be an original Nash flow for the instance with l_a replaced by $\hat{l}_a$. Then we have $C(z') \leq C(z)$ using the fact that series-parallel graphs are immune to the Braess paradox, see Milchtaich [12, Lemma 4]. Note that, in particular, we find $C(x)/C(z) \leq C(x)/C(z')$. By repeating this argument, we may without loss of generality assume that all latency functions l_a are constant between x_a and z_a for $a \in Z$. Afterwards, we can even replace the function $\hat{l}_a$ by a function that has the constant value of $l_a(x_a)$ everywhere.

In the remainder of the proof, we will denote P_j as a flow-carrying arc for sensitivity class $j \in [h]$ that maximizes the path latency amongst all flow-carrying path for sensitivity class $j \in [h]$, i.e., $P_j = \mathrm{argmax}_{P \in \mathcal{P} : x_{P,j} > 0} \{l_P(x)\}$. Moreover, there also exists a path P_0 with the property that $z_a \geq x_a$ and $z_a > 0$ for all arcs $a \in P_0$ (see, e.g., Lemma 2 [12]).

For fixed $a < b \in \{1, \ldots, h\}$, the Nash conditions imply that (these steps are of a similar nature as Lemma 1 [4])

$$l_{P_a}(x) + \gamma_a \cdot \delta_{P_a}(x) \leq l_{P_b}(x) + \gamma_a \cdot \delta_{P_b}(x)$$
$$l_{P_b}(x) + \gamma_b \cdot \delta_{P_b}(x) \leq l_{P_a}(x) + \gamma_b \cdot \delta_{P_a}(x).$$

Adding up these inequalities implies that $(\gamma_b - \gamma_a)\delta_{P_b}(x) \leq (\gamma_b - \gamma_a)\delta_{P_a}(x)$, which in turn yields that $\delta_{P_b}(x) \leq \delta_{P_a}(x)$ (using that $\gamma_a < \gamma_b$ if $a < b$). Furthermore, we also have

$$l_{P_1}(x) + \gamma_1 \delta_{P_1}(x) \leq l_{P_0}(x) + \gamma_1 \delta_{P_0}(x), \tag{4}$$

and $l_{P_0}(x) = l_{P_0}(z) \leq l_{P_1}(z) \leq l_{P_1}(x)$, which can be seen as follows. The equality follows from the fact that l_a is constant for all $a \in Z$ and, by choice, P_0 only consists of arcs in Z. The first inequality follows from the Nash conditions of the original Nash flow z, since there exists a flow-decomposition in which the path P_0 is used (since the flow on all arcs of P_0 is strictly positive in z). The second inequality follows from the fact that

$$\sum_{e \in P_1} l_e(z_e) = \sum_{e \in P_1 \cap X} l_e(z_e) + \sum_{e \in P_1 \cap Z} l_e(z_e) \leq \sum_{e \in P_1 \cap X} l_e(x_e) + \sum_{e \in P_1 \cap Z} l_e(x_e)$$

using that $z_e \leq x_e$ for $e \in X$ and the fact that latency functions for $e \in Z$ are constant. In particular, we find that $l_{P_0}(x) \leq l_{P_1}(x)$. Adding this inequality to (4), we obtain $\gamma_1 \delta_{P_1}(x) \leq \gamma_1 \delta_{P_0}(x)$ and therefore $\delta_{P_1}(x) \leq \delta_{P_0}(x)$. Thus $\delta_{P_h}(x) \leq \delta_{P_{h-1}}(x) \leq \cdots \leq \delta_{P_1}(x) \leq \delta_{P_0}(x)$. Moreover, by using induction it can be shown that

$$l_{P_j}(x) \leq l_{P_0}(x) + \gamma_1 \delta_{P_0}(x) + \left[\sum_{g=1}^{j-1} (\gamma_{g+1} - \gamma_g)\delta_{P_g}(x) \right] - \gamma_j \delta_{P_j}(x). \tag{5}$$

Using (5), we then have

$$C(x) \leq \sum_{j=1}^{h} r_j l_{P_j}(x) \quad \text{(by choice of the paths } P_j)$$

$$\leq \sum_{j=1}^{h} r_j \left(l_{P_0}(x) + \gamma_1 \delta_{P_0}(x) + \left[\sum_{g=1}^{j-1} (\gamma_{g+1} - \gamma_g) \delta_{P_g}(x) \right] - \gamma_j \delta_{P_j}(x) \right)$$

$$= l_{P_0}(x) + \gamma_1 \delta_{P_0}(x) + \sum_{j=1}^{h} (r_{j+1} + \cdots + r_h)(\gamma_{j+1} - \gamma_j) \delta_{P_j}(x) - r_j \gamma_j \delta_{P_j}(x)$$

$$\leq l_{P_0}(x) + \gamma_1 \delta_{P_0}(x)$$

$$+ \sum_{j=1}^{h-1} \left[(r_{j+1} + \cdots + r_h) \gamma_{j+1} - (r_j + r_{j+1} + \cdots + r_h) \gamma_j \right] \delta_{P_j}(x)$$

In the last inequality, we leave out the last negative term $-r_h \gamma_h \delta_{P_h}(x)$. Note that $\gamma_1 = (r_1 + \cdots + r_h)\gamma_1$ since we have normalized the demand to 1. We can then apply Lemma 1 with $\tau_i = \delta_{P_i}(x)$ for $i = 0, \ldots, h-1$ and $c_i = \gamma_i \cdot \sum_{p=i}^{h} r_p$ for $i = 1, \ldots, k$. Continuing the estimate, we get

$$C(x) \leq l_{P_0}(x) + \max_{j \in [h]} \left\{ \gamma_j \cdot \sum_{p=j}^{h} r_p \right\} \cdot \delta_{P_0}(x) \leq \left[1 + \beta \cdot \max_{j \in [h]} \left\{ \gamma_j \left(\sum_{p=j}^{h} r_p \right) \right\} \right] C(z)$$

where for the second inequality we use that $\delta_{P_0}(x) \leq \beta l_{P_0}(x)$, which holds by definition, and $l_{P_0}(x) = l_{P_0}(z) = C(z)$, which holds because z is an original Nash flow and all arcs in P_0 have strictly positive flow in z (and because of the fact that all arcs in P_0 have a constant latency functions).

To prove tightness, fix $j \in [h]$ and consider the following instance on two arcs. We take $(l_1(y), \delta_1(y)) = (1, \beta)$ and $(l_2(y), \delta_2(y))$ with $\delta_2(y) = 0$ and $l_2(y)$ a strictly increasing function satisfying $l_2(0) = 1 + \epsilon$ and $l_2(r_j + r_{j+1} + \cdots + r_h) = 1 + \gamma_j \beta$, where $\epsilon < \gamma_j \beta$. The (unique) original Nash flow is given by $z = (z_1, z_2) = (1, 0)$ with $C(z) = 1$. The (unique) β-deviated Nash flow x is given by $x = (x_1, x_2) = (r_1 + r_2 + \cdots + r_{j-1}, r_j + r_{j+1} + \cdots + r_h)$ with $C(x) = 1 + \beta \cdot \gamma_j (r_j + \cdots + r_h)$. Since this construction holds for all $j \in [h]$, we find the desired lower bound. □

Continuous Sensitivity Distributions. We obtain a similar result for more general (not necessarily discrete) sensitivity distributions. That is, we are given a Lebesgue integrable *sensitivity density function* $\psi : \mathbb{R}_{\geq 0} \to \mathbb{R}_{\geq 0}$ over the total demand. Since we can normalize the demand to 1, we have the condition that $\int_0^{\infty} \psi(y) dy = 1$. We then find the following natural generalizations of our upper bounds:

1. $\epsilon\text{-SR}(\mathcal{I}) \leq 1 + \beta \int_0^{\infty} y \cdot \psi(y) dy$, and
2. $\beta\text{-DR}(\mathcal{I}) \leq 1 + \beta \cdot \sup_{t \in \mathbb{R}_{\geq 0}} \left\{ t \cdot \int_t^{\infty} \psi(y) dy \right\}$.

These bounds are both asymptotically tight for all distributions. Details are given the full version [7].

4   Homogeneous Population

The reverse of Proposition 1 also holds for homogeneous players in single-commodity instances. As a consequence, the set of β-deviated Nash flows and the set of ϵ-approximate Nash flows with $\epsilon = \beta\gamma$ coincide in this case.

Recall that for homogeneous players we have $\gamma_{ij} = \gamma$ for all $i \in [k]$, $j \in [h_i]$ and some $\gamma \geq 0$.

Proposition 2. *Let $\mathcal{I}$ be a single-commodity congestion game with homogeneous players. f is an ϵ-approximate Nash flow for $\mathcal{I}$ if and only if f is a β-deviated Nash flow for $\mathcal{I}$ with $\epsilon = \beta\gamma$.*

Upper Bound on the Stability Ratio. Our main result in this section is an upper bound on the ϵ-stability ratio. Given the above equivalence, this bound also applies to the β-deviation ratio with $\epsilon = \beta\gamma$.

The following concept of alternating paths is crucial. For single-commodity instances an alternating path always exists (see, e.g., [13]) (Fig. 1).

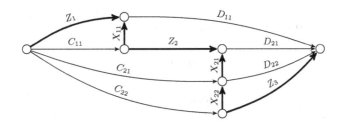

Fig. 1. Sketch of the situation in the proof of Theorem 2 with $q_1 = 1$ and $q_2 = 2$.

Definition 1 (Alternating path [10,13]). *Let $\mathcal{I}$ be a single-commodity network congestion game and let x and z be feasible flows. We partition the edges $E = X \cup Z$ such that $Z = \{a \in E : z_a \geq x_a \text{ and } z_a > 0\}$ and $X = \{a \in E : z_a < x_a \text{ or } z_a = x_a = 0\}$. We say that π is an alternating s,t-path if the arcs in $\pi \cap Z$ are oriented in the direction of t, and the arcs in $\pi \cap X$ are oriented in the direction of s. We call the number of backward arcs on π the backward length of π and refer to it by $q(\pi) = |\pi \cap X|$.*

Theorem 2. *Let $\mathcal{I}$ be a single-commodity network congestion game. Let $\epsilon \geq 0$ be fixed and consider an arbitrary alternating path π with backward length $q = q(\pi)$. If $\epsilon < 1/q$, then the ϵ-stability ratio is bounded by*

$$\epsilon\text{-}SR(\mathcal{I}) \leq \frac{1+\epsilon}{1-\epsilon \cdot q} \leq \frac{1+\epsilon}{1-\epsilon \cdot n}.$$

Note that the restriction on ϵ stated in the theorem always holds if $\epsilon < 1/n$. In particular, for $\epsilon \ll 1/n$ we roughly get $\epsilon\text{-}SR(\mathcal{I}) \leq 1+\epsilon n$. The proof of Theorem 2 is inspired by a technique of Nikolova and Stier-Moses [13], but technically more involved.

Proof. Let $x = f^\epsilon$ be an ϵ-approximate Nash flow and let $z = f^0$ an original Nash flow. Let $\pi = Z_1 X_1 Z_2 X_2 \ldots Z_{\eta-1} X_{\eta-1} Z_\eta$ be an alternating path for x and z, where Z_i and X_i are maximal sections consisting of consecutive arcs, respectively, in Z and X (i.e., $Z_i \subseteq Z$ and $X_i \subseteq X$ for all i). Furthermore, we let $q_i = |X_i|$ and write $X_i = (X_{iq_i}, \ldots, X_{i2}, X_{i1})$, where X_{ij} are the arcs in the section X_i. By definition, for every arc X_{ij} there exists a path $C_{ij} X_{ij} D_{ij}$ that is flow-carrying for x.[5]

For convenience, we define $C_{01} = D_{\eta,0} = \emptyset$. Furthermore, we denote $P^{\max}$ as a path maximizing $l_P(x)$ over all paths $P \in \mathcal{S}$. For convenience, we will abuse notation, and write $Q = Q(x) = \sum_{a \in Q} l_a(x)$ for $Q \subseteq E$.

Note that for all i, j:

$$C_{ij}(x) + X_{ij}(x) + D_{ij}(x) \leq P^{\max}(x). \tag{6}$$

Fix some $i \in \{1, \ldots, \eta-1\}$. Then we have $C_{i1} + X_{i1} + D_{i1} \leq (1+\epsilon)(C_{i-1,q_{i-1}} + Z_i + D_{i1})$ by definition of an ϵ-approximate Nash flow. This implies that (leaving out D_{i1} on both sides) $C_{i1} + X_{i1} \leq (1+\epsilon)Z_i + C_{i-1,q_{i-1}} + \epsilon(C_{i-1,q_{i-1}} + D_{i1})$. Furthermore, for all $j \in \{2, \ldots, q_i\}$, we have $C_{ij} + X_{ij} + D_{ij} \leq (1+\epsilon)(C_{i,j-1} + D_{ij})$ which implies (again leaving out D_{ij} on both sides)

$$C_{ij} + X_{ij} \leq C_{i,j-1} + \epsilon(C_{i,j-1} + D_{ij}).$$

Adding up these inequalities for $j \in \{1, \ldots, q_i\}$ and subtracting $\sum_{j=1}^{q_i-1} C_{ij}$ from both sides, we obtain for all $i \in \{1, \ldots, \eta-1\}$

$$C_{i,q_i} + \sum_{j=1}^{q_i} X_{ij} \leq C_{i-1,q_{i-1}} + (1+\epsilon)Z_i + \epsilon\left(\sum_{j=1}^{q_i} D_{ij} + C_{i-1,q_{i-1}} + \sum_{j=1}^{q_i-1} C_{ij}\right). \tag{7}$$

Moreover, we also have

$$P^{\max} \leq (1+\epsilon)(C_{\eta-1,\eta-1} + Z_\eta) = C_{\eta-1,\eta-1} + (1+\epsilon)Z_\eta + \epsilon C_{\eta-1,\eta-1}. \tag{8}$$

Adding up the inequalities in (7) for all $i \in \{1, \ldots, \eta-1\}$, and the inequality in (8), we obtain

$$P^{\max} + \sum_{i=1}^{\eta-1} C_{i,q_i} + \sum_{i=1}^{\eta-1}\sum_{j=1}^{q_i} X_{ij} \leq \sum_{i=1}^{\eta-1} C_{i,q_i} + (1+\epsilon)\sum_{i=1}^{\eta} Z_i + \epsilon\left(\sum_{i=1}^{\eta-1}\sum_{j=1}^{q_i} C_{ij} + D_{ij}\right)$$

which simplifies to

$$P^{\max} + \sum_{i=1}^{\eta-1}\sum_{j=1}^{q_i} X_{ij} \leq (1+\epsilon)\sum_{i=1}^{\eta} Z_i + \epsilon\left(\sum_{i=1}^{\eta-1}\sum_{j=1}^{q_i} C_{ij} + D_{ij}\right). \tag{9}$$

---

[5] Note that for a Nash flow one can assume that there is a flow-carrying path traversing all arcs $X_{iq_i}, \ldots, X_{i1}$; but this cannot be done for an approximate Nash flow.

Using (6), we obtain

$$\sum_{i=1}^{\eta-1}\sum_{j=1}^{q_i} C_{ij} + D_{ij} \le \sum_{i=1}^{\eta-1}\sum_{j=1}^{q_i} P^{\max} - X_{ij} = \left(\sum_{i=1}^{\eta-1} q_i\right) P^{\max} - \sum_{i=1}^{\eta-1}\sum_{j=1}^{q_i} X_{ij}.$$

Combining this with (9), and rearranging some terms, we get

$$(1 - \epsilon \cdot q)P^{\max} \le (1+\epsilon)\left[\sum_{i=1}^{\eta} Z_i - \sum_{i=1}^{\eta-1}\sum_{j=1}^{q_i} X_{ij}\right]$$

$$= (1+\epsilon)\left[\sum_{e \in Z \cap \pi} l_e(x_e) - \sum_{e \in X \cap \pi} l_e(x_e)\right]$$

where $q = q(\pi) = \sum_{i=1}^{\eta-1} q_i$ is the backward length of π.

Similarly (see also [13, Lemma 4.5]), it can be shown that

$$l_Q(z) \ge \sum_{e \in Z \cap \pi} l_e(z_e) - \sum_{e \in X \cap \pi} l_e(z_e) \qquad (10)$$

for any path Q with $z_Q > 0$ (these all have the same latency, since z is an original Nash flow). Using a similar argument as in [13, Theorem 4.6], we obtain

$$(1 - \epsilon \cdot q)l_{P^{\max}}(x) \le (1+\epsilon)\left[\sum_{e \in Z \cap \pi} l_e(x_e) - \sum_{e \in X \cap \pi} l_e(x_e)\right]$$

$$\le (1+\epsilon)\left[\sum_{e \in Z \cap \pi} l_e(z_e) - \sum_{e \in X \cap \pi} l_e(z_e)\right] \le (1+\epsilon)l_Q(z).$$

By multiplying both sides with the demand r, we obtain $(1 - \epsilon \cdot q)C(x) \le (1 - \epsilon \cdot q)r \cdot l_{P^{\max}}(x) \le (1+\epsilon)r \cdot l_Q(z) = (1+\epsilon)C(z)$ for $\epsilon < 1/q$, which proves the claim. $\square$

Tight Bound on the Stability Ratio. In this section, we consider instances for which all backward sections of the alternating path π consist of a single arc., i.e., $q_i = 1$ for all $i = 1, \dots, \eta - 1$. We then have $q = \sum_{i=1}^{\eta-1} q_i \le \lfloor n/2 \rfloor - 1$ since every arc in X must be followed directly by an arc in Z (and we can assume w.l.o.g. that the first and last arc are contained in Z). By Theorem 2, we obtain $\epsilon\text{-SR}(\mathcal{I}) \le (1+\epsilon)/(1 - \epsilon \cdot (\lfloor n/2 \rfloor - 1))$ for all $\epsilon < 1/(\lfloor n/2 \rfloor - 1)$. We show that this bound is tight. Further, we show that there exist instances for which $\epsilon\text{-SR}(\mathcal{I})$ is unbounded for $\epsilon \ge 1/(\lfloor n/2 \rfloor - 1)$. This completely settles the case of $q_i = 1$ for all i.

Our construction is based on the *generalized Braess graph* [14]. By construction, alternating paths for these graphs satisfy $q_i = 1$ for all i. See the full version [7] for details.

Theorem 3. *Let $n = 2m$ be fixed and let $\mathcal{B}^m$ be the set of all instances on the generalized Braess graph with n nodes. Then*

$$\sup_{\mathcal{I} \in \mathcal{B}^m} \epsilon\text{-SR}(\mathcal{I}) = \begin{cases} \frac{1+\epsilon}{1 - \epsilon \cdot (\lfloor n/2 \rfloor - 1)} & \text{if } \epsilon < \frac{1}{\lfloor n/2 \rfloor - 1}, \\ \infty & \text{otherwise.} \end{cases}$$

Non-symmetric Matroid Congestion Games. In the previous sections, we considered (symmetric) network congestion games only. It is interesting to consider other combinatorial strategy sets as well. In this section we make a first step in this direction by focusing on the bases of matroids as strategies.

A matroid congestion game is given by $\mathcal{J} = (E, (l_e)_{e \in E}, (\mathcal{S}_i)_{i \in [k]}, (r_i)_{i \in [k]})$, and matroids $\mathcal{M}_i = (E, \mathcal{I}_i)$ over the ground set E for every $i \in [k]$.[6] The strategy set $\mathcal{S}_i$ consists of the *bases* of the matroid $\mathcal{M}_i$, which are the independent sets of maximum size, e.g., spanning trees in an undirected graph. We refer the reader to Schrijver [16] for an extensive overview of matroid theory.

As for network congestion games, it can be shown that in general the ϵ-stability ratio can be unbounded (see Theorem 5 [7] in the appendix of [7]); this also holds for general path deviations because the proof of Proposition 2 in the appendix holds for arbitrary strategy sets. However, if we consider path deviations induced by the sum of edge deviations (as in [6,13]), then we can obtain a more positive result for general matroids.

Recall that for every resource $e \in E$ we have a deviation function $\delta_e : \mathbb{R}_{\geq 0} \to \mathbb{R}_{\geq 0}$ satisfying $0 \leq \delta_e(x) \leq \beta l_e(x)$ for all $x \geq 0$. The deviation of a basis B is then given by $\delta_B(f) = \sum_{e \in B} \delta_e(f_e)$.

Theorem 4. *Let $\mathcal{J} = (E, (l_e)_{e \in E}, (\mathcal{S}_i)_{i \in [k]}, (r_i)_{i \in [k]})$ be a matroid congestion game with homogeneous players. Let $\beta \geq 0$ be fixed and consider β-bounded basis deviations as defined above. Then the β-deviation ratio is upper bounded by β-DR$(\mathcal{J}) \leq 1 + \beta$. Further, this bound is tight already for 1-uniform matroid congestion games.*

Acknowledgements. We thank the anonymous referees for their very useful comments, and one reviewer for pointing us to Lemma 1 [4].

References

1. Chen, P.-A., Keijzer, B.D., Kempe, D., Schäfer, G.: Altruism and its impact on the price of anarchy. ACM Trans. Econ. Comput. **2**(4), 17:1–17:45 (2014)
2. Christodoulou, G., Koutsoupias, E., Spirakis, P.G.: On the performance of approximate equilibria in congestion games. Algorithmica **61**(1), 116–140 (2011)
3. Englert, M., Franke, T., Olbrich, L.: Sensitivity of Wardrop equilibria. In: Monien, B., Schroeder, U.-P. (eds.) SAGT 2008. LNCS, vol. 4997, pp. 158–169. Springer, Heidelberg (2008). doi:10.1007/978-3-540-79309-0_15
4. Fleischer, L.: Linear tolls suffice: new bounds and algorithms for tolls in single source networks. Theoret. Comput. Sci. **348**(2), 217–225 (2005)
5. Fujishige, S., Goemans, M.X., Harks, T., Peis, B., Zenklusen, R.: Matroids are immune to braess paradox. CoRR, abs/1504.07545 (2015)
6. Kleer, P., Schäfer, G.: The impact of worst-case deviations in non-atomic network routing games. CoRR, abs/1605.01510 (2016)

---

[6] A matroid over E is given by a collection $\mathcal{I} \subseteq 2^E$ of subsets of E (called *independent sets*). The pair $\mathcal{M} = (E, \mathcal{I})$ is a *matroid* if the following three properties hold: (i) $\emptyset \in \mathcal{I}$; (ii) If $A \in \mathcal{I}$ and $B \subseteq A$, then $B \in \mathcal{I}$. (iii) If $A, B \in \mathcal{I}$ and $|A| > |B|$, then there exists an $a \in A \setminus B$ such that $B + a \in \mathcal{I}$.

7. Kleer, P., Schäfer, G.: Path deviations outperform approximate stability in heterogeneous congestion games. CoRR, abs/1707.01278 (2017)
8. Koutsoupias, E., Papadimitriou, C.: Worst-case equilibria. In: Meinel, C., Tison, S. (eds.) STACS 1999. LNCS, vol. 1563, pp. 404–413. Springer, Heidelberg (1999). doi:10.1007/3-540-49116-3_38
9. Lianeas, T., Nikolova, E., Stier-Moses, N.E.: Asymptotically tight bounds for inefficiency in risk-averse selfish routing. In: Proceedings of the Twenty-Fifth International Joint Conference on Artificial Intelligence. IJCAI 2016, NY, USA, New York, pp. 338–344 (2016)
10. Lin, H., Roughgarden, T., Tardos, É., Walkover, A.: Stronger bounds on Braess's paradox and the maximum latency of selfish routing. SIAM J. Discrete Math. **25**(4), 1667–1686 (2011)
11. Meir, R., Parkes, D.C.: Playing the wrong game: smoothness bounds for congestion games with risk averse agents. CoRR, abs/1411.1751 (2017)
12. Milchtaich, I.: Network topology and the efficiency of equilibrium. Games Econ. Behav. **57**(2), 321–346 (2006)
13. Nikolova, E., Stier-Moses, N.E.: The burden of risk aversion in mean-risk selfish routing. In: Proceedings of the Sixteenth ACM Conference on Economics and Computation, EC 2015, pp. 489–506. ACM, New York (2015)
14. Roughgarden, T.: On the severity of Braess's paradox: designing networks for selfish users is hard. J. Comput. Syst. Sci. **72**(5), 922–953 (2006)
15. Roughgarden, T.: Intrinsic robustness of the price of anarchy. J. ACM **62**(5), 32 (2015)
16. Schrijver, A.: Combinatorial Optimization: Polyhedra and Efficiency Vol. B Matroids Trees Stable Sets Algorithms and Combinatorics. Springer, Heidelberg (2003). Chaps. 39–69
17. Wardrop, J.G.: Some theoretical aspects of road traffic research. In: Proceedings of the Institution of Civil Engineers, vol. 1, pp. 325–378 (1952)

Mechanism Design, Incentives and
Regret Minimization

Agent Incentives of Strategic Behavior in Resource Exchange

Zhou Chen[1], Yukun Cheng[2]([✉]), Xiaotie Deng[3], Qi Qi[1], and Xiang Yan[3]

[1] Department of Industrial Engineering and Logistics Management,
Hong Kong University of Science and Technology, Kowloon, Hong Kong
zchenaq@connect.ust.hk, kaylaqi@ust.hk
[2] School of Data Science, Zhejiang University of Finance and Economics,
Hangzhou, China
ykcheng@amss.ac.cn
[3] Department of Computer Science, Shanghai Jiao Tong University, Shanghai, China
deng-xt@cs.sjtu.edu.cn, xyansjtu@163.com

Abstract. In a resource exchange system, resources are shared among multiple interconnected peers. Peers act as both suppliers and customers of resources by making a certain amount of their resources directly available to other network participants. Their utilities are determined by the total amount of resources received from all neighbors. According to a preset mechanism, the allocation of the shared resources depends on the information that agents submit to the mechanism. The participating agents, however, may try to strategically manipulate its submitted information to influence the allocation with the expectation of its utility improvement. In this paper, we consider the tit-for-tat popular proportional response mechanism and discuss the incentives an agent may lie, by a *vertex splitting strategy*. We apply the concept of *incentive ratio* to characterize the multiplication factor by which utility of an agent can be increased with the help of the vertex splitting strategy. Because of the bounded rationality in the decentralized resource exchange system, a smaller incentive ratio makes the agents have the less incentive to play strategically. However the incentive ratio is proved to be unbounded in linear exchange market recently. In this paper we focus on the setting on trees, our linear exchange market proves to have the incentive ratio of exact two under the proportional response mechanism against the vertex splitting strategic behaviors of participating agents.

1 Introduction

The rapid growth of wireless and mobile Internet has led to wide applications of exchanging resources over networks. Participants in such networks act as both suppliers and customers of resources (such as processing power, disk storage or network bandwidth), and make their resources directly available to others according to preset rules [14]. To motivate sharing, [9] pioneered the use of incentive techniques to drive cooperation and to promote voluntary contributions

© Springer International Publishing AG 2017
V. Bilò and M. Flammini (Eds.): SAGT 2017, LNCS 10504, pp. 227–239, 2017.
DOI: 10.1007/978-3-319-66700-3_18

by participating agents. Taking this approach, the BitTorrent protocol, created by Cohen in 2001, has been well recognized as an Internet success that changed "the entertainment industry and the interchange of information in Web" [8]. In the past couple of years, the widespread applications of mobile technologies have further made sharing, together with the rising of mobile enabled companies such as Uber, a focal point in the interface of computer science and economics [15].

To capture the success of BitTorrent, Wu and Zhang [17] introduced the proportional response protocol, under which each peer contributes its resource to each neighbor in proportion to the amount received from its neighbors. They showed its economic efficiency by its convergence to the market equilibrium of a pure exchange economy. In recent works, [6,7] proved the incentive advantage of this protocol against strategic behavior of mis-reporting connectivity and agent capacities. In this paper, we further exploring its resistance to manipulative behavior by considering a setting where an agent disguises itself by creating several copied false nodes with its resources split among them, such as one may take new IP addresses in the network environment.

The problem is modeled by an undirected graph G, where each vertex v represents an agent with w_v units of divisible resources (or weight) for sharing among its neighbors. The utility U_v is the sum of resources obtained from its neighbors. The proportional response mechanism will send out each vertex's resource to its neighbors in proportion to the amount it receives from all its neighbors. The concern is associated with a cheating strategy of an agent v who splits itself into a subset of nodes such that each is connected to some of its original neighbors and is assigned a certain amount of resource of v (summing up to its original amount of resource) to manipulate the proportional response mechanism. The new utility of this agent is calculated by summing over all its copied nodes'. Conveniently, we name this strategic behavior as *vertex splitting*.

Example 1. *Consider graph G of 4 vertices in Fig. 1(a). At equilibrium, v_3 gets its utility 5 from v_4. In Fig. 1(b), v_3 splits itself into two copied nodes v_3^1 and v_3^2, and assigns resource 1 to each node respectively. The new utility of v_3 is 20/3, which is larger than its original utility 5. (In Sect. 2 we will explain how to compute the utility from the proportional response mechanism in details.)*

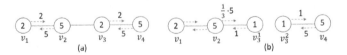

Fig. 1. Numbers in cycles are weights; dashed arrows indicate resource exchange amounts in the proportional sharing mechanism.

Example 1 exhibits a vulnerability of the proportional response mechanism against the vertex splitting strategy. The next immediate question is its level of significance for which is characterized by the concept of *incentive ratio*, introduced in [5], defined as the ratio of the largest possible utility that an agent can gain by cheating and its utility in honest behavior under a given mechanism. In the resource exchange environment, all agents only have limited knowledge due to the decentralization of the system. Therefore they need an incredible effort to know the full information of the game and do complicated computation as well. Thus a smaller incentive ratio implies that an agent has less incentive to play strategically as finding the optimal splitting strategy is too difficult.

Related Work. The automated process through information and communication technology for Internet applications has made their successes relied on the voluntary cooperations of participating agents. [9] pioneered the study of such incentive techniques in mechanism design and in performance analysis for such peer-to-peer resource sharing systems. The proportional response model for the BitTorrent protocol for resource exchange was shown to be equivalent to the market equilibrium solution in [17]. Agent strategic behavior in market equilibrium has been analyzed in the Fisher market equilibrium for linear markets [1] and for constant elasticity of substitution markets [3]. For the proportional response sharing protocol, focusing on two types of strategic behaviors for each agent: cheating on its connectivity with the rest of network and misreporting its own resource amount, the system is known to be stable [6,7].

The vertex splitting behavior considered here is motivated by and similar to false-name bid [19]. That is, the agent misrepresents itself by creating several fictitious identities. It is known that the VCG mechanism is not incentive compatible against the false-name bidding, as a result of the study on the false-name-proof auction mechanisms design [11,18], and on the efficiency guarantee of the VCG mechanism [2].

The concept of incentive ratio [5] is motivated by the concept of *price of anarchy* [12,16]. The former measures individual gains one may acquire in deviation from truthful behavior, but the latter models the loss of social efficiency in selfish Nash equilibrium in comparison to social optimality.

Main Results. We analyze incentive ratio to quantitatively measure the maximal magnitude of utility gain by vertex splitting followed a proportional response mechanism. We prove that the incentive ratios are exactly 2 on trees by proposing proper examples for lower bound and characterizing the worst case network structure for upper bound.

While the incentive ratio was known for the Fisher market with well characterized matching bound of two for linear utilities [4,5], it is not applicable to our setting of exchange market, a special case of the exchange economy. A recent result [13] claims an unbounded incentive ratios in linear exchange economy. Our results of bounded incentive ratio places the practical network sharing economy with a market equilibrium solution more rational in terms of truthful behavior.

2 Preliminary

Our resource exchange system is based on a connected and undirected network $G = (V, E)$. Each vertex $v \in V$ represents an agent with an upload resource amount (weight) $w_v > 0$ for exchange with its neighbors, where $\Gamma(v) = \{u : (v, u) \in E\}$ is the neighborhood of v. Let x_{vu} be the amount of resource v allocates to neighbor u ($0 \leq x_{vu} \leq w_v$) and $X = \{x_{uv}\}$ be an allocation. The utility of agent v is defined as $U_v(X) = \sum_{u \in \Gamma(v)} x_{uv}$, i.e. all received resource from its neighbors. In the resource exchange environment, one of critical issues is how to design an allocation mechanism to maintain the agents participation, i.e., ensuring that agents will share their resources in a fair fashion. Wu and Zhang [17] pioneered the concept of "proportional response" inspired by the idea of "tit-for-tat" for the consideration of fairness.

Proportional Response. A mechanism is called *proportional response* if an allocation X from this mechanism satisfies $x_{vu} = \frac{x_{uv}}{\sum_{k \in \Gamma(v)} x_{kv}} w_v$, that is the allocation of each agent's resource is proportional to what it receives from its neighbors.

To achieve a proportional response mechanism, a combinatorial structure, called *bottleneck decomposition* is derived in [17]. For set $S \subseteq V$, define $w(S) = \sum_{v \in S} w_v$ and $\Gamma(S) = \cup_{v \in S} \Gamma(v)$. It is possible that $S \cap \Gamma(S) \neq \emptyset$. Denote $\alpha(S) = \frac{w(\Gamma(S))}{w(S)}$ to be the inclusive expansion ratio of S, or the α-*ratio* of S for short. A set $B \subseteq V$ is called a *bottleneck* of G if $\alpha(B) = \min_{S \subseteq V} \alpha(S)$. A bottleneck with the maximal size is called the *maximal bottleneck*.

Bottleneck Decomposition. Given $G = (V, E; w)$. Start with $V_1 = V$, $G_1 = G$ and $i = 1$. Find the maximal bottleneck B_i of G_i and let G_{i+1} be the induced subgraph on the vertex set $V_{i+1} = V_i - (B_i \cup C_i)$, where $C_i = \Gamma(B_i) \cap V_i$, the neighbor set of B_i in G_i. Repeat if $G_{i+1} \neq \emptyset$ and set $k = i$ if $G_{i+1} = \emptyset$. Then we call $\mathcal{B} = \{(B_1, C_1), \cdots, (B_k, C_k)\}$ the bottleneck decomposition of G, (B_i, C_i) the i-th bottleneck pair and $\alpha_i = w(C_i)/w(B_i)$ the α-ratio of (B_i, C_i).

Proposition 1 (Wu and Zhang [17]). *Given a graph G, the bottleneck decomposition of G is unique and*

(1) $0 < \alpha_1 < \alpha_2 < \cdots < \alpha_k \leq 1$;
(2) if $\alpha_i = 1$, then $i = k$ and $B_i = C_i$; otherwise B_i is an independent set and $B_i \cap C_i = \emptyset$;
(3) there is no edge between B_i and B_j, $i \neq j \in \{1, \cdots, k\}$;
(4) if there is an edge between B_i and C_j, then $j \leq i$.

B-class and C-class. Given $\mathcal{B} = \{(B_1, C_1), \cdots, (B_k, C_k)\}$. For pair (B_i, C_i) with $\alpha_i < 1$, each vertex in B_i (or C_i) is called a *B*-class (or *C*-class) vertex. For the special case $B_k = C_k = V_k$, i.e., $\alpha_k = 1$, all vertices in B_k are categorized as both *B*-class and *C*-class.

BD Mechanism. Given the bottleneck decomposition $\mathcal{B}$, an allocation (Wu and Zhang [17]) can be determined by distinguishing three cases. We name it as BD Mechanism for short.

- For (B_i, C_i) with $\alpha_i < 1$, consider the bipartite graph $\widehat{G} = (B_i, C_i; E_i)$ with $E_i = B_i \times C_i$. Construct a network $N = (V_N, E_N)$ with $V_N = \{s, t\} \cup B_i \cup C_i$ and directed edges (s, u) with capacity w_u for $u \in B_i$, (v, t) with capacity w_v/α_i for $v \in C_i$ and (u, v) with capacity $+\infty$ for $(u, v) \in E_i$. The max-flow min-cut theorem ensures a maximal flow $\{f_{uv}\}$, $u \in B_i$ and $v \in C_i$, such that $\sum_{v \in \Gamma(u) \cap C_i} f_{uv} = w_u$ and $\sum_{u \in \Gamma(v) \cap B_i} f_{uv} = w_v/\alpha_i$. Let the allocation be $x_{uv} = f_{uv}$ and $x_{vu} = \alpha_i f_{uv}$ implying $\sum_{u \in \Gamma(v) \cap B_i} x_{vu} = \sum_{u \in \Gamma(v) \cap B_i} \alpha_i \cdot f_{vu} = w_v$.

- For $\alpha_k = 1$ (i.e., $B_k = C_k$), construct a bipartite graph $\widehat{G} = (B_k, B'_k; E'_k)$ where B'_k is a copy of B_k, there is an edge $(u, v') \in E'_k$ if and only if $(u, v) \in E[B_k]$. Construct a network by the above method, for any edge $(u, v') \in E'_k$, there exists flow $f_{uv'}$ such that $\sum_{v' \in \Gamma(u) \cap B'_k} f_{uv'} = w_u$. Let the allocation be $x_{uv} = f_{uv'}$.

- For any other edge $(u, v) \notin B_i \times C_i$, $i = 1, 2, \cdots, k$, define $x_{uv} = 0$.

Proposition 2 [17]. *BD Mechanism is a proportional response mechanism.*

On the other hand the resource exchange system can be modeled as a pure exchange economy, for which an efficient allocation is the market equilibrium.

Market Equilibrium. In the exchange economy, price vector $p = (p_v)_{v \in V}$ together with an allocation X is called a *market equilibrium* if for any agent $v \in V$ the following holds, 1. $\sum_{u \in \Gamma(v)} x_{vu} = w_v$ (market clearance); 2. $\sum_{u \in \Gamma(v)} p_u \frac{x_{uv}}{w_u} \leq p_v$ (budget constraint); 3. $X = (x_{vu})$ maximizes the utility $U_v = \sum_{u \in \Gamma(v)} x_{uv}$ subject to the budget constraint (individual optimality).

BD Mechanism is not only fair as stated above but also efficient, since the proportional response allocation from it is also a market equilibrium. Given a bottleneck decomposition, if a price vector p is well defined as: $p_u = \alpha_i w_u$, if $u \in B_i$; and $p_u = w_u$ otherwise, then

Proposition 3 [17]. (p, X) *is a market equilibrium. Furthermore, each agent u's utility is $U_u = w_u \cdot \alpha_i$ if $u \in B_i$; $U_u = \frac{w_u}{\alpha_i}$ if $u \in C_i$.*

Note that $U_u \geq w_u$ if u is in C_i and $U_u \leq w_u$ if u is in B_i, as $\alpha_i \leq 1$ by Proposition 1-(1). For convenience, we let α_u be the α-ratio of u where $\alpha_u = \alpha_i$, if $u \in B_i \cup C_i$. Recall Example 1. The bottleneck decompositions, α-ratios and utilities before and after vertex splitting are listed in the following table.

$B_1 = \{v_2, v_4\}$, $C_1 = \{v_1, v_3\}$	$\alpha_{v_3} = \frac{2}{5}$	$U_{v_3} = 5$
$B'_1 = \{v_4\}$, $C'_1 = \{v_3^2\}$	$\alpha'_{v_3^2} = \frac{1}{5}$	$U'_{v_3^2} = 5$
$B'_2 = \{v_2\}$, $C'_2 = \{v_3^1\}$	$\alpha'_{v_3^1} = \frac{3}{5}$	$U'_{v_3^1} = \frac{5}{3}$

From a system design point of view, although BD Mechanism shall allocate resource among interconnected participants fairly and efficiently, a problem occurs that, an agent may or may not follow BD Mechanism at the execution level. Can agents make strategic moves for gains in their utilities? We call such a problem with incentive compatibility consideration the *resource exchange game*.

In this paper, we consider a strategic move, called *vertex splitting strategy*, that is one agent may split itself into several copied nodes, and assign a weight to each node. Formally, in a resource exchange network G, the collection $\mathbf{w} = (w_1, \cdots, w_n) \in R^n$ is referred as the *weight profile*. For agent v, let $\mathbf{w}_{-v}$ be the weight profile without v. Since the utility of agent v depends on G and $\mathbf{w}$, it is written as $U_v(G; \mathbf{w})$. After splitting into m nodes, $1 \leq m \leq d_v$ (d_v is the degree of v), and assigning an amount of resource to each node, agent v's new utility is denoted by $U_v'(G'; w_{v^1}, \cdots, w_{v^m}, \mathbf{w}_{-v})$, where G' is the resulting graph after vertex splitting and $w_{v^i} \in [0, w_v]$ is the amount of resource assigned to node v^i with $\sum_{i=1}^m w_{v^i} = w_v$.

Definition 1 (Incentive Ratio). *In a resource exchange game, the incentive ratio of agent v under BD Mechanism for the vertex splitting strategy is*

$$\zeta_v = \max_{1 \leq m \leq d_v} \max_{w_{v^i} \in [0, w_v], \sum_{i=1}^m w_{v^i} = w_v; \mathbf{w}_{-v}; G'} \frac{U_v'(G'; w_{v^1}, \cdots, w_{v^m}, \mathbf{w}_{-v})}{U_v(G; \mathbf{w}_v)}.$$

The incentive ratio of BD mechanism in resource exchange game is defined to be $\zeta = \max_{v \in V} \zeta_v$.

There is a special case that a strategic agent v splits itself into d_v nodes and each node is connected to one of neighbors. The following proposition shows that this cheating way would not reduce the generality of the result.

Proposition 4. *In a resource exchange game, the incentive ratio of BD mechanism with respect to vertex splitting strategy can be achieved by splitting d_v nodes and each node is connected to one neighbor for any vertex v, where d_v is the degree of agent v.*

The proof of Proposition 4 is left in the full version. Therefore we only consider the cheating way that v splits d_v nodes. Conveniently, let the copied nodes set be $\Lambda(v) = \{v^1, \cdots, v^{d_v}\}$ and the neighborhood be $\Gamma(v) = \{u^1, \cdots, u^{d_v}\}$, where v^j is adjacent to u^j in G'.

3   Incentive Ratios of BD Mechanism on Trees

In this section, we focus on the resource exchange game in which the underlying network is a tree. Our main result in this paper is the following.

Theorem 1. *If the network G of resource exchange system is a tree, then the incentive ratio of BD Mechanism for the vertex splitting strategy is exactly 2.*

Before providing the details of discussion, let us introduce some additional notations. Recall neighborhood is $\Gamma(v) = \{u^1, u^2, \cdots, u^{d_v}\}$. In the original network G with weight profile $\mathbf{w}$, the resource allocation from BD Mechanism is

$X = (x_{vu})$ and the bottleneck decomposition is $\mathcal{B}$. we partition neighborhood $\Gamma(v)$ into two disjoint subsets: $\Gamma_1(v) = \{u^i | x_{vu^i} > 0\}$ and $\Gamma_2(v) = \{u^i | x_{vu^i} = 0\}$. Hence, v only exchanges the resource with the neighbors in $\Gamma_1(v)$. W.l.o.g., let index sets $I_1(v) = \{i | u^i \in \Gamma_1(v)\}$ and $I_2(v) = \{i | u^i \in \Gamma_2(v)\}$.

If agent v plays the vertex splitting strategy, then the original network G shall be decomposed into d_v connected subtrees, because G is a tree and the degree of v is d_v. Likewise, we label all subtrees as $G^1, \cdots, G^{d_v}$, each containing one copied node v^i and its unique neighbor u^i, $i = 1, \cdots, d_v$. For any given weight assignment $(w_{v^1}, \cdots, w_{v^{d_v}})$ among all copied nodes, the utility of v^i depends on the structure of G^i and its weight w_{v^i}. So we denote it by $U^i(w_{v_i})$. In addition, based on the partition of index set, the subtrees $\{G^i\}$, copied node set $\Lambda(v)$ and weights $\{w_{v^i}\}$ are also divided into corresponding two subsets.

Recently Cheng *et al.*, raised the issue of agent deviation from the BD Mechanism by the acts of cheating on the resource amount it owns [7]. They confirmed the incentive compatibility of BD Mechanism with respect to such a so called weight misreporting manipulative move. A mechanism is called *incentive compatible* if truthful revelation is a best response for each agent under such a mechanism, irrespective of what is reported by the other agents.

Proposition 5 [7]. *For any agent u in a resource exchange game, the utility of u is continuous and monotonically nondecreasing on it reported value x. As $x \leq w_u$, the dominant strategy of agent u is to report it true weight.*

Since the bottleneck decomposition of G, the utility and α-ratio of agent v depend on its reported weight x given the other agents' weights, we can view its utility and α-ratio as functions on x, denoted by $U_v(x)$ and $\alpha_v(x)$. In [7], Cheng *et al.* characterized $\alpha_v(x)$ as follows.

Proposition 6 [7]. *For any agent v and any other agents' reported weights, let $\alpha_v(x)$ be v's α-function of its reported weight $x \in [0, w_v]$. Then there exist three cases*

(1). $\alpha_v(x)$ is non-decreasing and v is in C-class for all $x \in [0, w_v]$;
(2). $\alpha_v(x)$ is non-increasing and v is in B-class for all $x \in [0, w_v]$;
(3). there is a number $x^* \in (0, w_v]$ with $\alpha_v(x^*) = 1$, and
    (3.1). $\alpha_v(x)$ is non-decreasing and v is in C-class if $0 \leq x \leq x^*$;
    (3.2). $\alpha_v(x)$ is non-increasing and v is in B-class if $x^* \leq x \leq w_v$.

From Proposition 6, we note that if v is in B-class when its report weight $x = 0$, then it must keep to be a B-class vertex and $\alpha_v(x)$ is non-increasing. But if v is in C-class when $x = 0$, then it may be in C-class or B-class during $x \in [0, w_v]$.

In the following, we first show the lower bound of the incentive ratio is equal to 2 by proposing an example.

Example 2. *There is a path G containing 5 vertices, shown in Fig. 2(a). The weights of all vertices are $w_{v_1} = w_{v_3} = M/4$, $w_{v_2} = M$, $w_{v_4} = 2$ and $w_{v_5} = 1$, where M is a large enough number. The bottleneck decomposition of G is*

$\{(B_1, C_1)\}$, where $B_1 = \{v_2, v_4\}$ and $C_1 = \{v_1, v_3, v_5\}$ with $\alpha_1 = 1/2$. The utility of v_4 is $U_{v_4} = 2\alpha_1 = 1$.

Now vertex v_4 strategically splits itself into v_4^1 and v_4^2 and assigns $2 - \varepsilon$ and ε resources to each node, respectively, where ε is an arbitrarily small positive number as shown in Fig. 2(b). The new bottleneck decomposition changes to be $\{(B_1', C_1'), (B_2', C_2')\}$, where $B_1' = \{v_5\}$, $C_1' = \{v_4^2\}$ and $B_2' = \{v_2, v_4^1\}$, $C_2' = \{v_1, v_3\}$. The α-ratios are $\alpha_1' = \varepsilon$ and $\alpha_2' = \frac{M/2}{M+2-\varepsilon}$, respectively. At this time $\lim_{M\to\infty} U'_{v_4^1} = 1$ and $\lim_{M\to\infty} U'_{v_4^2} = 1$ which means $\lim_{M\to\infty} U'_{v_4} = 2$.

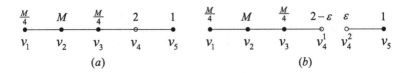

Fig. 2. An example that v_4 can improve twice of its utility by vertex splitting strategy.

Hence to obtain Theorem 1, we only need to prove that the upper bound of incentive ratio is 2, too. For this purpose, we discuss this issue from two-fold aspects corresponding to the partition of $\Gamma(v)$. On the one hand, by applying the property of proportional response from BD Mechanism and the monotonic property of α-function in Proposition 6, we prove the upper bound of the sum of utilities of all nodes $\{v^i\}_{i\in I_1}$ is $(1+\delta)U_v$, where $\delta = (\sum_{i\in I_1} w_{v^i})/w_v$ and U_v is the utility of v in G.

Lemma 1. *If G is a tree and the other agents' weight profile $\mathbf{w}_{-v}$ is given, then the sum of utilities of v^i, $i \in I_1$, satisfies $\sum_{i\in I_1} U^i(w_{v^i}) \leq (1+\delta)U_v$, for any weight assignment $(w_{v^1}, \cdots, w_{v^{d_v}})$, where $\delta = (\sum_{i\in I_1} w_{v^i})/w_v$.*

On the other hand, by applying the bottleneck decomposition of G^i, $i \in I_2$, and the monotonic property of α-function, we show the upper bound of the utility of each copied node v^i, $i \in I_2$.

Lemma 2. *If the underlying network G is a tree and the other agents' weight profile $\mathbf{w}_{-v}$ is given, then utility $U^i(w_{v^i}) \leq \frac{w_{v^i}}{w_v} U_v$, $i \in I_2$, for any weight assignment $(w_{v^1}, \cdots, w_{v^{d_v}})$.*

Combining above two lemmas, the upper bound of 2 can be deduced directly.

Theorem 2. *If the underlying network G is a tree, then the incentive ratio of BD Mechanism for the vertex splitting strategy is at most 2, i.e. $\zeta \leq 2$.*

Proof. Since $\delta = (\sum_{i \in I_1} w_{v^i})/w_v$, then $(\sum_{i \in I_2} w_{v^i})/w_v = 1 - \delta$. For any weight assignment $(w_{v^1}, \cdots, w_{v^{d_v}})$,

$$U'_v(w_{v^1}, \cdots, w_{v^{d_v}}) = \sum_{i \in I_1} U^i(w_{v^i}) + \sum_{i \in I_2} U^i(w_{v^i}) \leq (1 + \delta)U_v + \sum_{i \in I_2} \frac{w_{v^i}}{w_v} U_v$$

$$= (1 + \delta)U_v + (1 - \delta)U_v = 2U_v.$$

Because of the arbitrariness of $(w_{v^1}, \cdots, w_{v^{d_v}})$ and $\mathbf{w}_{-v}$, we conclude $\zeta \leq 2$. This completes the proof.

Discussion for Lemma 1. Before proceeding with the proof of Lemma 1, some other notations and propositions are necessary to introduce. In this subsection, we are interested in another ratio β_v, named as *exchange ratio* [10] of vertex v, which quantifies the aggregate amount of resource that vertex v receives per unit of resource that offers to its neighbors, i.e., $\beta_v = U_v/w_v$. From Proposition 3, we know $\beta_v = \alpha_v$, if v is in B-class, and $\beta_v = 1/\alpha_v$, if v is in C-class. Based on this relationship of β_v and α_v, β-ratio also can be written as a function on x, the reported weight by v given the other agents' weights. That is $\beta_v(x) = U_v(x)/x$. By the monotonicity of α-function in Proposition 6, $\beta_v(x)$ has the property as

Proposition 7. *For any agent v and any other agents' reported weights, the exchange ratio function $\beta_v(x) = U_v(x)/x$ is non-increasing with $x \in [0, w_v]$.*

Now let us pay attention to an initial state in which the weight assignment is $w_{v^i} = x_{vu^i}$, $i = 1, 2, \cdots, d_v$. As stated before, x_{vu^i} is the amount of resource allocated from v to u^i under BD Mechanism in original network G. Further, because $X = (x_{vu})$ obtained from BD Mechanism has the property of proportional response (Proposition 2), then $\frac{x_{vu^i}}{w_v} = \frac{x_{u^iv}}{\sum_{l=1}^{d_v} x_{u^lv}} = \frac{x_{u^iv}}{U_v}$. Hence

Proposition 8. *For any vertex v, its exchange ratio β_v satisfies*

$$\beta_v = \frac{U_v}{w_v} = \frac{x_{u^iv}}{x_{vu^i}}, \ i \in I_1, \tag{1}$$

If vertex v splits itselt with weight assignment $(x_{vu^1}, \cdots, x_{vu^{d_v}})$, it is not hard to see that the allocation in subtree G^i is the same as the sub-allocation of $X = (x_{vu})$ restricted in G^i, and the exchange ratio of each copied node v^i in G^i is also equal to β_v. In addition, v^i is a leaf in G^i, which makes its utility only be from its unique neighbor u^i. Thus the utility of v^i when its weight is x_{vu^i} is $U^i(x_{vu^i}) = x_{u^iv} = \beta_v \cdot x_{vu^i}$.

On the other hand, let us focus on the connected subtree G^i, $i \in I_1$, which can be viewed as a sub-resource exchange system. So all results for original G are suitable for each G^i. Suppose all vertex weights are given. Because G^i only contains one copied node v^i, the β-ratio of v^i only depends on w_{v^i} for any weight assignment $(w_{v^1}, \cdots, w_{v^{d_v}})$. So such a ratio can be written as a function of w_{v^i}, i.e., $\beta_{v^i}(w_{v^i}) = U^i(w_{v^i})/w_{v^i}$. It is observed that $\beta_{v^i}(x_{vu^i}) = \beta_v$.

Proof of Lemma 1: For any weight assignment $(w_{v^1}, \cdots, w_{v^{d_v}})$, in which $\sum_{i \in I_1} w_{v^i} = \delta w_v$,

$$
\begin{aligned}
\sum_{i \in I_1} U^i(w_{v^i}) - U_v &= \sum_{i \in I_1} U^i(w_{v^i}) - \sum_{i \in I_1} U^i(x_{vu^i}) \\
&= \sum_{w_{v^i} \leq x_{vu^i}} [U^i(w_{v^i}) - U^i(x_{vu^i})] + \sum_{w_{v^i} > x_{vu^i}} [U^i(w_{v^i}) - U^i(x_{vu^i})] \\
&\leq \sum_{w_{v^i} > x_{vu^i}} [U^i(w_{v^i}) - U^i(x_{vu^i})] \\
&= \sum_{w_{v^i} > x_{vu^i}} [w_{v^i} \cdot \beta_{v^i}(w_{v^i}) - x_{vu^i} \cdot \beta_v] \\
&= \sum_{w_{v^i} > x_{vu^i}} [x_{vu^i}(\beta_{v^i}(w_{v^i}) - \beta_v) + (w_{v^i} - x_{vu^i})\beta_{v^i}(w_{v^i})] \\
&\leq \sum_{w_{v^i} > x_{vu^i}} (w_{v^i} - x_{vu^i})\beta_v \leq \sum_{i \in I_1} w_{v^i}\beta_v = \delta U_v
\end{aligned}
$$

in which, the first equality comes from $U_v = \sum_{i \in I_1} x_{u^i v} = \sum_{i \in I_1} U^i(x_{vu^i})$; the monotonically nondecreasing property of utility function on weight in Proposition 5 promises the first inequality and the second inequality is right since the β-function is non-increasing on weight in Proposition 7. □

Discussion for Lemma 2. To prove Lemma 2, some additional notations and propositions are necessary to be introduced. As defined before, **w** is the weight profile and $\mathcal{B} = \{(B_1, C_1), \cdots, (B_k, C_k)\}$ is the bottleneck decomposition of G. For each subtree G^i, the vertex set of G^i is denoted by $V(G^i)$ and vertex set V of G can be decomposed in another way that $V = \cup_{i=1}^{d_v} [V(G^i) - \{v^i\}] \cup \{v\}$. For convenience, we denote $V^i = V(G^i) - \{v^i\}$. Let us focus on the sub-decomposition $\mathcal{B}^i$ obtained by restricting $\mathcal{B}$ into subset V^i, $i \in I_2$. W.l.o.g., $\mathcal{B}^i$ is written as $\mathcal{B}^i = \{(B_1^i, C_1^i), \cdots, (B_{k^i}^i, C_{k^i}^i)\}$. By the acyclic structure of G, we have the following proposition whose proof will be in the full version.

Proposition 9. *For each pair (B_h^i, C_h^i) in $\mathcal{B}^i$, $h = 1, \cdots, k^i$ and $i \in I_2$, there is a pair (B_f, C_f) in $\mathcal{B}$, such that either $(B_f, C_f) = (B_h^i, C_h^i)$ or $B_h^i \subset B_f$, $C_h^i \subset C_f$ and $\frac{w(C_h^i)}{w(B_h^i)} = \frac{w(C_f)}{w(B_f)} = \alpha_f$.*

Now let us focus on the bottleneck decomposition of G^i, $i \in I_2$. Of course, its decomposition depends on w_{v^i}, if the weights of other vertices in G^i are fixed. W.l.o.g, we denote it by $\mathcal{B}^i(w_{v^i})$, $w_{v^i} \in [0, w_v]$. Similarly, we start our discussion from the initial state that $w_{v^i} = x_{vu^i}$, $i = 1, \cdots, d_v$, where $x_{vu^i} = 0$, for each $i \in I_2$. Please pay attention that $\mathcal{B}^i(0)$ is the bottleneck decomposition of G^i when $w_{v^i} = 0$ and $\mathcal{B}^i$ is the sub-decomposition of $\mathcal{B}$ restricted in $G^i - \{v^i\}$. Because v^i is a leaf in G^i with unique neighbor u^i and $w_{v^i} = 0$, we can derive $\mathcal{B}^i(0)$ from $\mathcal{B}^i$ directly as the following.

$$(B_h^i(0), C_h^i(0)) = \begin{cases} (B_h^i, C_h^i \cup \{v^i\}), & \text{if } u^i \in B_h^i; \\ (B_h^i \cup \{v^i\}, C_h^i), & \text{if } u^i \in C_h^i; \\ (B_h^i, C_h^i), & \text{otherwise.} \end{cases} \tag{2}$$

Further, the condition of $w_{v^i} = 0$ makes $\alpha_h^i(0) = \alpha_h^i$, $h = 1, \cdots, k^i$.

Proof of Lemma 2: W.l.o.g., we suppose $v \in B_j \cup C_j$, $u^i \in B_f \cup C_f$ in $\mathcal{B}$ and $u^i \in B_h^i \cup C_h^i$ in $\mathcal{B}^i$. So (B_h^i, C_h^i) is the pair obtained by restricting (B_f, C_f) in to V^i and $\alpha_h^i = \alpha_f$ by Proposition 9. Moreover, $v^i \in B_h^i(0) \cup C_h^i(0)$. Then its α-ratio is $\alpha_{v^i}(0) = \alpha_h^i(0) = \alpha_h^i = \alpha_f$.

If $v \in B_j$, then u^i must be in C_f (also in C_h^i) with $f \leq j$ by Property 1. Under this case, $\alpha_f \leq \alpha_j = \alpha_v = \beta_v = \frac{U_v}{w_v}$. The fact $u^i \in C_f$ makes $u^i \in C_h^i$, inducing $u^i \in C_h^i(0)$ and $v^i \in B_h^i(0)$ by (2). Proposition 6 tells us if v^i is a B-class vertex when its weight is zero, then v^i must always be in B-class and $\alpha_{v^i}(w_{v^i})$ is non-increasing for all $w_{v^i} \in [0, w_v]$. Hence, for any $w_{v^i} \in [0, w_v]$

$$U^i(w_{v^i}) = w_{v^i} \cdot \alpha_{v^i}(w_{v^i}) \leq w_{v^i} \cdot \alpha_{v^i}(0) = w_{v^i} \cdot \alpha_f \leq w_{v^i} \cdot \alpha_j = \frac{w_{v^i}}{w_v} U_v.$$

If $v \in C_j$, then $\frac{U_v}{w_v} = \beta_v = \frac{1}{\alpha_j} \geq 1$ and u^i may be in B_f or C_f. If $u \in C_f$ (also in C_h^i), then v^i must be a B-class vertex when $w_{v^i} = 0$. By the same reason, we know v^i is a B-class vertex for all $w_{v^i} \in [0, w_v]$ and

$$U^i(w_{v^i}) \leq w_{v^i} \leq w_{v^i} \frac{1}{\alpha_j} = \frac{w_{v^i}}{w_v} U_v. \tag{3}$$

If $u^i \in B_f$ (also in B_h^i), then $v^i \in C_h^i(0)$. By Proposition 6, it is possible that there is $x^* \in (0, w_v]$, such that $\alpha_{v^i}(x^*) = 1$. In addition, v^i is in B-class when $w_{v^i} \in [x^*, w_v]$ and v is in C-class with $\alpha_{v^i}(w_{v^i})$ non-decreasing when $0 \leq x \leq x^*$. Clearly, if $w_{v^i} \in [x^*, w_v]$, then the fact that v^i is a B-class vertex results in $U^i(w_{v^i}) \leq \frac{w_{v^i}}{w_v} U_v$ by (3). If $w_{v^i} \in [0, x^*]$, then v^i is in C-class and

$$U^i(w_{v^i}) = w_{v^i} \cdot \frac{1}{\alpha_{v^i}(w_{v^i})} \leq w_{v^i} \cdot \frac{1}{\alpha_{v^i}(0)} = w_{v^i} \cdot \frac{1}{\alpha_f} = \frac{w_{v^i}}{w_v} U_v, \tag{4}$$

where the inequality comes from the non-decreasing property of $\alpha_{v^i}(w_{v^i})$ when $w_{v^i} \in [0, x^*]$. If x^* does not exist, then there is only one situation that v is in C-class for all $w_{v^i} \in [0, w_v]$. Under this situation, we know $\alpha_{v^i}(w_{v^i})$ is non-decreasing in $[0, w_v]$ by Proposition 6. Applying the above deduction in (4), we conclude $U^i(w_{v^i}) \leq \frac{w_{v^i}}{w_v} U_v$ for all $w_{v^i} \in [0, w_v]$. This completes the lemma. $\quad\square$

4 Conclusion

This paper discusses the issue of possible strategic manipulations of agents with respect to a resource allocation mechanism, BD Mechanism, for the application of resource exchange. We study the incentives of a vertex splitting strategy of agents and characterize how much utility can be increased by such plays for resource exchange game, establishing new understandings on the strategic stability of BD Mechanism from the mechanism design perspective. We prove that the incentive ratios of BD Mechanism for the vertex splitting strategy on trees are exactly 2. Compared with a recent work [13] proving that the incentive ratio for the linear exchange economy in Arrow-Debreu markets is unbounded, our results of bounded incentive ratio place the practical resource exchange with a market equilibrium solution more rational in terms of truthful behavior.

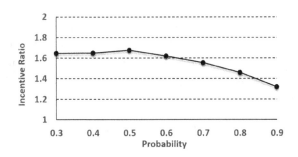

Fig. 3. The numerical experiment results on random graphs.

Though we have completed the study of incentive ratio on trees, there is a space left that how to explore the incentive ratio of BD Mechanism on general graphs. Fig. 3 shows the average incentive ratio among 1000 numerical experiments on random graphs, in which each edge is generated with probability $p \in \{0.1, 0.2, \cdots, 0.9\}$. Further, in our all simulations, the incentive ratio of BD Mechanism for the vertex splitting strategy on any graph never exceeds 2. Inspired by these simulation results, we conjecture the incentive ratio is no more than 2 for any graph which will be left as future work.

Acknowledgments. This research was partially supported by the National Nature Science Foundation of China (Nos. 11301475, 11426026, 61632017, 61173011), by a Project 985 grant of Shanghai Jiao Tong University, and by the Research Grant Council of Hong Kong (ECS Project No. 26200314, GRF Project No. 16213115 and GRF Project No. 16243516).

References

1. Adsul, B., Babu, C.S., Garg, J., Mehta, R., Sohoni, M.: Nash equilibria in fisher market. In: Kontogiannis, S., Koutsoupias, E., Spirakis, P.G. (eds.) SAGT 2010. LNCS, vol. 6386, pp. 30–41. Springer, Heidelberg (2010). doi:10.1007/978-3-642-16170-4_4
2. Alkalay, C., Vetta, A.: False-name bidding and economic efficiency in combinatorial auctions. In: AAAI, pp. 538–544 (2014)
3. Braanzei, S., Chen, Y.L., Deng, X.T., Filos-Ratsikas, A., Kristoffer, S., Frederiksen, S., Zhang, J.: The fisher market game: equilibrium and welfare. In: AAAI (2014)

4. Chen, N., Deng, X., Zhang, H., Zhang, J.: Incentive ratios of fisher markets. In: Czumaj, A., Mehlhorn, K., Pitts, A., Wattenhofer, R. (eds.) ICALP 2012. LNCS, vol. 7392, pp. 464–475. Springer, Heidelberg (2012). doi:10.1007/978-3-642-31585-5_42
5. Chen, N., Deng, X., Zhang, J.: How profitable are strategic behaviors in a market? In: Demetrescu, C., Halldórsson, M.M. (eds.) ESA 2011. LNCS, vol. 6942, pp. 106–118. Springer, Heidelberg (2011). doi:10.1007/978-3-642-23719-5_10
6. Cheng, Y., Deng, X., Pi, Y., Yan, X.: Can bandwidth sharing be truthful? In: Hoefer, M. (ed.) SAGT 2015. LNCS, vol. 9347, pp. 190–202. Springer, Heidelberg (2015). doi:10.1007/978-3-662-48433-3_15
7. Cheng, Y., Deng, X., Qi, Q., Yan, X.: Truthfulness of a proportional sharing mechanism in resource exchange. In: IJCAI, pp. 187–193 (2016)
8. Dalakov, G.: History of Computers and Computing, Internet, Internet conquers the world, BitTorrent. http://historycomputer.com/Internet/Conquering/BitTorrent.html
9. Feldman, M., Lai, K., Stoica, I., et al.: Robust incentive techniques for peer-to-peer networks. In: Proceedings of the 5th ACM conference on Electronic commerce, pp. 102–111. ACM (2004)
10. Georgiadis, L., Iosifidisy, G., Tassiulas, L.: Exchange of services in networks: competition, cooperation, and fairness. In: Proceedings of the 2015 ACM International Conference on Measurement and Modeling of Computer Systems (SIGMETRICS 2015), pp. 43–56 (2015)
11. Iwasaki, A., Conitzer, V., Omori, Y., et al.: Worst-case efficiency ratio in false-name-proof combinatorial auction mechanisms. In: Proceedings of the 9th International Conference on Autonomous Agents and Multiagent Systems, vol. 1. International Foundation for Autonomous Agents and Multiagent Systems, pp. 633–640 (2010)
12. Koutsoupias, E., Papadimitriou, C.: Worst-case equilibria. In: Meinel, C., Tison, S. (eds.) STACS 1999. LNCS, vol. 1563, pp. 404–413. Springer, Heidelberg (1999). doi:10.1007/3-540-49116-3_38
13. Polak, I.: The incentive ratio in exchange economies. In: Chan, T.-H.H., Li, M., Wang, L. (eds.) COCOA 2016. LNCS, vol. 10043, pp. 685–692. Springer, Cham (2016). doi:10.1007/978-3-319-48749-6_49
14. Schollmeier, R.: A definition of peer-to-peer networking for the classification of peer-to-peer architectures and applications. In: 2001 Proceedings of First International Conference on Peer-to-Peer Computing, pp. 101–102. IEEE (2001)
15. Schor, J.: Debating the sharing economy. J. Self-Gov. Manag. Econ. 4(3), 7–22 (2016)
16. Roughgarden, T., Tardos, E.: How bad is selfish routing. J. ACM 49(2), 236–259 (2002)
17. Wu, F., Zhang, L.: Proportional response dynamics leads to market equilibrium. In: STOC, pp. 354–363 (2007)
18. Yokoo, M.: False-name bids in combinatorial auctions. ACM SIGecom Exchanges 7(1), 1–4 (2007)
19. Yokoo, M., Sakurai, Y., Matsubara, S.: The effect of false-name bids in combinatorial auctions: new fraud in Internet auctions. Games Econ. Behav. 46, 174–188 (2004)

A 3-Player Protocol Preventing Persistence in Strategic Contention with Limited Feedback

George Christodoulou[1], Martin Gairing[1], Sotiris Nikoletseas[2,3],
Christoforos Raptopoulos[2,3(✉)], and Paul Spirakis[1,2,3]

[1] Department of Computer Science, University of Liverpool, Liverpool, UK
{G.Christodoulou,gairing,P.Spirakis}@liverpool.ac.uk
[2] Computer Engineering and Informatics Department,
University of Patras, Patras, Greece
nikole@cti.gr, raptopox@ceid.upatras.gr
[3] Computer Technology Institute and Press "Diophantus", Patras, Greece

Abstract. In this paper, we study contention resolution protocols from a game-theoretic perspective. In a recent work [8], we considered *acknowledgment-based* protocols, where a user gets feedback from the channel only when she attempts transmission. In this case she will learn whether her transmission was successful or not. One of the main results of [8] was that no acknowledgment-based protocol can be in equilibrium. In fact, it seems that many natural acknowledgment-based protocols fail to prevent users from unilaterally switching to persistent protocols that always transmit with probability 1. It is therefore natural to ask how powerful a protocol must be so that it can beat persistent deviators.

In this paper we consider *age-based* protocols, which can be described by a sequence of probabilities of transmitting in each time step. Those probabilities are given beforehand and do not change based on the transmission history. We present a 3-player age-based protocol that can prevent users from unilaterally deviating to a persistent protocol in order to decrease their expected transmission time. It is worth noting that the answer to this question does not follow from the results and proof ideas of [8]. Our protocol is non-trivial, in the sense that, when all players use it, finite expected transmission time is guaranteed. In fact, we show that this protocol is preferable to any deadline protocol in which, after some fixed time, attempt transmission with probability 1 in every subsequent step. An advantage of our protocol is that it is very simple to describe, and users only need a counter to keep track of time. Whether there exist n-player age-based protocols that do not use counters and can prevent persistence is left as an open problem for future research.

Keywords: Contention resolution · Age-based protocol · Persistent deviator · Game theory

1 Introduction

A fundamental problem in networks is *contention resolution* in multiple access channels. In such a setting there are multiple users that want to communicate

V. Bilò and M. Flammini (Eds.): SAGT 2017, LNCS 10504, pp. 240–251, 2017.
DOI: 10.1007/978-3-319-66700-3_19

with each other by sending messages into a multiple access channel (or broadcast channel). The channel is not centrally controlled, so two or more users can transmit their messages at the same time, in which case there is a collision and no transmission is successful. The objective in contention resolution is the design of *distributed protocols* for resolving such conflicts, while simultaneously optimizing some performance measure, like channel utilization or average throughput.

Following the standard assumption in this area, we assume that time is discrete and messages are broken up into fixed sized packets, which fit exactly into one time slot. In fact, we consider one of the simplest possible scenarios where each user only needs to send a single packet through the channel. Most studies on distributed contention resolution protocols (see Sect. 1.2) are based on the assumption that users will always follow the algorithm. In this paper, following [10] we drop this assumption, and we assume that a player will only obey a protocol if it is in her best interest, given the other players stick to the protocol. Therefore, we model the situation from a game-theoretic perspective, i.e. as a stochastic game with the users as selfish *players*.

One of the main results of Fiat et al. [10] was the design of an incentive-compatible transmission protocol which guarantees that (with high probability) all players will transmit successfully in time linear in the number of players n. The authors assume a *ternary* feedback channel, i.e. each player receives feedback of the form $0/1/2^+$ after each time step, indicating whether zero, one, or more than one transmission was attempted. In a related paper, Christodoulou et al. [9] designed efficient ϵ-equilibrium protocols under a stronger assumption that each player receives as feedback the number of players that attempted transmission; this is called *multiplicity* feedback. They also assume non-zero transmission costs, in which case the protocols of [10] do not apply.

All of the protocols defined in the above two works belong to the class of *full-sensing* protocols [14], in which the channel feedback is broadcasted to all sources. However, in wireless channels, there are situations where full-sensing is not possible because of the *hidden-terminal problem* [24]. In a previous work [8], we considered *acknowledgment-based* protocols, which use a more limited feedback model – the only feedback that a user gets is whether her transmission was successful or not. A user that does not transmit cannot "listen" to the channel and therefore does not get any feedback. In other words, the only information that a user has is the history of her own transmission attempts. Acknowledgment-based protocols have been extensively studied in the literature (see e.g. [14] and references therein).

Our main concern in [8] was the existence of acknowledgment-based protocols that are in equilibrium. For $n = 2$ players, we showed that there exists such a protocol, which guarantees finite expected transmission time. Even though the general question for more than 2 players was left open in [8], we ruled out that such a protocol can be *age-based*. Age-based protocols are a special case of acknowledgment-based protocols and can be described by a sequence of probabilities (one for each time-step) of transmitting in each time step. Those probabilities are given beforehand and do not change based on the transmission history.

The well known ALOHA protocol [1] is a special age-based protocol, where – except for the first round – users always transmit with the same probability. Since an age-based protocol $\mathcal{P}$ cannot be in equilibrium, it is beneficial for players to deviate from $\mathcal{P}$ to some other protocol. In fact, most natural acknowledgment-based protocols fail to prevent users from unilaterally switching to the persistent protocol that always transmits with probability 1. It is therefore natural to ask how powerful a protocol must be with respect to memory (and feedback) in order to be able to prevent persistent deviators.

1.1   Our Contribution

The question that we consider in this paper is whether there exist age-based protocols that can prevent users from unilaterally deviating to a persistent protocol (in which they attempt a transmission in every step until they successfully transmit) in order to decrease their expected transmission time. In particular, such protocols should be non-trivial, in the sense that using the protocol should guarantee a finite expected transmission time for the users. It is worth noting that the answer to this question does not follow from the results and proof ideas of [8]. We give a positive answer for the case of 3 players (users), by presenting and analyzing such a protocol (see definition below). In particular, we show that this protocol is preferable to any deadline protocol in which, after some fixed time, attempt transmission with probability 1 in every subsequent step.

Let $c \geq 1$ and $p \in [0,1]$ be constants. We define the protocol $\mathcal{P} = \mathcal{P}(c,p)$ as follows: the transmission probability $\mathcal{P}_t$ at any time t is equal to p if $t = \sum_{j=0}^{k} \lfloor 2c^j \rfloor$, for some $k = 0, 1, \dots$ and it is equal to 1 otherwise. The intuition behind this protocol is that with every collision it is increasingly harder for remaining users to successfully transmit, and thus "aggressive" protocols are suboptimal.

Our main result is the following:

Theorem 1. *Assume there are 3 players in the system, two of which use protocol $\mathcal{P}(1.1, 0.75)$. Then, the third player will prefer using protocol $\mathcal{P}(1.1, 0.75)$ over any deadline protocol $\mathcal{D}$.*

In addition, we show that the expected transmission time of a fixed player when all players use $\mathcal{P}(1.1, 0.75)$ is upper bounded by 2759 and is thus finite. We believe that our ideas can be used to give a positive answer also for the case of $n > 3$ players, but probably not for too large values of n.

An advantage of our protocol is that it is very simple to describe, and users only need a counter to keep track of time. Whether there exist n-player age-based protocols using finite memory that can prevent persistence is left as an open problem for future research.

Due to lack of space some proofs are omitted but can be found in the complete version of our paper [7].

1.2   Other Related Work

Perhaps the most famous multiple-access communication protocol is the (slotted) ALOHA protocol [1,22]. Follow-up papers study the efficiency of multiple-access protocols for packets that are generated by some stochastic process (see e.g. [12,13,21]), or worst-case scenarios of bursty inputs [5].

The main focus of many contention resolution protocols is on actual *conflict resolution*. In such a scenario, it is assumed that there are n users in total, and k of them collide. In such an event, a resolution algorithm is called, which ensures that all the colliding packets are successfully transmitted [6,16,25]. There is extensive study on the efficiency of protocols under various information models (see [14] for an overview). When k is known, [11] provides an $O(k + \log k \log n)$ *acknowledgment-based* algorithm, while [19] provides a matching lower bound. For the ternary model, [15] provides a bound of $\Omega(k(\log n/\log k))$ for all deterministic algorithms.

There are various game theoretic models of slotted ALOHA that have been studied in the literature, apart from the ones mentioned in the introduction; see for example [2,3,18]. However, in most of these models only transmission protocols that always transmit with the same probability are considered. There has been also research on pricing schemes [26] as well as on cases in which the channel quality changes dynamically with time and players must choose their transmission levels accordingly [4,20,27]. An interesting game-theoretic model that lies between the contention and congestion model was studied in [17]; where decisions of *when* to submit is part of the action space of the players.

2   Model

Game Structure. Let $N = \{1, 2, \ldots, n\}$ be the set of agents, each one of which has a single packet that he wants to send through a common channel. All players know n. We assume time is discretized into slots $t = 1, 2, \ldots$. The players that have not yet successfully transmitted their packet are called *pending* and initially all n players are pending. At any given time slot t, a pending player i has two available actions, either to *transmit* his packet or to *remain quiet*. In a *(mixed) strategy*, a player i transmits his packet at time t with some probability that potentially depends on information that i has gained from the channel based on previous transmission attempts. If exactly one player transmits in a given slot t, then his transmission is *successful*, the successful player exits the game (i.e. he is no longer pending), and the game continues with the rest of the players. On the other hand, whenever two or more agents try to access the channel (i.e. transmit) at the same slot, a *collision* occurs and their transmissions fail, in which case the agents remain in the game. Therefore, in case of collision or if the channel is idle (i.e. no player attempts to transmit) the set of pending agents remains unchanged. The game continues until all players have successfully transmitted their packets.

Transmission Protocols. Let $X_{i,t}$ be the indicator variable that indicates whether player i attempted transmission at time t. For any $t \geq 1$, we denote by $\boldsymbol{X}_t$ the transmission vector at time t, i.e. $\boldsymbol{X}_t = (X_{1,t}, X_{2,t}, \ldots, X_{n,t})$. An *acknowlegment-based* protocol, uses very limited channel feedback. After each time step t, only players that attempted a transmission receive feedback, and the rest get no information. In fact, the information received by a player i who transmitted during t is whether his transmission was successful (in which case he gets an acknowledgement and exits the game) or whether there was a collision.

Let $\boldsymbol{h}_{i,t}$ be the vector of the *personal transmission history* of player i up to time t, i.e. $\boldsymbol{h}_{i,t} = (X_{i,1}, X_{i,2}, \ldots, X_{i,t})$. We also denote by $\boldsymbol{h}_t$ the transmission history of all players up to time t, i.e. $\boldsymbol{h}_t = (\boldsymbol{h}_{1,t}, \boldsymbol{h}_{2,t}, \ldots, \boldsymbol{h}_{n,t})$. In an acknowledgement-based protocol, the actions of player i at time t depend only (a) on his personal history $\boldsymbol{h}_{i,t-1}$ and (b) on whether he is pending or not at t. A *decision rule* $f_{i,t}$ for a pending player i at time t, is a function that maps $\boldsymbol{h}_{i,t-1}$ to a probability $\Pr(X_{i,t} = 1|\boldsymbol{h}_{i,t-1})$. For a player $i \in N$, a *(transmission) protocol* f_i is a sequence of decision rules $f_i = \{f_{i,t}\}_{t \geq 1} = f_{i,1}, f_{i,2}, \cdots$.

A transmission protocol is *anonymous* if and only if the decision rule assigns the same transmission probability to all players with the same personal history. In particular, for any two players $i \neq j$ and any $t \geq 0$, if $\boldsymbol{h}_{i,t-1} = \boldsymbol{h}_{j,t-1}$, it holds that $f_{i,t}(\boldsymbol{h}_{i,t-1}) = f_{j,t}(\boldsymbol{h}_{j,t-1})$. In this case, we drop the subscript i in the notation, i.e. we write $f = f_1 = \cdots = f_n$.

We call a protocol f_i for player i *age-based* if and only if, for any $t \geq 1$, the transmission probability $\Pr(X_{i,t} = 1|\boldsymbol{h}_{i,t-1})$ depends only (a) on time t and (b) on whether player i is pending or not at t. In this case, we will denote the transmission probability by $p_{i,t} \stackrel{def}{=} \Pr(X_{i,t} = 1|\boldsymbol{h}_{i,t-1}) = f_{i,t}(\boldsymbol{h}_{i,t-1})$.

We call a transmission protocol f_i *non-blocking* if and only if, for any $t \geq 1$ and any transition history $\boldsymbol{h}_{i,t-1}$, the transmission probability $\Pr(X_{i,t} = 1|\boldsymbol{h}_{i,t-1})$ is always smaller than 1. A protocol f_i for player i is a *deadline protocol with deadline* $t_0 \in \{1, 2, \ldots\}$ if and only if $f_{i,t}(\boldsymbol{h}_{i,t-1}) = 1$, for any player i, any time slot $t \geq t_0$ and any transmission history $\boldsymbol{h}_{i,t-1}$. A *persistent player* is one that uses the deadline protocol with deadline 1.

Individual Utility. Let $\boldsymbol{f} = (f_1, f_2, \ldots, f_n)$ be such that player i uses protocol $f_i, i \in N$. For a given transmission sequence $\boldsymbol{X}_1, \boldsymbol{X}_2, \ldots$, which is consistent with $\boldsymbol{f}$, define the *latency* or *success time* of agent i as $T_i \stackrel{def}{=} \inf\{t : X_{i,t} = 1, X_{j,t} = 0, \forall j \neq i\}$. That is, T_i is the time at which i successfully transmits. Given a transmission history $\boldsymbol{h}_t$, the n-tuple of protocols $\boldsymbol{f}$ induces a probability distribution over sequences of further transmissions. In that case, we write $C_i^{\boldsymbol{f}}(\boldsymbol{h}_t) \stackrel{def}{=} \mathbb{E}[T_i|\boldsymbol{h}_t, \boldsymbol{f}] = \mathbb{E}[T_i|\boldsymbol{h}_{i,t}, \boldsymbol{f}]$ for the expected latency of agent i incurred by a sequence of transmissions that starts with $\boldsymbol{h}_t$ and then continues based on $\boldsymbol{f}$. For anonymous protocols, i.e. when $f_1 = f_2 = \cdots = f_n = f$, we will simply write $C_i^f(\boldsymbol{h}_t)$ instead[1].

Equilibria. The objective of every agent is to minimize her expected latency. We say that $\boldsymbol{f} = \{f_1, f_2, \ldots, f_n\}$ is in *equilibrium* if for any transmission history

---

[1] Abusing notation slightly, we will also write $C_i^{\boldsymbol{f}}(\boldsymbol{h}_0)$ for the *unconditional* expected latency of player i induced by $\boldsymbol{f}$.

h_t the agents cannot decrease their expected latency by unilaterally deviating after t; that is, for all agents i, for all time slots t, and for all decision rules f_i' for agent i, we have

$$C_i^{\boldsymbol{f}}(\boldsymbol{h}_t) \leq C_i^{(\boldsymbol{f}_{-i}, f_i')}(\boldsymbol{h}_t),$$

where $(\boldsymbol{f}_{-i}, f_i')$ denotes the protocol profile[2] where every agent $j \neq i$ uses protocol f_j and agent i uses protocol f_i'.

3   A 3-Player Protocol that Prevents Persistence

In this section we prove that there is an anonymous age-based protocol $\mathcal{P}(c, p)$ for 3 players that has finite expected latency and prevents players from unilaterally switching to any deadline protocol. In what follows, Alice is one of the three players in the system.

For some parameters $c \geq 1$ and $p \in [0, 1]$, which will be specified later, we define the protocol $\mathcal{P} = \mathcal{P}(c, p)$ as follows:

$$\mathcal{P}_t = \begin{cases} p, & \text{if } t = \sum_{j=0}^{k} \lfloor 2c^j \rfloor, \text{ for some } k = 0, 1, \ldots \\ 1, & \text{otherwise.} \end{cases} \tag{1}$$

For $k = 0, 1, 2, \ldots$, define the k-th non-trivial transmission time s_k to be the time step on which the decision rule for a pending player using $\mathcal{P}$ is to transmit with probability p. In particular, $s_k \overset{def}{=} \sum_{j=0}^{k} \lfloor 2c^j \rfloor$, by definition of the protocol. For technical reasons, we set $s_k = 0$, for any $k < 0$. Furthermore, for $k = 1, 2, \ldots$, define the k-th (non-trivial) inter-transmission time x_k as the time between the k-th and $(k-1)$-th non-trivial transmission time, i.e. $x_k \overset{def}{=} s_k - s_{k-1} = \lfloor 2c^k \rfloor$. The following elementary result will be useful for the analysis of the protocol. The proof can be found in [7].

Lemma 1. *For any $k, k', j \in \{0, 1, \ldots\}$, such that $k' > k$, and any $c \in [1, 2]$, we have that*

$$c^{k'-k-1}(c-1)x_{k+j} \leq x_{k'+j} \leq c^{k'-k-1}(c+1)x_{k+j}.$$

3.1   Expected Latency for a Persistent Player

Assume that Alice is a persistent player, i.e. she uses the deadline protocol g with deadline 1, i.e. $g_t = 1$, for all $t \geq 1$, while both other players use protocol $\mathcal{P}$. For $n \in \{1, 2, 3\}, k \in \{0, 1, \ldots\}$, let $Y_{n,k}'$ be the additional time after s_{k-1} that Alice needs to successfully transmit when there are n pending players. It is evident that Alice will be the first player to successfully transmit, so there will be no need to calculate $\mathbb{E}[Y_{2,k}']$ or $\mathbb{E}[Y_{1,k}']$.

The proof of the following Theorem can be found in [7].

---

[2] For an anonymous protocol f, we denote by (f_{-i}, f_i') the profile where agent $j \neq i$ uses protocol f and agent i uses protocol f_i'.

Theorem 2. *If $\frac{1}{1-(1-p)^2} < c \leq 2$, then $\mathbb{E}[Y'_{3,0}] = \infty$. That is, the expected latency for Alice when she is persistent and both other players use protocol $\mathcal{P}(c,p)$ is infinity.*

Remark 1. At first glance, the above result may seem surprising. Indeed, let Z denote the number of times that the persistent player has a collision whenever the other players transmit with probability p (i.e. we do not count collissions when the other two players transmit with probability 1, which causes certain collision). It is easy to see that $Z + 1$ is a geometric random variable with probability of success $(1 - p)^2$. Therefore, $\mathbb{E}[Z + 1] = \frac{1}{(1-p)^2}$ is finite! On the other hand, it is not hard to see that the (actual) time $Y'_{3,0}$ needed for the persistent player to successfully transmit is given by $Y'_{3,0} = \sum_{j=0}^{Z} \lfloor 2c^j \rfloor$. In particular, $Y'_{3,0}$ is a strictly convex function of Z, for any $c > 1$, and so, by Jensen's inequality (see e.g. [23]) $\mathbb{E}[Y'_{3,0}] > \sum_{j=0}^{\mathbb{E}[Z]} \lfloor 2c^j \rfloor$.

3.2   Expected Latency When All Players Use $\mathcal{P}(c,p)$

Assume that all three players use protocol $\mathcal{P}$. For $n \in \{1, 2, 3\}, k \in \{0, 1, \ldots\}$, let $Y_{n,k}$ be the additional time after s_{k-1} that Alice needs to successfully transmit, when there are n pending players. The following corollary, which is a direct consequence of Lemma 1, will be useful for our analysis.

Corollary 1. *For any $n \in \{1, 2, 3\}$, any $k, k' \in \{0, 1, \ldots\}$ with $k' > k$, and $c \in [1, 2]$ we have that*

$$c^{k'-k-1}(c-1)\mathbb{E}[Y_{n,k}] \leq \mathbb{E}[Y_{n,k'}] \leq c^{k'-k-1}(c+1)\mathbb{E}[Y_{n,k}].$$

The main purpose of this section is to prove Theorem 3. To do this, we need to consider $\mathbb{E}[Y_{n,k}]$, for all values of $n \in \{1, 2, 3\}, k \in \{0, 1, \ldots\}$.

The Case $n = 1$. When only Alice is pending, we have

$$\mathbb{E}[Y_{1,k}] = \lfloor 2c^k \rfloor p + \left(\sum_{j=k}^{k+1} \lfloor 2c^j \rfloor\right) p(1 - p) + \ldots$$

$$= \sum_{\ell=k}^{\infty} \left(\left(\sum_{j=k}^{\ell} \lfloor 2c^j \rfloor\right) p(1 - p)^{\ell-k}\right)$$

$$\leq \sum_{\ell=k}^{\infty} \left(\left(\sum_{j=k}^{\ell} 2c^j\right) p(1 - p)^{\ell-k}\right)$$

$$= \sum_{\ell=k}^{\infty} \left(\left(2\frac{c^{\ell+1} - c^k}{c - 1}\right) p(1 - p)^{\ell-k}\right)$$

$$\leq \frac{2cp}{(c - 1)(1 - p)^k} \sum_{\ell=k}^{\infty} \left(c^\ell (1 - p)^\ell\right). \tag{2}$$

In particular, by the above inequality, we have the following:

Lemma 2. *If* $1 < c < \frac{1}{1-p}$, *then* $\mathbb{E}[Y_{1,k}]$ *is finite, for any (finite)* k.

The Case $n = 2$. Fix $k'_1 > 0$ and assume $c \in [1, 2]$ (so that we can apply Corollary 1). When two players are pending (i.e. Alice and one other player), we have, for all $i = 0, 1, 2, \ldots, k'_1 - 1$,

$$\mathbb{E}[Y_{2,i}] = \lfloor 2c^i \rfloor + p(1 - p)\mathbb{E}[Y_{1,i+1}] + (1 - 2p(1 - p))\mathbb{E}[Y_{2,i+1}]$$

Set $\delta = 1 - 2p(1 - p)$. Multiplying the corresponding equation for $\mathbb{E}[Y_{2,i}]$ by δ^i, for each $i = 0, 1, 2, \ldots, k'_1 - 1$ and adding up, we get

$$\mathbb{E}[Y_{2,0}] = \sum_{i=0}^{k'_1-1} \delta^i \lfloor 2c^i \rfloor + p(1 - p) \sum_{i=0}^{k'_1-1} \delta^i \mathbb{E}[Y_{1,i+1}] + \delta^{k'_1} \mathbb{E}[Y_{2,k'_1}]. \qquad (3)$$

By the second inequality of Corollary 1 for $n = 2$ and $k = 0$, we get

$$\mathbb{E}[Y_{2,0}] \leq \sum_{i=0}^{k'_1-1} \delta^i \lfloor 2c^i \rfloor + p(1 - p) \sum_{i=0}^{k'_1-1} \delta^i \mathbb{E}[Y_{1,i+1}] + \delta^{k'_1} c^{k'_1-1}(c + 1)\mathbb{E}[Y_{2,0}]. \qquad (4)$$

Observe now that, if we have $1 < c < \frac{1}{1-p}$, then, by Lemma 2, the terms $\sum_{i=0}^{k'_1-1} \delta^i \lfloor 2c^i \rfloor + p(1 - p) \sum_{i=0}^{k'_1-1} \delta^i \mathbb{E}[Y_{1,i+1}]$ in the above inequality are finite and strictly positive. Therefore, $\mathbb{E}[Y_{2,0}]$ (which is also strictly positive), will be finite if, in addition to $c < \frac{1}{1-p}$ and $c \in [1, 2]$, the following inequality holds:

$$\delta^{k'_1} c^{k'_1-1}(c + 1) < 1. \qquad (5)$$

Taking $k'_1 \to \infty$ (in fact, given p, c, we can choose a minimum, finite value for k'_1 so that the above inequality holds, see also [7]), we have that, if c satisfies $c < \frac{1}{1-2p(1-p)}$, and also $c < \frac{1}{1-p}$ (so that $\mathbb{E}[Y_{1,i}]$ is finite for all finite i), and $c \in (0, 2]$ (so that we can apply Corollary 1), then $\mathbb{E}[Y_{2,0}]$ is finite. In fact, we can prove the following more general result:

Lemma 3. *If* $1 < c < \min\left\{\frac{1}{1-p}, \frac{1}{1-2p(1-p)}, 2\right\}$, *then* $\mathbb{E}[Y_{2,k}]$ *is finite, for any (finite)* k.

Proof. By the above arguments, when $1 < c < \min\left\{\frac{1}{1-p}, \frac{1}{1-2p(1-p)}, 2\right\}$, $\mathbb{E}[Y_{2,0}]$ is finite. But, by the second inequality of Corollary 1, we also have that $\mathbb{E}[Y_{2,k}] \leq c^{k-1}(c + 1)\mathbb{E}[Y_{2,0}]$, which completes the proof. □

The Case $n = 3$. Fix $k'_2 > 0$ and assume $c \in [1, 2]$. When all three players are pending, we have, for all $i = 0, 1, 2, \ldots, k'_2 - 1$,

$$\mathbb{E}[Y_{3,i}] = \lfloor 2c^i \rfloor + 2p(1 - p)^2 \mathbb{E}[Y_{2,i+1}] + (1 - 3p(1 - p)^2)\mathbb{E}[Y_{3,i+1}].$$

Set $\beta = 1 - 3p(1-p)^2$. Multiplying the corresponding equation for $\mathbb{E}[Y_{3,i}]$ by β^i, for all $i = 0, 1, 2, \ldots, k_2' - 1$ and adding up, we get

$$\mathbb{E}[Y_{3,0}] = \sum_{i=0}^{k_2'-1} \beta^i \lfloor 2c^i \rfloor + 2p(1-p)^2 \sum_{i=0}^{k_2'-1} \beta^i \mathbb{E}[Y_{2,i+1}] + \beta^{k_2'} \mathbb{E}[Y_{3,k_2'}].$$

By the second inequality of Corollary 1 for $n = 3$ and $k = 0$, we get

$$\mathbb{E}[Y_{3,0}] \le \sum_{i=0}^{k_2'-1} \beta^i \lfloor 2c^i \rfloor + 2p(1-p)^2 \sum_{i=0}^{k_2'-1} \beta^i \mathbb{E}[Y_{2,i+1}] + \beta^{k_2'} c^{k_2'-1}(c+1)\mathbb{E}[Y_{3,0}]. \quad (6)$$

Observe now that, if we have $1 < c < \min\left\{\frac{1}{1-p}, \frac{1}{1-2p(1-p)}, 2\right\}$, then, by Lemma 3, the terms $\sum_{i=0}^{k_2'-1} \beta^i \lfloor 2c^i \rfloor + 2p(1-p)^2 \sum_{i=0}^{k_2'-1} \beta^i \mathbb{E}[Y_{2,i+1}]$ in the above inequality are finite and strictly positive. Therefore, $\mathbb{E}[Y_{3,0}]$ (which is also strictly positive), will be finite if, in addition to $1 < c < \min\left\{\frac{1}{1-p}, \frac{1}{1-2p(1-p)}, 2\right\}$, the following inequality holds:

$$\beta^{k_2'} c^{k_2'-1}(c+1) < 1. \quad (7)$$

Taking $k_2' \to \infty$ (in fact, given p, c, we can choose a minimum, finite value for k_2' so that the above inequality holds, see also [7]), we have that, if c satisfies $c < \frac{1}{1-3p(1-p)^2}$, and also $c < \min\left\{\frac{1}{1-p}, \frac{1}{1-2p(1-p)}, 2\right\}$ (so that $\mathbb{E}[Y_{2,i}]$ is finite for all finite i), then $\mathbb{E}[Y_{3,0}]$ is finite. Similarly to the proof of Lemma 3, we can prove the following more general result:

Theorem 3. *If $1 < c < \min\left\{\frac{1}{1-p}, \frac{1}{1-2p(1-p)}, \frac{1}{1-3p(1-p)^2}, 2\right\}$, then $\mathbb{E}[Y_{3,k}]$ is finite, for any (finite) k. In particular, the expected latency of Alice when all players (including Alice herself) use protocol $\mathcal{P}(c,p)$ is finite.*

3.3   Feasibility

We first show that there are values for p and c, such that the following inequalities hold at the same time:

$$1 < c < \min\left\{\frac{1}{1-p}, \frac{1}{1-2p(1-p)}, \frac{1}{1-3p(1-p)^2}\right\}$$

and

$$\frac{1}{1-(1-p)^2} < c \le 2.$$

By Theorems 2 and 3, if all the above inequalities hold, then $\mathbb{E}[Y_{3,0}]$ is finite, while $\mathbb{E}[Y_{3,0}']$ is infinite.

For $p = 3/4$, the above inequalities become: $1 < c < \min\{4, 8/5, 64/55\} \approx$ 1.163 and $1.066 \approx 16/15 < c \leq 2$. Therefore, selecting $p = 3/4$ and $c = 1.1$, we have an anonymous age-based protocol that has finite expected latency and that prevents players from unilaterally switching to a persistent protocol. In fact we prove a slightly more general result:

Theorem 4 (restatement of Theorem 1). *Assume there are 3 players in the system, two of which use protocol $\mathcal{P}(1.1, 0.75)$. Then, the third player will prefer using protocol $\mathcal{P}(1.1, 0.75)$ over any deadline protocol $\mathcal{D}$.*

Proof. Extending the notation used in the previous sections, let $Y_{3,0}^{\mathcal{D}}$ (respectively $Y_{3,0}$) be the time needed for the third player to successfully transmit when she uses protocol $\mathcal{D}$ (respectively protocol $\mathcal{P}(1.1, 0.75)$). Furthermore, let $Y'_{3,k}$, $k \in \{0, 1, \ldots\}$, be the additional time after s_{k-1} (i.e. the $(k-1)$-th non-trivial transmission time) that the third player needs to successfully transmit when she uses a deadline protocol with deadline 1.

Since $c = 1.1$ and $p = 0.75$, we have that $1 < c < \min\left\{\frac{1}{1-p}, \frac{1}{1-2p(1-p)}, \frac{1}{1-3p(1-p)^2}\right\}$ and $\frac{1}{1-(1-p)^2} < c \leq 2$. Therefore, by Theorems 2 and 3, we have that $\mathbb{E}[Y_{3,0}]$ is finite and $\mathbb{E}[Y_{3,0}^{\mathcal{D}}] = \infty$, which means that the third player prefers using $\mathcal{P}(1.1, 0.75)$ over a deadline protocol with deadline 1.

We now prove that the third player prefers using $\mathcal{P}(1.1, 0.75)$ over any deadline protocol $\mathcal{D}$ with deadline $t_0 = t_0(\mathcal{D})$ as well. Let $\mathcal{E}$ be the event that none of the first two players has successfully transmitted before t_0. Let also $\xi = \xi(t_0)$ be the number of times t such that $\mathcal{P}(1.1, 0.75)_t = p = 0.75$ (i.e. the protocol $\mathcal{P}(1.1, 0.75)$ suggests transmitting with probability less than 1) before time t_0 (i.e. $\xi(t_0)$ is the number of non-trivial transmissions before t_0). We can see that

$$\Pr(\mathcal{E}) \geq (1 - 2p(1-p))^\xi.$$

In fact, this lower bound is quite crude, since it does not take into account the third player, so the probability that one of the first two players succesffully transmits during a non-trivial transmission time step when both are pending is $2p(1-p)$. We now have the following:

$$\mathbb{E}\left[Y_{3,0}^{\mathcal{D}}\right] = \sum_{t=0}^{\infty} t \Pr\left(Y_{3,0}^{\mathcal{D}} = t\right)$$

$$\geq \Pr(\mathcal{E}) \sum_{t=\tau}^{\infty} t \Pr\left(Y_{3,0}^{\mathcal{D}} = t | \mathcal{E}\right) = \Pr(\mathcal{E}) \sum_{t=\tau}^{\infty} t \Pr(Y'_{3,\xi} = t)$$

$$\geq \Pr(\mathcal{E}) \sum_{t=0}^{\infty} t \Pr(Y'_{3,\xi} = t) - t_0^2 = \Pr(\mathcal{E})\mathbb{E}[Y'_{3,\xi}] - t_0^2$$

$$\geq \Pr(\mathcal{E})c^{\xi-1}(c-1)\mathbb{E}[Y'_{3,0}] - t_0^2 = \infty$$

where in the last inequality we used the first inequality of Corollary 1. Therefore, the third player prefers using $\mathcal{P}(1.1, 0.75)$ over $\mathcal{D}$ as well. Since $\mathcal{D}$ is arbitrary, the proof is complete. $\square$

In [7], we show that when all three players use the protocol $\mathcal{P}(1.1, 0.75)$, the expected latency of a fixed player is upper bounded by 2759. It is worth noting that a naive protocol where each player transmits with constant probability, say $\frac{1}{3}$, at any time t, has a better expected latency than that of $\mathcal{P}(c, p)$, but on the other hand it does not prevent players from unilaterally switching to some deadline protocol.

References

1. Abramson, N.: The ALOHA system: another alternative for computer communications. In: Proceedings of the Fall Joint Computer Conference, 17–19 November 1970, pp. 281–285. ACM New York (1970)
2. Altman, E., El Azouzi, R., Jiménez, T.: Slotted aloha as a game with partial information. Comput. Netw. **45**(6), 701–713 (2004)
3. Altman, E., Barman, D., Benslimane, A., Azouzi, R.: Slotted aloha with priorities and random power. In: Boutaba, R., Almeroth, K., Puigjaner, R., Shen, S., Black, J.P. (eds.) NETWORKING 2005. LNCS, vol. 3462, pp. 610–622. Springer, Heidelberg (2005). doi:10.1007/11422778_49
4. Auletta, V., Moscardelli, L., Penna, P., Persiano, G.: Interference games in wireless networks. In: Papadimitriou, C., Zhang, S. (eds.) WINE 2008. LNCS, vol. 5385, pp. 278–285. Springer, Heidelberg (2008). doi:10.1007/978-3-540-92185-1_34
5. Bender, M., Farach-Colton, M., He, S., Kuszmaul, B., Leiserson, C.: Adversarial contention resolution for simple channels. In: SPAA 2005, pp. 325–332. ACM (2005)
6. Capetanakis, J.: Tree algorithms for packet broadcast channels. IEEE Trans. Inf. Theory **25**(5), 505–515 (1979)
7. Christodoulou, G., Gairing, M., Nikoletseas, S.E., Raptopoulos, C., Spirakis, P.G.: A 3-player protocol preventing persistence in strategic contention with limited feedback. arXiv:1707.01439 [cs.GT]
8. Christodoulou, G., Gairing, M., Nikoletseas, S.E., Raptopoulos, C., Spirakis, P.G.: Strategic contention resolution with limited feedback. In: Proceedings of the 24th Annual European Symposium on Algorithms (ESA), pp. 30:1–30:16 (2016)
9. Christodoulou, G., Ligett, K., Pyrga, E.: Contention resolution under selfishness. Algorithmica **70**(4), 675–693 (2014)
10. Fiat, A., Mansour, Y., Nadav, U.: Efficient contention resolution protocols for selfish agents. In: SODA 2007, pp. 179–188. SIAM, Philadelphia (2007)
11. Geréb-Graus, M., Tsantilas, T.: Efficient optical communication in parallel computers. In: SPAA 1992, pp. 41–48. ACM, New York (1992)
12. Goldberg, L.A., MacKenzie, P.D.: Analysis of practical backoff protocols for contention resolution with multiple servers. J. Comput. Syst. Sci. **58**(1), 232–258 (1999)
13. Goldberg, L.A., Mackenzie, P.D., Paterson, M., Srinivasan, A.: Contention resolution with constant expected delay. J. ACM **47**(6), 1048–1096 (2000)
14. Goldberg, L.A.: Notes on contention resolution (2002). http://www.cs.ox.ac.uk/people/leslieann.goldberg/contention.html
15. Greenberg, A., Winograd, S.: A lower bound on the time needed in the worst case to resolve conflicts deterministically in multiple access channels. J. ACM **32**(3), 589–596 (1985)
16. Hayes, J.: An adaptive technique for local distribution. IEEE Trans. Commun. **26**(8), 1178–1186 (1978)

17. Koutsoupias, E., Papakonstantinopoulou, K.: Contention issues in congestion games. In: Czumaj, A., Mehlhorn, K., Pitts, A., Wattenhofer, R. (eds.) ICALP 2012. LNCS, vol. 7392, pp. 623–635. Springer, Heidelberg (2012). doi:10.1007/978-3-642-31585-5_55

18. Ma, R.T., Misra, V., Rubenstein, D.: Modeling and analysis of generalized slotted-aloha MAC protocols in cooperative, competitive and adversarial environments. In: ICDCS 2006, p. 62. IEEE, Washington, DC, USA (2006)

19. MacKenzie, P.D., Plaxton, C.G., Rajaraman, R.: On contention resolution protocols and associated probabilistic phenomena. J. ACM 45(2), 324–378 (1998)

20. Menache, I., Shimkin, N.: Efficient rate-constrained nash equilibrium in collision channels with state information. In: INFOCOM 2008, pp. 403–411 (2008)

21. Raghavan, P., Upfal, E.: Stochastic contention resolution with short delays. Technical report, Weizmann Science Press of Israel, Jerusalem, Israel (1995)

22. Roberts, L.: Aloha packet system with and without slots and capture. SIGCOMM Comput. Commun. Rev. 5(2), 28–42 (1975)

23. Sheldon, R.: A First Course in Probability. Pearson, London (2012)

24. Tobagi, F.A., Kleinrock, L.: Packet switching in radio channels: part II-the hidden terminal problem in carrier sense multiple-access and the busy-tone solution. IEEE Trans. Commun. 23(12), 1417–1433 (1975)

25. Tsybakov, B.S., Mikhailov, V.A.: Free synchronous packet access in a broadcast channel with feedback. Probl. Inf. Transm. 14(4), 259–280 (1978)

26. Wang, D., Comaniciu, C., Tureli, U.: Cooperation and fairness for slotted aloha. Wirel. Pers. Commun. 43(1), 13–27 (2007)

27. Zheng, D., Ge, W., Zhang, J.: Distributed opportunistic scheduling for ad-hoc communications: an optimal stopping approach. In: MobiHoc 2007, pp. 1–10. ACM (2007)

Hedging Under Uncertainty: Regret Minimization Meets Exponentially Fast Convergence

Johanne Cohen[1], Amélie Héliou[2,3], and Panayotis Mertikopoulos[4(✉)]

[1] LRI-CNRS, Université de Paris-Sud, Université Paris-Saclay, Orsay, France
johanne.cohen@lri.fr
[2] AMIB Project, Inria Saclay, 91120 Palaiseau, France
[3] LIX CNRS UMR 7161, Ecole Polytechnique, Université Paris-Saclay, 91120
Palaiseau, France
amelie.heliou@polytechnique.edu
[4] Univ. Grenoble Alpes, CNRS, Grenoble INP, Inria, LIG, 38000 Grenoble, France
panayotis.mertikopoulos@imag.fr

Abstract. This paper examines the problem of multi-agent learning in
N-person non-cooperative games. For concreteness, we focus on the so-
called "hedge" variant of the (EW) algorithm, one of the most widely
studied algorithmic schemes for regret minimization in online learning. In
this multi-agent context, we show that (a) dominated strategies become
extinct (a.s.); and (b) in generic games, pure Nash equilibria are attracting with high probability, even in the presence of uncertainty and noise of
arbitrarily high variance. Moreover, if the algorithm's step-size does not
decay too fast, we show that these properties occur at a quasi-exponential
rate – that is, much faster than the algorithm's $\mathcal{O}(1/\sqrt{T})$ worst-case
regret guarantee would suggest.

Keywords: Dominated strategies · Exponential weights · Nash
equilibrium · No-regret learning

1 Introduction

In its most basic form, the prototypical framework of online learning can be
summarized as follows: at each instance $t = 1, 2, \ldots$ of a repeated decision
process, a player selects an action α_t from some finite set $\mathcal{A}$, and they obtain
a reward $u_t(\alpha_t)$ based on an a priori unknown payoff function $u_t \colon \mathcal{A} \to \mathbb{R}$.
Subsequently, the player observes some problem-specific feedback (for instance,
the resulting payoff vector or some estimate thereof), and selects a new action
seeking to maximize their reward over time. In the absence of any other considerations, this objective is usually quantified by asking that the player's *regret*
$\mathrm{Reg}(T) \equiv \max_{\alpha \in \mathcal{A}} \sum_{t=1}^{T} [u_t(\alpha) - u_t(\alpha_t)]$ grow sublinearly in T, a property
known as "no regret".

Game-theoretic learning is a multi-agent extension of the above in which
every player's payoffs are determined by the actions of all players via a fixed

© Springer International Publishing AG 2017
V. Bilò and M. Flammini (Eds.): SAGT 2017, LNCS 10504, pp. 252–263, 2017.
DOI: 10.1007/978-3-319-66700-3_20

mechanism – the *game*. Of course, in many applications, this mechanism may be unknown and/or opaque to the players (who, conceivably, might not even know that they are playing a game). As a result, we are led to the following key questions: if the players of a repeated game agnostically update their strategies following an algorithm that minimizes their individual regret, does the induced sequence of play converge to a Nash equilibrium (or some other rationally justifiable solution concept)? And if so, does this still hold true if the players' observations are contaminated by noise and/or uncertainty?

On the positive side, under no-regret learning, the players' empirical frequencies of play converge to the game's *Hannan set* [12], also known as the set of *coarse correlated equilibria* (CCE) [13].[1] As such, a partial answer to the first question is that coarse correlated equilibria are indeed learnable via no-regret learning. In general however, the Hannan set may contain highly non rationalizable outcomes, so the real answer to this question is "no". For instance, Viossat and Zapechelnyuk [26] recently constructed an example of a 4 × 4 symmetric game with a coarse correlated equilibrium that assigns positive weight *only* on strictly dominated strategies – an "equilibrium" in name only.

In view of this, our aim in this paper is to examine when no-regret learning leads to a set of rationally justifiable strategies – such as the game's Nash set or the set of undominated strategies. To that end, we focus on the widely used "hedge" variant [9] of the *exponential weights* (EW) algorithm [19,27], where the probability of choosing an action is proportional to the exponential of its cumulative payoff over time (so better-performing actions are employed exponentially more often). As is well known, "hedging" is min-max optimal in terms of the achieved regret minimization rate [5]. Nonetheless, beyond Hannan consistency, few conclusions can be drawn from this property in a game-theoretic environment where finer convergence criteria apply.

A further complication that arises in game-theoretic learning is that it is often important to establish the convergence of the *actual* sequence of play generated by a learning process – as opposed to its time average. In online learning, averaging comes up naturally because the focus is on the player's regret. In a game-theoretic context however, even if time averages converge, the actual sequence of play may fail to converge altogether (or may do so at a completely different rate), so the players' actual behavior (and the payoffs they obtain) could be drastically different in the two regimes. Indeed, regret-based results offer little insight on the "last iterate" of the process, so the analysis of the latter requires a completely different set of tools and techniques.

1.1 Our Results

In their recent paper, Viossat and Zapechelnyuk [26] showed that, in general, the set of coarse correlated equilibria may contain non rationalizable strategies supported exclusively on strictly dominated strategies. Nevertheless, by

---

[1] As the second name suggests, this set contains the game's set of correlated equilibria (CE) – and hence, the game's set of Nash equilibria as well.

leveraging the consistent negative reinforcement of actions that perform badly in past instances of play, we show that hedging eliminates dominated strategies with probability 1. Moreover, we show that the rate of elimination is $\mathcal{O}(\exp(-c\sum_{t=1}^{T}\gamma_t))$ for some positive constant $c > 0$, where γ_t is a variable step-size parameter; in other words, dominated strategies become extinct exponentially fast if γ_t decays slower than $1/t$.

With respect to equilibrium convergence, we show that hedging in generic games converges locally to pure Nash equilibria with high probability, and the convergence is global with probability 1 if the game's equilibrium satisfies a certain variational inequality. The only added caveat for this result is that the algorithm's step-size parameter must satisfy a summability condition which precludes the use of very aggressive step-size policies. Nevertheless, the algorithm still achieves an exponential $\mathcal{O}(e^{-cT^{1-b}})$ convergence rate for step-size sequences of the form $\gamma_t \sim 1/t^b$, $b \in (1/2, 1)$.

To account for the fact that players may not have access to perfect payoff observations, we assume throughout that players can only estimate their payoff vectors up to a possibly unbounded error with arbitrarily high variance. This uncertainty is countered by means of a judicious choice of the algorithm's step-size parameter γ_t which can be used to control the weight with which new observations enter the algorithm at a given stage. This is made possible by exploiting results from martingale limit theory and the theory of stochastic approximation.

1.2   Related Work

Algorithms and dynamics for learning in games have received considerable attention over the last few decades. Such procedures can be divided into two broad categories, depending on whether they evolve in continuous or discrete time: the former includes the numerous dynamics for learning and evolution (see [23] for a survey), whereas the latter focuses on learning algorithms for infinitely iterated games (such as fictitious play and its variants). In this paper, we focus exclusively on discrete-time algorithms.

In this framework, it is natural to consider agents who learn from their experience by small adjustments in their behavior based on local – and possibly imperfect – information. Several such approaches in the literature can be viewed as *decentralized no-regret dynamics* – for example the multiplicative/exponential weights algorithm and its variants [9,19,27], Follow the Regularized/Perturbed Leader [14,21], etc. Indeed, regret bounds can be used to guarantee that each player's utility approaches long-term optimality in adversarial environments, a natural first step towards long-term rational behavior. For example, it has been shown in [2,22] that the sum of utilities approaches an approximate optimum, and there is convergence of time averages towards an equilibrium in two-player zero-sum games [3,7,9]. In all these examples, the players' average regret vanishes at the worst-case rate of $\mathcal{O}(1/\sqrt{T})$ where T denotes the play horizon. This convergence rate was recently improved by Syrgkanis *et al.* [24] for a wide class of N-player normal form games using a natural class of regularized learning algorithms. However, the convergence results established in [24] concerned (a)

the set of coarse correlated equilibria (which may contain highly non rationalizable strategies); and (*b*) the "long-run average" $\bar{x}_T = T^{-1} \sum_{t=1}^{T} x_t$ of the actual sequence of play. By contrast, our paper focuses squarely on the algorithm's "last iterate" (which determines the players' rewards at each stage), and finer rationality properties (such as the elimination of dominated strategies or convergence to pure Nash equilibria) that cannot be deduced from coarse equilibrium convergence results.

The HEDGE algorithm received much attention in various fields such as optimization [1], multi-armed bandit problems [4], and in general algorithmic game theory. In particular, the number of payoff queries needed to compute approximate correlated equilibria has been studied in [10]: upper and lower bounds have been derived, as well as reductions between problems, such as the reduction of the problem of verifying an approximate well supported Nash equilibrium to the problem of computing a well supported Nash equilibrium under some assumptions.

Kleinberg et al. [15] studied the behavior of the dynamic of the HEDGE algorithm for some particular load balancing games (the so-called atomic load balancing games) in the "*bulletin board*" model. In this latter model, players know the actual payoff of each strategy according to the actual strategies played. They proved that if all players play according to the same mixed strategy, the dynamics of the history of play converge, and the limit is necessarily a stable distribution over states, such as a mixed Nash or a correlated equilibrium. Furthermore, the average performance of the dynamics has been analyzed in atomic load balancing games. Recently, in a similar spirit, Foster et al. [8] showed that some variants of HEDGE algorithms are such that the average of the outcome converge rapidly to an approximation of the optimum in smooth games. In parallel, Krichene *et al.* [16] extended the result to congestion games, and proved that a discounted variant of the HEDGE algorithm converges to the set of Nash equilibria in the sense of Cesàro means (time averages), while strong convergence can be guaranteed with some additional conditions. Their proof is based on the so-called *Kullback–Leibler* (KL) divergence. Finally, Coucheney *et al.* [6] also showed that a "penalty-regulated" variant of the HEDGE algorithm with bandit feedback converges to logit equilibria in congestion games, but their techniques do not extend to actual Nash equilibria. In the current paper, we do not restrict ourselves to congestion games; instead, we consider generic games that admit a Nash equilibrium in pure strategies, of which congestion and potential games are a special case.

2   Preliminaries

Throughout the paper, we focus on games that are played by a (finite) set $\mathcal{N} = \{1, \ldots, N\}$ of N *players* (or *agents*). Each player $i \in \mathcal{N}$ is assumed to have a finite set of *actions* (or *pure strategies*) $\mathcal{A}_i$, and the players' preferences for one action over another are represented by each action's *utility* (or *payoff*). Specifically, as players interact with each other, the individual payoff of each player is given by a function $u_i \colon \mathcal{A} \equiv \prod_i \mathcal{A}_i \to \mathbb{R}$ of all players' actions, and each agent seeks to maximize the utility $u_i(\alpha_i; \alpha_{-i})$ of their chosen action $\alpha_i \in \mathcal{A}_i$ against the action profile α_{-i} of his opponents.[2] A game is then called (weakly) *generic* if there are no unilateral payoff ties, i.e. if $u_i(\alpha_i; \alpha_{-i}) \neq u_i(\beta_i; \alpha_{-i})$ for all $\alpha_i, \beta_i \in \mathcal{A}_i$, $i \in \mathcal{N}$.

Players can also use *mixed strategies* by playing probability distributions $x_i = (x_{i\alpha_i})_{\alpha_i \in \mathcal{A}_i} \in \Delta(\mathcal{A}_i)$ over their action sets $\mathcal{A}_i$. The resulting probability vector x_i is called the *mixed strategy* of the i-th player and the set $\mathcal{X}_i = \Delta(\mathcal{A}_i)$ is the corresponding mixed strategy space of player i; aggregating over players, we also write $\mathcal{X} = \prod_i \mathcal{X}_i$ for the game's *strategy space*, i.e. the space of all mixed strategy profiles $x = (x_i)_{i \in \mathcal{N}}$.

In this context (and in a slight abuse of notation), the expected payoff of the i-th player in the mixed strategy profile $x = (x_1, \ldots, x_N)$ is

$$u_i(x) = \sum_{\alpha_1 \in \mathcal{A}_1} \cdots \sum_{\alpha_N \in \mathcal{A}_N} u_i(\alpha_1, \ldots, \alpha_N) x_{1\alpha_1} \cdots x_{N\alpha_N}. \tag{1}$$

Accordingly, if player i plays the pure strategy $\alpha_i \in \mathcal{A}_i$, we will write

$$v_{i\alpha_i}(x) = u_i(\alpha_i; x_{-i}) = u_i(x_1, \ldots, \alpha_i, \ldots, x_N) \tag{2}$$

for the payoff corresponding to α_i, and $v_i(x) = (v_{i\alpha_i}(x))_{\alpha_i \in \mathcal{A}_i}$ for the resulting *payoff vector* of player i. A player's expected payoff may thus be written as

$$u_i(x) = \sum_{\alpha_i \in \mathcal{A}_i} x_{i\alpha_i} v_{i\alpha_i}(x) = \langle v_i(x) | x_i \rangle, \tag{3}$$

where $\langle v_i | x_i \rangle$ denotes the canonical bilinear pairing between v_i and x_i.

A fundamental rationality principle in game theory is that, assuming full knowledge of the game, a player would have no incentive to play an action that always yields suboptimal payoffs with respect to another. To formalize this, $\alpha_i \in \mathcal{A}_i$ is called *(strictly) dominated* by β_i (and we write $\alpha_i \prec \beta_i$) if

$$u_i(\alpha_i; \alpha_{-i}) < u_i(\beta_i; \alpha_{-i}) \quad \text{for all } \alpha_{-i} \in \mathcal{A}_{-i} \equiv \prod_{j \neq i} \mathcal{A}_j, \ i \in \mathcal{N}. \tag{4}$$

Extending the notion of strategic dominance, the most widely used solution concept in game theory is that of a *Nash equilibrium* (NE), i.e. a state $x^* \in \mathcal{X}$ which is unilaterally stable in the sense that

$$u_i(x_i^*; x_{-i}^*) \geq u_i(x_i; x_{-i}^*) \quad \text{for all } x_i \in \mathcal{X}_i, \ i \in \mathcal{N}, \tag{NE}$$

or, equivalently, writing $\operatorname{supp}(x)$ for the support of x:

$$v_{i\alpha_i}(x^*) \geq v_{i\beta_i}(x^*) \quad \text{for all } \alpha_i \in \operatorname{supp}(x_i^*) \text{ and all } \beta_i \in \mathcal{A}_i, \ i \in \mathcal{N}. \tag{5}$$

If equilibrium x^* is *pure* (i.e. $\operatorname{supp}(x_i^*) = \{\alpha_i^*\}$ for some $\alpha_i^* \in \mathcal{A}_i$ and all $i \in \mathcal{N}$), then it is called a *pure equilibrium*. In generic games, a pure equilibrium satisfies (5) as a strict inequality for all $\beta_i \notin \operatorname{supp}(x_i^*)$, $i \in \mathcal{N}$, so we sometimes refer to

---

[2] In the above $(\alpha_i; \alpha_{-i})$ is shorthand for $(\alpha_1, \ldots, \alpha_i, \ldots, \alpha_N)$, used here to highlight the action of player i against that of all other players.

pure equilibria in generic games as *strict*. Such equilibria will play a key role in our analysis, so we provide a convenient variational characterization below:

Proposition 1. *In generic games, x^* is a pure equilibrium if and only if*

$$\langle v(x)|x - x^* \rangle \leq -\tfrac{1}{2}\mu\|x - x^*\| \quad \text{for some } \mu > 0 \text{ and for all } x \text{ near } x^*, \quad (6)$$

where $\|x\| = \sum_i \sum_{\alpha \in \mathcal{A}_i} |x_{i\alpha_i}|$ denotes the L^1-norm of x.

There are two remarks to be made here. First, if (6) holds for all $x \in \mathcal{X}$, then x^* is the unique Nash equilibrium of the game: indeed, if $x' \neq x^*$ is another Nash equilibrium of the game, we would have $0 \leq \langle v(x')|x' - x^* \rangle \leq -\tfrac{1}{2}\mu\|x' - x^*\| < 0$, a contradiction. On the other hand, (6) does not imply that x^* is in any way "dominant", locally otherwise: for instance, it is easy to verify that the unique Nash equilibrium of the Prisoner's Dilemma satisfies (6), even though it is not Pareto efficient (let alone dominant). In general, (6) could seen as a "finite games" variant of the notion of evolutionary stability in population games [23]; as such, it is easy to verify that (6) is satisfied in the Prisoner's Dilemma and its variants, generic competition games, potential games with a unique equilibrium, etc.

3   Hedging Under Uncertainty

The algorithm that we examine is the so-called "hedge" variant of the exponential weights algorithm [9]. In a nutshell, the main idea of the algorithm is as follows: At each stage $t = 1, 2, \ldots$ of the process, players maintain and update a "performance score" for each of their actions (pure strategies) based on each action's cumulative payoff up to stage t. These scores are then converted to mixed strategies by assigning exponentially higher probability to actions with higher scores, and a new action is drawn based on these mixed strategies.

More precisely, this iterative process can be encoded as follows:

Algorithm 1.1. HEDGE with variable step-size γ_t

1   Each player $i \in \mathcal{N}$ chooses an initial score vector $y_i(0)$
2   **for** each round t
3   Each player $i \in \mathcal{N}$ plays $x_i(t) = \Lambda_i(y_i(t-1))$ where the *logit map* Λ_i is defined as

$$\Lambda_i(y_i) = \frac{1}{\sum_{\alpha \in \mathcal{A}_i} \exp(y_{i\alpha})}(\exp(y_{i\alpha}))_{\alpha \in \mathcal{A}_i}. \quad (7)$$

4   Each player $i \in \mathcal{N}$ draws a pure strategy $\alpha_i(t)$ according to $x_i(t)$
5   Each player $i \in \mathcal{N}$ gets an estimate $\hat{v}_i(t)$ of their payoff vector $v_i(\alpha(t))$
6   Each player $i \in \mathcal{N}$ updates their score vectors as

$$y_i(t) = y_i(t-1) + \gamma_t \hat{v}_i(t), \quad (8)$$

    end for

Mathematically, the above algorithm can be expressed as

$$x_i(t) = \Lambda_i(y_i(t-1)),$$
$$y_i(t) = y_i(t-1) + \gamma_t \hat{v}(t), \tag{HEDGE}$$

with $y_i(0)$ initialized arbitrarily. Motivated by practical implementation issues (especially in large networks and telecommunication systems), this formulation further assumes that players have *imperfect knowledge* of their payoff vectors $v_i(\alpha(t))$ at each iteration of the algorithm – for instance, contaminated by measurement errors or other uncertainty factors. To formalize this, we will consider a general feedback model of the form

$$\hat{v}_i(t) = v_i(\alpha(t)) + \xi_i(t), \tag{9}$$

where the error process $\xi = (\xi_i)_{i \in \mathcal{N}}$ is a L^2-bounded martingale difference sequence with respect to the history $\mathcal{F}_t$ of the process $(x(t), \alpha(t), \hat{v}(t), y(t))$. In other words, we assume that $\xi(t)$ satisfies the statistical hypotheses

1. *Zero-mean:*

$$\mathbb{E}[\xi(t) \mid \mathcal{F}_{t-1}] = 0 \quad \text{for all } t = 1, 2, \dots \text{(a.s.);} \tag{H1}$$

2. *Finite mean squared error:* there exists some $\sigma > 0$ such that

$$\mathbb{E}[\|\xi(t)\|_\infty^2 \mid \mathcal{F}_{t-1}] \leq \sigma^2 \quad \text{for all } t = 1, 2, \dots \text{(a.s.).} \tag{H2}$$

Put differently, Hypotheses (H1) and (H2) simply mean that the payoff vector estimates $\hat{v}_i$ are *conditionally unbiased and bounded in mean square*, i.e.

$$\mathbb{E}[\hat{v}(t) \mid \mathcal{F}_{t-1}] = v(x(t)), \tag{10a}$$

$$\mathbb{E}[\|\hat{v}(t)\|_\infty^2 \mid \mathcal{F}_{t-1}] \leq V^2, \tag{10b}$$

where $V > 0$ is a finite positive constant (in the noiseless case $\xi = 0$ the constant V is simply a bound on the players' maximum absolute payoff).[3] Thus, Hypotheses (H1) and (H2) allow for a broad range of noise distributions, including all compactly supported, (sub-)Gaussian, (sub-)exponential and log-normal distributions.

---

[3] Note here that (10a) is phrased in terms of the players' *mixed* strategy profile $x(t)$, not the action profile $\alpha(t) = (\alpha_i(t); \alpha_{-i}(t))$ which is chosen based on $x(t)$ at stage t. To see that (H1) indeed implies (10a), simply recall that $x_i(t) = \Lambda_i(y_i(t-1))$ so

$$\mathbb{E}[\hat{v}_{i\alpha_i}(t) \mid \mathcal{F}_{t-1}] = \sum_{\alpha_{-i} \in \mathcal{A}_{-i}} [u_i(\alpha_i; \alpha_{-i}) x_{\alpha_{-i}}(t) + \mathbb{E}[\xi_{i\alpha_i}(t) \mid \mathcal{F}_{t-1}]]$$

$$= u_i(\alpha_i; x_{-i}(t)) = v_{i\alpha_i}(x(t)). \tag{11}$$

4 Rationality Analysis and Results

In this section, we present our main convergence results for (HEDGE). We begin with the fact that hedging bypasses the negative result of Viossat and Zapachelnyuk [26], and does not lead to non-rationalizable, strategically dominated outcomes:

Theorem 1. *Suppose that* (HEDGE) *is run with a step-size sequence of the form* $\gamma_t \propto 1/t^b$ *for some* $b \leq 1$ *(not necessarily positive), and noisy payoff observations satisfying Hypotheses* (H1) *and* (H2). *If* $\alpha_i \in \mathcal{A}_i$ *is dominated, there exists some* $c > 0$ *such that*

$$x_{i\alpha_i}(T+1) = \mathcal{O}(\exp(-c\textstyle\sum_{t=1}^T \gamma_t)) \quad \text{with probability 1.} \quad (12)$$

In particular, if $b < 1$, α_i *becomes extinct exponentially fast (a.s.).*

Proof. Suppose that $\alpha_i \prec \beta_i$ for some $\beta_i \in \mathcal{A}_i$. Then, suppressing the player index i for simplicity, we get

$$y_\beta(T) - y_\alpha(T) = c_{\beta\alpha} + \sum_{t=1}^T \gamma_t \left[\hat{v}_\beta(t) - \hat{v}_\alpha(t)\right]$$

$$= c_{\beta\alpha} + \sum_{t=1}^T \gamma_t \left[v_\beta(x(t)) - v_\alpha(x(t))\right] + \sum_{t=1}^T \gamma_t \zeta_t, \quad (13)$$

where we set $c_{\beta\alpha} = y_\beta(0) - y_\alpha(0)$ and $\zeta_t = \hat{v}_\beta(t) - v_\beta(x(t)) - [\hat{v}_\alpha(t) - v_\alpha(x(t))]$. Since $\alpha \prec \beta$, there exists some $\mu > 0$ such that $v_\beta(x) - v_\alpha(x) \geq \mu$ for all $x \in \mathcal{X}$. Then, (13) yields

$$y_\beta(T) - y_\alpha(T) \geq c_{\beta\alpha} + \theta_T \left[\mu + \frac{\sum_{t=1}^T \gamma_t \zeta_t}{\theta_T}\right], \quad (14)$$

with $\theta_T = \sum_{t=1}^T \gamma_t$.

Since $\mathbb{E}[\zeta_t \mid \mathcal{F}_{t-1}] = 0$ and $\sup_t \mathbb{E}[\zeta_t^2 \mid \mathcal{F}_{t-1}] < \infty$ by the reformulation (10a) and (10b) of Hypotheses (H1) and (H2) respectively, the law of large numbers for martingale difference sequences [11, Theorem 2.18] gives $\theta_T^{-1} \sum_{t=1}^T \gamma_t \zeta_t \to 0$ (a.s.), provided that $\sum_{t=1}^\infty (\gamma_t/\theta_t)^2 < \infty$ and $\sum_{t=1}^\infty \gamma_t = \infty$. This last assumption is satisfied for all $b \leq 1$, so we readily get $\theta_T^{-1} \sum_{t=1}^T \gamma_t \zeta_t \to 0$ with probability 1. As a result, for all $c \in (0, \mu)$, there exists some random (but a.s. finite) T_0 such that if $T \geq T_0$, then $y_\beta(T) - y_\alpha(T) \geq c\theta_T$. Thus, with $x_\alpha(T+1) = \Lambda_\alpha(y(T))$ by definition, we get

$$x_\alpha(T+1) = \frac{e^{y_\alpha(T)}}{\sum_\kappa e^{y_\kappa(T)}} \leq \frac{e^{y_\alpha(T)}}{e^{y_\beta(T)}} = e^{y_\alpha(T) - y_\beta(T)} \leq e^{-c\sum_{t=1}^T \gamma_t} \quad \text{(a.s.),} \quad (15)$$

and our proof is complete. $\qquad\square$

Remark 1. It should be noted here that the elimination of dominated strategies with imperfect knowledge of the game's payoffs is by no means a given.[4] For instance, if players play a greedy best response scheme at each round and the payoff observation errors are not supported on a small, compact set, dominated strategies will be played infinitely often (simply because at each round, any strategy could be erroneously perceived as a best response).[5] With this in mind, the fact that the rate of elimination (12) *improves* with more aggressive – even *increasing* – step-size sequences γ_t is somewhat surprising because it suggests that players can employ (HEDGE) in a very greedy fashion and achieve fast dominated strategy extinction rates, even in the presence of arbitrarily high estimation errors.

Remark 2. Regarding the number of players and actions per player, our proof reveals that c depends only on the player's payoffs – specifically, we can take $c = \frac{1}{2}\min_{\alpha_{-i} \in \mathcal{A}_{-i}}[u_i(\beta_i; \alpha_{-i}) - u_i(\alpha_i; \alpha_{-i})] > 0$. In other words, the algorithm's half-life is asymptotically *independent* of the size of the game, and only depends on the players' relative payoff differences.

Remark 3. Finally, we should note that Theorem 1 also extends to the case of iteratively dominated strategies.[6] As such, Theorem 1 implies that the sequence of play induced by (HEDGE) actually converges to the set of iteratively undominated strategies of the game.

We now turn to the convergence properties of (HEDGE) in generic games that admit pure Nash equilibria:

Theorem 2. *Fix a confidence level $\varepsilon > 0$ and suppose that (HEDGE) is run with a small enough (depending on ε) step-size γ_t satisfying $\sum_{t=1}^{\infty} \gamma_t^2 < \infty$ and $\sum_{t=1}^{\infty} \gamma_t = \infty$, and imperfect payoff observations satisfying Hypotheses (H1) and (H2). If x^* is a pure equilibrium of a generic game and (HEDGE) is initialized not too far from x^*, we have*

$$\mathbb{P}\left(\|x(T+1) - x^*\| \le C' e^{-c\sum_{t=1}^{T} \gamma_t} \text{ for all } t\right) \ge 1 - \varepsilon, \qquad (16)$$

where $c > 0$ is a constant that only depends on the game and $C' > 0$ is a (random) constant that depends on the initialization of (HEDGE). In particular, under the stated assumptions, $x(t) \to x^*$ with probability at least $1 - \varepsilon$.

Corollary 1. *With assumptions as above, if (HEDGE) is run with a step-size of the form $\gamma_t = \gamma/t^b$ for some sufficiently small $\gamma > 0$ and $b \in (1/2, 1)$, we have*

$$\mathbb{P}\left(\|x(T+1) - x^*\| = \mathcal{O}\left(e^{-\frac{c\gamma}{1-b}T^{1-b}}\right)\right) \ge 1 - \varepsilon \qquad (17)$$

for some positive constant $c > 0$.

---

[4] To the best of our knowledge, the closest result is the elimination of dominated strategies under exponential learning in *continuous* time [20]; for a survey of the relevant literature, see [25].

[5] Recall also the counterexample of [26] discussed in the introduction.

[6] This can be shown by an induction argument on the rounds of elimination of dominated strategies as in [18].

Relegating the (somewhat involved) proof of Theorem 2 to the long version of the paper, we note here that, in contrast to Theorem 1, the "$L^2 - L^1$" summability requirement $\sum_{t=1}^{\infty} \gamma_t^2 < \infty$ and $\sum_{t=1}^{\infty} \gamma_t = \infty$ constrains the admissible step-size policies that lead to pure equilibrium (for instance, constant step-size policies are no longer admissible). In particular, the most aggressive step-size that satisfies the assumptions of Theorem 2 is $\gamma_t \propto t^{-b}$ for some b close (but not equal) to $1/2$, leading to a convergence rate of $\lambda^{t^{1-b}}$ for some real $\lambda < 1$ (cf. Corollary 1). This bound on b is due to the second moment growth bound required by Doob's maximal inequality; if there is finer control on the moments of the noise process ξ (for instance, if the noise is sub-exponential), the lower bound $b > 1/2$ can be pushed all the way down to $b > 0$, implying a quasi-linear convergence rate.

As was hinted above, the main idea behind the proof of Theorem 2 is to use Doob's maximal inequality for martingales to show that the probability of $x(t)$ escaping the basin of attraction of a pure Nash equilibrium x^* can be made arbitrarily small if the algorithm's step-size is chosen appropriately. Building on this, if x^* satisfies the variational inequality (6) throughout $\mathcal{X}$, we have the stronger result:

Theorem 3. *Suppose that (HEDGE) is run with a step-size sequence γ_t such that $\sum_{t=1}^{\infty} \gamma_t^2 < \infty$ and $\sum_{t=1}^{\infty} \gamma_t = \infty$ and imperfect payoff observations satisfying Hypotheses (H1) and (H2). If x^* satisfies (6) for all $x \in \mathcal{X}$, then:*

1. $\lim_{T \to \infty} x(T) = x^*$ *(a.s.).*
2. *There exists a (deterministic) constant $c > 0$ such that*

$$\|x(T+1) - x^*\| = \mathcal{O}\left(e^{-c\sum_{t=1}^{T} \gamma_t}\right). \tag{18}$$

Corollary 2. *With assumptions as above, if (HEDGE) is run with a step-size of the form $\gamma_t = \gamma/t^b$ some $b \in (1/2, 1)$, we have*

$$\|x(T+1) - x^*\| = \mathcal{O}(e^{-\frac{\mu\gamma}{1-b}T^{1-b}}). \tag{19}$$

In contrary to Theorems 1 and 2, the proof of Theorem 3 relies heavily on the so-called *Kullback–Leibler* (KL) divergence [17], defined here as

$$D_{\mathrm{KL}}(x^*, x) = \sum_{i \in \mathcal{N}} \sum_{\alpha_i \in \mathcal{A}_i} x^*_{i\alpha_i} \log \frac{x^*_{i\alpha_i}}{x_{i\alpha_i}} \quad \text{for all } x \in \mathcal{X}^\circ. \tag{20}$$

The KL divergence is a positive-definite, asymmetric distance measure that is tailored to the analysis of the replicator dynamics [18,23,28]. By using this divergence as a discrete-time Lyapunov function, we show that x^* is a *recurrent point* of the process $x(t)$, i.e. $x(t)$ visits any neighborhood of x^* infinitely many times. We then use a stochastic approximation argument to show that the process actually converges to x^* at an asymptotic rate of $\mathcal{O}(e^{-c\sum_{t=1}^{T} \gamma_t})$.

262 J. Cohen et al.

The step-size assumption in the statement of Theorem 3 is key in achieving this, but it is important to note it can be relaxed to $\sum_{t=1}^{T} \gamma_t^2 / \sum_{t=1}^{T} \gamma_t \to 0$ if the players' feedback noise is bounded (for instance, if players have access to their actual pure payoff information). When this is the case, it is possible to achieve an $\mathcal{O}(e^{-cT^{1-b}})$ convergence rate for any $b > 0$ by using a step-size sequence of the form $\gamma_t \propto 1/t^b$.

5 Conclusions

Our main goal is to analyse the convergence properties of the "hedge" variant of the exponential weights algorithm [9] in generic N-player games that admit a Nash equilibrium in pure strategies. Using techniques drawn from the theory of stochastic approximation and martingale limit theory, we showed that (i) dominated strategies become extinct (a.s.); (ii) pure equilibria are locally attracting with high probability; and (iii) pure equilibria that satisfy a certain variational stability condition are globally attracting with probability 1. Moreover, despite the uncertainty in the players' payoff observations, we showed that the elimination of dominated strategies is exponentially fast – in stark contrast to more unrefined best response schemes which, under uncertainty, may lead to playing dominated strategies infinitely often. Using the theory of stochastic approximation and discrete-time martingale processes, we showed that the algorithm's convergence (local or global, depending on the context) to a pure Nash equilibrium is also exponentially fast, even in the presence of uncertainty and noise of arbitrary magnitude. These results apply to all generic games that admit a Nash equilibrium in pure strategies, and not only to the class of coordination and anti-coordination (or congestion) games that have been the traditional focus of the game-theoretic learning literature.

Acknowledgment. This work was partially supported from the French National Research Agency (ANR) under grant no. ANR–16–CE33–0004–01 (ORACLESS) and the Huawei HIRP FLAGSHIP project ULTRON. ·

References

1. Arora, S., Hazan, E., Kale, S.: The multiplicative weights update method: a meta-algorithm and applications. Theory Comput. **8**(1), 121–164 (2012)
2. Blum, A., Hajiaghayi, M.T., Ligett, K., Roth, A.: Regret minimization and the price of total anarchy. In: STOC 2008: Proceedings of the 40th Annual ACM Symposium on the Theory of Computing, pp. 373–382. ACM (2008)
3. Blum, A., Mansour, Y.: Learning, regret minimization, and equilibria (Chap. 4). In: Nisan, N., Roughgarden, T., Tardos, E., Vazirani, V.V. (eds.) Algorithmic Game Theory. Cambridge University Press, Cambridge (2007)
4. Bubeck, S., Cesa-Bianchi, N.: Regret analysis of stochastic and nonstochastic multi-armed bandit problems. Found. Trends Mach. Learn. **5**(1), 1–122 (2012)
5. Cesa-Bianchi, N., Lugosi, G.: Prediction, Learning, and Games. Cambridge University Press, Cambridge (2006)

6. Coucheney, P., Gaujal, B., Mertikopoulos, P.: Penalty-regulated dynamics and robust learning procedures in games. Math. Oper. Res. **40**(3), 611–633 (2015)
7. Foster, D., Vohra, R.V.: Calibrated learning and correlated equilibrium. Games Econ. Behav. **21**(1), 40–55 (1997)
8. Foster, D.J., Lykouris, T., Sridharan, K., Tardos, E.: Learning in games: robustness of fast convergence. In: Advances in Neural Information Processing Systems, pp. 4727–4735 (2016)
9. Freund, Y., Schapire, R.E.: Adaptive game playing using multiplicative weights. Games Econ. Behav. **29**, 79–103 (1999)
10. Goldberg, P.W., Roth, A.: Bounds for the query complexity of approximate equilibria. ACM Trans. Econ. Comput. **4**(4), 24:1–24:25 (2016)
11. Hall, P., Heyde, C.C.: Martingale Limit Theory and Its Application. Probability and Mathematical Statistics. Academic Press, New York (1980)
12. Hannan, J.: Approximation to Bayes risk in repeated play. In: Dresher, M., Tucker, A.W., Wolfe, P. (eds.) Contributions to the Theory of Games. Annals of Mathematics Studies, vol. 39, pp. 97–139. Princeton University Press, Princeton (1957)
13. Hart, S., Mas-Colell, A.: A simple adaptive procedure leading to correlated equilibrium. Econometrica **68**(5), 1127–1150 (2000)
14. Kalai, A., Vempala, S.: Efficient algorithms for online decision problems. J. Comput. Syst. Sci. **71**(3), 291–307 (2005)
15. Kleinberg, R., Piliouras, G., Tardos, É.: Load balancing without regret in the bulletin board model. Distrib. Comput. **24**(1), 21–29 (2011)
16. Krichene, W., Drighès, B., Bayen, A.M.: Learning Nash equilibria in congestion games. arXiv preprint arXiv:1408.0017 (2014)
17. Kullback, S., Leibler, R.A.: On information and sufficiency. Ann. Math. Stat. **22**(1), 79–86 (1951)
18. Laraki, R., Mertikopoulos, P.: Higher order game dynamics. J. Econ. Theory **148**(6), 2666–2695 (2013)
19. Littlestone, N., Warmuth, M.K.: The weighted majority algorithm. Inf. Comput. **108**(2), 212–261 (1994)
20. Mertikopoulos, P., Moustakas, A.L.: The emergence of rational behavior in the presence of stochastic perturbations. Ann. Appl. Probab. **20**(4), 1359–1388 (2010)
21. Mertikopoulos, P., Sandholm, W.H.: Learning in games via reinforcement and regularization. Math. Oper. Res. **41**(4), 1297–1324 (2016)
22. Roughgarden, T.: Intrinsic robustness of the price of anarchy. J. ACM (JACM) **62**(5), 32 (2015)
23. Sandholm, W.H.: Population Games and Evolutionary Dynamics. Economic Learning and Social Evolution. MIT Press, Cambridge (2010)
24. Syrgkanis, V., Agarwal, A., Luo, H., Schapire, R.E.: Fast convergence of regularized learning in games. In: Advances in Neural Information Processing Systems, pp. 2989–2997 (2015)
25. Viossat, Y.: Evolutionary dynamics and dominated strategies. Econ. Theory Bull. **3**(1), 91–113 (2015)
26. Viossat, Y., Zapechelnyuk, A.: No-regret dynamics and fictitious play. J. Econ. Theory **148**(2), 825–842 (2013)
27. Vovk, V.G.: Aggregating strategies. In: COLT 1990: Proceedings of the 3rd Workshop on Computational Learning Theory, pp. 371–383 (1990)
28. Weibull, J.W.: Evolutionary Game Theory. MIT Press, Cambridge (1995)

Resource Allocation

Tradeoffs Between Information and Ordinal Approximation for Bipartite Matching

Elliot Anshelevich and Wennan Zhu$^{(\boxtimes)}$

Rensselaer Polytechnic Institute, Troy, NY, USA
eanshel@cs.rpi.edu, zhuw5@rpi.edu

Abstract. We study ordinal approximation algorithms for maximum-weight bipartite matchings. Such algorithms only know the ordinal preferences of the agents/nodes in the graph for their preferred matches, but must compete with fully omniscient algorithms which know the true numerical edge weights (utilities). Ordinal approximation is all about being able to produce good results with only limited information. Because of this, one important question is how much better the algorithms can be as the amount of information increases. To address this question for forming high-utility matchings between agents in $\mathcal{X}$ and $\mathcal{Y}$, we consider three ordinal information types: when we know the preference order of only nodes in $\mathcal{X}$ for nodes in $\mathcal{Y}$, when we know the preferences of both $\mathcal{X}$ and $\mathcal{Y}$, and when we know the total order of the edge weights in the entire graph, although not the weights themselves. We also consider settings where only the top preferences of the agents are known to us, instead of their full preference orderings. We design new ordinal approximation algorithms for each of these settings, and quantify how well such algorithms perform as the amount of information given to them increases.

1 Introduction

Many important settings involve agents with preferences for different outcomes. Such settings include, for example, social choice and matching problems. Although the quality of an outcome to an agent may be measured by a numerical utility, it is often not possible to obtain these exact utilities when forming a solution. This can occur because eliciting numerical information from the agents may be too difficult, the agents may not want to reveal this information, or even because the agents themselves do not know the exact numerical values. On the other hand, eliciting *ordinal* information (i.e., the preference ordering of each agent over the outcomes) is often much more reasonable. Because of this, there has been a lot of recent work on *ordinal approximation algorithms*: these are algorithms which only use ordinal preference information as their input, and yet return a solution provably close to the optimum one (e.g., [3–5,9–12,17]).

This work was partially supported by NSF award CCF-1527497.

V. Bilò and M. Flammini (Eds.): SAGT 2017, LNCS 10504, pp. 267–279, 2017.
DOI: 10.1007/978-3-319-66700-3_21

In other words, these are algorithms which only use limited ordinal information, and yet can compete in the quality of solution produced with omniscient algorithms which know the true (possibly latent) numerical utility information.

Ordinal approximation is all about being able to produce good results with only limited information. Because of this, it is important to quantify how well algorithms can perform as more information is given. If the quality of solutions returned by ordinal algorithms greatly improves when they are provided more information, then it may be worthwhile to spend a lot of resources in order to acquire such more detailed information. If, on the other hand, the improvement is small, then such an acquisition of more detailed information would not be worth it. Thus the main question we consider in this paper is: *How does the quality of ordinal algorithms improve as the amount of information provided increases?*

In this paper, we specifically consider this question in the context of computing a maximum-utility matching in a metric space. Matching problems, in which agents have preferences for which other agents they want to be matched with, are ubiquitous. The maximum-weight metric matching problem specifically provides solutions to important applications, such as forming diverse teams and matching in friendship networks (see [4,5] for much more discussion of this). Formally, there exists a complete undirected bipartite graph for two sets of agents $\mathcal{X}$ and $\mathcal{Y}$ of size N, with an edge weight $w(x,y)$ representing how much utility $x \in \mathcal{X}$ and $y \in \mathcal{Y}$ derive from their match; these edge weights satisfy the triangle inequality. The algorithms we consider, however, do not have access to such numerical edge weights: they are only given ordinal information about the agent preferences. The goal is to form a perfect matching between $\mathcal{X}$ and $\mathcal{Y}$, in order to approximate the maximum weight matching as much as possible using only the given ordinal information. We compare the weight of the matching returned by our algorithms with the true maximum-weight perfect matching in order to quantify the performance of our ordinal algorithms.

Types of Ordinal Information. Ordinal approximation algorithms for maximum weight matching have been considered before in [4,5], although only for complete graphs; algorithms for bipartite graphs require somewhat different techniques. Our main contribution, however, lies in considering many types of ordinal information, forming different algorithms for each, and quantifying how much better types of ordinal information improve the quality of the matching formed. Specifically, we consider the following types of ordinal information.

- The most restrictive model we consider is *one-sided preferences*. That is, only preferences for agents in $\mathcal{X}$ over agents in $\mathcal{Y}$ are given to our algorithm. These preferences are assumed to be consistent with the (hidden) agent utilities, i.e., if x prefers y_1 to y_2, then it must be that $w(x, y_1) \geq w(x, y_2)$. Such one-sided preferences may occur, for example, when $\mathcal{X}$ represents people and $\mathcal{Y}$ represents houses. People have preferences over different houses, but houses do not have preferences over people. These types of preferences also apply to settings in which both sides have preferences, but we only have access to the preferences of $\mathcal{X}$, e.g., because the agents in $\mathcal{Y}$ are more secretive.

– The next level of ordinal information we consider is *two-sided preferences*, that is, both preferences for agents in $\mathcal{X}$ over $\mathcal{Y}$ and agents in $\mathcal{Y}$ over $\mathcal{X}$ are given. This setting could apply to the situation that two sets of people are collaborating, and they have preferences over each other, or of a matching between job applicants and possible employers. As we consider the model in a metric space, the distance (weight) between two people could represent the diversity of their skills, and a person prefers someone with most diverse skills from him/her in order to achieve the best results of collaboration.

– The most informative model which we consider in this paper is that of *total-order*. That is, the order of all the edges in the bipartite graph is given to us, instead of only local preferences for each agent. In this model, global ordinal information is available, compared to the preferences of each agent in the previous two models. Studying this setting quantifies how much efficiency is lost due to the fact that we only know ordinal information, as opposed to the fact that we only know *local* information given to us by each agent.

Comparing the results for the above three information types allows us to answer questions like: "Is it worth trying to obtain two-sided preference information or total order information when only given one-sided preferences?" However, above we always assumed that for an agent x, we are given their entire preferences for all the agents in $\mathcal{Y}$. Often, however, an agent would not give their preference ordering for all the agents they could match with, and instead would only give an ordered list of their top preferences. Because of this, in addition to the three models described above, we also consider the case of *partial* ordinal preferences, in which only the top α fraction of a preference list is given by each agent of $\mathcal{X}$. Thus for $\alpha = 0$ no information at all is given to us, and for $\alpha = 1$ the full preference ordering of an agent is given. Considering partial preferences tells us when, if there is a cost to buying information, we might choose to buy only part of the ordinal preferences. We establish tradeoffs between the percentage of available preferences and the possible approximation ratio for all three models of information above, and thus quantify when a specific amount of ordinal information is enough to form a high-quality matching.

Our Contributions. We show that as we obtain more ordinal information about the agent preferences, we are able to form better approximations to the maximum-utility matching, even without knowing the true numerical edge weights. Our main results are shown in Fig. 1.

Using only one-sided preference information, with only the order of top αN preferences given for agents in $\mathcal{X}$, we are able to form a $(3 - (2 - \sqrt{2})\alpha)$-approximation. We do this by combining random serial dictatorship with purely random matchings. When $\alpha = 1$, the algorithm yields a $(\sqrt{2}+1)$-approximation. This is the first non-trivial analysis for the performance of RSD on maximum bipartite matching in a metric space, and this analysis is one of our main contributions.

Given two-sided information, with the order of top αN preferences for agents in $\mathcal{X}$ and $\mathcal{Y}$, we can do significantly better. When $\alpha \geq \frac{1}{2}$, adopting an exist-

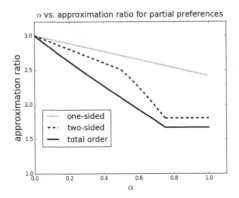

Fig. 1. α vs. approximation ratio for partial information. As we obtain more information about the agent preferences (α increases), we are able to form better approximation to the maximum-weight matching. The tradeoff for one-sided preferences is linear, while it is more complex for two-sided and total order.

ing framework in [4], by mixing greedy and random algorithms, and adjusting it for bipartite graphs, we get a $\frac{(3-2\alpha)(3-\alpha)}{2\alpha^2-3\alpha+3}$-approximation. When $\alpha \leq \frac{1}{2}$, the framework would still work, but would not produce a good approximation. We instead design a different algorithm to get better results. Inspired by RSD, we take advantage of the information of preferences from both sets of agents, adjust RSD to obtain "undominated" edges in each step, and finally combine it with random matchings to get a $(3 - \alpha)$-approximation. When $\alpha \geq \frac{3}{4}$, the algorithm yields a 1.8-approximation.

For the total-ordering model, the order of top αN^2 heaviest edges in the bipartite graph is given. We use the framework in [4] again to obtain a $\frac{2+\sqrt{1-\alpha}}{2-\sqrt{1-\alpha}}$-approximation. Here we must re-design the framework to deal with the cases that $\alpha \leq \frac{3}{4}N$, which is not a straight-forward adjustment. When $\alpha \geq \frac{3}{4}N$ the algorithm yields a $\frac{5}{3}$-approximation.

Finally, in Sect. 6 we analyze the case when edge weights cannot be too different: the highest weight edge is at most β times the lowest weight edge in one-sided model. When the edge weights have this relationship, we can extend our analysis to give a $(\sqrt{\beta - \frac{3}{4}} + \frac{1}{2})$-approximation, even without assuming that edge weights form a metric.

Discussion and Related Work. Previous work on forming good matchings can largely be classified into the following classes. First, there is a large body of work assuming that numerical weights or utilities don't exist, only ordinal preferences. Such work studies many possible objectives, such as forming stable matchings (see e.g., [15,16]), or maximizing objectives determined only by the ordinal preferences (e.g., [2,8]). Second, there is work assuming that numerical utilities or weights exist, and are *known* to the matching designer. Unlike the above two settings, we consider the case when numerical weights *exist*, but are latent or *unknown*, and yet the goal is to approximate the true social welfare, i.e., maximum weight of a perfect matching. Note that although some previous work assumes that all numerical utilities are known, they often still use algorithms which only require ordinal information, and thus fit into our framework; we discuss some of these results below.

Similar to our one-sided model, house allocation [1] is a popular model of assigning n agents to n items. [6] studied the ordinal welfare factor and the linear welfare factor of RSD and other ordinal algorithms. [14] studied both maximum matching and maximum vertex weight matching using an extended RSD algorithm. These either used objectives depending only on ordinal preferences, such as the size of the matching formed, or used node weights (as opposed to edge weights). [9,11] assumed the presence of numerical agent utilities and studied the properties of RSD. Crucially, this work assumed normalized agent utilities, such as unit-sum or unit-range. This allowed [9,11] to prove approximation ratios of $\Theta(\sqrt{n})$ for RSD. Instead of assuming that agent utilities are normalized, we consider agents in a metric space; this different correlation between agent utilities allows us to prove much stronger results, including a constant approximation ratio for RSD. Kalyanasundaram et al. studied serial dictatorship for maximum weight matching in a metric space [13], and gave a 3-approximation for RSD in this, while we are able to get a tighter bound of 2.41-approximation.[1]

Besides maximizing social welfare, minimizing the social cost of a matching is also popular. [7] studied the approximation ratio of RSD and augmentation of serial dictatorship (SD) for minimum weight matching in a metric space. Their setting is very similar to ours, except that we consider the maximization problem, which has different applications [4,5], and allows for a much better approximation factor (constant instead of linear in n) using different techniques.

Another area studying ordinal approximation algorithms is social choice, where the goal is to decide a single winner in order to maximize the total social welfare. This is especially related to our work when the hidden utilities of voters are in a metric space (see e.g., [3,10,12,17]).

The work most related to ours is [4,5]. As mentioned above, we use an existing framework [4] for the two-sided and the total-order model. While the goal is the same: to approximate the maximum weight matching using ordinal information, this paper is different from [4] in several aspects. [4] only considered approximating the true maximum weight matching for non-bipartite complete graphs. We instead focus on bipartite graphs, and especially on considering different levels of ordinal information by analyzing three models with increasing amount of information, and also consider partial preferences. Although we use similar techniques for parts of two-sided and total-order model analysis, they need significant adjustments to deal with bipartite graphs and partial prefer-

---

[1] Note that many of the papers mentioned here specifically attempt to form *truthful* algorithms. While RSD is certainly truthful, in this paper we attempt to quantify what can be done using ordinal information in the presence of latent numerical utilities, and leave questions of truthfulness to future work.

ences; moreover, the method used for analyzing the one-sided model is quite different from [4].

2   Model and Notation

For all the problems studied in this paper, we are given as input two sets of agents $\mathcal{X}$ and $\mathcal{Y}$ with $|\mathcal{X}| = |\mathcal{Y}| = N$. $G = (\mathcal{X}, \mathcal{Y}, E)$ is an undirected complete bipartite graph with weights on the edges. We assume that the agent preferences are derived from a set of underlying hidden edge weights $w(x, y)$ for each edge (x, y), $x \in \mathcal{X}, y \in \mathcal{Y}$. $w(x, y)$ represents the utility of the match between x and y, so if x prefers y_1 to y_2, then it must be that $w(x, y_1) \geq w(x, y_2)$. Let $OPT(G)$ denote the complete bipartite matching that gives the maximum total edge weights. $w(G)$ of any bipartite graph G is the total edge weight of the graph, and $w(M)$ of any matching M is the total weight of edges in the matching. The agents lie in a metric space, by which we will only mean that, $\forall x_1, x_2 \in \mathcal{X}, \forall y_1, y_2 \in \mathcal{Y}, w(x_1, y_1) \leq w(x_1, y_2) + w(x_2, y_1) + w(x_2, y_2)$. We assume this property in all sections except for Sect. 6.

For the setting of one-sided preferences, $\forall x \in \mathcal{X}$, we are given a strict preference ordering P_x over the agents in $\mathcal{Y}$. When dealing with partial preferences, only top αN agents in P_x are given to us in order. We assume αN is an integer, $\alpha \in [0, 1]$. Of course, when $\alpha = 0$, nothing can be done except to form a completely random matching. For two-sided partial preferences, we are given both the top α fraction of preferences P_x of agents x in $\mathcal{X}$ over those in $\mathcal{Y}$, and vice versa. For the total order setting, we are given the order of the highest-weight αN^2 edges in the complete bipartite graph $G = (\mathcal{X}, \mathcal{Y}, E)$.

3   One-Sided Ordinal Preferences

For one-sided preferences, our problem becomes essentially a house allocation problem to maximize social welfare, see e.g., [9,11,14]. Before we proceed, it is useful to establish a baseline for what approximation factor is reasonable. Simply picking a matching uniformly at random immediately results in a 3-approximation (see Theorem 2), and there are examples showing that this bound is tight. Other well-known algorithms, such as Top Trading Cycle, also cannot produce better than a 3-approximation to the maximum weight matching for our setting. Serial Dictatorship, which uses only one-sided ordinal information, is also known to give a 3-approximation to the maximum weight matching for our problem [13]. Serial Dictatorship simply takes an arbitrary agent from $x \in \mathcal{X}$, assigns it x's favorite unallocated agent from $\mathcal{Y}$, and repeats. Unfortunately, it is not difficult to show that this bound of 3 is tight. Our first major result in this paper is to prove that *Random* Serial Dictatorship always gives a $(\sqrt{2} + 1)$-approximation in expectation, no matter what the true numerical weights are, thus giving a significant improvement to all the algorithms mentioned above.

Algorithm 1. *Random Serial Dictatorship (RSD)* Given bipartite graph $G = (\mathcal{X}, \mathcal{Y}, E)$, initialize a matching $M = \emptyset$. Pick an agent x uniformly at random from $\mathcal{X}$. Let y denote x's most preferred agent in $\mathcal{Y}$, take $e = (x, y)$ from E and add it to M. Remove x, y, and all edges containing x or y. Repeat until no agent is left, return M.

Theorem 1. *Suppose* $G = (\mathcal{X}, \mathcal{Y}, E)$ *is a complete bipartite graph on the set of nodes* $\mathcal{X}, \mathcal{Y}$ *with* $|\mathcal{X}| = |\mathcal{Y}| = N$. *Then, the expected weight of the perfect matching* M *returned by Algorithm 1 is* $E[w(M)] \geq \frac{1}{\sqrt{2}+1} w(OPT(G))$.

Proof Sketch We give a proof sketch here; full proofs for all our results can be found in the full version of this paper at http://www.cs.rpi.edu/~eanshel/. Let $Min(G)$ denote a *minimum* weight perfect matching on G, and $RSD(G)$ denote the expected weight returned by Algorithm 1 on graph G. For any $x \in \mathcal{X}$, we use $\lambda(x)$ to denote the edge between x and its most preferred agent in $\mathcal{Y}$. Define $R(x)$ as the remaining graph after removing x, x's most preferred agent, and all the edges containing x or x's most preferred agent from G. We now state the main technical lemma which allows us to prove the result. This lemma gives a bound on the maximum weight matching in terms of the quantities defined above.

Lemma 1. *For any given graph* $G = (\mathcal{X}, \mathcal{Y}, E)$, *one of the following two cases must be true:*

> **Case 1:** $w(OPT(G)) \leq \frac{1}{|\mathcal{X}|} \sum_{x \in \mathcal{X}} w(OPT(R(x))) + \frac{\sqrt{2}+1}{|\mathcal{X}|} \sum_{x \in \mathcal{X}} w(\lambda(x))$
> **Case 2:** $w(OPT(G)) \leq (\sqrt{2}+1) w(Min(G))$

We will prove this lemma below, but first we discuss how the rest of the proof proceeds. When Case 1 above holds, we know that at any step of the algorithm, the change in the weight of the optimum solution in the remaining graph is not that different from the weight of the edge selected by our algorithm. This allows us to compare the weight of OPT with the weight of the matching returned by our algorithm. In fact, this is the technique used in a previous paper [5] to analyze RSD for complete graphs (i.e., non-bipartite graphs), and show that RSD gives a 2-approximation for perfect matching on complete graphs. It is important to note here that this *does not* work for bipartite graphs. In bipartite matching, there are examples in which using only this method will not give an approximation ratio better than 3. We get around this problem by adding Case 2 to our lemma, and then using this to prove the theorem.

Proof Sketch of Lemma 1. For any fixed $x \in \mathcal{X}$, denote x's most preferred agent in $\mathcal{Y}$ as y (so $\lambda(x) = (x, y)$). In $OPT(G)$, suppose x is matched to $b \in \mathcal{Y}$, and y is matched to $a \in \mathcal{X}$. In $Min(G)$, suppose b is matched to $m \in \mathcal{X}$. $\forall x \in \mathcal{X}$, there exist y, a, b, m as described above. Denote edge (x, y) by $\lambda(x)$, (x, b) by $P(x)$, (a, y) by $\bar{P}(x)$, and (a, b) by $D(x)$.

We'll prove Lemma 1 by showing that if **Case 2** is not true, then **Case 1** must be true. Suppose **Case 2** is not true, i.e., $w(OPT(G)) > (\sqrt{2}+1) w(Min(G))$.

Suppose that random serial dictatorship picks $x \in \mathcal{X}$. Then $OPT(R(x))$ is at least as good as the matching obtained by removing $P(x)$ and $\bar{P}(x)$, and adding $D(x)$ to $OPT(G)$ (the rest stay the same):

$$w(OPT(R(x))) \geq w(OPT(G)) - w(P(x)) - w(\bar{P}(x)) + w(D(x))$$

Summing this up over all nodes x, we obtain:

$$\frac{1}{|\mathcal{X}'|} \sum_{x \in \mathcal{X}'} w(OPT(R(x))) \geq (1 - \frac{1}{|\mathcal{X}'|})w(OPT(G)) - \frac{1}{|\mathcal{X}'|} \sum_{x \in \mathcal{X}'} (w(\bar{P}(x)) - w(D(x))) \tag{1}$$

By the triangle inequality, we know that $w(a, y) \leq w(a, b) + w(m, b) + w(m, y)$ Because $\lambda(m)$ is the edge to m's most preferred agent, $w(m, y) \leq w(\lambda(m))$, and thus $w(\bar{P}(x)) \leq w(D(x)) + w(m, b) + w(\lambda(m)))$.

Summing this up for all $x \in \mathcal{X}$, note that each x is matched to a unique b in $OPT(G)$, and each b is matched to a unique m in $Min(G)$, so each agent in $\mathcal{Y}$ appears as b exactly once and each agent in $\mathcal{X}$ appears as m exactly once.

$$\sum_{x \in \mathcal{X}} (w(\bar{P}(x)) - w(D(x))) \leq w(Min(G)) + \sum_{x \in \mathcal{X}} w(\lambda(x))) \tag{2}$$

Combining Inequalitys 1 and 2,

$$\frac{1}{|\mathcal{X}|} \sum_{x \in \mathcal{X}} w(OPT(R(x))) \geq (1 - \frac{1}{|\mathcal{X}|})w(OPT(G)) - \frac{1}{|\mathcal{X}|}[w(Min(G)) + \sum_{x \in \mathcal{X}} w(\lambda(x))] \tag{3}$$

$w(P(x)) \leq w(\lambda(x))$ since $\lambda(x)$ is the most preferred edge of x, so it is obvious that $w(OPT(G)) \leq \sum_{x \in \mathcal{X}} w(\lambda(x))$. Combining this with our assumption about $Min(G)$, we obtain the desired result. For detailed proof, see the full version. □

Partial One-Sided Ordinal Preferences

In this section, we consider the case when we are given even less information than in the previous one, i.e., only partial preferences. We begin by establishing the following easy result for the completely random algorithm.

Theorem 2. *The uniformly random perfect matching is a 3-approximation to the maximum-weight matching.*

Algorithm 2. *Run Algorithm 1, stop when* $|M| = \alpha N$*, then form random matches until all agents are matched. Return M.*

Theorem 3. *Suppose $G = (\mathcal{X}, \mathcal{Y}, E)$ is a complete bipartite graph on the set of nodes $\mathcal{X}, \mathcal{Y}$ with $|\mathcal{X}| = |\mathcal{Y}| = N$. There is a strict preference ordering P_x over the agents in $\mathcal{Y}$ for each agent $x \in \mathcal{X}$. We are only given top αN agents in P_x in order. Then, the expected weight of the perfect matching M returned by Algorithm 2 is $E[w(M)] \geq \frac{1}{3 - (2 - \sqrt{2})\alpha} w(OPT(G))$, as shown in Fig. 1.*

Proof Sketch. We establish a linear tradeoff as α increases. Note that this would not work for combining any two arbitrary algorithms. The key insight which makes this proof work is that, at every step, the expected weight of RSD is higher than in the following step, and that RSD always produces an edge weight which is better than random in expectation.   □

4   Two-Sided Ordinal Preferences

For two-sided preferences, we give separate algorithms for the cases when $\alpha \geq \frac{1}{2}$ and when $\alpha \leq \frac{1}{2}$, as these require somewhat different techniques.

$\boldsymbol{\alpha \geq \frac{1}{2}}$ While for the case when $\alpha < \frac{1}{2}$ new techniques are necessary to obtain a good approximation, the approach for the case when $\alpha \geq \frac{1}{2}$ is essentially the same as the one used in [4]. We adopt this approach to deal with bipartite graphs and with partial preferences, giving us a 1.8-approximation for $\alpha = 1$. To do this, we re-state the definition of Undominated Edges from [4], and a standard greedy algorithm for forming a matching of size k.

Definition 1. *(Undominated Edges) Given a set E of edges, $(x, y) \in E$ is said to be an undominated edge if for all (x, a) and (y, b) in E, $w(x, y) \geq w(x, a)$ and $w(x, y) \geq w(y, b)$.*

Note that an undominated edge must always exist: either there are two nodes x and y such that they are each other's top preferences (and so (x, y) is undominated), or there is a cycle $x_1, x_2, \ldots$ in which x_{i+1} is the top preference of x_i, in which case all edges in the cycle must be the same weight, and thus all edges in the cycle are undominated. This also gives us an algorithm for determining if an edge (x, y) is undominated: either x and y prefer each other over all other agents, or it is part of such a cycle of top preferences.

Algorithm 3 (Undominated Greedy). *Given bipartite graph $G = (\mathcal{X}, \mathcal{Y}, E)$, initialize a matching $M = \emptyset$. Pick an arbitrary undominated edge $e = (x, y)$ from E and add it to M. Remove x, y, and all edges containing x or y from E. Repeat until $|M| = k$. Return M.*

Algorithm 4. Given bipartite graph $G = (\mathcal{X}, \mathcal{Y}, E)$, and top αN of $P(\mathcal{X})$, top αN of $P(\mathcal{Y})$. Let M_0 be the output returned by Algorithm 3 for E, $k = \alpha N$. Let $M_1 = M_0 \cup$ (Uniformly random matching on the rest of the agents). We get M_2 in the following way: randomly choose $(2\alpha - 1)N$ edges from M_0 that stay matched, and unmatch all other agents in M_0. Then form a random bipartite matching between all the agents which were not matched in M_0, and the nodes which we chose from M_0 to become unmatched. Return M_1 with probability $\frac{3-2\alpha}{3-\alpha}$ and M_2 with probability $\frac{\alpha}{3-\alpha}$.

Note that for $\alpha > \frac{3}{4}$ this algorithm does not seem to provide better guarantees than for $\alpha = \frac{3}{4}$. Because of this, for $\alpha > \frac{3}{4}$, we simply run the same algorithm for $\alpha = \frac{3}{4}$.

Theorem 4. *Algorithm 4 returns a $\frac{(3-2\alpha)(3-\alpha)}{2\alpha^2-3\alpha+3}$-approximation to the maximum-weight perfect matching given two-sided ordering when $\frac{1}{2} \leq \alpha \leq \frac{3}{4}$.*

$\alpha \leq \frac{1}{2}$ Unlike the case for $\alpha \geq \frac{1}{2}$, this case requires different techniques than in [4]. While the techniques above would still work, they will not give us a bound as good as the one we form below. The idea in this section is to do something similar to our one-sided algorithm for partial preferences: run the greedy algorithm for a while, and then switch to random. Unfortunately, if we simply run the greedy Algorithm 3 and then switch to random, this will not form a good approximation. The reason why this is true is that an undominated edge which is picked by the greedy algorithm may be much worse than the average weight of an edge, and so the approximation factor of the random algorithm will dominate, giving only a 3-approximation. Even taking an undominated edge uniformly at random has this problem. We can fix this, however, by picking each undominated edge with an appropriate probability, as described below. Such an algorithm results in matchings which are guaranteed to be better than either RSD or Random, thus allowing us to prove the result.

Algorithm 5. Given bipartite graph $G = (\mathcal{X}, \mathcal{Y}, E)$, and top αN of $P(\mathcal{X})$, top αN of $P(\mathcal{Y})$. Pick an agent x uniformly at random from $\mathcal{X}$, let x's most preferred agent be y in $\mathcal{Y}$. We choose an undominated edge by the following method: if (x, y) is an undominated edge, take it and continue to pick the next agent. Otherwise check whether the edge between y and its most preferred node is undominated, keep doing this until we find an undominated edge (x_1, y_1) and add it to M. Remove x_1, y_1, and all edges containing x_1 or y_1. Repeat until $|M| = \alpha N$. Form a uniformly random matching for the remaining graph G, add the edges returned by the algorithm to M, and return M.

This algorithm guarantees that an undominated edge is chosen for any x in any bipartite graph G. Now, before we reach an undominated edge, the weights of edges are non-decreasing in the order they are checked. Thus whenever a node x is picked, the algorithm adds an undominated edge (x_1, y_1) to the matching which is guaranteed to have higher weight than all edges leaving x.

Theorem 5. *Algorithm 5 returns a $(3 - \alpha)$-approximation to the maximum-weight perfect matching given two-sided ordering when $0 \leq \alpha \leq \frac{1}{2}$.*

Proof Sketch. We use a similar method and the same notation as in Sect. 3 to proof this theorem. Essentially, because we are always picking undominated edges, we can form a linear interpolation between a factor of 2 and a factor of 3 for random matching, instead of between factors $\sqrt{2} + 1$ and 3 as for one-sided preferences. The reason why we are able to form such an interpolation is entirely because of the probabilities with which we choose the undominated edges; if we simply chose arbitrary undominated edges or choose them uniformly at random, then there are examples where the random edge weights will dominate and result in a poor approximation, since undominated edges are only guaranteed to be within a factor of 3 of the average edge weight.   □

5 Total Ordering of Edge Weights

For the setting in which we are given the top αN^2 edges of G in order, we prove that for $\alpha = \frac{3}{4}$, we can obtain an approximation of $\frac{5}{3}$ in expectation. For larger α, however, more information does not seem to help, and so we simply use the algorithm for $\alpha = \frac{3}{4}$ for any $\alpha > \frac{3}{4}$.

Algorithm 6. Given bipartite graph $G = (\mathcal{X}, \mathcal{Y}, E)$, initialize a matching $M = \emptyset$. Pick the heaviest edge $e = (x, y)$ from E and add it to M. Remove x, y, and all edges containing x or y from E. Repeat until $|M| = k$. Return M.

The algorithm for bipartite matching with partial ordinal information is similar to that with partial two-sided ordinal information, except that we only need to consider the case that $k \leq \frac{1}{2}N$, $\alpha \leq \frac{3}{4}$.

Algorithm 7. Given bipartite graph $G = (\mathcal{X}, \mathcal{Y}, E)$, and order of the top αN^2 edges in the graph. Let M_0 be the output returned by Algorithm 6 for E, $k = (1 - \sqrt{1 - \alpha})N$. Let $M_1 = M_0 \cup$ (Uniformly random matching on the rest of the agents). We get M_2 in the following way: let B denote the complete bipartite graph on the set of nodes not matched in M_0. Randomly choose $(2\sqrt{1 - \alpha} - 1)N$ nodes from both sets of agents in B, get the perfect matching output by random Algorithm and add to M_2, then unmatched all agents in M_0, form a random bipartite matching between agents in M_0 and the agents not chosen in B and add to M_2. Return M_1 with probability $\frac{2}{2 + \sqrt{1 - \alpha}}$ and M_2 with probability $\frac{\sqrt{1 - \alpha}}{2 + \sqrt{1 - \alpha}}$.

Theorem 6. *Algorithm 7 returns a $\frac{2 + \sqrt{1 - \alpha}}{2 - \sqrt{1 - \alpha}}$-approximation to the maximum-weight matching in expectation for $\alpha \leq \frac{3}{4}$, as shown in Fig. 1.*

6 One-Sided Preferences with Restricted Edge Weights

In previous sections, we made the assumption that the agents lie in a metric space, and thus the edge weights, although unknown to us, must follow the triangle inequality. In this section we once again consider the most restrictive type of agent preferences—that of one-sided preferences—but now instead of assuming that agents lie in a metric space, we instead consider settings where edges weights cannot be infinitely different from each other. This applies to settings where the agents are at least somewhat indifferent and the items are somewhat similar; the least-preferred agent and the most-preferred items differ only by a constant factor to any agent. Indeed, when for example purchasing a house in a reasonable market (i.e., once houses that almost no one would buy have been removed from consideration), it is unlikely that any agent would like house x so much more than house y that they would be willing to pay hundreds of times more for x than for y.

More formally, for each agent $i \in \mathcal{X}$, we are given a strict preference ordering P_i over the agents in $\mathcal{Y}$. In this section we assume that the highest weight edge e_{max} is at most β times of the lowest weight edge e_{min}. We normalize the lowest weight edge e_{min} in the graph to $w(e_{min}) = 1$; then for any edge $e \in E$, $w(e) \le \beta$. We use similar analysis as in Sect. 3, except that instead of getting bounds by using the triangle inequality, the relationships among edge weights are bounded by our assumption of the highest and lowest weight edge ratio. As stated above, we no longer assume the agents lie in a metric space in this section.

Theorem 7. *Suppose $G = (\mathcal{X}, \mathcal{Y}, E)$ is a complete bipartite graph on the set of nodes $\mathcal{X}, \mathcal{Y}$ with $|\mathcal{X}| = |\mathcal{Y}| = N$. $w(e_{min}) = 1$, $\forall e \in E$, $w(e) \le \beta$. The expected weight of the perfect matching returned by Algorithm 1 is $w(M) \ge \frac{1}{\sqrt{\beta - \frac{3}{4}} + \frac{1}{2}} w(OPT)$ (see plot and proof in the full version).*

References

1. Abdulkadiroğlu, A., Sönmez, T.: Random serial dictatorship and the core from random endowments in house allocation problems. Econometrica **66**(3), 689–701 (1998)
2. Abraham, D.J., Irving, R.W., Kavitha, T., Mehlhorn, K.: Popular matchings. SIAM J. Comput. **37**(4), 1030–1045 (2007)
3. Anshelevich, E., Bhardwaj, O., Postl, J.: Approximating optimal social choice under metric preferences. In: AAAI (2015)
4. Anshelevich, E., Sekar, S.: Blind, greedy, and random: algorithms for matching and clustering using only ordinal information. In: AAAI (2016)
5. Anshelevich, E., Sekar, S.: Truthful mechanisms for matching and clustering in an ordinal world. In: Cai, Y., Vetta, A. (eds.) WINE 2016. LNCS, vol. 10123, pp. 265–278. Springer, Heidelberg (2016). doi:10.1007/978-3-662-54110-4_19
6. Bhalgat, A., Chakrabarty, D., Khanna, S.: Social welfare in one-sided matching markets without money. In: Goldberg, L.A., Jansen, K., Ravi, R., Rolim, J.D.P. (eds.) APPROX/RANDOM -2011. LNCS, vol. 6845, pp. 87–98. Springer, Heidelberg (2011). doi:10.1007/978-3-642-22935-0_8
7. Caragiannis, I., Filos-Ratsikas, A., Frederiksen, S.K.S., Hansen, K.A., Tan, Z.: Truthful facility assignment with resource augmentation: an exact analysis of serial dictatorship. In: Cai, Y., Vetta, A. (eds.) WINE 2016. LNCS, vol. 10123, pp. 236–250. Springer, Heidelberg (2016). doi:10.1007/978-3-662-54110-4_17
8. Chakrabarty, D., Swamy, C.: Welfare maximization and truthfulness in mechanism design with ordinal preferences. In: ITCS (2014)
9. Christodoulou, G., Filos-Ratsikas, A., Frederiksen, S.K.S., Goldberg, P.W., Zhang, J., Zhang, J.: Social welfare in one-sided matching mechanisms. In: Osman, N., Sierra, C. (eds.) AAMAS 2016. LNCS (LNAI), vol. 10002, pp. 30–50. Springer, Cham (2016). doi:10.1007/978-3-319-46882-2_3
10. Feldman, M., Fiat, A., Golomb, I.: On voting and facility location. In: EC (2016)
11. Filos-Ratsikas, A., Frederiksen, S.K.S., Zhang, J.: Social welfare in one-sided matchings: random priority and beyond. In: Lavi, R. (ed.) SAGT 2014. LNCS, vol. 8768, pp. 1–12. Springer, Heidelberg (2014). doi:10.1007/978-3-662-44803-8_1
12. Goel, A., Krishnaswamy, A.K., Munagala, K.: Metric distortion of social choice rules: lower bounds and fairness properties. In: EC (2017)
13. Kalyanasundaram, B., Pruhs, K.: On-line weighted matching. In: SODA, vol. 91, pp. 234–240 (1991)
14. Krysta, P., Manlove, D., Rastegari, B., Zhang, J.: Size versus truthfulness in the house allocation problem. In: EC (2014)

15. Rastegari, B., Condon, A., Immorlica, N., Leyton-Brown, K.: Two-sided matching with partial information. In: EC (2013)
16. Roth, A.E., Sotomayor, M.: Two-sided matching. Handb. Game Theory Econ. Appl. **1**, 485–541 (1992)
17. Skowron, P., Elkind, E.: Social choice under metric preferences: scoring rules and STV. In: AAAI (2017)

Group Strategyproof Pareto-Stable Marriage with Indifferences via the Generalized Assignment Game

Nevzat Onur Domaniç$^{(\boxtimes)}$, Chi-Kit Lam$^{(\boxtimes)}$, and C. Gregory Plaxton$^{(\boxtimes)}$

University of Texas at Austin, Austin, TX 78712, USA
{onur,geocklam,plaxton}@cs.utexas.edu

Abstract. We study the variant of the stable marriage problem in which the preferences of the agents are allowed to include indifferences. We present a mechanism for producing Pareto-stable matchings in stable marriage markets with indifferences that is group strategyproof for one side of the market. Our key technique involves modeling the stable marriage market as a generalized assignment game. We also show that our mechanism can be implemented efficiently. These results can be extended to the college admissions problem with indifferences.

1 Introduction

The stable marriage problem was first introduced by Gale and Shapley [13]. The stable marriage market involves a set of men and women, where each agent has ordinal preferences over the agents of the opposite sex. The goal is to find a disjoint set of man-woman pairs, called a *matching*, such that no other man-woman pair prefers each other to their partners in the matching. Such matchings are said to be *stable*. When preferences are strict, a unique man-optimal stable matching exists and can be computed by the man-proposing deferred acceptance algorithm of Gale and Shapley [13]. A mechanism is said to be *group strategyproof for the men* if no coalition of men can be simultaneously matched to strictly preferred partners by misrepresenting their preferences. Dubins and Freedman [8] show that the mechanism that produces man-optimal matchings is group strategyproof for the men when preferences are strict. In our work, we focus on group strategyproofness for the men, since no stable mechanism is strategyproof for both men and women [19].

We remark that the notion of group strategyproofness used here assumes no side payments within the coalition of men. It is known that group strategyproofness for the men is impossible for the stable marriage problem with strict preferences when side payments are allowed [21, Chap. 4]. This notion of group strategyproofness is also different from strong group strategyproofness, in which at least one man in the coalition gets matched to a strictly preferred partner while the other men in the coalition get matched to weakly preferred

This research was supported by NSF Grant CCF–1217980.

© Springer International Publishing AG 2017
V. Bilò and M. Flammini (Eds.): SAGT 2017, LNCS 10504, pp. 280–291, 2017.
DOI: 10.1007/978-3-319-66700-3_22

partners. It is known that strong group strategyproofness for the men is impossible for the stable marriage problem with strict preferences [8, attributed to Gale].

Indifferences in the preferences of agents arise naturally in real-world applications such as school choice [1, 10, 11]. For the marriage problem with indifferences, Sotomayor [24] argues that Pareto-stability is an appropriate solution concept. A matching is said to be *weakly stable* if no man-woman pair strictly prefers each other to their partners in the matching. A matching is said to be *Pareto-optimal* if there is no other matching that is strictly preferred by some agent and weakly preferred by all agents. If a matching is both weakly stable and Pareto-optimal, it is said to be *Pareto-stable*.

Weakly stable matchings, unlike strongly stable or super-stable matchings [14], always exist. However, not all weakly stable matchings are Pareto-optimal [24]. Pareto-stable matchings can be obtained by applying successive Pareto-improvements to weakly stable matchings. Erdil and Ergin [10, 11] show that this procedure can be carried out efficiently. Pareto-stable matchings also exist and can be computed in strongly polynomial time for many-to-many matchings [2] and multi-unit matchings [3]. Instead of using the characterization of Pareto-improvement chains and cycles, Kamiyama [15] gives another efficient algorithm for many-to-many matchings based on rank-maximal matchings. However, none of these mechanisms addresses strategyproofness.

We remark that the notion of Pareto-optimality here is different from *man-Pareto-optimality*, which only takes into account the preferences of the men. It is known that man-Pareto-optimality is not compatible with strategyproofness for the stable marriage problem with indifferences [10, 16]. The notion of Pareto-optimality here is also different from *Pareto-optimality in expected utility*, which permits Pareto-domination by non-pure outcomes. A result of Zhou [25] implies that Pareto-optimality in expected utility is not compatible with strategyproofness for the stable marriage problem with indifferences.

Until recently, it was not known whether a strategyproof Pareto-stable mechanism exists. In our recent workshop paper [7], we present a generalization of the deferred acceptance mechanism that is Pareto-stable and strategyproof for the men. If the market has n agents, our implementation of this mechanism runs in $O(n^4)$ time, matching the time bound of the algorithm of Erdil and Ergin [10, 11][1]. The proof of strategyproofness relies on reasoning about a certain threshold concept in the stable marriage market, and this approach seems difficult to extend to address group strategyproofness.

In this paper, we introduce a new technique useful for investigating incentive compatibility for coalitions of men. We present a Pareto-stable mechanism for the stable marriage problem with indifferences that is provably group strategyproof

---

[1] The algorithm of Erdil and Ergin proceeds in two phases. In the first phase, ties are broken arbitrarily and the deferred acceptance algorithm is used to obtain a weakly stable matching. In the second phase, a sequence of Pareto-improvements are applied until a Pareto-stable matching is reached. In App. A in the full version of [7], we show that this algorithm does not provide a strategyproof mechanism.

for the men, by modeling the stable marriage market as an appropriate form of the generalized assignment game. In Sect. 4 and the full version [6, App. B and C] of this paper, we show that this mechanism coincides with the generalization of the deferred acceptance mechanism presented in [7]. Thus we obtain an $O(n^4)$-time group strategyproof Pareto-stable mechanism.

The Generalized Assignment Game. The assignment game, introduced by Shapley and Shubik [22], involves a two-sided matching market with monetary transfer in which agents have unit-slope linear utility functions. This model has been generalized to allow agents to have continuous, invertible, and increasing utility functions [4,5,18]. Some models that generalize both the assignment game and the stable marriage problems have also been developed, but their models are not concerned with the strategic behavior of agents [12,23]. The formulation of the generalized assignment game in this paper follows the presentation of Demange and Gale [5].

In their paper, Demange and Gale establish various elegant properties of the generalized assignment game, such as the lattice property and the existence of one-sided optimal outcomes. (One-sided optimality or man-optimality is a stronger notion than one-sided Pareto-optimality or man-Pareto-optimality.) These properties are known to hold for the stable marriage market in the case of strict preferences [17, attributed to Conway], but fail in the case of weak preferences [21, Chap. 2]. Given the similarities between stable marriage markets and generalized assignment games, it is natural to ask whether stable marriage markets can be modeled as generalized assignment games. Demange and Gale discuss this question and state that "the model of [Gale and Shapley] is not a special case of our model". The basic obstacle is that it is unclear how to model an agent's preferences within the framework of a generalized assignment game: on the one hand, even though ordinal preferences can be converted into numeric utility values, such preferences are expressed in a manner that is independent of any monetary transfer; on the other hand, the framework demands that there is an amount of money that makes an agent indifferent between any two agents on the other side of the market.

In Sect. 2, we review key concepts in the work of Demange and Gale, and introduce the *tiered-slope market* as a special form of the generalized assignment game in which the slopes of the utility functions are powers of a large fixed number. Then, in Sect. 3, we describe our approach for converting a stable marriage market with indifferences into an associated tiered-slope market. While these are both two-sided markets that involve the same set of agents, the utilities achieved under an outcome in the associated tiered-slope market may not be equal to the utilities under a corresponding solution in the stable marriage market. Nevertheless, we are able to establish useful relationships between certain sets of solutions to these two markets.

Our first such result, Theorem 2, shows that Pareto-stability in the stable marriage market with indifferences follows from stability in the associated tiered-slope market, even though it does not follow from weak stability in the stable marriage market with indifferences. This can be seen as a partial analogue to the case of strict preferences, in which stability in the stable marriage market implies Pareto-stability [13]. This also demonstrates that, in addition to using the deferred acceptance procedure to solve the generalized assignment game [4], we can use the generalized assignment game to solve the stable marriage problem with indifferences.

In Lemma 5, we establish that the utility achieved by any man in a man-optimal solution to the associated tiered-slope market uniquely determines the tier of preference to which that man is matched in the stable marriage market with indifferences. Another consequence of this lemma is that any matched man in a man-optimal outcome of the associated tiered-slope market receives at least one unit of money from his partner. We can then deduce that if a man strictly prefers his partner to a woman, then the woman has to offer a large amount of money in order for the man to be indifferent between her offer and that of his partner. Since individual rationality prevents any woman from offering such a large amount of money, this explains how we overcome the obstacle of any man being matched with a less preferred woman in exchange for a sufficiently large payment.

A key result established by Demange and Gale is that the man-optimal mechanism is group strategyproof for the men. Using this result and Lemma 5, we are able to show in Theorem 3 that group strategyproofness for the men in the stable marriage market with indifferences is achieved by man-optimality in the associated tiered-slope market, even though it is incompatible with man-Pareto-optimality in the stable marriage market with indifferences [10,16]. This can be seen as a partial analogue to the case of strict preferences, in which man-optimality implies group strategyproofness [8].

Extending to the College Admissions Problem. We also consider the settings of incomplete preference lists and one-to-many matchings, in which efficient Pareto-stable mechanisms are known to exist [2,3,10,11,15]. Preference lists are incomplete when an agent declares another agent of the opposite sex to be unacceptable. Our mechanism is able to support such incomplete preference lists through an appropriate choice of the reserve utilities of the agents in the associated tiered-slope market. In fact, our mechanism also supports indifference between being unmatched and being matched to some partner.

The one-to-many variant of the stable marriage problem with indifferences is the college admissions problem with indifferences. In this model, students and colleges play the roles of men and women, respectively, and colleges are allowed to be matched with multiple students, up to their capacities. We provide the formal definition of the model in the full version [6, App. D] of this paper. By a simple reduction from college admissions markets to stable marriage markets, our mechanism is group strategyproof for the students[2] and produces a Pareto-stable matching in polynomial time.

---

[2] A stable mechanism can be strategyproof only for the side having unit demand, namely the students [20].

Organization of this Paper. In Sect. 2, we review the generalized assignment game and define the tiered-slope market. In Sect. 3, we introduce the tiered-slope markets associated with the stable marriage markets with indifferences, and use them to obtain a group strategyproof, Pareto-stable mechanism. In Sect. 4 and the full version [6, App. B and C] of this paper, we discuss efficient implementations of the mechanism and its relationship with the generalization of the deferred acceptance algorithm presented in [7]. Due to space limitations, some of the proofs are omitted from this paper. See the full version [6] for all of the proof details.

2   Tiered-Slope Market

The generalized assignment game studied by Demange and Gale [5] involves two disjoint sets I and J of agents, which we call *men* and *women* respectively. We assume that the sets I and J do not contain the element 0, which we use to denote being unmatched. For each man $i \in I$ and woman $j \in J$, the compensation function $f_{i,j}(u_i)$ represents the compensation that man i needs to receive in order to attain utility u_i when he is matched to woman j. Similarly, for each man $i \in I$ and woman $j \in J$, the compensation function $g_{i,j}(v_j)$ represents the compensation that woman j needs to receive in order to attain utility v_j when she is matched to man i. Moreover, each man $i \in I$ has a reserve utility r_i and each woman $j \in J$ has a reserve utility s_j.

In this paper, we assume that the compensation functions are of the form

$$f_{i,j}(u_i) = u_i \lambda^{-a_{i,j}} \quad \text{and} \quad g_{i,j}(v_j) = v_j - (b_{i,j}N + \pi_i)$$

and the reserve utilities are of the form

$$r_i = \pi_i \lambda^{a_{i,0}} \quad \text{and} \quad s_j = b_{0,j}N,$$

where

$$\pi \in \mathbb{Z}^I; \quad N \in \mathbb{Z}; \quad \lambda \in \mathbb{Z}; \quad a \in \mathbb{Z}^{I \times (J \cup \{0\})}; \quad b \in \mathbb{Z}^{(I \cup \{0\}) \times J}$$

such that $N > \max_{i \in I} \pi_i \geq \min_{i \in I} \pi_i \geq 1$ and

$$\lambda \geq \max_{(i,j) \in (I \cup \{0\}) \times J} (b_{i,j} + 1)N \geq \min_{(i,j) \in (I \cup \{0\}) \times J} (b_{i,j} + 1)N \geq N.$$

We denote this *tiered-slope market* as $\mathcal{M} = (I, J, \pi, N, \lambda, a, b)$. When $a_{i,j} = 0$ for every man $i \in I$ and woman $j \in J \cup \{0\}$, this becomes a *unit-slope market* $(I, J, \pi, N, \lambda, 0, b)$. Notice that the compensation functions in a unit-slope market coincide with those in the assignment game [22] where buyer $j \in J$ has a valuation of $b_{i,j}N + \pi_i$ on house $i \in I$. For better readability, we write $\exp_\lambda(\xi)$ to denote λ^ξ.

A *matching* is a function $\mu \colon I \to J \cup \{0\}$ such that for any woman $j \in J$, we have $\mu(i) = j$ for at most one man $i \in I$. Given a matching μ and a woman $j \in J$, we denote

$$\mu(j) = \begin{cases} i & \text{if } \mu(i) = j \\ 0 & \text{if there is no man } i \in I \text{ such that } \mu(i) = j \end{cases}$$

An *outcome* is a triple (μ, u, v), where μ is a matching, $u \in \mathbb{R}^I$ is the utility vector of the men, and $v \in \mathbb{R}^J$ is the utility vector of the women. An outcome (μ, u, v) is *feasible* if the following conditions hold for every man $i \in I$ and woman $j \in J$.

1. If $\mu(i) = j$, then $f_{i,j}(u_i) + g_{i,j}(v_j) \leq 0$.
2. If $\mu(i) = 0$, then $u_i = r_i$.
3. If $\mu(j) = 0$, then $v_j = s_j$.

A feasible outcome (μ, u, v) is *individually rational* if $u_i \geq r_i$ and $v_j \geq s_j$ for every man $i \in I$ and woman $j \in J$. An individually rational outcome (μ, u, v) is *stable* if $f_{i,j}(u_i) + g_{i,j}(v_j) \geq 0$ for every man $i \in I$ and woman $j \in J$.

A stable outcome (μ, u, v) is *man-optimal* if for any stable outcome (μ', u', v') we have $u_i \geq u_i'$ for every man $i \in I$. It has been shown that man-optimal outcomes always exist [5, Property 2]. Theorem 1 below provides a useful group strategyproofness result for man-optimal outcomes.

Theorem 1. *Let (μ, u, v) and (μ', u', v') be man-optimal outcomes of tiered-slope markets $(I, J, \pi, N, \lambda, a, b)$ and $(I, J, \pi, N, \lambda, a', b)$, respectively. If $a \neq a'$, then there exists a man $i_0 \in I$ and a woman $j_0 \in J \cup \{0\}$ with $a_{i_0,j_0} \neq a'_{i_0,j_0}$ such that $u_{i_0} \geq u'_{i_0} \exp_\lambda(a_{i_0,\mu'(i_0)} - a'_{i_0,\mu'(i_0)})$.*

Proof. This follows directly from [5, Theorem 2], which establishes group strategyproofness for the men in the generalized assignment game with no side payments. Notice that the value $u'_{i_0} \exp_\lambda(a_{i_0,\mu'(i_0)} - a'_{i_0,\mu'(i_0)})$ is the true utility of man i_0 under matching μ' as defined in their paper, both in the case of being matched to $\mu'(i_0) \neq 0$ with compensation $u'_{i_0} \exp_\lambda(-a'_{i_0,\mu'(i_0)})$ and in the case of being unmatched. □

3   Stable Marriage with Indifferences

The stable marriage market involves a set I of men and a set J of women. We assume that the sets I and J are disjoint and do not contain the element 0, which we use to denote being unmatched. The preference relation of each man $i \in I$ is specified by a binary relation $\succeq_i$ over $J \cup \{0\}$ that satisfies transitivity and totality. To allow indifferences, the preference relation is not required to satisfy anti-symmetry. Similarly, the preference relation of each woman $j \in J$ is specified by a binary relation $\succeq_j$ over $I \cup \{0\}$ that satisfies transitivity and totality. We denote this *stable marriage market* as $(I, J, (\succeq_i)_{i \in I}, (\succeq_j)_{j \in J})$.

A *matching* is a function $\mu: I \to J \cup \{0\}$ such that for any woman $j \in J$, we have $\mu(i) = j$ for at most one man $i \in I$. Given a matching μ and a woman $j \in J$, we denote

$$\mu(j) = \begin{cases} i & \text{if } \mu(i) = j \\ 0 & \text{if there is no man } i \in I \text{ such that } \mu(i) = j \end{cases}$$

A matching μ is *individually rational* if $j \succeq_i 0$ and $i \succeq_j 0$ for every man $i \in I$ and woman $j \in J$ such that $\mu(i) = j$. An individually rational matching μ is *weakly stable* if for any man $i \in I$ and woman $j \in J$, either $\mu(i) \succeq_i j$ or $\mu(j) \succeq_j i$. (Otherwise, such a man i and woman j form a *strongly blocking pair*.)

For any matchings μ and μ', we say that the binary relation $\mu \succeq \mu'$ holds if $\mu(i) \succeq_i \mu'(i)$ and $\mu(j) \succeq_j \mu'(j)$ for every man $i \in I$ and woman $j \in J$. A weakly stable matching μ is *Pareto-stable* if for any matching μ' such that $\mu' \succeq \mu$, we have $\mu \succeq \mu'$. (Otherwise, the matching μ is not *Pareto-optimal* because it is *Pareto-dominated* by the matching μ'.)

A *mechanism* is an algorithm that, given a stable marriage market $(I, J, (\succeq_i)_{i \in I}, (\succeq_j)_{j \in J})$, produces a matching μ. A mechanism is said to be *group strategyproof (for the men)* if for any two different preference profiles $(\succeq_i)_{i \in I}$ and $(\succeq'_i)_{i \in I}$, there exists a man $i_0 \in I$ with preference relation $\succeq_{i_0}$ different from $\succeq'_{i_0}$ such that $\mu(i_0) \succeq_{i_0} \mu'(i_0)$, where μ and μ' are the matchings produced by the mechanism given $(I, J, (\succeq_i)_{i \in I}, (\succeq_j)_{j \in J})$ and $(I, J, (\succeq'_i)_{i \in I}, (\succeq_j)_{j \in J})$ respectively. (Such a man i_0 belongs to the coalition but is not matched to a strictly preferred woman by expressing preference relation $\succeq'_{i_0}$ instead of his true preference relation $\succeq_{i_0}$.)

3.1 The Associated Tiered-Slope Market

We construct the *tiered-slope market* $\mathcal{M} = (I, J, \pi, N, \lambda, a, b)$ *associated with* stable marriage market $(I, J, (\succeq_i)_{i \in I}, (\succeq_j)_{j \in J})$ as follows. We take $N \geq |I| + 1$ and associate with each man $i \in I$ a fixed and distinct priority $\pi_i \in \{1, 2, \ldots, |I|\}$. We convert the preference relations $(\succeq_i)_{i \in I}$ of the men to integer-valued (nontransferable) utilities $a \in \mathbb{Z}^{I \times (J \cup \{0\})}$ such that for every man $i \in I$ and women $j_1, j_2 \in J \cup \{0\}$, we have $j_1 \succeq_i j_2$ if and only if $a_{i,j_1} \geq a_{i,j_2}$. Similarly, we convert the preference relations $(\succeq_j)_{j \in J}$ of the women to integer-valued (nontransferable) utilities $b \in \mathbb{Z}^{(I \cup \{0\}) \times J}$ such that for every woman $j \in J$ and men $i_1, i_2 \in I \cup \{0\}$, we have $i_1 \succeq_j i_2$ if and only if $b_{i_1,j} \geq b_{i_2,j} \geq 1$. Finally, we take

$$\lambda = \max_{\substack{i \in I \cup \{0\} \\ j \in J}} (b_{i,j} + 1)N.$$

In order to achieve group strategyproofness, we require that N and π should not depend on the preferences $(\succeq_i)_{i \in I}$ of the men. We further require that b does not depend on the preferences $(\succeq_i)_{i \in I}$ of the men, and that a_{i_0,j_0} does not depend on the other preferences $(\succeq_i)_{i \in I \setminus \{i_0\}}$ for any man $i_0 \in I$ and woman $j_0 \in J \cup \{0\}$. In other words, a man $i_0 \in I$ is only able to manipulate his own utilities $(a_{i_0,j})_{j \in J \cup \{0\}}$. One way to satisfy these conditions is by taking a_{i_0,j_0} to be the number of women $j \in J \cup \{0\}$ such that $j_0 \succeq_{i_0} j$ for every man $i_0 \in I$ and woman $j_0 \in J \cup \{0\}$, and taking b_{i_0,j_0} to be the number of men $i \in I \cup \{0\}$ such that $i_0 \succeq_{j_0} i$ for every man $i_0 \in I \cup \{0\}$ and woman $j_0 \in J$. (These conditions are not used until Sect. 3.3, where we prove group strategyproofness.)

Intuitively, each woman has a compensation function with the same form as a buyer in the assignment game [22]. The valuation $b_{i,j}N + \pi_i$ that woman j assigns to man i has a first-order dependence on the preferences over the men and a second-order dependence on the priorities of the men, which are used to break any ties in her preferences. From the perspective of man i, if he highly prefers a woman j, he assigns a large exponent $a_{i,j}$ in the slope associated with woman j, and thus expects only a small amount of compensation.

3.2 Pareto-Stability

In this subsection, we study the Pareto-stability of matchings in the stable marriage market that correspond to stable outcomes in the associated tiered-slope market. We first show that individual rationality in the associated tiered-slope market implies individual rationality in the stable marriage market (Lemmas 1 and 2). Then, we show that stability in the associated tiered-slope market implies weak stability in the stable marriage market (Lemma 3). Finally, we show that stability in the associated tiered-slope market is sufficient for Pareto-stability in the stable marriage market (Lemma 4 and Theorem 2). The proof of Lemma 4 is given in [6, App. A].

Lemma 1. *Let (μ, u, v) be an individually rational outcome in tiered-slope market $\mathcal{M} = (I, J, \pi, N, \lambda, a, b)$. Let $i \in I$ be a man and $j \in J$ be a woman. Then*

$$0 < \exp_\lambda(a_{i,0}) \le u_i < \exp_\lambda(a_{i,\mu(i)}+1) \quad and \quad 0 \le b_{0,j}N \le v_j < (b_{\mu(j),j}+1)N.$$

Proof. The lower bounds

$$u_i \ge \pi_i \exp_\lambda(a_{i,0}) \ge \exp_\lambda(a_{i,0}) > 0 \quad and \quad v_j \ge b_{0,j}N \ge 0$$

follow directly from individual rationality. If $\mu(i) = 0$, then feasibility implies $u_i = \pi_i \exp_\lambda(a_{i,0}) < \exp_\lambda(a_{i,0} + 1)$. If $\mu(j) = 0$, then feasibility implies $v_j = b_{0,j}N < (b_{0,j} + 1)N$. It remains to show that the upper bounds hold when $\mu(i) \ne 0$ and $\mu(j) \ne 0$. Without loss of generality, we may assume that $\mu(i) = j$, so feasibility implies

$$u_i \exp_\lambda(-a_{i,j}) - (b_{i,j}N + \pi_i - v_j) \le 0.$$

Since $u_i \ge 0$ and $v_j \ge 0$, we have

$$u_i \exp_\lambda(-a_{i,j}) - (b_{i,j}N + \pi_i) \le 0 \quad and \quad -(b_{i,j}N + \pi_i - v_j) \le 0.$$

Since $\pi_i < N$ and $b_{i,j}N + \pi_i < \lambda$, we have

$$u_i \exp_\lambda(-a_{i,j}) - \lambda < 0 \quad and \quad -(b_{i,j}N + N - v_j) < 0.$$

Thus $u_i < \exp_\lambda(a_{i,\mu(i)} + 1)$ and $v_j < (b_{\mu(j),j} + 1)N$. □

Lemma 2 (Individual Rationality). *Let (μ, u, v) be an individually rational outcome in the tiered-slope market $\mathcal{M} = (I, J, \pi, N, \lambda, a, b)$ associated with stable marriage market $(I, J, (\succeq_i)_{i \in I}, (\succeq_j)_{j \in J})$. Then μ is an individually rational matching in the stable marriage market.*

Proof. Let $i \in I$ be a man and $j \in J$ be a woman. Then, by Lemma 1, we have

$$\exp_\lambda(a_{i,0}) < \exp_\lambda(a_{i,\mu(i)} + 1) \qquad \text{and} \qquad b_{0,j} N < (b_{\mu(j),j} + 1)N.$$

Thus $a_{i,\mu(i)} + 1 > a_{i,0}$ and $b_{\mu(j),j} + 1 > b_{0,j}$, and hence $a_{i,\mu(i)} \geq a_{i,0}$ and $b_{\mu(j),j} \geq b_{0,j}$. We conclude that $\mu(i) \succeq_i 0$ and $\mu(j) \succeq_j 0$. □

Lemma 3 (Stability). *Let (μ, u, v) be a stable outcome in the tiered-slope market $\mathcal{M} = (I, J, \pi, N, \lambda, a, b)$ associated with stable marriage market $(I, J, (\succeq_i)_{i \in I}, (\succeq_j)_{j \in J})$. Then μ is a weakly stable matching in the stable marriage market.*

Proof. Since the outcome (μ, u, v) is individually rational in market $\mathcal{M}$, Lemma 2 implies that the matching μ is individually rational in the stable marriage market. It remains to show that there is no strongly blocking pair.

For the sake of contradiction, suppose there exists a man $i \in I$ and a woman $j \in J$ such that neither $\mu(i) \succeq_i j$ nor $\mu(j) \succeq_j i$. Then $a_{i,j} > a_{i,\mu(i)}$ and $b_{i,j} > b_{\mu(j),j}$. Hence $a_{i,j} \geq a_{i,\mu(i)} + 1$ and $b_{i,j} \geq b_{\mu(j),j} + 1$. Since (μ, u, v) is a stable outcome in $\mathcal{M}$, we have

$$
\begin{aligned}
0 &\leq u_i \exp_\lambda(-a_{i,j}) - (b_{i,j} N + \pi_i - v_j) \\
&< \exp_\lambda(a_{i,\mu(i)} + 1) \exp_\lambda(-a_{i,j}) - (b_{i,j} N + \pi_i - (b_{\mu(j),j} + 1)N) \\
&\leq 1 - \pi_i,
\end{aligned}
$$

where the second inequality follows from Lemma 1. Thus, $\pi_i < 1$, a contradiction. □

Lemma 4. *Let (μ, u, v) be a stable outcome in the tiered-slope market $\mathcal{M} = (I, J, \pi, N, \lambda, a, b)$. Let μ' be an arbitrary matching. Then*

$$\sum_{i \in I} \left(u_i \exp_\lambda(-a_{i,\mu'(i)}) - \pi_i \right) \geq \sum_{j \in J} \left(b_{\mu'(j),j} N - v_j \right).$$

Furthermore, the inequality is tight if and only if the outcome (μ', u, v) is stable.

Theorem 2 (Pareto-stability). *Let (μ, u, v) be a stable outcome in the tiered-slope market $\mathcal{M} = (I, J, \pi, N, \lambda, a, b)$ associated with stable marriage market $(I, J, (\succeq_i)_{i \in I}, (\succeq_j)_{j \in J})$. Then μ is a Pareto-stable matching in the stable marriage market.*

Proof. Since the outcome (μ, u, v) is stable in market $\mathcal{M}$, Lemma 3 implies that the matching μ is weakly stable in the stable marriage market. It remains to show that the matching μ is not Pareto-dominated.

Let μ' be a matching of the stable marriage market such that $\mu' \succeq \mu$. Then $\mu'(i) \succeq_i \mu(i)$ and $\mu'(j) \succeq_j \mu(j)$ for every man $i \in I$ and woman $j \in J$. Hence $a_{i,\mu'(i)} \geq a_{i,\mu(i)}$ and $b_{\mu'(j),j} \geq b_{\mu(j),j}$ for every man $i \in I$ and woman $j \in J$. Since $a_{i,\mu'(i)} \geq a_{i,\mu(i)}$ for every man $i \in I$, we have

$$\sum_{i \in I} \left(u_i \exp_\lambda(-a_{i,\mu'(i)}) - \pi_i \right) \leq \sum_{i \in I} \left(u_i \exp_\lambda(-a_{i,\mu(i)}) - \pi_i \right).$$

Applying Lemma 4 to both sides, we get

$$\sum_{j \in J} \left(b_{\mu'(j),j} N - v_j \right) \leq \sum_{j \in J} \left(b_{\mu(j),j} N - v_j \right).$$

Since $b_{\mu'(j),j} \geq b_{\mu(j),j}$ for every woman $j \in J$, the inequalities are tight. Hence $a_{i,\mu'(i)} = a_{i,\mu(i)}$ and $b_{\mu'(j),j} = b_{\mu(j),j}$ for every man $i \in I$ and woman $j \in J$. Thus $\mu(i) \succeq_i \mu'(i)$ and $\mu'(j) \succeq_j \mu(j)$ for every man $i \in I$ and woman $j \in J$. We conclude that $\mu \succeq \mu'$.   □

3.3   Group Strategyproofness

In this subsection, we study the group strategyproofness of matchings in the stable marriage market that correspond to man-optimal outcomes in the associated tiered-slope market. We first show that the utilities of the men in man-optimal outcomes in the associated tiered-slope market reflect the utilities of the men in the stable marriage market (Lemma 5). Then we prove group strategyproofness in the stable marriage market using group strategyproofness in the associated tiered-slope market (Theorem 3).

Lemma 5. *Let (μ, u, v) be a man-optimal outcome in the tiered-slope market $\mathcal{M} = (I, J, \pi, N, \lambda, a, b)$ associated with stable marriage market $(I, J, (\succeq_i)_{i \in I}, (\succeq_j)_{j \in J})$. Then $\exp_\lambda(a_{i,\mu(i)}) \leq u_i < \exp_\lambda(a_{i,\mu(i)} + 1)$ for every man $i \in I$.*

The proof of Lemma 5 is given in [6, App. A]. Since the compensation received by a man $i \in I$ matched with a woman $\mu(i) \neq 0$ is given by $u_i \exp_\lambda(-a_{i,\mu(i)})$, Lemma 5 implies that the amount of compensation in man-optimal outcomes is at least 1 and at most λ. In fact, no woman is willing to pay more than λ under any individual rational outcome.

Theorem 3 (Group strategyproofness). *If a mechanism produces matchings that correspond to man-optimal outcomes of the tiered-slope markets associated with the stable marriage markets, then it is group strategyproof and Pareto-stable.*

Proof. We have shown Pareto-stability in the stable marriage market in Theorem 2. It remains only to show group strategyproofness.

Let $(I, J, (\succeq_i)_{i \in I}, (\succeq_j)_{j \in J})$ and $(I, J, (\succeq'_i)_{i \in I}, (\succeq_j)_{j \in J})$ be stable marriage markets where $(\succeq_i)_{i \in I}$ and $(\succeq'_i)_{i \in I}$ are different preference profiles. Let $(I, J,$

$\pi, N, \lambda, a, b)$ and $(I, J, \pi, N, \lambda, a', b)$ be the tiered-slope markets associated with stable marriage markets $(I, J, (\succeq_i)_{i \in I}, (\succeq_j)_{j \in J})$ and $(I, J, (\succeq'_i)_{i \in I}, (\succeq_j)_{j \in J})$, respectively. Let (μ, u, v) and (μ', u', v') be man-optimal outcomes of the tiered-slope markets $(I, J, \pi, N, \lambda, a, b)$ and $(I, J, \pi, N, \lambda, a', b)$, respectively.

Since the preference profiles $(\succeq_i)_{i \in I}$ and $(\succeq'_i)_{i \in I}$ are different, we have $a \neq a'$. So, by Theorem 1, there exists a man $i_0 \in I$ and a woman $j_0 \in J \cup \{0\}$ with $a_{i_0, j_0} \neq a'_{i_0, j_0}$ such that $u_{i_0} \geq u'_{i_0} \exp_\lambda(a_{i_0, \mu'(i_0)} - a'_{i_0, \mu'(i_0)})$. Hence

$$\exp_\lambda(a_{i_0, \mu(i_0)} + 1) > u_{i_0}$$

$$\geq \frac{u'_{i_0}}{\exp_\lambda(a'_{i_0, \mu'(i_0)})} \exp_\lambda(a_{i_0, \mu'(i_0)})$$

$$\geq \exp_\lambda(a_{i_0, \mu'(i_0)}),$$

where the first and third inequalities follow from Lemma 5. This shows that $a_{i_0, \mu(i_0)} + 1 > a_{i_0, \mu'(i_0)}$. Hence $a_{i_0, \mu(i_0)} \geq a_{i_0, \mu'(i_0)}$, and we conclude that $\mu(i_0) \succeq_{i_0} \mu'(i_0)$. Also, since $a_{i_0, j_0} \neq a'_{i_0, j_0}$, the preference relations $\succeq_{i_0}$ and $\succeq'_{i_0}$ are different. Therefore, the mechanism is group strategyproof. $\square$

4   Efficient Implementation

The implementation of our group strategyproof Pareto-stable mechanism for stable marriage with indifferences amounts to computing a man-optimal outcome for the associated tiered-slope market. Since all utility functions in the tiered-slope market are linear functions, we can perform this computation using the algorithm of Dütting et al. [9], which was developed for multi-item auctions. If we model each woman j as a non-dummy item in the multi-item auction with price given by utility v_j, then the utility function of each man on each non-dummy item is a linear function of the price with a negative slope. Using the algorithm of Dütting et al., we can compute a man-optimal (envy-free) outcome using $O(n^5)$ arithmetic operations, where n is the total number of agents. Since poly(n) precision is sufficient, our mechanism admits a polynomial-time implementation.

For the purpose of solving the stable marriage problem, it is actually sufficient for a mechanism to produce the matching without the utility vectors u and v of the associated tiered-slope market. In [6, App. B and C], we show that the generalization of the deferred acceptance algorithm presented in [7] can be used to compute a matching that corresponds to a man-optimal outcome for the associated tiered-slope markets. The proof of Theorem 4 is given in [6, App. C.3].

Theorem 4. *There exists an $O(n^4)$-time algorithm that corresponds to a group strategyproof Pareto-stable mechanism for the stable marriage market with indifferences, where n is the total number of men and women.*

References

1. Abdulkadiroğlu, A., Pathak, P.A., Roth, A.E.: Strategy-proofness versus efficiency in matching with indifferences: redesigning the NYC high school match. Am. Econ. Rev. **99**, 1954–1978 (2009)

2. Chen, N.: On computing Pareto stable assignments. In: Proceedings of the 29th International Symposium on Theoretical Aspects of Computer Science, pp. 384–395 (2012)
3. Chen, N., Ghosh, A.: Algorithms for Pareto stable assignment. In: Proceedings of the Third International Workshop on Computational Social Choice, pp. 343–354 (2010)
4. Crawford, V.P., Knoer, E.M.: Job matching with heterogeneous firms and workers. Econometrica **49**, 437–450 (1981)
5. Demange, G., Gale, D.: The strategy structure of two-sided matching markets. Econometrica **53**, 873–888 (1985)
6. Domaniç, N.O., Lam, C.K., Plaxton, C.G.: Group strategyproof Pareto-stable marriage with indifferences via the generalized assignment game, July 2017. https://arxiv.org/abs/1707.01496
7. Domaniç, N.O., Lam, C.K., Plaxton, C.G.: Strategyproof Pareto-stable mechanisms for two-sided matching with indifferences. In: Fourth International Workshop on Matching under Preferences, April 2017. https://arxiv.org/abs/1703.10598
8. Dubins, L.E., Freedman, D.A.: Machiavelli and the Gale-Shapley algorithm. Am. Math. Mon. **88**, 485–494 (1981)
9. Dütting, P., Henzinger, M., Weber, I.: An expressive mechanism for auctions on the web. ACM Trans. Econ. Comput. **4**, 1:1–1:34 (2015)
10. Erdil, A., Ergin, H.: What's the matter with tie-breaking? Improving efficiency in school choice. Am. Econ. Rev. **98**, 669–689 (2008)
11. Erdil, A., Ergin, H.: Two-sided matching with indifferences (2015). working paper
12. Eriksson, K., Karlander, J.: Stable matching in a common generalization of the marriage and assignment models. Discrete Math. **217**, 135–156 (2000)
13. Gale, D., Shapley, L.S.: College admissions and the stability of marriage. Am. Math. Mon. **69**, 9–15 (1962)
14. Irving, R.W.: Stable marriage and indifference. Discrete Appl. Math. **48**, 261–272 (1994)
15. Kamiyama, N.: A new approach to the Pareto stable matching problem. Math. Oper. Res. **39**, 851–862 (2014)
16. Kesten, O.: School choice with consent. Q. J. Econ. **125**, 1297–1348 (2010)
17. Knuth, D.: Marriages Stables. Montreal University Press, Montreal (1976)
18. Quinzii, M.: Core and competitive equilibria with indivisibilities. Int. J. Game Theory **13**, 41–60 (1984)
19. Roth, A.E.: The economics of matching: stability and incentives. Math. Oper. Res. **7**, 617–628 (1982)
20. Roth, A.E.: The college admissions problem is not equivalent to the marriage problem. J. Econ. Theory **36**, 277–288 (1985)
21. Roth, A.E., Sotomayor, M.: Two-sided Matching: A Study in Game-Theoretic Modeling and Analysis. Cambridge University Press, New York (1990)
22. Shapley, L.S., Shubik, M.: The assignment game I: the core. Int. J. Game Theory **1**, 111–130 (1971)
23. Sotomayor, M.: Existence of stable outcomes and the lattice property for a unified matching market. Math. Soc. Sci. **39**, 119–132 (2000)
24. Sotomayor, M.: The Pareto-stability concept is a natural solution concept for discrete matching markets with indifferences. Int. J. Game Theory **40**, 631–644 (2011)
25. Zhou, L.: On a conjecture by gale about one-sided matching problems. J. Econ. Theory **52**, 123–135 (1990)

The Spectrum of Equilibria for the Colonel Blotto and the Colonel Lotto Games

Marcin Dziubiński[✉]

Institute of Informatics, University of Warsaw, Banacha 2, 02-097 Warsaw, Poland
m.dziubinski@mimuw.edu.pl

Abstract. We study Nash equilibria of a symmetric Colonel Blotto and Colonel Lotto games. In these game two players, with $N \geq 1$ units of resources each, distribute their resources simultaneously across $K \geq 2$ battlefields. We introduce a characteristic of equilibria in this game called *spectrum* which stands for the fraction of battlefields receiving given numbers of units. We provide complete characterization of spectra of Nash equilibria in Colonel Lotto game as well as necessary conditions on spectra of Nash equilibria in Colonel Blotto game for the case of $K \mid N$.

1 Introduction

The Colonel Blotto game, introduced by Borel [5], is a prime example of a conflict with multiple battlefields [13], where two parties distribute their limited resources across a number of fronts aiming to beat the opponent at every front. A player wins a battlefield if the number of her resources is strictly larger than the number of resources of the opponent. This is a natural problem that arises in numerous applications, such as military conflicts [4], political competition [14], or network security [7]. Examples of conflicts with multiple battlefields include security games [12,18], duelling algorithms [11], and chopstick auctions [17]. This paper addresses the problem of how to distribute the available resources in a competitive setting. This should be seen as a more complex set up than the one used in conflicts on networks (c.f. [2,6,8]), where the users decide whether they use or do not use a single protection resource (for instance, an antivirus software) to prevent, potentially contagious, attacks on the network.

A continuous variant of the game, where each party can assign any fraction of her total resources to a battlefield, was solved by Gross and Wagner in [9] (the symmetric variant) and by Roberson [16] (asymmetric variant).[1] In addition, a complete characterization of Nash equilibria of the continuous variant on two battlefields was obtained by Macdonell and Mastronardi [15]. The discrete variant is much less understood (for many cases of the asymmetric variant we do not even know the value of the game). The most comprehensive results for this

---

This work was supported by Polish National Science Centre through grant nr 2014/13/B/ST6/01807.
[1] Colonel Blotto game is symmetric when both players have the same amount of resources and it is asymmetric otherwise.

V. Bilò and M. Flammini (Eds.): SAGT 2017, LNCS 10504, pp. 292–306, 2017.
DOI: 10.1007/978-3-319-66700-3_23

variant were obtained by Hart [10]. Hart found Nash equilibria for the symmetric variant of the game, as well as for some cases of the asymmetric variant. To obtain these results, Hart introduces a new game, called Colonel Lotto, where the players choose partitions of their resources only and then one part from each partition is drawn equilprobably and then the chosen parts are matched. He shows that equilibria in this game correspond to equilibria in Colonel Blotto game which are symmetric across battlefields, i.e. are invariant under battlefields relabelling.

This paper considers the symmetric variant of the Colonel Blotto game and the Colonel Lotto game. Strategies are probability distributions over the set of distributions of resources across the battlefields. We introduce a characteristic of such strategies called *spectrum*. It stands for a fraction of battlefields receiving given numbers of units of resources. We provide full characterization of spectra of equilibrium strategies for the case where the number of battlefields divides the number of units of resources. To achieve this goal, in particular, we construct new, previously unknown, equilibrium strategies of the game. The notion of spectrum is more general and can be used to characterize equilibria in other models of conflict with multiple battlefields. Our work builds on [10], but we extend it towards providing full characterization of a non-trivial characteristic of equilibria in Colonel Lotto game and we provide necessary conditions for spectra of all equilibrium strategies in Colonel Blotto game (not only those which are symmetric across battlefields). To our knowledge, this is the first result of this sort for the discrete variant of the game.

It should be noted that although explicit characterization of Nash equilibria in discrete Colonel Blotto is known only partially, the computational side of the problem is understood much better. In recent papers, [1, 3] efficient, polynomial time, algorithms to compute Nash equilibria are proposed for a more general variant of the Colonel Blotto and the Colonel Lotto games, where both the players budgets as well as the battlefields values may be heterogeneous.

To summarize, the contributions of this paper are as follows: (i) we introduce the notion of spectrum of mixed strategies for conflicts with multiple battlefields, (ii) we provide full characterization of Nash equilibria (in terms of spectrum) in symmetric Colonel Lotto game and provide necessary conditions on spectrum of equilibria in symmetric Colonel Blotto game, both for the case where the number of battlefields divides the number of units of resources each player has, (iii) we construct new equilibria for the game and we develop a notation for representing complex equilibria. The first contribution is a general concept which is applicable to other games of conflict with multiple battlefields. The second contribution allows for an efficient verification of whether a given strategy is an equilibrium strategy in the game or not (in the case of Colonel Lotto game) or for partial verification of optimality of given strategies (in the case of Colonel Blotto game). Notice that using the algorithm of Ahmedinejad et al. [1] for that purpose is not straightforward, because, in particular, it only finds a subset of equilibria in the games (all the equilibria it computes have support of polynomial size). It may also be useful when the problem of finding equilibria satisfying additional constraints is considered (e.g. equilibria which provide best expected payoffs against all strategies). The third contribution extends the set of equilibria constructed in [10].

1.1 Spectrum: Informal Introduction

Before defining the game and the concept of spectrum formally, we provide an informal definition and discuss its more general usefulness. A conflict with multiple battlefields requires the players to distribute their limited resources across a number of battlefields. Hence a pure strategy in any such game is a vector that specifies the number of units of resources received by each battlefield. A typical feature of such conflicts is that optimal strategies involve using mixed strategies to 'confuse' the opposition. Characterizing such strategies, especially finding their full characterization, is challenging, because the number of allocations they could randomize over is exponential. One way forward is to fully characterize non-trivial characteristics of such strategies. One such characteristic is spectrum and it can be applied to any conflict with multiple battlefields where players distribute different numbers of resources across a number of battlefields (c.f. [13] for an overview of such games).

Spectrum of a mixed strategy for a conflict with multiple battlefields is a probability distribution on the set of possible numbers of units of resources, $\{0, \ldots, N\}$ that a player can use. Given a number of units M it provides the fraction of battlefields that receive M under the given mixed strategy. For example, given a mixed strategy where one of allocations $[2, 0, 1]$, $[0, 3, 0]$ and $[1, 0, 2]$ is used with probability $1/3$ in a conflict with $K = 3$ battlefields by a player with 3 units, the spectrum is $(\frac{4}{9}, \frac{2}{9}, \frac{2}{9}, \frac{1}{9})$, that is on average, $4/9$ of battlefields receive 0 units, $2/9$ of battlefields receive 1 unit, $2/9$ of battlefields receive 2 units, and $1/9$ of battlefields receives 3 units.

In this paper, we demonstrate usefulness of spectrum by providing full characterization, in terms of it, of equilibria in symmetric Colonel Blotto game where $K \mid N$. Such characterization is useful when the problem of finding equilibria satisfying additional constraints is considered (e.g. equilibria which provide best expected payoffs against all strategies). It can also be used to verify algorithms for computing equilibria for the game. Spectrum could also be useful to obtain some information about equilibria for conflicts with multiple battlefields where finding any equilibria is an open problem, e.g. chopstick auctions [17]. Obtaining spectra of equilibrium strategies could help in finding the actual strategies.

2   Colonel Blotto, Colonel Lotto, and General Lotto Games

In this section we introduce three games that are useful in the analysis of Colonel Blotto game (but are also of interest on their own). These are Colonel Blotto, Colonel Lotto and General Lotto games. Since the focus of this paper are symmetric variants of these games, where both players have the same strength, for the remaining part of the paper we restrict attention to symmetric variants of the games only (and simplify the definitions accordingly).

Given two natural numbers, $N \geq 1$ and $K \geq 2$, the *Colonel Blotto game*, $\mathcal{B}(N, K)$, is defined as follows. There are two players A and B having N units of resources (or army), each, to distribute simultaneously over K battlefields. A pure strategy of a player is an ordered K-partition, $\boldsymbol{x} = (x_1, \ldots, x_K)$, of N, so that $x_1 + \ldots + x_K = N$ and each x_i is a natural number. The set of strategies of each player is

$$S = \left\{ \boldsymbol{x} \in \mathbb{Z}_+^K : \sum_{i=1}^K x_i = N \right\}. \tag{1}$$

A mixed strategy, $\boldsymbol{\xi}$, of $\mathcal{B}(N, K)$ is a probability distribution on S. Let $\Pi(K)$ be the set of all permutations on $\{1, \ldots, K\}$. Any permutation $\pi \in \Pi(k)$ gives rise to a bijection on S which maps any ordered partition $\boldsymbol{x} = (x_1, \ldots, x_K) \in S$ to $\pi(\boldsymbol{x}) = (x_{\pi(1)}, \ldots, x_{\pi(K)})$. A mixed strategy $\boldsymbol{\xi}$ is *symmetric across battlefields* if for any permutation $\pi \in \Pi(K)$, $\boldsymbol{\xi} = \boldsymbol{\xi} \circ \pi$.

After the units are distributed, the payoff of each player is determined as follows. For each battlefield where a player has a strictly larger number of units of resources placed she receives the score 1, while for each battlefield where a player has a strictly smaller number of units of resources placed, he receives the score -1. The score on the tied battlefields is 0 for each player. The overall payoff is the average of payoffs obtained for all battlefields, that is, given the strategies $\boldsymbol{x}$ and $\boldsymbol{y}$ of A and B, respectively, it is

$$h_{\mathcal{B}}(\boldsymbol{x}, \boldsymbol{y}) = \frac{1}{K} \sum_{i=1}^K \operatorname{sign}(x_i - y_i).$$

Colonel Blotto is a zero-sum game.

In this paper we follow the approach of [10]. To study the Colonel Blotto game, [10] proposed a symmetrized-across-battlefields variant of this game called the *Colonel Lotto* game. In this game, denoted by $\mathcal{L}(N, K)$, the battlefields are indistinguishable and players simultaneously divide their units into K groups, which are then randomly paired. A pure strategy of each player A is an unordered K-partition, $\boldsymbol{x} = \langle x_1, \ldots, x_K \rangle$, of N. The payoff of each player is an average over all possible pairings, that is, given the strategies x and y of A and B, respectively, it is

$$h_{\mathcal{L}}(\boldsymbol{x}, \boldsymbol{y}) = \frac{1}{K^2} \sum_{i=1}^K \sum_{j=1}^K \operatorname{sign}(x_i - y_j).$$

To see the connection between the Colonel Blotto and Colonel Lotto games, given a pure strategy $\boldsymbol{x}$ of player A in $\mathcal{B}(N, K)$, let $\sigma(\boldsymbol{x})$ denote a mixed strategy that assigns equal probability, $\frac{1}{K!}$, to each permutation of $\boldsymbol{x}$. Similarly, given a mixed strategy $\boldsymbol{\xi}$ of player A, let $\sigma(\boldsymbol{\xi})$ denote a mixed strategy obtained by replacing each pure strategy $\boldsymbol{x}$ in the support of $\boldsymbol{\xi}$ by $\sigma(\boldsymbol{x})$. The strategies $\sigma(\boldsymbol{x})$ and $\sigma(\boldsymbol{\xi})$ are symmetric across battlefields. As was observed by [10], $h_{\mathcal{B}}(\sigma(\boldsymbol{\xi}), \boldsymbol{y}) = h_{\mathcal{L}}(\boldsymbol{\xi}, \boldsymbol{y})$, for any pure strategy $\boldsymbol{y}$ of player B. Consequently,

$h_{\mathcal{B}}(\sigma(\xi), \eta) = h_{\mathcal{L}}(\xi, \eta)$, for any mixed strategy η of player B. Analogously for the strategies of player B. Hence the following observation can be made

Observation 1 [10]. *The Colonel Blotto game $\mathcal{B}(N, K)$ and the Colonel Lotto game $\mathcal{L}(N, K)$ have the same value. Moreover, the mapping σ maps the optimal strategies in the Colonel Lotto game onto the optimal strategies in the Colonel Blotto game that are symmetric across battlefields.*

Following [10], we define yet another game called *General Lotto*. Given a positive rational number $n \in \mathbb{Q}_{++}$, the General Lotto game $\mathcal{G}(n)$ is defined as follows. There are two players, A and B, who simultaneously choose discrete probability distributions over non-negative natural numbers such that the expectations under the distributions proposed must be n. This restriction could be seen as a budget constraint, restricting the average amount of resources that can be assigned to a single battlefield. The set of strategies of a player

$$T = \left\{ \boldsymbol{p} \in [0, 1]^{\mathbb{Z}_+} : \sum_{i=0}^{+\infty} p_i = 1 \text{ and } \sum_{i=0}^{+\infty} i p_i = n \right\}.$$

Let X and Y be integer valued random variables distributed according to the distributions $\boldsymbol{x}$ and $\boldsymbol{y}$, proposed by A and B, respectively (i.e. $\mathbf{Pr}(X = i) = x_i$ and $\mathbf{Pr}(Y = i) = y_i$, for all $i \in \mathbb{Z}_+$). Then the payoff of player A is

$$H(X, Y) = \mathbf{Pr}(X > Y) - \mathbf{Pr}(X < Y) = \sum_{i=0}^{+\infty} x_i (\mathbf{Pr}(Y < i) - \mathbf{Pr}(Y > i)). \quad (2)$$

General Lotto is a zero sum game.

[10] provided full characterization of equilibria in this game. We start with introducing some useful strategies. Given $m \geq 1$, let U_O^m, U_E^m, and U^m be vectors defined as follows:[2]

$$U_O^m = \underbrace{[0, 1, \ldots, 0, 1, 0]^T}_{2m+1}, \quad U_E^m = \underbrace{[1, 0, \ldots, 1, 0, 1]^T}_{2m+1}, \quad U^m = \underbrace{[1, 1, \ldots, 1, 1, 1]^T}_{2m+1}.$$

Given $m \geq 1$, let $\boldsymbol{u}_O^m$, $\boldsymbol{u}_E^m$, and $\boldsymbol{u}^m$, be stochastic vectors defined as follows:

$$\boldsymbol{u}_O^m = \left(\frac{1}{m}\right) U_O^m, \quad \boldsymbol{u}_E^m = \left(\frac{1}{m+1}\right) U_E^m, \quad \boldsymbol{u}^m = \left(\frac{1}{2m+1}\right) U^m.$$

Notice that probability distribution represented by $\boldsymbol{u}^m$ is a uniform distribution on the set $\{0, \ldots, 2m\}$. Similarly, the probability distribution represented by $\boldsymbol{u}_O^m$ is a uniform distribution on the set of odd numbers from $\{0, \ldots, 2m\}$ and $\boldsymbol{u}_E^m$ is a uniform distribution on the set of even numbers from $\{0, \ldots, 2m\}$. Clearly, $\boldsymbol{u}^m$ is a convex combination of $\boldsymbol{u}_O^m$ and $\boldsymbol{u}_E^m$,

$$\boldsymbol{u}^m = \left(\frac{m}{2m+1}\right) \boldsymbol{u}_O^m + \left(\frac{m+1}{2m+1}\right) \boldsymbol{u}_E^m. \quad (3)$$

---

[2] For convenience reasons, throughout the paper we number the rows of matrices starting for 0 and we number the columns starting from 1.

We will use $\mathrm{conv}(\boldsymbol{u}_O^m, \boldsymbol{u}_E^m)$ to denote the set of all convex combinations of $\boldsymbol{u}_O^m$ and $\boldsymbol{u}_E^m$. The following theorem characterizes equilibrium strategies of General Lotto games.

Theorem 1 [10]. *Let $n > 0$ be an integer. Then the value of General Lotto game, val $\mathcal{G}(n) = 0$. Moreover, a strategy $\boldsymbol{x}$ is an equilibrium strategy of player A or B iff $\boldsymbol{x} \in \mathrm{conv}(\boldsymbol{u}_O^n, \boldsymbol{u}_E^n)$.*

Equilibria of General Lotto game can be used to construct equilibria of Colonel Lotto game and, consequently, symmetric across battlefields equilibria of Colonel Blotto game. To see that, notice that any K-partition $\langle z_1, \ldots, z_K \rangle$ of a natural number C can be seen as a discrete random variable Z with values in the set $\{z_1, \ldots, z_K\}$ and the distribution obtained by assigning to each $z_1, \ldots, z_K$ the probability $\frac{1}{K}$. The expected value of Z is then $\mathbf{Ex}(Z) = \frac{C}{K}$, which is the average number of units per battlefield. This construction links the pure strategies $\boldsymbol{x}$ and $\boldsymbol{y}$ of players A and B in Colonel Lotto game with discrete integer valued random variables X and Y. The strategies of players A and B in Colonel Lotto game could be seen as non-negative, integer valued random variables bounded by N and having expectations N/K. The payoff $h_{\mathcal{L}}(x, y)$ can be then written as

$$h_{\mathcal{L}}(x, y) = H(X, Y) = \mathbf{Pr}(X > Y) - \mathbf{Pr}(X < Y). \tag{4}$$

General Lotto game could be seen as a modification of Colonel Lotto game where, on one hand, the strategies of the players are unbounded random variables, and on the other hand, the units of resources are distributed on the continuum of battlefields. In a strategy $\boldsymbol{x}$ of General Lotto the probability, x_i corresponds to the fraction of battlefields that receive i units and this fraction can be any number in the interval $[0, 1]$. A mixed strategy $\boldsymbol{\xi}$ of Colonel Lotto game could be represented in a similar way, by providing the fraction of battlefields each number of units of resources is assigned to. The fact that there are finitely many battlefields restricts the set of fractions that can appear in such a strategy. Thus every mixed strategy $\boldsymbol{\xi}$ in the Colonel Lotto game $\mathcal{L}(N, K)$ corresponds to a strategy $\boldsymbol{x}$ in the General Lotto game $\mathcal{G}(N/K)$. In this case we say that $\boldsymbol{\xi}$ (N, K)-*implements* $\boldsymbol{x}$. A strategy in $\mathcal{G}(N/K)$ for which an (N, K)-implementing strategy exists is called (N, K)-*feasible*. Every equilibrium strategy in a General Lotto game $\mathcal{G}(N/K)$ which is (N, K)-feasible is an equilibrium strategy in Colonel Lotto game $\mathcal{L}(N, K)$. Using this approach, [10] found equilibrium strategies in Colonel Lotto game. [10] gave necessary and sufficient conditions for (N, K)-feasibility of some of the equilibrium strategies in General Blotto game $\mathcal{G}(N/K)$.

Proposition 1 [10]. *Let $N \geq 1$ and $K \geq 2$ be natural numbers such that $K \,|\, N$ and let $n = N/K$*

1. *$\boldsymbol{u}_O^n$ is (N, K)-feasible if and only if $N \equiv K \pmod 2$.*
2. *$\boldsymbol{u}_E^n$ is (N, K)-feasible if and only if N is even.*

Proof of this proposition is constructive. Hart constructs strategies in Colonel Lotto games which (N, K)-implement the corresponding vectors in the proposition, thus providing equilibrium strategies for Colonel Lotto game and symmetric across battlefields equilibrium strategies for Colonel Blotto game.

3   Analysis

We start with some key definitions. Given a mixed strategy $\boldsymbol{\xi}$ in $\mathcal{B}(N, K)$ and a battlefield $i \in \{1, \dots, K\}$, a *marginal distribution of* $\boldsymbol{\xi}$ *corresponding to* i is the probability distribution $\boldsymbol{\xi}^i$ such that for each number of units of resources $y \in \{0, \dots, N\}$,

$$\xi_y^i = \mathbf{Pr}(i \text{ gets } y \text{ units under } \boldsymbol{\xi}) = \sum_{\boldsymbol{x} \in S} \xi(\boldsymbol{x})[x_i = y], \tag{5}$$

where, given a condition φ, $[\varphi]$ is the Iverson bracket that takes value 1 when φ holds and value 0 otherwise. The concept of spectrum of a mixed strategy in $\mathcal{B}(N, K)$ is defined as follows.

Definition 1 (Spectrum). *Let* $\boldsymbol{\xi}$ *be a mixed strategy in* $\mathcal{B}(N, K)$. *A spectrum of* $\boldsymbol{\xi}$ *is the probability distribution* $\mathsf{spec}(\boldsymbol{\xi})$ *over the set of possible numbers of units of resources such that for each number of units of resources* $y \in \{0, \dots, N\}$,

$$\mathsf{spec}_y(\boldsymbol{\xi}) = \left(\frac{1}{K}\right) \sum_{i=1}^{K} \sum_{\boldsymbol{x} \in S} \xi(\boldsymbol{x})[x_i = y] = \left(\frac{1}{K}\right) \sum_{i=1}^{K} \xi_y^i, \tag{6}$$

i.e. it is the probability that an equiprobably picked battlefield is allocated y *units of resources under* $\boldsymbol{\xi}$.

Marginal distributions and spectrum of mixed strategies in Colonel Lotto game are defined analogously. Notice that in the case of mixed strategies symmetric across battlefields in $\mathcal{B}(N, K)$ as well as mixed strategies in $\mathcal{L}(N, K)$, all marginal distributions are the same and are equal to the spectrum of the distribution. Notice also that spectrum $\mathsf{spec}(\boldsymbol{\xi})$ of a mixed strategy $\boldsymbol{\xi}$ in Colonel Lotto game $\mathcal{L}(N, K)$ is a strategy in $\mathcal{G}(N/K)$ (N, K)-implemented by $\boldsymbol{\xi}$ and hence it is (N, K)-feasible.

In this section we provide a full characterization of spectra of equilibrium strategies in Colonel Blotto game $\mathcal{B}(N, K)$ and of marginal distributions of equilibrium strategies in Colonel Lotto game $\mathcal{L}(N, K)$. We start with the following observation, which implies that the two objects are the same (the proof is moved to the Appendix).

Observation 2. *For any equilibrium strategy* $\boldsymbol{\xi}$ *in* $\mathcal{B}(N, K)$, $\sigma(\boldsymbol{\xi})$ *is an equilibrium strategy in* $\mathcal{B}(N, K)$ *as well. Moreover,* $\mathsf{spec}(\boldsymbol{\xi}) = \mathsf{spec}(\sigma(\boldsymbol{\xi}))$.

It follows from Observation 2 that to characterize spectra of equilibrium strategies in $\mathcal{B}(N, K)$ we can restrict attention to equilibrium strategies which are symmetric across battlefields. Clearly spectrum of such strategies is the same as their marginal distributions. Hence spectra of equilibrium strategies of $\mathcal{B}(N, K)$ are the same as marginal distributions of equilibrium strategies in $\mathcal{L}(N, K)$.

We start with providing full characterization of the equilibrium strategies in the General Lotto game $\mathcal{G}(N/K)$ that are (N, K)-implementable by strategies in Colonel Lotto game $\mathcal{L}(N, K)$, for $K \mid N$. In further analysis we will use the strategies defined by [10] to prove Proposition 1 along with new strategies introduced below. To allow for representation of more complex strategies given in the remaining part of the paper, we develop some auxiliary notation.

A pure strategy in Colonel Lotto game $\mathcal{L}(N, K)$ can be represented as a $1 \times K$ row vector of natural numbers that sum up to N. A mixed strategy $\sum_{i=1}^{r} \lambda_i X_i$, where probability of each partition in the support is a rational number, $\lambda_i = n_i/d_i$, can be represented as a $l \times K$ matrix, where $l = \sum_{i=1}^{r} l_i = \sum_{i=1}^{r} q \operatorname{lcm}(d_1, \ldots, d_r) \frac{n_i}{d_i}$, $q \geq 1$ is a natural number, and each row representing partition X_i is repeated l_i times.[3]

Given a $r \times c$ matrix of natural numbers, $\mathbf{X}$, and a natural number $i \in \mathbb{Z}_+$, let $\#\mathbf{X}(i)$ be the number of appearances of i in $\mathbf{X}$. A vector $\boldsymbol{x}$ such that $x_i = \#\mathbf{X}(i)$ is called the *cardinality vector* of $\mathbf{X}$ and a vector $\boldsymbol{z}$ such that $z_i = \#\mathbf{X}(i)/(rc)$ is called the *spectrum* of $\mathbf{X}$. We use $\mathsf{card}(\mathbf{X})$ and $\mathsf{spec}(\mathbf{X})$ to denote the cardinality vector and the spectrum, respectively, for a matrix $\mathbf{X}$. As we observed above, a mixed strategy (matrix) $\mathbf{X}$ of a player in Colonel Lotto game $\mathcal{L}(N, K)$ gives rise to a strategy $\boldsymbol{x} = \mathsf{spec}(\mathbf{X})$ in General Lotto game $\mathcal{G}(N/K)$. In this case we say that $\mathbf{X}$ (N, K)-*implements* $\boldsymbol{x}$. We use the same terminology with regard to cardinality vectors. Given an optimal strategy $\boldsymbol{x}$ of a player in $\mathcal{G}(N/K)$, any mixed strategy $\mathbf{X}$ that (N, K)-implements $\boldsymbol{x}$ is an optimal strategy for a player in $\mathcal{L}(N, K)$. Hence to find optimal strategies (and values) of Colonel Lotto game $\mathcal{L}(N, K)$ we need to try to (N, K)-implement an optimal strategy from General Lotto game $\mathcal{G}(N/K)$. As it turns out, all (N, K)-feasible corner strategies in General Lotto game can be implemented by mixed strategies in Colonel Lotto game for which all the probabilities are rational numbers. To find such rational (N, K)-implementations we need to solve the following problem: *Given a spectrum $\boldsymbol{x}$ and numbers N and K find a $l \times K$ matrix $\mathbf{X}$ such that $l \geq 1$, every row of $\mathbf{X}$ sums up to N, and $\mathsf{spec}(\mathbf{X}) = \boldsymbol{x}$.* If such a matrix exists for given $\boldsymbol{x}$, N and K, then $\boldsymbol{x}$ is (N, K)-*feasible*.

Matrices representing equilibrium strategies will be constructed from simpler matrices by means of horizontal and vertical composition. Given a $r \times c_1$ matrix $\mathbf{X}$ and a $r \times c_2$ matrix $\mathbf{Y}$, $\mathbf{X}|\mathbf{Y} = [\mathbf{X}|\mathbf{Y}]$ denotes a $r \times (c_1 + c_2)$ matrix being the horizontal composition of matrices $\mathbf{X}$ and $\mathbf{Y}$. Similarly, given a $r_1 \times c$ matrix $\mathbf{X}$ and a $r_2 \times c$ matrix $\mathbf{Y}$, we use

$$\mathbf{X}/\!\!/\mathbf{Y} = \begin{bmatrix} \mathbf{X} \\ \mathbf{Y} \end{bmatrix}$$

to denote a $(r_1 + r_2) \times c$ matrix being the vertical composition of $\mathbf{X}$ and $\mathbf{Y}$. We will also use the standard generalizations of these operators to multiple

---

[3] Given natural numbers $c_1, \ldots, c_r$, $\operatorname{lcm}(c_1, \ldots, c_m)$ is their least common multiple.

arguments, $|_{i=1}^z \mathbf{X}_i$ and $/\!/_{i=1}^z \mathbf{X}_i$, where $\mathbf{X}_i$ are matrices satisfying the required dimensional constraints. Clearly $\mathsf{card}(\mathbf{X}|\mathbf{Y}) = \mathsf{card}(\mathbf{X}) + \mathsf{card}(\mathbf{Y})$ as well as $\mathsf{card}(\mathbf{X}/\!/\mathbf{Y}) = \mathsf{card}(\mathbf{X}) + \mathsf{card}(\mathbf{Y})$. Similarly, $\mathsf{spec}(|_{i=1}^z \mathbf{X}) = \mathsf{spec}(\mathbf{X})$ and $\mathsf{spec}(/\!/_{i=1}^z \mathbf{X}) = \mathsf{spec}(\mathbf{X})$.

Given a natural number $m \geq 1$, let $\mathbf{E}(m)$ be $(m+1) \times 2$ matrix defined as follows:

$$\mathbf{E}(m) = \begin{bmatrix} 0, & 2m \\ 2, & 2m-2 \\ \vdots & \vdots \\ 2m, & 0 \end{bmatrix},$$

that is the i'th row, $e_i = \begin{bmatrix} 2i, & 2(m-i) \end{bmatrix}$. Each row of matrix $\mathbf{E}(m)$ represents a bipartition, a strategy in Colonel Lotto game with $K = 2$ battlefields and $2m$ units. The whole matrix represents a mixed strategy, where each partition (row) is chosen with probability $1/(m+1)$.

Given an even natural number $m \geq 2$, let $\mathbf{RE}(m) = \mathbf{RE}^{\mathrm{I}}(m)/\!/\mathbf{RE}^{\mathrm{II}}(m)$ be a $(m+1) \times 3$ matrix consisting of two blocks, $\mathbf{RE}^{\mathrm{I}}(m)$ and $\mathbf{RE}^{\mathrm{II}}(m)$, where

$$\mathbf{RE}^{\mathrm{I}}(m) = \begin{bmatrix} 0, & m, & 2m \\ 2, & m+2, & 2m-4 \\ \vdots & \vdots & \vdots \\ m, & 2m, & 0 \end{bmatrix}, \quad \mathbf{RE}^{\mathrm{II}}(m) = \begin{bmatrix} 0, & m+2, & 2m-2 \\ 2, & m+4, & 2m-6 \\ \vdots & \vdots & \vdots \\ m-2, & 2m, & 2 \end{bmatrix},$$

that is the i'th row, $re_i^{\mathrm{I}} = \begin{bmatrix} 2i, & m+2i, & 2m-4i \end{bmatrix}$, and the i'th row $re_i^{\mathrm{II}} = \begin{bmatrix} 2i, & m+2i+2, & 2m-4i-2 \end{bmatrix}$.

Similarly, given $m \geq 2$, let $\mathbf{O}(m)$ be a $m \times 2$ matrix defined as $\mathbf{O}(m) = \mathbf{E}(m-1) + 1$, that is row $o_i = \begin{bmatrix} 2i+1, & 2(m-i-1)+1 \end{bmatrix}$, and, given an odd $m \geq 3$, let $\mathbf{RO}(m) = \mathbf{RE}(m-1) + 1$.

Notice that $\mathsf{card}(\mathbf{E}(m)) = 2U_{\mathrm{E}}^m$ and $\mathsf{card}\left(|_{i=1}^l \mathbf{E}(m)\right) = 2lU_{\mathrm{E}}^m$, for all $l \in \mathbb{Z}_+$. Clearly $\mathsf{spec}\left(|_{i=1}^l \mathbf{E}(m)\right) = \boldsymbol{u}_{\mathrm{E}}^m$, for all $l \in \mathbb{Z}_+$. Thus $|_{i=1}^{\frac{K}{2}} \mathbf{E}(m)$ (mK, K)-implements $\boldsymbol{u}_{\mathrm{E}}^m$ in the case of K being even. It can be easily checked that $\mathsf{card}(\mathbf{RE}(m)) = 3U_{\mathrm{E}}^m$ and $\mathsf{spec}(\mathbf{RE}(m)) = \boldsymbol{u}_{\mathrm{E}}^m$. Hence $\mathsf{spec}\left(\mathbf{RE}(m)|\left(|_{i=1}^l \mathbf{E}(m)\right)\right) = \boldsymbol{u}_{\mathrm{E}}^m$, for all $l \in \mathbb{Z}_+$. Thus $\left(\mathbf{RE}(m)|\left(|_{i=1}^{\frac{K-3}{2}} \mathbf{E}(m)\right)\right)$ (mK, K)-implements $\boldsymbol{u}_{\mathrm{E}}^m$ in the case of K being odd. The cardinality vector for $\mathbf{O}(m)$ is the cardinality vector for $\mathbf{E}(m-1)$, $2U_{\mathrm{E}}^{m-1}$, 'shifted up' by one place, which is exactly $2U_{\mathrm{O}}^m$. Similarly, $\mathsf{card}(\mathbf{RO}(m)) = 3U_{\mathrm{O}}^m$. Thus $|_{i=1}^{\frac{K}{2}} \mathbf{O}(m)$ (mK, K)-implements $\boldsymbol{u}_{\mathrm{O}}^m$ in the case of K being even and $\left(\mathbf{RO}(m)|\left(|_{i=1}^{\frac{K-3}{2}} \mathbf{O}(m)\right)\right)$ (mK, K)-implements $\boldsymbol{u}_{\mathrm{O}}^m$ in the case of K being odd.

Proposition 1 implies that in the case of $K \mid N$ and K being even both U_{O}^n and U_{E}^n, where $n = N/K$, are (N, K)-feasible. Thus any convex combination of these vectors is (N, K)-feasible as well, as it is a spectrum of convex combinations of strategies implementing $\boldsymbol{u}_{\mathrm{O}}^n$ and $\boldsymbol{u}_{\mathrm{E}}^n$. Proposition below provides the complete characterization of (N, K)-feasible General Lotto equilibrium strategies in the case of $K \mid N$ and K being odd.[4]

Proposition 2. *Let $N \geq 1$ and $K \geq 2$ be natural numbers such that $K \mid N$ and let $n = N/K$.*

1. *If K is odd and N is odd, then $\lambda \boldsymbol{u}_O^n + (1 - \lambda)\boldsymbol{u}_E^n$ is (N, K)-feasible if and only if $\lambda \in \left[\frac{1}{K}, 1\right]$.*
2. *If K is odd and N is even, then $(1 - \lambda)\boldsymbol{u}_O^n + \lambda \boldsymbol{u}_E^n$ is (N, K)-feasible if and only if $\lambda \in \left[\frac{1}{K}, 1\right]$.*

Proof. For point 1, suppose that K is odd and N is odd (in this case n is odd as well). For the left to right implication notice that any K-partition of N must contain at least one odd number. This, together with the fact that $\boldsymbol{u}_E^n$ puts positive probabilities on even numbers only implies that $\lambda \in [1/K, 1]$. For the right to left notice that, by point 1 of Proposition 1, $\boldsymbol{u}_O^n$ is (N, K)-feasible. Thus it is enough to show that $\boldsymbol{z} = \frac{1}{K}\boldsymbol{u}_O^n + \frac{K-1}{K}\boldsymbol{u}_E^n$ is (N, K)-feasible. Vector $\boldsymbol{z}$ can be rewritten as $\boldsymbol{z} = \boldsymbol{r} + \left(\frac{K-3}{K}\right)\boldsymbol{u}_E^n$, where

$$\boldsymbol{r} = \left(\frac{1}{n(n+1)K}\right)((n+1)U_O^n + 2nU_E^n). \tag{7}$$

Let $\mathbf{R}(n)$ be a $n(n + 1) \times 3$ matrix consisting of four blocks, $\mathbf{R}(n) = \mathbf{R}^{\mathrm{I}}(n) /\!\!/ \mathbf{R}^{\mathrm{II}}(n) /\!\!/ \mathbf{R}^{\mathrm{III}}(n) /\!\!/ \mathbf{R}^{\mathrm{IV}}(n)$, defined below.

$$\mathbf{R}^{\mathrm{I}}(n) = \begin{bmatrix} \{ & 0, & 2n-1, n+1 & \}^{n-1} \\ \{ & 2, & 2n-5, n+3 & \}^{n-1} \\ \vdots & \vdots & \vdots & \\ \{ & n-1, & 1, & 2n & \}^{n-1} \end{bmatrix}, \quad \mathbf{R}^{\mathrm{II}}(n) = \begin{bmatrix} \{ & 2, & 2n-3, n+1 & \}^{n-1} \\ \{ & 4, & 2n-7, n+3 & \}^{n-1} \\ \vdots & \vdots & \vdots & \\ \{ & n-1, & 3, & 2n-2 & \}^{n-1} \end{bmatrix},$$

$$\mathbf{R}^{\mathrm{III}}(n) = \begin{bmatrix} \{ & 2, & n-2, 2n & \}^{2} \\ \{ & 4, & n-4, 2n & \}^{2} \\ \vdots & \vdots & \vdots & \\ \{ & n-1, & 1, & 2n & \}^{2} \end{bmatrix}, \quad \mathbf{R}^{\mathrm{IV}}(n) = \begin{bmatrix} \{ & 0, 2n-1, n+1 & \}^{2} \\ \{ & 0, 2n-3, n+3 & \}^{2} \\ \vdots & \vdots & \vdots & \\ \{ & 0, & n, & 2n & \}^{2} \end{bmatrix},$$

(in the definitions, the notation $\left[\{ a_1, \ldots, a_l \}^t\right]$ means that row $\left[a_1, \ldots, a_l\right]$ is repeated t times). $\mathbf{R}^{\mathrm{I}}(n)$ is a $\frac{(n+1)(n-1)}{2} \times 3$ block with the i'th part of the form $\left[2i, 2n - 4i - 1, n + 2i + 1\right]$. $\mathbf{R}^{\mathrm{II}}(n)$ is a $\frac{(n-1)^2}{2} \times 3$ block with the i'th part of the form $\left[2i + 2, 2n - 4i - 3, n + 2i + 1\right]$. $\mathbf{R}^{\mathrm{III}}(n)$ is a $(n - 1) \times 3$ block with the i'th part of the form $\left[2i + 2, n - 2i - 2, 2n\right]$. $\mathbf{R}^{\mathrm{IV}}(n)$ is a $(n + 1) \times 3$ block with the i'th part of the form $\left[0, 2n - 2i - 1, n + 2i + 1\right]$.

Each row of $\mathbf{R}(n)$ sums up to $3n$. Moreover, it is easy to verify that $\mathrm{card}(\mathbf{R}(n)) = Z(n)$ where $Z(n) = (n + 1)U_O^n + 2nU_E^n = [2n, n + 1, 2n, \ldots, n + 1, 2n]$, that is $z_{2i} = 2n$ and $z_{2i+1} = n+1$. Let $\mathbf{T}(n) = \mathbf{R}(n) | \left(|_{i=1}^{(K-3)/2} /\!\!/_{j=1}^{n} \mathbf{E}(n)\right)$.

---

[4] If $K \mid N$ and K is even, then N must be even as well, so K being odd is the only remaining case.

$\mathbf{T}(n)$ is a $n(n+1) \times K$ matrix with each row summing up to nK. By observations above, $\mathsf{card}(\mathbf{T}(n)) = (n+1)U_O^n + (K-1)nU_E^n$ and $\mathsf{spec}(\mathbf{T}(n)) = \boldsymbol{z}$. This shows that $\mathbf{T}(n)$ (N, K)-implements $\boldsymbol{z}$ and so $\boldsymbol{z}$ is (N, K)-feasible.

For point 2, suppose that K is odd and N is even (in this case n is even). For the left to right implication notice that any K-partition of N must contain at least one even number. This, together with the fact that $\boldsymbol{u}_O^n$ puts positive frequencies on odd numbers only implies that $\lambda \in [1/K, 1]$. For the right to left notice that, by point 1 or Proposition 1, $\boldsymbol{u}_E^n$ is (N, K)-feasible. Thus it is enough to show that $\boldsymbol{z} = \frac{1}{K}\boldsymbol{u}_E^n + \frac{K-1}{K}\boldsymbol{u}_O^n$ is (N, K)-feasible. Vector $\boldsymbol{z}$ can be rewritten as $\boldsymbol{z} = \boldsymbol{s} + \left(\frac{K-3}{K}\right)\boldsymbol{u}_O^n$, where

$$\boldsymbol{s} = \left(\frac{1}{n(n+1)K}\right)\left(nU_E^n + 2(n+1)U_O^n\right). \tag{8}$$

Let $\mathbf{S}(n)$ be a $n(n+1) \times 3$ matrix consisting of three blocks, $\mathbf{S}(n) = \mathbf{S}^I(n)/\!/\mathbf{S}^{II}(n)/\!/\mathbf{S}^{III}(n)$, defined below.

$$\mathbf{S}^{II}(n) = \begin{bmatrix} \{ & 1, & n-1, 2n & \}^2 \\ \{ & 3, & n-3, 2n & \}^2 \\ & \vdots & \vdots & \vdots \\ \{ n-1, & 1, & 2n & \}^2 \end{bmatrix}, \quad \mathbf{S}^{III}(n) = \begin{bmatrix} \{ & 0, 2n-1, & n+1 & \}^2 \\ \{ & 0, 2n-1, & n+3 & \}^2 \\ & \vdots & \vdots & \vdots \\ \{ & 0, & n+1, 2n-1 & \}^2 \end{bmatrix},$$

A $n(n-1) \times 3$ block $\mathbf{S}^I(n) = \mathbf{R}(n-1) + 1$. A $n \times 3$ block $\mathbf{S}^{II}(n)$ with the i'th part of the form $[2i+1, n-2i-1, 2n]$. A $n \times 3$ block $\mathbf{S}^{III}(n)$ with i'th part of the form $[0, 2n-2i-1, n+2i+1]$.

Each row of $\mathbf{S}(n)$ sums up to $3n$. Moreover, it is easy to verify that $\mathsf{card}(\mathbf{S}^{II}(n)/\!/\mathbf{S}^{III}(n)) = Y(n)$ where $Y(n) = [n, 4, 0, \ldots, 4, n]$, that is $y_0 = y_{2n} = n$, $z_{2i} = 0$ and $z_{2i-1} = 4$. The cardinality vector for $\mathbf{S}^I(n)$ is the cardinality vector for $\mathbf{R}(n-1)$, $Z(n)$, shifted up by one position. Adding it to $Y(n)$ we obtain the vector $W(n) = 2(n+1)U_O^n + nU_E^n = [n, 2(n+1), n, \ldots, 2(n+1), n]$, that is $w_{2i} = n$ and $w_{2i+1} = 2(n+1)$. Hence $\mathsf{card}(\mathbf{S}(n)) = \mathsf{card}(\mathbf{S}^I(n)) + \mathsf{card}(\mathbf{S}^{II}(n)/\!/\mathbf{S}^{III}(n)) = W(n)$.

Let $\mathbf{T}(n) = \mathbf{S}(n)| \left(|_{i=1}^{(K-3)/2}/\!/_{j=1}^{n+1}\mathbf{O}(n)\right)$. $\mathbf{T}(n)$ is a $n(n+1) \times K$ matrix with each row summing up to nK. By observations above, $\mathsf{card}(\mathbf{T}(n)) = nU_E^n + (K-1)(n+1)U_O^n$ and $\mathsf{spec}(\mathbf{T}(n)) = \boldsymbol{z}$. This shows that $\mathbf{T}(n)$ (N, K)-implements $\boldsymbol{z}$ and so $\boldsymbol{z}$ is (N, K)-feasible. $\qquad\square$

As a corollary of Propositions 1 and 2, we have that the strategy $\boldsymbol{u}^n$ is (N, K)-feasible if $K \mid N$ and $n = N/K$ (the proof is moved to the Appendix).

Corollary 1. *Let $N \geq 1$ and $K \geq 2$ be natural numbers such that $K \mid N$ and let $n = N/K$. Strategy $\boldsymbol{u}^n$ is (N, K)-feasible.*

Propositions 1 and 2 provide a full characterization of equilibrium strategies in the General Lotto game, $\mathcal{G}(N/K)$, that are implementable by equilibrium strategies in Colonel Lotto game $\mathcal{L}(N, K)$. It turns out that every equilibrium strategy of Colonel Lotto game $\mathcal{L}(N, K)$ (N, K)-implements an equilibrium strategy of General Lotto game $\mathcal{G}(N/K)$. Hence we have the following theorem, providing full characterization of equilibrium marginal distributions in Colonel Lotto game.

Theorem 2. *Let $N \geq 1$, $K \geq 3$, and $K \mid N$, and let $n = N/K$. A (mixed) strategy $\boldsymbol{\xi}$ of $\mathcal{L}(N, K)$ is an equilibrium strategy in $\mathcal{L}(N, K)$ if and only if for all $i \in \{1, \ldots, K\}$, $\boldsymbol{\xi}^i = \hat{\boldsymbol{\xi}}$ where*

1. $\hat{\boldsymbol{\xi}} \in \text{conv}(\boldsymbol{u}_O^n, \boldsymbol{u}_E^n)$ *(in the case of K even).*
2. $\hat{\boldsymbol{\xi}} \in \{\lambda \boldsymbol{u}_O^n + (1 - \lambda)\boldsymbol{u}_E^n : \lambda \in [1/K, 1]\}$ *(in the case of K odd and N odd).*
3. $\hat{\boldsymbol{\xi}} \in \{(1 - \lambda)\boldsymbol{u}_O^n + \lambda \boldsymbol{u}_E^n : \lambda \in [1/K, 1]\}$ *(in the case of K odd and N even).*

Proof. Fix $N \geq 1$ and $K \geq 3$ such that $K \mid N$. It is enough to show that every equilibrium strategy in Colonel Lotto game $\mathcal{L}(N, K)$ (N, K)-implements an equilibrium strategy of General Lotto game $\mathcal{G}(N/K)$. Then, by Proposition 2, the result follows.

Take any equilibrium strategy $\boldsymbol{\xi}$ in $\mathcal{L}(N, K)$ and let $\boldsymbol{X}$ be the strategy in $\mathcal{G}(N/K)$ that $\boldsymbol{\xi}$ (N, K)-implements, so that $\text{spec}(\boldsymbol{\xi}) = \boldsymbol{X}$. It is enough to show that $\boldsymbol{X} \in \text{conv}(\boldsymbol{u}_O^n, \boldsymbol{u}_E^n)$. Let $p_i = \mathbf{Pr}(\boldsymbol{X} = i)$. By Eq. (2) and by $\mathbf{Ex}(\boldsymbol{X}) = \sum_{i=1}^{+\infty} \mathbf{Pr}(\boldsymbol{X} \geq i)$,

$$H(\boldsymbol{X}, \boldsymbol{u}^n) = -\frac{\mathbf{Pr}(\boldsymbol{X} \geq 2n + 1)}{2n + 1} - \left(\frac{2}{2n + 1}\right) \sum_{i=2n+2}^{+\infty} \mathbf{Pr}(\boldsymbol{X} \geq i). \qquad (9)$$

Since $\boldsymbol{\xi}$ is an equilibrium strategy and, by Corollary 1, $\boldsymbol{u}^n$ is (N, K)-feasible so $h_{\mathcal{L}}(\boldsymbol{\xi}, \boldsymbol{v}) = 0$, where $\boldsymbol{v}$ is a strategy in $\mathcal{L}(N, K)$ that (N, K)-implements $\boldsymbol{u}^n$. Hence, by Eq. (4), $H(\boldsymbol{X}, \boldsymbol{u}^n) = 0$. Thus it must be that $\mathbf{Pr}(\boldsymbol{X} \geq 2n + 1) = 0$ and so $p_i = 0$ for all $i \geq 2n + 1$. This implies that $\boldsymbol{X}$ can be represented by a $(2n + 1)$-dimensional stochastic vector with mean n. Notice that in the case of $n = N/K = 1$, any $(2n+1)$-dimensional stochastic vector with mean n is a convex combination of $\boldsymbol{u}_O^1 = [0, 1, 0]$ and $\boldsymbol{u}_E^1 = [1/2, 0, 1/2]$. Hence $\boldsymbol{X} \in \text{conv}(\boldsymbol{u}_O^1, \boldsymbol{u}_E^1)$.

Assume that $n \geq 2$. We will construct (N, K)-feasible strategies such that any strategy yielding payoff greater or equal to 0 against them in $\mathcal{G}(N/K)$ must be a convex combination of $\boldsymbol{u}_O^n$ and $\boldsymbol{u}_E^n$. Let

$$\boldsymbol{S}^i = \left(\frac{1}{2K}\right)((K - 1)\mathbf{1}_{n-i} + 2\mathbf{1}_n + (K - 1)\mathbf{1}_{n+i}), \qquad i \in \{0, \ldots, n\},$$

$$\boldsymbol{L}^i = \left(\frac{1}{2K}\right)((K - 3)\mathbf{1}_{n-i} + 2\mathbf{1}_{n-i+1} + 2\mathbf{1}_{n-1} + (K - 1)\mathbf{1}_{n+i}), \quad i \in \{2, \ldots, n\},$$

$$\boldsymbol{R}^i = \left(\frac{1}{2K}\right)((K - 1)\mathbf{1}_{n-i} + 2\mathbf{1}_{n+1} + 2\mathbf{1}_{n+i-1} + (K - 3)\mathbf{1}_{n+i}), \quad i \in \{2, \ldots, n\}.$$

be strategies in $\mathcal{G}(N/K)$ where, given $j \in \mathbb{Z}_+$, $\mathbf{1}_j$ is the degenerate probability distribution on $\mathbb{Z}_+$ assigning probability 1 to j. Let $\mathcal{S}(K, n) = \{\boldsymbol{S}^i\}_{i=0}^n \cup \{\boldsymbol{L}^i, \boldsymbol{R}^i\}_{i=2}^n$. The following lemma is needed (the proof is moved to the Appendix).

Lemma 1. *For all integer $K \geq 3$ and $n \geq 2$, and any $\boldsymbol{Y} \in \mathcal{S}(K, n)$*

1. *$\boldsymbol{Y}$ is (nK, K)-feasible, and*
2. *There exists $\lambda \in (0, 1)$ and a (nK, K)-feasible strategy $\boldsymbol{T}$ in $\mathcal{G}(n)$ such that $\boldsymbol{u}^n = \lambda \boldsymbol{Y} + (1 - \lambda)\boldsymbol{T}$.*

By Corollary 1, $\boldsymbol{u}^n$ is (N, K)-feasible and, by Lemma 1, $\boldsymbol{S}^i$ is (N, K)-feasible (for any $i \in \{0, \ldots, n\}$), and there exists $\lambda > 0$ and (N, K)-feasible $\boldsymbol{T}$ such that that $\boldsymbol{u}^n = \lambda \boldsymbol{S}^i + (1 - \lambda)\boldsymbol{T}$. Let $\boldsymbol{v}$ be a strategy that (N, K)-implements $\boldsymbol{u}^n$ and $\boldsymbol{\theta}$ be a strategy that (N, K)-implements $\boldsymbol{T}$. Since $\boldsymbol{\xi}$ is an equilibrium strategy in $\mathcal{L}(N, K)$ and the value of the game is 0 so is must be that $h_{\mathcal{L}}(\boldsymbol{\xi}, \boldsymbol{s}^i) \geq 0$, $h_{\mathcal{L}}(\boldsymbol{\xi}, \boldsymbol{\theta}) \geq 0$, and $h_{\mathcal{L}}(\boldsymbol{\xi}, \boldsymbol{v}) \geq 0$. On the other hand, $\boldsymbol{v}$ is an equilibrium strategy, so $h_{\mathcal{L}}(\boldsymbol{\xi}, \boldsymbol{v}) = 0$. Hence $h_{\mathcal{L}}(\boldsymbol{\xi}, \boldsymbol{s}^i) = 0$ and, by Eq. (4), $H(\boldsymbol{X}, \boldsymbol{S}^i) = 0$. Similarly, $H(\boldsymbol{X}, \boldsymbol{L}^i) = 0$ and $H(\boldsymbol{X}, \boldsymbol{R}^i) = 0$, for all $i = \{2, \ldots, n\}$. Let $w_i = H(\boldsymbol{X}, \boldsymbol{1}_i)$. Since $H(\boldsymbol{X}, \boldsymbol{S}^0) = 0$ and $H(\boldsymbol{X}, \boldsymbol{S}^1) = 0$ so

$$w_n = 0 \quad \text{and} \quad w_{n-1} = -w_{n+1}. \tag{10}$$

Since $H(\boldsymbol{X}, \boldsymbol{S}^i) = H(\boldsymbol{X}, \boldsymbol{L}^i)$ so

$$w_{n-i} + w_n = w_{n-i+1} + w_{n-1}, \quad \text{for } i \in \{2, \ldots, n\}. \tag{11}$$

Similarly, since $H(\boldsymbol{X}, \boldsymbol{S}^i) = H(\boldsymbol{X}, \boldsymbol{R}^i)$ so

$$w_n + w_{n+i} = w_{n+1} + w_{n+i-1}, \quad \text{for } i \in \{2, \ldots, n\}. \tag{12}$$

By Eqs. (10) and (12),

$$w_{i+1} - w_i = w_{n+1}, \quad \text{for } i \in \{0, \ldots, 2n - 1\}. \tag{13}$$

Since $w_{i+1} - w_i = -(p_i + p_{i+1})$ so, by (13), $p_i + p_{i+1} = p_{i+1} + p_{i+2}$ and, consequently, $p_i = p_{i+2}$, for all $i \in \{0, \ldots, 2n - 2\}$. Hence $\boldsymbol{X} \in \text{conv}(\boldsymbol{u}_O^n, \boldsymbol{u}_E^n)$. □

An immediate corollary from Theorem 2 and Observation 2 is the following result providing a necessary property of equilibrium strategies in Colonel Blotto game in terms of their spectra.

Corollary 2. *Let $N \geq 1$, $K \geq 3$, and $K \mid N$, and let $n = N/K$. If a (mixed) strategy $\boldsymbol{\xi}$ of $\mathcal{B}(N, K)$ is an equilibrium strategy of $\mathcal{B}(N, K)$ then*

1. $\text{spec}(\boldsymbol{\xi}) \in \text{conv}(\boldsymbol{u}_O^n, \boldsymbol{u}_E^n)$ *(in the case of K even).*
2. $\text{spec}(\boldsymbol{\xi}) \in \{\lambda \boldsymbol{u}_O^n + (1 - \lambda)\boldsymbol{u}_E^n : \lambda \in [1/K, 1]\}$ *(in the case of K odd and N odd).*
3. $\text{spec}(\boldsymbol{\xi}) \in \{(1 - \lambda)\boldsymbol{u}_O^n + \lambda \boldsymbol{u}_E^n : \lambda \in [1/K, 1]\}$ *(in the case of K odd and N even).*

In particular, it follows from Corollary 2 that a probability that an equiprobably picked battlefield receives a given number of units of resources must the same for all even numbers in $\{0, \ldots, 2N/K\}$ and must be the same for all odd numbers in $\{0, \ldots, 2N/K\}$. Moreover, the maximal number of units of resources assigned with positive probability to a battlefield by an equilibrium strategy is $2N/K$. Given a strategy, checking whether this criteria is satisfied can be done in linear time (with respect to the size of the strategy). Moreover, in the case of Colonel Lotto game, the criteria is both sufficient and necessary. Hence, in this case, we have a simple method for deciding whether a given strategy is an equilibrium strategy in the game. Notice that such a method is not provided by the algorithm of Ahmedinejad et al. [1], because the algorithm finds only some equilibria of the game (in particular, their support must be polynomial). Hence to use the algorithm for this purpose, we would have to find the equilibrium strategies, use them to compute the value of the game, and then check whether the given strategy guarantees the payoff not less than the value against all the strategies of the opponent.

4 Conclusions

This paper studied equilibrium strategies of (symmetric and discrete) Colonel Blotto game. We introduced the notion of spectrum of strategies in this game and provided complete characterization of spectra of equilibria for the cases where the number of battlefields divides the number of units of resources a player has. We showed that the spectra are suitable convex combinations of two extreme distributions: one, that mixes uniformly on odd numbers of units of resources and another one, that mixes uniformly on even numbers.

References

1. Ahmadinejad, A., Dehghani, S., Hajiaghayi, M., Lucier, B., Mahini, H., Seddighin, S.: From duels to battlefields: computing equilibria of Blotto and other games. In: Schuurmans, D., Wellman, M.P. (eds.) Proceedings of the Thirtieth AAAI Conference on Artificial Intelligence, 12–17 February 2016, Phoenix, Arizona, USA, pp. 376–382. AAAI Press (2016)
2. Aspnes, J., Chang, K., Yampolskiy, A.: Inoculation strategies for victims of viruses and the sum-of-squares partition problem. J. Comput. Syst. Sci. **72**(6), 1077–1093 (2006)
3. Behnezhad, S., Dehghani, S., Derakhshan, M., HajiAghayi, M., Seddighin, S.: Faster and simpler algorithm for optimal strategies of blotto game. In: Singh, S.P., Markovitch, S. (eds.) Proceedings of the Thirty First AAAI Conference on Artificial Intelligence, 4–9 February 2017, San Francisco, California, USA, pp. 369–375. AAAI Press (2017)
4. Blackett, D.: Some blotto games. Naval Res. Logist. Q. **1**(1), 55–60 (1954)
5. Borel, É.: La Théorie du Jeu et les Équations Intégrales à Noyau Symétrique. Comptes Rendus de l'Académie des Sci. **173**, 1304–1308 (1921). Translated by Savage, L.J.: The theory of play and integral equations with skew symmetric kernels. Econometrica **21** 97–100 (1953)
6. Cerdeiro, D., Dziubiński, M., Goyal, S.: Individual security and network design. In: Proceedings of the Fifteenth ACM Conference on Economics and Computation, EC 2014, pp. 205–206. ACM, New York (2014)

7. Chia, P.H., Chuang, J.: Colonel Blotto in the phishing war. In: Baras, J.S., Katz, J., Altman, E. (eds.) GameSec 2011. LNCS, vol. 7037, pp. 201–218. Springer, Heidelberg (2011). doi:10.1007/978-3-642-25280-8_16
8. Dziubiński, M., Goyal, S.: How do you defend a network? Theor. Econ. **12**(1), 331–376 (2017)
9. Gross, O., Wagner, A.: A continuous Colonel Blotto game. RM-408, RAND Corporation, Santa Monica, CA (1950)
10. Hart, S.: Discrete Colonel Blotto and General Lotto games. Int. J. Game Theory **36**, 441–460 (2008)
11. Immorlica, N., Kalai, A., Lucier, B., Moitra, A., Postlewaite, A., Tennenholtz, M.: Dueling algorithms. In: Proceedings of the Forty-Third Annual ACM Symposium on Theory of Computing, STOC 2011, pp. 215–224. ACM, New York (2011)
12. Kiekintveld, C., Jain, M., Tsai, J., Pita, J., Ordóñez, F., Tambe, M.: Computing optimal randomized resource allocations for massive security games. In: Proceedings of the 8th International Conference on Autonomous Agents and Multiagent Systems, AAMAS 2009, vol. 1, pp. 689–696. International Foundation for Autonomous Agents and Multiagent Systems, Richland, SC (2009)
13. Kovenock, D., Roberson, B.: Conflicts with multiple battlefields. In: Garfinkel, M., Skaperdas, S. (eds.) Oxford Handbook of the Economics of Peace and Conflict. Oxford University Press, Oxford (2012)
14. Laslier, J.F., Picard, N.: Distributive politics and electoral competition. J. Econ. Theory **103**(1), 106–130 (2002)
15. Macdonell, S., Mastronardi, N.: Waging simple wars: a complete characterization of two-battlefield blotto equilibria. Econ. Theor. **58**(1), 183–216 (2015)
16. Roberson, B.: The Colonel Blotto game. Econ. Theor. **29**(1), 1–24 (2006)
17. Szentes, B., Rosenthal, R.: Three-object two-bidder simultaneous auctions: chopsticks and tetrahedra. Games Econ. Behav. **44**(1), 114–133 (2003)
18. Tambe, M.: Security and Game Theory: Algorithms, Deployed Systems, Lessons Learned. Cambridge University Press, Cambridge (2011)

On Proportional Allocation in Hedonic Games

Martin Hoefer[1] and Wanchote Jiamjitrak[2]([⊠])

[1] Institute for Computer Science, Goethe University Frankfurt, Frankfurt, Germany
mhoefer@cs.uni-frankfurt.de
[2] Department of Computer Science, Aalto University, Espoo, Finland
wanchote.jiamjitrak@aalto.fi

Abstract. Proportional allocation is an intuitive and widely applied mechanism to allocate divisible resources. We study proportional allocation for profit sharing in coalition formation games. Here each agent has an impact or reputation value, and each coalition represents a joint project that generates a total profit. This profit is divided among the agents involved in the project based on their reputation. We study existence, computational complexity, and social welfare of core-stable states with proportional sharing.

Core-stable states always exist and can be computed in time $O(m \log m)$, where m is the total number of projects. Moreover, when profits have a natural monotonicity property, there exists a reputation scheme such that the price of anarchy is 1, i.e., every core-stable state is a social optimum. However, these schemes exhibit a strong inequality in reputation of agents and thus imply a lacking fairness condition. Our main results show a tradeoff between reputation imbalance and the price of anarchy. Moreover, we show lower bounds and computational hardness results on the reputation imbalance when prices of anarchy and stability are small.

1 Introduction

Profit sharing is a central domain in game theory and has attracted a large amount of interest, mostly as cooperative transferable-utility (TU) games. Usually, there are n agents, and a characteristic function specifies the profit for each subset of agents. The goal is to divide the profit of the grand coalition in a fair and stable way. There has been particular interest in TU games resulting from combinatorial optimization problems. For example, in the matching game [20] each agent is a node in an edge-weighted graph, and the profit of a subset of agents is the max-weight matching in the induced subgraph. For these games, there is a large variety of stability and fairness concepts, most prominently variants of the core. In fact, the core of a matching game might be empty, and the (approximate) core enjoys a close connection to the natural integer program of max-weight matching [9].

An underlying assumption is that (deviating) subsets of agents can freely negotiate shares and distribute profit. In many application contexts, however,

This work was done while the authors were at Saarland University. Supported by DFG Cluster of Excellence MMCI at Saarland University.

© Springer International Publishing AG 2017
V. Bilò and M. Flammini (Eds.): SAGT 2017, LNCS 10504, pp. 307–319, 2017.
DOI: 10.1007/978-3-319-66700-3_24

profit shares are less directly negotiable, e.g., when allocating credit for joint work. For example, in scientific publishing, credit is assigned based on a variety of aspects and rules, such as reputation, visibility, previous achievements, etc. Moreover, collaborative online platforms (recommendation systems, Wikis, etc.) can design and implement centralized rules for credit allocation among the users.

In this paper, we study natural and simple *proportional allocation* rules to distribute profit or credit among agents that engage in joint projects. Proportional allocation is a central approach in a variety of contexts and has been studied, e.g., for allocating divisible goods in mechanism design [6,8,15]. It can be used to express consequences of rich-get-richer-phenomena (also termed *Matthew effect*), was studied to distribute profits in stable matching [1], or appeared in probabilistic models for allocating scientific credit [16]. Moreover, proportional response dynamics are a successful method to compute market equilibria [4,22].

We study the properties of the proportional allocation mechanism in matching and coalition formation games. In this scenario, coalitions represent joint projects that agents can engage in. Each project yields a profit value, which is shared among the involved agents in proportion to an agent-specific parameter r_u. Intuitively, this parameter specifies the *influence of the agent*. Depending on the application context, it captures its, e.g., importance, visibility, or reputation. It might result from previous achievements (e.g., by reputation in societies) or be subject to design (e.g., by assignment in collaborative online systems).

Given such a profit sharing scheme, agents strategically choose the projects to engage in. More formally, a given set projects, profit values, and agent reputations constitutes a hedonic coalition formation game [10], where the proportional allocation mechanism yields the agent utilities. Our goal is to shed light on the equilibria of such games, i.e., existence, structure and social welfare of core-stable states. More precisely, we are interested in the structure of reputation values and the resulting prices of anarchy and stability.

Contribution and Overview. In Sect. 2, we observe that in our games, a core-stable state always exists. This is mostly a consequence of earlier work on stable matching [1]. Let k and $k_{\min}$ be the size of the largest and smallest project with non-zero profit in the game, respectively. For equal sharing (all influences the same), prices of anarchy and stability are $\Theta(k)$. In fact, if profits and influences are misaligned in a worst-case fashion, it is known that prices of anarchy and stability can be unbounded, even for the special case of matching [1].

In Sect. 3, we consider games with a natural monotonicity condition (termed inclusion-monotone), where we find an interesting trade-off between the required difference in influence and price of anarchy. When the ratio of maximum and minimum influence is bounded by α, there are reputations such that the price of anarchy is bounded by $\max\{1, k^2/(k-1+\alpha^{k_{\min}/n})\}$, where $k_{\min}$ is the size of the smallest project with non-zero profit in the game. When $\alpha = (k^2 - k + 1)^{n/k_{\min}}$, the price of anarchy drops to – with a suitable assignment of influence, we can eliminate any inefficiency in the game. For environments, in which assigning influence values is possible, we also provide an efficient algorithm that, given an optimum state S^*, computes influence values that achieve this bound on the price of anarchy. While finding an optimum solution can be NP-hard, our algorithm can also work with an arbitrary ρ-approximative state S, and the price of anarchy bound increases by a factor ρ. Note that the natural representation of our games is linear in the number of agents n and the number of projects/coalitions m, since we must specify a possibly arbitrary positive profit value for each possible project. Our algorithm runs in time polynomial in n and m. Consequently, it is strongest if $m = poly(n)$ (which is often the case, e.g., for matching games).

On the downside, when approaching a price of anarchy of 1, the society becomes extremely hierarchical – the maximum difference in influences grow exponentially large. In Sect. 4 we show that for inclusion-monotone games a factor difference of $n-1$ in influences can be required to obtain a price of stability of 1. If the profits of projects in a core-stable optimum should be shared equally, we strengthen this to an exponential lower bound of $(k+1)^{n/k_{\min}-1}$. Moreover, for games that are not inclusion-monotone, inefficiency of all core-stable states can be unavoidable.

Finally, in Sect. 5 we discuss computational hardness results. For a given optimal state S^* and a given upper bound α on the ratio of influences, it is NP-hard to decide whether we can make S^* stable, even if every project has size $k = 2$. Note that this hardness does not stem from computing S^*, since S^* is a max-weight matching when $k = 2$, which can be computed in polynomial time. We also show lower bounds and hardness results for the case with influence values in $\{1, x\}$ with $x > 1$, where agents have either "low" or "high" influence.

Due to space constraints, further proofs can be found in the appendix of the full version of this paper.

Further Related Work. Computing stability concepts in hedonic coalition formation games is a recent line of research [7,12,17–19]. Many stability concepts are NP- or PLS-hard to compute. This holds even in the case of additive-separable coalition profits, which can be interpreted by an underlying graph structure with weighted edges, and the profit of a coalition is measured by the total edge weights covered by the coalition [3,11,21]. The price of anarchy was studied, e.g., in [5].

Our work is inspired by proportional allocation mechanisms. The model we study was proposed for stable matching in [1] under the name *Matthew-effect sharing*. We study general hedonic coalition formation games. While the results in [1] bound prices of anarchy and stability for worst-case profit and influence values, our approach here is to study the necessary inherent trade-offs in social welfare in equilibrium and inequality of influence in the population. In this sense, our paper is closely related to [14], who study the trade-offs between social welfare and difference in profit shares. A drawback of [14] is that it allows to design arbitrary profit shares for every coalition and every agent. Thus, it allows a designer an unnatural amount of freedom when assigning credit to stabilize good states. In contrast, our approach with proportional sharing based on influence and reputation represents a more restricted, structured and arguably realistic way of how credit from joint projects might be allocated to agents.

2   Model and Preliminaries

A *proportional coalition formation game* is a hedonic coalition formation game based on a weighted hypergraph $G = (V, C, w)$. There is a set V of *agents* (or vertices) and a set of weighted *coalitions* (or hyperedges) $C \subseteq 2^V$, where $|c| \geq 2$ for every $c \in C$. Let $w : C \to \mathbb{R}^+$ represent the positive *weight* or *total profit* of each coalition $c \in C$. Unless stated otherwise, we use $n = |V|$, $m = |C|$, $k = \max_{c \in C} |c|$ (the maximum size of any coalition in C), and $k_{\min} = \min_{c \in C} |c|$ (the minimum size of any coalition in C).

Each agent $v \in V$ has a *reputation* $r_v > 0$, which we scale throughout to satisfy $\min_{v \in V} r_v = 1$. When a coalition $c \in C$ is formed, the profit $w(c)$ is shared among the $v \in c$ proportionally to r_v. We define a *reputation scheme* as a vector of reputations for the agents $R = (r_v)_{v \in V}$.

A *coalition structure* or *state* $S \subseteq C$ is a collection of pairwise disjoint coalitions c from C, i.e., for each $v \in V$ we have $|\{c \mid c \in S, v \in c\}| \leq 1$. For each coalition $c \in S$, the *profit* of agent $u \in c$ is a proportional share of weight $w(c)$:

$$p_u(S) = p_u(c) = \frac{r_u}{\sum_{v \in c} r_v} \cdot w(c) \tag{1}$$

Note that $p_u(c) > 0$ for all $c \in C$ and $u \in c$ by definition. For every agent $u \in V$ such that S contains no coalition that includes u, we assume $p_u(S) = 0$.

For a coalition structure S, a *blocking coalition* $c \in C \setminus S$ is a coalition such that, for each $v \in c$, $p_v(c) > p_v(S)$. Every agent in c gains strictly more profit than he currently obtains in S if they deviate to c instead. In the case of matching and $k = 2$, we speak of a *blocking pair*. A coalition structure S is termed *core-stable* if there is no blocking coalition in[1] $C \setminus S$.

To assess the quality of reputation schemes, we quantify the social welfare of the resulting core-stable coalition structures. The *social welfare* of a coalition structure S is $w(S) := \sum_{c \in S} w(c)$. We call the coalition structure S^* with the highest social welfare the *optimal coalition structure*. We measure the quality of reputation schemes using the prices of anarchy (PoA) and stability (PoS) of core-stable coalition structures. While the core-stable coalition structures depend on reputations, the optimal coalition structures (and hence, optimal social welfare) do not. Consequently, our goal is to measure the quality of reputation schemes based on PoA and PoS. In particular, we strive to design reputations to maximize social welfare of the resulting core-stable coalition structures.

It turns out that we can obtain very small PoA and PoS using a hierarchy of reputations with extremely large differences, which is often undesirable due to reasons of fairness and equality. As a consequence, we try to limit unequal reputations and, in particular, strive to quantify the tension between efficiency and equality. We measure the degree of equality using a parameter as follows. A reputation scheme R is α-*bounded* if $\alpha \geq \frac{\max_{v \in V} r_v}{\min_{v \in V} r_v} = \max_{v \in V} r_v$. Intuitively, a smaller α indicates that reputation is more uniform reputation.

---

[1] Core-stability usually means that no subset of agents wants to deviate. We recover this interpretation when we assume all coalitions $c \in 2^V \setminus C$ have profit $w(c) = -1$.

A simple solution to achieve perfect equality is when every agent has the same reputation. This results in *equal sharing*, and it results in PoA (and PoS) of at most k, where k is the size of the largest coalition. The following result is shown, e.g., in [2, Theorem 2.9, Corollary 8.2]. In the full version of this paper, we include a proof for completeness and discuss an example game.

Proposition 1. *The PoA and the PoS in hedonic coalition formation games with equal sharing is exactly k.*

Some of our results apply to instances with an additional property. A game $G = (V, C, w)$ is *inclusion monotone* if for any $c, c' \in C$ with $c' \subsetneq c$, we have $\frac{w(c)}{|c|} > \frac{w(c')}{|c'|}$. Note that, trivially, every instance of matching with $k = 2$ is inclusion monotone.

3   Existence and Computation

Let us first discuss our existence and computational results. We define an *improvement step* for a coalition structure S by adding a blocking coalition to S while removing all coalitions that intersect with it from S. It can be seen rather directly that every game has a (strong) lexicographical potential function. As a consequence, a core-stable coalition structure exists in every game and for every reputation scheme, and every sequence of improvement steps always converges. By considering coalitions in non-increasing order of $\frac{w(c)}{\sum_{u \in c} r_u}$, it is possible to arrive at a core-stable structure from any initial structure in at most n steps. The proof is a rather direct extension of [1, Theorem 8], and we include it in the full version of this paper for completeness.

Proposition 2. *For any game $G = (V, C, w)$ and proportional sharing based on reputation scheme R, there always exists a core-stable coalition structure. Given any initial coalition structure, we need at most $O(n)$ improvement steps to reach a core-stable coalition structure.*

Hence, for any game and any reputation scheme we have both *existence* and *convergence*, but it might be the case that every core-stable coalition structure has small social welfare or reputations are extremely different. The subsequent algorithm shows how reputation schemes can provide a trade-off between α-boundedness and the PoA. For a given inclusion-monotone instance and a parameter $\alpha > 1$, the algorithm provides a reputation scheme that is α-bounded and guarantees a PoA of strictly better than k.

When $\alpha = 1$, we have equal sharing, the price of anarchy is at most k (due to Proposition 1) and a greedy procedure computes the $O(n)$ improvement steps to reach a core-stable state (due to Proposition 2). Algorithm 1 generalizes this approach to obtain improved bounds for $\alpha > 1$. It uses a similar structure as a corresponding algorithm in [14]. In each iteration, we choose one coalition to be a part of our solution and assign the reputation to each agent in this coalition, then remove the agents from consideration. Let c be a coalition with the largest ratio of $\frac{w(c)}{|c|-1+x}$, where $x = \alpha^{k_{\min}/n}$. There are three cases in the ith iteration:

Algorithm 1. Computing a reputation scheme for given α

Input: Inclusion monotone $G = (V, C, w)$, optimal structure S^*, bound α
Output: α-bounded reputation scheme R, core-stable coalition structure S

1   Initialize $i \leftarrow 0$, $C_0 \leftarrow C$, $S \leftarrow \emptyset$, $x \leftarrow \alpha^{k_{\min}/n}$ and $r_v \leftarrow 0$ for all $v \in V$
2   **while** $C_i \neq \emptyset$ **do**
3      $c \leftarrow \arg\max_{c \in C_i} \left(\frac{w(c)}{|c|-1+x} \right)$
4   **if** $c \in S^*$ **then** $s_i^* \leftarrow c$
5   **else if** $c \notin S^*$ and $\frac{w(c)}{|c|-1+x} < \frac{w(c')}{|c'|}$ for some $c' \in S^*$ that $c' \cap c \neq \emptyset$ **then**
6         $c' \leftarrow \arg\max_{c' \in S^*} \left(\frac{w(c')}{|c'|} \right)$
7         $s_i^* \leftarrow c'$
8   **else** $s_i^* \leftarrow c$
9   **for** $u \in s_i^*$ **do** $r_u \leftarrow x^i$
10     $S \leftarrow S \cup s_i^*$
11     $C_{i+1} \leftarrow C_i$
12   **for** $c \in C_i$ with $c \cap s_i^* \neq \emptyset$ **do** $C_{i+1} \leftarrow C_{i+1} \setminus \{c\}$
13     $i \leftarrow i+1$
14   **for** $v \in V$ with $r_v = 0$ **do** $r_v \leftarrow x^i$

---

1. If c is a part of the optimal coalition structure S^*, we call it s_i^*.
2. If c is not in S^* and is overlapping with some coalitions in S^*, then we consider c' in S^* such that c' has the highest ratio of $\frac{w(c')}{|c'|}$ among all overlapping coalitions. If c' is large enough to stabilize, we will choose c' instead of c in order to make our solution closest to S^* as much as possible. So we call c' as s_i^*.
3. If c is not in S^*, but c has a high ratio of $\frac{w(c)}{|c|-1+x}$ so that we should stabilize c instead of stabilizing a coalition from the optimal coalition structure, then we choose c to be s_i^*.

Then, we stabilize s_i^* by assigning the same reputation to each included agent. This reputation increases by the factor of x in the next iteration. Then we remove all the agents in s_i^* and their incident coalitions from consideration. The algorithm terminates when there is no coalition left to consider.

Theorem 1. *For a given inclusion-monotone instance, and given $\alpha > 1$ and any optimal coalition structure S^*, Algorithm 1 computes in polynomial time an α-bounded reputation scheme R with PoA at most $\max\{1, k^2/(k-1+\alpha^{k_{\min}/n})\}$ and a core-stable coalition structure S that achieves both bounds.*

Proof. We first consider the running time. The algorithm sorts all coalitions by the ratio of $\frac{w(c)}{|c|-1+x}$, which takes $O(m \log m)$ time. Then, in each iteration we only consider one coalition and its overlapping coalitions, which can be done in $O(m)$ time. In total, the running time is bounded by $O(m^2)$. Recall from the discussion in the introduction that the input size is $\Omega(n + m)$, hence the algorithm runs in polynomial time.

We now show core-stability of S. As the invariant of the algorithm we maintain that coalitions dropped from consideration will never form any blocking coalition. This holds since we assign reputations that increase by a factor of x in every iteration. Consider three cases as in the algorithm,

1. In the first case, since c is the coalition that has the maximum ratio of $\frac{w(c)}{|c|-1+x}$, we assign the same reputation to each agent in c, and each agent gets a profit of $\frac{w(c)}{|c|}$. Consider an overlapping coalition c', and let $u \in c' \cap c$. There are two subcases: (1) c' is a proper subset of c. Since the instance is inclusion monotone and we share profit equally in the coalition, we have

$$p_u(c') = \frac{w(c')}{|c'|} < \frac{w(c)}{|c|} = p_u(c).$$

(2) There is an agent $v \in c' \setminus c$ who has a reputation $r_v \geq xr_u$. Then the profit $u \in c \cap c'$ gains from c' is

$$p_u(c') \leq \frac{w(c')}{|c'|-1+x} \leq \frac{w(c)}{|c|-1+x} < \frac{w(c)}{|c|} = p_u(c).$$

This shows that every $u \in c \cap c'$ gains more profit by staying with c.
2. In the second case, we choose c' that is in S^* instead of c, each agent in c' gains a profit of $\frac{w(c')}{|c'|}$. Consider an overlapping coalition c'', and let $u \in c'' \cap c'$. There are two subcases: (1) c'' is a proper subset of c'. We apply the same argument as in the first subcase of the first case. (2) There is an agent in $v \in c'' \setminus c'$ who has a reputation $r_v \geq xr_u$. Then, the profit $u \in c' \cap c''$ gains from c'' is

$$p_u(c'') \leq \frac{w(c'')}{|c''|-1+x} \leq \frac{w(c)}{|c|-1+x} < \frac{w(c')}{|c'|} = p_u(c').$$

This shows that every $u \in c' \cap c''$ gains more profit by staying with c.
3. In the third case, we can use the same analysis as in the first case because we choose the coalition that has the maximum ratio of $\frac{w(c)}{|c|-1+x}$.

This concludes that the resulting state S is core-stable.

Now consider any arbitrary core-stable state S' and coalition c added to S in the first round of the algorithm. Then agents u with $r_u = 1$ are exactly the ones in c, so every overlapping $c' \in S'$ is either a subset of c or has at least one $v \in c' \setminus c$ with $r_v \geq xr_u$ and no agent with reputation less than 1. Hence, the strict inequalities above apply to all agents in c and imply that c is blocking.

Now suppose all coalitions added to S by our algorithm up to round i are in S', but c added in round $i+1$ is not. Then the agents with smaller reputation are exactly the ones in the coalitions added in the first i rounds. As such, they are part of S and do not overlap with c. Hence, every overlapping coalition $c' \in S'$ is either a subset of c or has only agents with same or higher reputation and at least one agent with $r_v \geq xr_u$. Therefore, the strict inequalities above apply to all agents in c and imply that c is blocking. By induction, every core-stable state must contain all coalitions of S, and S is the unique core-stable state.

For α-boundedness, observe that the minimum reputation in R is always 1. In each iteration, we add one coalition to S with size at least $k_{\min}$, so there are at most $n/k_{\min}$ iterations. As a consequence, the maximum reputation is at most $x^{n/k_{\min}} = \alpha$, i.e., R is α-bounded.

Finally, for the PoA, we see that the solution S of Algorithm 1 deviates from S^* only in iterations that apply the third case, when $c \notin S^*$ and $\frac{w(c)}{|c|-1+x} \geq \frac{w(c')}{|c'|}$ for all $c' \in S^*$. c can intersect at most $|c|$ other coalitions $c' \in S^*$, hence

$$\text{PoA} \leq \frac{|c| \cdot w(c')}{w(c)} \leq \frac{|c| \cdot w(c')}{(|c|-1+x) \cdot \frac{w(c')}{|c'|}} = \frac{|c| \cdot |c'|}{|c|-1+x} \leq \frac{k^2}{k-1+\alpha^{k_{\min}/n}}.$$

This proves the theorem.   □

The algorithm reveals a trade-off between α and PoA. By increasing α, the guaranteed PoA decreases and vice-versa. While the algorithm itself runs in polynomial time, it uses S^* as input, which is NP-hard to compute (finding S^* trivially generalizes, e.g., the standard SET-PACKING problem). Hence, the above trade-off mostly applies in terms of existence.

Interestingly, the algorithm also yields a trade-off in terms of (overall) efficient computation. Our analysis of the social welfare of the output structure applies w.r.t. to the social welfare of the input structure. Consequently, if Algorithm 1 is given any input structure S', it will output a core-stable coalition structure S with social welfare at least $w(S) \geq w(S') \cdot (k-1+\alpha^{k_{\min}/n})/k^2$.

Corollary 1. *If Algorithm 1 is applied using any coalition structure S' that represents a ρ-approximation to the optimal social welfare, it computes an α-bounded reputation scheme with PoA at most $\rho \cdot \max\{1, k^2/(k-1+\alpha^{k_{\min}/n})\}$.*

4   Lower Bounds

In this subsection, we will show a number of lower bounds. Algorithm 1 applies to games that are inclusion monotone, and it shows that we can always reduce PoA to 1 if α is chosen large enough. Next, we will show that there are instances that are not inclusion monotone, where for arbitrarily large α we cannot stabilize an optimal coalition structure.

Proposition 3. *There are classes of non-inclusion monotone instances such that (1) every reputation scheme yields a PoS of at least $2 - \frac{4}{n+2}$; (2) every α-bounded reputation scheme yields a PoS of at least $(n-1+\alpha)/(1+\alpha)$.*

The previous proposition shows that the trade-off shown in Theorem 1 does not apply in instances that are not inclusion monotone. The next result complements the bound on α in Theorem 1 when PoS is 1.

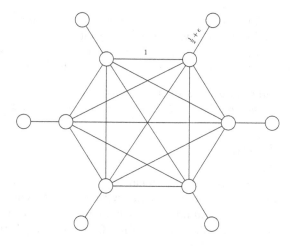

Fig. 1. An instance that requires $\alpha \geq n - 1$ whenever PoS is 1 (in this case, $n = 12$)

Proposition 4. *There is a class of inclusion-monotone instances where every reputation scheme with PoS of 1 has $\alpha \geq n - 1$.*

Proof. Consider an instance of the type depicted in Fig. 1. The instance $G = (V, C, w)$ consists of a clique of size $\frac{n}{2}$ (denoted by $K_{n/2}$). Every coalition/edge c in $K_{n/2}$ has $w(c) = 1$. For each agent, we create an additional agent (called "*leaf agent*") and include a coalition with a clique agent of weight $\frac{1}{2} + \epsilon$. The optimal social welfare is $\frac{n}{2}(\frac{1}{2} + \epsilon)$, and S^* is composed of exactly the $n/2$ coalitions with leaf agents. For PoS 1, we need to maximize the profit of clique agents in S^*, since they are the only ones with deviations. Hence, we can w.l.o.g. assign the minimum reputation of $r_u = 1$ to all leaf agents.

Let $r_1, r_2, \ldots, r_{n/2}$ be the reputations of clique agents. Consider r_i and r_j; in order to avoid a blocking coalition with agents i and j, at least one of them must gain at least as much profit as in S^*. Assume i is such an agent, then $(\frac{1}{2} + \epsilon) \cdot \frac{r_i}{r_i + 1} \geq 1 \cdot \frac{r_i}{r_i + r_j}$. For $\epsilon \to 0$, this implies $r_j \geq r_i + 2$. Since an inequality of this form must hold for every pair $\{i, j\}$ of clique agents, we have

$$\max_{i,j \in [\frac{n}{2}]} \{r_i - r_j\} \geq 2(n/2 - 1) = n - 2.$$

Since $r_i \geq 1$ for all $i = 1, \ldots, n$, this implies $\alpha = \max_i r_i \geq n - 1$.   □

This lower bound for α is linear in n, but if we apply Algorithm 1 with $k = k_{\min} = 2$ and postulate a PoA of 1, then we can only guarantee $\alpha \leq 3^{n/2}$. Hence, in general, our results leave significant room for improvement. Note that the output of Algorithm 1 has the property that in some (in fact, the unique) core-stable coalition structure the profits in each coalition are shared equally. For schemes with this property we can show a drastically improved lower bound, which is asymptotically tight for constant k. Hence, to show existence and computation of schemes with smaller inequality, we need substantially different techniques.

A scheme R has *equal sharing in stability* if there is a core-stable coalition structure S such that for every $c \in S$ we have $r_i = r_j$, for every $i, j \in c$.

Proposition 5. *There is a class of inclusion-monotone instances where every scheme R with equal sharing in stability and PoS of 1 requires $\alpha \geq (k+1)^{\frac{n}{k_{\min}} - 1}$.*

5 Hardness Results

In this section, we consider computational hardness results that complement our upper bounds in Sect. 3. Even in games with $k = 2$, in which an optimum coalition structure S^* is a maximum-weight matching that can be computed in polynomial time, there is no efficient algorithm for computing R that makes S^* core-stable and minimizes the inequality α.

Theorem 2. *Given an optimal coalition structure S^* and given $\alpha \geq 1$, it is NP-hard to decide whether there is an α-bounded reputation scheme such that S^* is core-stable. It remains NP-hard even if every coalition has size exactly k, for any $k \geq 2$.*

Proof. We will show a reduction from the GRAPH COLORING problem. First consider the case when $k = 2$. For an instance of GRAPH COLORING given by an unweighted graph $G = (V, E)$ with $V = \{v_1, \ldots, v_n\}$, we construct a game $G' = (V', E', w)$ as follows. Let $V' = V_1 \cup V_2$ with $V_1 = V$ and $V_2 = \{v_{n+1}, \ldots, v_{2n}\}$, and $E' = E_1 \cup E_2$ with $E_1 = E$ and $E_2 = \{\{v_i, v_{n+i}\} \mid i = 1, \ldots, n\}$. We set $w(e) = 1$ if $e \in E_1$ and $w(e) = \frac{1}{2} + \epsilon$ if $e \in E_2$, for an arbitrarily small constant $\epsilon > 0$. Hence, G' is similar in spirit to Fig. 1 except we replace the clique by the coloring instance G. The optimal coalition structure $S^* = E_2$.

First assume G is ℓ-colorable. We show that there is an $(2\ell - 1)$-bounded reputation scheme that can stabilize S^* in G'. By Proposition 4, any two adjacent vertices in V_1 must have a difference in reputations of at least 2, otherwise the edge will be a blocking pair. So, if a vertex has i^{th} color class in G, then assign the corresponding agent a reputation of $2i - 1$ in G'. Finally, assign reputation 1 to all agents in V_2. This reputation scheme makes S^* core-stable, and the proof is identical to the one in Proposition 4.

Now assume there is a α^*-bounded reputation scheme R that makes S^* core-stable. We show that G is $\frac{\alpha^*+1}{2}$-colorable. First, we convert R as follows:

1. Normalize all the reputations to satisfy $\min_i r_i = 1$.
2. Change the reputations of all the leaf nodes to be 1.
3. For every normalized reputation value, if it is not an odd integer, decrease it down to next lower odd integer.

It is obvious that after these three conversions, the scheme is still α^*-bounded. Let us argue that the conversion also keeps S^* core-stable. Step 1 does not change anything because scaling all reputations does not change any profit shares. After step 2, agents in V_1 receive more profit in S^*, so S^* remains core-stable. In step 3, any two agents that have a difference in their reputations of at least two, the difference still remains at least two. Thus, S^* remains core-stable.

After the conversion, every agent in V_1 has an odd integer as reputation. Now, we just identify each odd integer with a color class. Since S^* is core-stable, adjacent vertices in G must differ in reputation by at least 2, i.e., belong to different color classes. Hence, G is $\frac{\alpha^*+1}{2}$-colorable. The result follows for $k = 2$.

To show that it remains NP-hard even if every coalition in the instance has size exactly k, we reduce the coloring problem on graph G to hypergraph G' as in the previous reduction. Here, however, for each coalition in G' we add $k-2$ more agents that only belong to that coalition. Every coalition has the same weight as in previous case. To stabilize S^*, we need $(\frac{1}{2} + \epsilon) \cdot \frac{r_i}{r_i+k-1} \geq 1 \cdot \frac{r_i}{r_i+r_j+k-2}$ for every $v_i, v_j \in V_1$. This leads to $r_j \geq r_i + k$ for any $j > i$. So, any two adjacent vertices in V must have the difference in reputations of at least k (instead of 2 as above). Applying the reduction as above, we can stabilize S^* in G' with an α^*-bounded reputation scheme if and only if G is $(\frac{\alpha^*+k-1}{k})$-colorable. $\qquad\square$

It shows that even approximating α^* is extremely hard, since the reduction preserves the well-known approximation hardness of GRAPH COLORING [13].

Corollary 2. *For any constant $\epsilon > 0$, α^* cannot be efficiently approximated within $n^{1-\epsilon}$ unless* NP = ZPP.

Let us also examine an interesting special case reputation scheme where we are allowed only to assign "high" and "low" reputations. More formally, let $R \in \{1, x\}^n$ for some $x > 1$, where we call such schemes *"restricted reputations"*. Unfortunately, the next theorem shows that finding a scheme with optimal α remains NP-hard when we are given a bound W on social welfare of a core-stable coalition structure (but not the exact optimum S^*). This is a weaker assumption than providing an optimal coalition structure directly as in the previous theorem. However, the result applies even for matching with $k = 2$, where existence of a solution with welfare at least W can be decided in polynomial time. Hence, the difficulty does not lie in *finding a good coalition structure* but it is again *inherent to the correct assignment of reputations*.

Theorem 3. *Given a positive rational number $x > 1$ and a bound on social welfare $W > 0$, it is* NP*-hard to decide whether there exist restricted reputations that results in a core-stable coalition structure S with $w(S) \geq W$. This holds even for instances with $k = 2$.*

Corollary 3. *For both restricted and general reputations, the following problems are* NP*-hard: (1) Given $W > 0$, find the α-bounded core-stable coalition structure S with minimum α such that $w(S) \geq W$; and (2) given $\alpha > 0$, find the α-bounded core-stable coalition structure with maximum social welfare.*

For restricted reputations, we might not be able to stabilize the optimal coalition structure. The final result lower bounds the PoS in terms of parameter x.

Proposition 6. *For $x > 1$, there are classes of instances such that for every restricted reputation scheme $R \in \{1, x\}^n$ (1) the PoS is at least $\frac{4}{x+1}$; and (2) the PoS is at least $2 - \frac{4}{x+3}$.*

References

1. Anshelevich, E., Bhardwaj, O., Hoefer, M.: Friendship and stable matching. In: Bodlaender, H.L., Italiano, G.F. (eds.) ESA 2013. LNCS, vol. 8125, pp. 49–60. Springer, Heidelberg (2013). doi:10.1007/978-3-642-40450-4_5
2. Anshelevich, E., Hoefer, M.: Contribution games in networks. Algorithmica **63**(1–2), 51–90 (2012)
3. Aziz, H., Brandt, F., Seedig, H.G.: Computing desirable partitions in additively separable hedonic games. Artif. Intell. **195**, 316–334 (2013)
4. Birnbaum, B., Devanur, N., Xiao, L.: Distributed algorithms via gradient descent for fisher markets. In: Proceedings of 12th Conference Electronic Commerce (EC), pp. 127–136 (2011)
5. Branzei, S., Larson, K.: Coalitional affinity games and the stability gap. In: Proceedings of 21st International Joint Conference on Artificial Intelligence (IJCAI), pp. 79–84 (2009)
6. Caragiannis, I., Voudouris, A.: Welfare guarantees for proportional allocations. Theory Comput. Syst. **59**(4), 581–599 (2016)
7. Cechlárova, K.: Stable partition problem. In: Kao, M.-Y. (ed.) Encyclopedia of Algorithms, pp. 1–99. Springer, New York (2008)
8. Christodoulou, G., Sgouritsa, A., Tang, B.: On the efficiency of the proportional allocation mechanism for divisible resources. Theory Comput. Syst. **59**(4), 600–618 (2016)
9. Deng, X., Ibaraki, T., Nagamochi, H.: Algorithmic aspects of the core of combinatorial optimization games. Math. Oper. Res. **24**(3), 751–766 (1999)
10. Dreze, J., Greenberg, J.: Hedonic coalitions: optimality and stability. Econometrica **48**(4), 987–1003 (1980)
11. Gairing, M., Savani, R.: Computing stable outcomes in hedonic games. In: Kontogiannis, S., Koutsoupias, E., Spirakis, P.G. (eds.) SAGT 2010. LNCS, vol. 6386, pp. 174–185. Springer, Heidelberg (2010). doi:10.1007/978-3-642-16170-4_16
12. Hajduková, J.: Coalition formation games: a survey. Int. Game Theory Rev. **8**(4), 613–641 (2006)
13. Håstad, J.: Clique is hard to approximate within $n^{1-\varepsilon}$. Acta Math. **182**(1), 105–142 (1999)
14. Hoefer, M., Wagner, L.: Designing profit shares in matching and coalition formation games. In: Chen, Y., Immorlica, N. (eds.) WINE 2013. LNCS, vol. 8289, pp. 249–262. Springer, Heidelberg (2013). doi:10.1007/978-3-642-45046-4_21
15. Johari, R., Tsitsiklis, J.: Efficiency loss in a resource allocation game. Math. Oper. Res. **29**(3), 407–435 (2004)
16. Kleinberg, J., Oren, S.: Mechanisms for (mis)allocating scientific credit. In: Proceedings of 43rd Symposium on Theory of Computing (STOC), pp. 529–538 (2011)
17. Peters, D.: Graphical hedonic games of bounded treewidth. In: Proceedings of 23rd Conference on Artificial Intelligence (AAAI), pp. 586–593 (2016)
18. Peters, D.: Towards structural tractability in hedonic games. In: Proceedings of 23rd Conference on Artificial Intelligence (AAAI), pp. 4252–4253 (2016)

19. Peters, D., Elkind, E.: Simple causes of complexity in hedonic games. In Proceedings of 24th International Joint Conference on Artificial Intelligence (IJCAI), pp. 617–623 (2015)
20. Shapley, L., Scarf, H.: On cores and indivisibility. J. Math. Econ. **1**(1), 23–37 (1974)
21. Sung, S.-C., Dimitrov, D.: Computational complexity in additive hedonic games. Eur. J. Oper. Res. **203**(3), 635–639 (2010)
22. Zhang, L.: Proportional response dynamics in the fisher market. Theor. Comput. Sci. **412**(24), 2691–2698 (2011)

Stable Marriage with Covering Constraints–A Complete Computational Trichotomy

Matthias Mnich[1,2] and Ildikó Schlotter[3(✉)]

[1] Universität Bonn, Bonn, Germany
mmnich@uni-bonn.de
[2] Maastricht University, Maastricht, The Netherlands
[3] Budapest University of Technology and Economics, Budapest, Hungary
ildi@cs.bme.edu

Abstract. We consider STABLE MARRIAGE WITH COVERING CONSTRAINTS (SMC): in this variant of STABLE MARRIAGE, we distinguish a subset of women as well as a subset of men, and we seek a matching with fewest number of blocking pairs that matches all of the distinguished people. We investigate how a set of natural parameters, namely the maximum length of preference lists for men and women, the number of distinguished men and women, and the number of blocking pairs allowed determine the computational tractability of this problem.

Our main result is a complete complexity trichotomy that, for each choice of the studied parameters, classifies SMC as polynomial-time solvable, NP-hard and fixed-parameter tractable, or NP-hard and W[1]-hard. We also classify all cases of one-sided constraints where only women may be distinguished.

1 Introduction

The STABLE MARRIAGE (SM) problem is a fundamental problem first studied by Gale and Shapley [16] in 1962. An instance of SM consists of a set $\mathcal{M}$ of men, a set $\mathcal{W}$ of women, and a preference list for each person ordering members of the opposite sex. We aim to find a stable matching, i.e., a matching for which there exists no pair of a man and a woman who prefer each other to their partners given by the matching; such a pair is called a blocking pair.

We consider a problem that we call STABLE MARRIAGE WITH COVERING CONSTRAINTS (SMC). Here, a set $\mathcal{W}^\star$ of women and a set $\mathcal{M}^\star$ of men are distinguished, and a feasible matching is one where each person in $\mathcal{W}^\star \cup \mathcal{M}^\star$ gets matched. By the Rural Hospitals Theorem [17] we know that the set of unmatched men and women is the same in all stable matchings, so clearly, feasible stable matchings may not exist. Thus, we define the task in SMC as finding a feasible matching with a minimum number of blocking pairs. Somewhat surprisingly, this natural extension of SM has not been considered before.

M. Mnich—Supported by ERC Starting Grant 306465 (BeyondWorstCase).
I. Schlotter—Supported by the Hungarian National Research Fund (OTKA grants no. K-108383 and no. K-108947).

© Springer International Publishing AG 2017
V. Bilò and M. Flammini (Eds.): SAGT 2017, LNCS 10504, pp. 320–332, 2017.
DOI: 10.1007/978-3-319-66700-3_25

Motivation. Our main motivation for studying SMC—apart from its natural definition—is its close relationship with the HOSPITALS/RESIDENTS WITH LOWER QUOTA (HRLQ) problem, modelling a situation where medical residents apply for jobs in hospitals: residents rank hospitals and vice versa, and hospitals declare both *lower* and *upper quotas* which bound the number of residents they can accept; the task is to find an assignment with a minimum number of blocking pairs. By "cloning" hospitals, HRLQ reduces to the case where each hospital has unit upper quota. In fact, this is equivalent to the special case of SMC where only women (or, equivalently, men) are distinguished. We refer to this problem with one-sided covering constraints, linking SMC and HRLQ, as SMC-1.

The HRLQ problem and its variants have recently gained quite some interest from the algorithmic community [4, 7, 13, 18, 20, 21, 25, 33, 37]. In his book, Manlove [29, Chap. 5.2] devotes an entire chapter to the algorithmics of HRLQ.

The reason for this interest in HRLQ is explained by its importance in several real-world matching markets [14, 15, 35] such as school admission systems, centralized assignment of residents to hospitals, or of cadets to military branches. Lower quotas are a common feature of such admission systems. Their purpose is often to remedy the effects of understaffing that are explained by the well-known Rural Hospitals Theorem [17]: as an example, governments usually want to assign at least a small number of medical residents to each rural hospital to guarantee a minimum service level. Minimum quotas also appear in controlled school choice programs [11, 28, 36] where students belong to a small number of types; schools set lower bounds for each type enact affirmative actions, such as admitting a certain number of minority students [11]. Another example is the German university admission system for admitting students to highly oversubscribed subjects, where a certain percentage of study places are assigned according to high school grades or waiting time [36]. But lower quotas may also arise due to financial considerations: for instance, a business course with too few (tuition-paying) attendees may not be profitable. Certain aspects of airline preferences for seat upgrade allocations can be also modelled by lower quotas [28].

Another motivation for studying SMC comes from the following scenario that we dub *Control for Stable Marriage*. Consider a two-sided market where each participant of the market expresses its preferences over members of the other party, and some central agent (e.g., a government) performs the task of finding a stable matching in the market. It might happen that this central agency wishes to apply a certain *control* on the stable matching produced: it may favour some participants by trying to assign them a partner in the resulting matching. Such a behaviour might be either malicious (e.g., the central agency may accept bribes and thus favour certain participants) or beneficial (e.g., it may favour those who are at disadvantage, like handicapped or minority participants). However, there might not be a stable matching that covers all participants the agency wants to favour; thus arises the need to produce a matching that is *as stable as possible* among those that fulfill our constraints—the most natural aim in such a case is to minimize the number of blocking pairs in the produced matching, which yields exactly the SMC problem. Similar control problems for voting systems have been extensively studied in social choice following the work initiated by Bartholdi III. [3], but have not yet been considered in connection to stable matchings.

Our Results. We provide an extensive algorithmic analysis of the SMC problem, and its special case SMC-1. In our analysis, we examine the influence of different aspects on the tractability of these problems. The aspects we consider are

- the number b of blocking pairs allowed,
- the number $|\mathcal{W}^\star|$ of women with covering constraint,
- the number $|\mathcal{M}^\star|$ of men with covering constraint,
- the maximum length $\Delta_\mathcal{W}$ of women's preference lists, and
- the maximum length $\Delta_\mathcal{M}$ of men's preference lists.

To investigate how these aspects affect the complexity of SMC, we use the framework of parameterized complexity, which deals with computationally hard problems and focuses on how certain *parameters* of a problem instance influence its tractability. We aim to design *fixed-parameter algorithms*, which perform well in practice if the value of the parameter on hand is small[1].

The choice of the above aspects (or parameters) is motivated by the aforementioned applications. For instance, we seek matchings where ideally no blocking pairs at all or at least only few of them appear, to ensure stability of the matching and happiness of those getting matched. The number of women/men with covering constraints corresponds, for instance, to the number of rural hospitals for which a minimum quota specifically must be enforced, which we can expect to be small among the set of all hospitals accepting medical residents. Finally, preference lists of hospitals and residents can be expected to be small, as each hospital might not rank many more candidates than positions it has to fill, whereas residents might rank only their top choices of hospitals. Hence, it is reasonable to assume that these parameters take small values in certain applications, and thus suitable fixed-parameter algorithms may be highly efficient in practice.

We draw a detailed landscape of the influence of each aspect, and *all* their combinations, on the complexity of the SMC problem. To this end, we consider all choices of aspects in $A = \{b, |\mathcal{W}^\star|, |\mathcal{M}^\star|, \Delta_\mathcal{M}, \Delta_\mathcal{W}\}$ as either being restricted to some constant integer, or regarded as a parameter, or left unbounded. Intuitively, these different choices for elements of A correspond to their expected "range" in applications, from very small to mid-range to large (compared to the size of the entire system). By considering all combinations, we can model all applications.

Our main result classifies the SMC problem for all such combinations into one of three cases, as being either "easy", "moderate", or "hard" to solve:

---

[1] For background on parameterized complexity, we refer to the recent monograph [9].

Theorem 1. *For each choice of aspects in $A = \{b, |\mathcal{W}^\star|, |\mathcal{M}^\star|, \Delta_{\mathcal{M}}, \Delta_{\mathcal{W}}\}$, SMC is in P, or NP-hard and fixed-parameter tractable, or NP-hard and W[1]-hard with the given parameterization[2], and is covered by one of the results in Table 1.*

In particular, SMC is W[1]-hard parameterized by $b + |\mathcal{W}^\star|$, even if there are no distinguished men (i.e., $|\mathcal{M}^\star| = 0$), $\Delta_{\mathcal{M}} = 3$, $\Delta_{\mathcal{W}} = 3$ and each distinguished woman finds only a single man acceptable.

Table 1 summarizes our results on the complexity of SMC. Note that some results are implied directly by the symmetrical roles of men and women in SMC, and thus are not stated explicitly. Proofs marked by ($\star$), as well as a decision diagram showing that the presented results indeed cover all restrictions of SMC with respect to $\{b, |\mathcal{W}^\star|, |\mathcal{M}^\star|, \Delta_{\mathcal{M}}, \Delta_{\mathcal{W}}\}$, can be found in the full version [32].

Table 1. Summary of our results for STABLE MARRIAGE WITH COVERING CONSTRAINTS. Here, $\Delta^\star$ denotes the maximum length of the preference list of any distinguished person.

Constants	Parameters	Complexity
$|\mathcal{M}^\star| = 0, |\mathcal{W}^\star| = 0$		P (Gale-Shapley alg.)
$|\mathcal{M}^\star| = 0, |\mathcal{W}^\star|, \Delta_{\mathcal{M}}$		P (Theorem 8)
$|\mathcal{M}^\star|, |\mathcal{W}^\star|, \Delta_{\mathcal{M}}, \Delta_{\mathcal{W}}$		P (Theorem 8)
$|\mathcal{M}^\star| = 0, \Delta_{\mathcal{M}} \leq 2$		P (Theorem 9)
$\Delta_{\mathcal{W}} \leq 2, \Delta_{\mathcal{M}} \leq 2$		P (Observation 11)
b		P (Observation 5)
$|\mathcal{M}^\star| = 0, \Delta_{\mathcal{W}} \leq 2, \Delta_{\mathcal{M}} \geq 3$		NP-hard (Theorem 10)
$|\mathcal{W}^\star| = 1, \Delta_{\mathcal{W}} \leq 2, \Delta_{\mathcal{M}} \geq 3$		NP-hard (Theorem 12)
$|\mathcal{M}^\star| = 0, \Delta_{\mathcal{W}} \geq 3, \Delta_{\mathcal{M}} \geq 3, \Delta^\star = 1$	$b + |\mathcal{W}^\star|$	W[1]-hard (Theorem 2)
$|\mathcal{M}^\star| = 0, |\mathcal{W}^\star| \geq 1, \Delta_{\mathcal{W}} \geq 3, \Delta^\star = 1$	$b + \Delta_{\mathcal{M}}$	W[1]-hard (Theorem 6)
$\Delta_{\mathcal{W}} \leq 2$	$|\mathcal{W}^\star| + |\mathcal{M}^\star|$	FPT (Theorem 13)
$\Delta_{\mathcal{W}} \leq 2$	b	FPT (Corollary 2)

As a special case, we answer a question by Hamada et al. [20] who gave an exponential-time algorithm that in time $O(|I|^{b+1})$ decides for a given instance I of HRLQ whether it admits a feasible matching with at most b blocking pairs[3]; the authors asked whether HRLQ is fixed-parameter tractable parameterized by b. As shown by Theorem 1, SMC-1 and therefore also HRLQ is W[1]-hard when parameterized by b, already in a very restricted setting. Thus, the answer to the question by Hamada et al. [20] is negative: SMC-1, and hence HRLQ, admits no fixed-parameter algorithm with parameter b unless FPT = W[1].

---

[2] Restrictions without parameters are classified as polynomial-time solvable or NP-hard.

[3] Hamada et al. claim only a run time $O((|\mathcal{W}||\mathcal{M}|)^{b+1})$, but their algorithm can easily be implemented to run in time $O(|I|^{b+1})$.

Related Work. There is a fast-growing body of literature on matching markets with lower quotas [4,7,13–15,18,20,21,25,33,37]. These papers study several variants of HRLQ, adapting the general model to the various specialties of practical problems. However, there are only a few papers which consider the problem of minimizing the number of blocking pairs [14,20]. The most closely related work to ours is that of Hamada et al. [20]: they prove that the HRLQ problem is NP-hard and give strong inapproximability results; they also consider the SMC-1 problem directly and propose an $O(|I|^{b+1})$ time algorithm for it.

A different line of research connected to SMC is the problem of *arranged marriages*, an early extension of SM suggested by Knuth [26] in 1976. Here, a set $Q^\star$ of man-woman pairs is distinguished, and we seek a stable matching that contains $Q^\star$ as a subset. Thus, as opposed to SMC, we not only require that each distinguished person is assigned *some* partner, but instead prescribe its partner exactly. Initial work on arranged marriages [19,26] was extended by Dias et al. [10] to consider also *forbidden marriages*, and was further generalized by Fleiner et al. [12] and Cseh and Manlove [8]. Despite the similar flavour, none of these papers have a direct consequence on the complexity of SMC.

Our work also fits the research line that addresses computationally hard stable matching problems by focusing on instances with bounded preference lists [6,22,24,27,34] or by studying their parameterized complexity [1,2,5,30,31].

2   Preliminaries

An instance I of the STABLE MARRIAGE (SM) problem consists of a set $\mathcal{M}$ of men and a set $\mathcal{W}$ of women. Each person $x \in \mathcal{M} \cup \mathcal{W}$ has a preference list $L(x)$ that strictly orders the members of the other party acceptable for x. We thus write $L(x)$ as a vector $L(x) = (y_1, \ldots, y_t)$, denoting that y_i is (strictly) *preferred* by x over y_j for each i and j with $1 \leq i < j \leq t$. A *matching* M for I is a set of man-woman pairs appearing in each other's preference lists such that each person is contained in at most one pair of M; some persons may be left unmatched by M. For each person x we denote by $M(x)$ the person assigned by M to x. For a matching M, a man m and a woman w included in each other's preference lists form a *blocking pair* if (i) m is either unmatched or prefers w to $M(m)$, and (ii) w is either unmatched or prefers m to $M(w)$. In the STABLE MARRIAGE WITH COVERING CONSTRAINTS (SMC) problem, we are given additional subsets $\mathcal{W}^\star \subseteq \mathcal{W}$ and $\mathcal{M}^\star \subseteq \mathcal{M}$ of *distinguished* people that must be matched; a matching M is *feasible* if it matches everybody in $\mathcal{W}^\star \cup \mathcal{M}^\star$. The objective of SMC is to find a feasible matching for I with minimum number of blocking pairs. If only people from one gender are distinguished, then w.l.o.g., we assume these to be women; this special case will be denoted by SMC-1.

The many-to-one extension of SMC-1 is the HOSPITALS/RESIDENTS WITH LOWER QUOTAS (HRLQ) problem whose input consists of a set $\mathcal{R}$ of residents and a set $\mathcal{H}$ of hospitals that have ordered preferences over the acceptable members of the other party. Each hospital $h \in \mathcal{H}$ has a quota lower bound $\underline{q}(h)$ and a quota upper bound $\overline{q}(h)$. One seeks an *assignment* M that maps a subset of the residents to hospitals that respects acceptability and is *feasible*, that is, $\underline{q}(h) \leq |M(h)| \leq \overline{q}(h)$ for each hospital h. Here, $M(h)$ is the set of residents assigned to some $h \in \mathcal{H}$ by M. We say that a hospital h is *under-subscribed* if $|M(h)| < \overline{q}(h)$. For an assignment M of an instance of HRLQ, a pair $\{r, h\}$ of a resident r and a hospital h is *blocking* if (i) r is unassigned or prefers h to the hospital assigned to r by M, and (ii) h is under-subscribed or prefers r to one of the residents in $M(h)$. The task in HRLQ is to find a feasible assignment with minimum number of blocking pairs.

Some instances of SMC may admit a *master list* over women, which is a total ordering $L_\mathcal{W}$ of all women, such that for each man $m \in \mathcal{M}$, the preference list $L(m)$ is the restriction of $L_\mathcal{W}$ to those women that m finds acceptable. Similarly, we consider master lists over men.

With each instance I of SMC (or HRLQ) we can naturally associate a bipartite graph G_I whose vertex partitions correspond to $\mathcal{M}$ and $\mathcal{W}$ (or $\mathcal{R}$ and $\mathcal{H}$, respectively), and there is an edge between a man $m \in \mathcal{M}$ and a woman $w \in \mathcal{W}$ (or between a resident $r \in \mathcal{R}$ and a hospital $h \in \mathcal{H}$, respectively) if they appear in each other's preference lists. We may refer to entities of I as vertices, or a pair of entities as edges, without mentioning G_I explicitly.

3 Strong Parameterized Intractability of SMC

This section shows parameterized intractability and inapproximability results for finding feasible matchings with minimum number of blocking pairs. A fundamental hypothesis about the complexity of NP-hard problems is the *Exponential Time Hypothesis* (ETH), which stipulates that algorithms solving all SATISFIABILITY instances in subexponential time cannot exist [23].

Theorem 2 ($\star$). *SMC-1 is W[1]-hard parameterized by $b + |\mathcal{W}^\star|$, and cannot be solved in time $f'(b) \cdot n^{o(\sqrt{b})}$ for any computable function f' unless ETH fails, even if there is a master list over men as well as one over women, all preference lists are of length at most 3, and $|L(w)| = 1$ for each woman $w \in \mathcal{W}^\star$.*

4 Polynomial-Time Approximation

Theorem 3 ($\star$). *Let I be an instance I of HRLQ, $\underline{q}_\Sigma$ the sum of lower quota bounds taken over all hospitals in I, and $\Delta_\mathcal{R}$ the maximum length of residents' preference lists. There is an algorithm that in polynomial time either outputs a feasible assignment for I with at most $(\Delta_\mathcal{R} - 1)\underline{q}_\Sigma$ blocking pairs, involving only $\underline{q}_\Sigma$ residents, or concludes that no feasible assignment exists.*

If both $\Delta_\mathcal{R}$ and $\underline{q}_\Sigma$ are constant, then Theorem 3 implies that HRLQ becomes polynomial-time solvable: if $b \geq (\Delta_\mathcal{R} - 1)\underline{q}_\Sigma$, then we apply Theorem 3 directly; if $b < (\Delta_\mathcal{R} - 1)\underline{q}_\Sigma$, then we use the algorithm by Hamada et al. [20] running in time $O(|I|^{b+1})$ which is polynomial, since b is upper-bounded by a constant.

Corollary 1. *HRLQ with both the maximum length $\Delta_{\mathcal{R}}$ of residents' prefer-ence lists and the total sum q_{Σ} of all lower quotas constant, is polynomial-time solvable.*

Another application of Theorem 3 is an approximation algorithm that works regardless of whether $\Delta_{\mathcal{R}}$ or q_{Σ} is a constant. In fact, the algorithm of Theorem 3 can be turned into a $(\Delta_{\mathcal{R}}-1)q_{\Sigma}$-factor approximation algorithm as follows. First, find a stable assignment M_s for I in polynomial time using the extension of the Gale-Shapley algorithm for the HOSPITALS/RESIDENTS problem. If M_s is not feasible, then by the Rural Hospitals Theorem [17], any feasible assignment for I must admit at least one blocking pair; hence, the algorithm of Theorem 3 yields an approximation with (multiplicative and also additive) factor $(\Delta_{\mathcal{R}} - 1)q_{\Sigma}$.

We also state an analogue of Theorem 3 that deals with SMC: it handles covering constraints on both sides, but assumes that all quota upper bounds are 1:

Theorem 4 ($\star$). *There is an algorithm that in polynomial time either outputs a feasible matching for an instance I of SMC with at most $(\Delta_{\mathcal{W}}-1)|\mathcal{M}^\star|+(\Delta_{\mathcal{M}}-1)|\mathcal{W}^\star|$ blocking pairs, or concludes that no feasible matching exists for I.*

5   SMC with Bounded Number of Distinguished Persons or Blocking Pairs

In Theorem 2 we proved W[1]-hardness of SMC-1 for the case where $\Delta_{\mathcal{M}} = \Delta_{\mathcal{W}} = 3$, with parameter $b + |\mathcal{W}^\star|$. Here we investigate those instances of SMC and SMC-1 where the length of preference lists may be unbounded, but either b, or the number of distinguished persons is constant.

First, if the number b of blocking pairs allowed is constant, then SMC can be solved by simply running the extended Gale-Shapley algorithm after guessing and deleting all blocking pairs. This complements the result by Hamada et al. [20].

Observation 5. *SMC can be solved in time $O(|I|^{b+1})$, where b denotes the number of blocking pairs allowed in the input instance I.*

In Theorem 6 we prove hardness of SMC-1 even if only one woman must be covered. If we require preferences to follow master lists, then a slightly weaker version of Theorem 6, where $|\mathcal{W}^\star| = 2$, still holds.

Theorem 6 ($\star$). *SMC-1 is W[1]-hard parameterized by $b + \Delta_{\mathcal{M}}$, even if $\mathcal{W}^\star = \{s\}$, $\Delta_{\mathcal{W}} = 3$, and $|L(s)| = 1$.*

Theorem 7 ($\star$). *SMC-1 is W[1]-hard parameterized by $b + \Delta_{\mathcal{M}}$, even if there is a master list over men as well as one over women, $|\mathcal{W}^\star| = 2$, $\Delta_{\mathcal{W}} \leq 3$, and $|L(w)| = 1$ for each $w \in \mathcal{W}^\star$.*

To contrast our intractability results, we show next that if each of $|\mathcal{W}^\star|$, $|\mathcal{M}^\star|$, $\Delta_\mathcal{W}$, and $\Delta_\mathcal{M}$ is constant, then SMC becomes polynomial-time solvable. Our algorithm relies on the observation that in this case, the number of blocking pairs in an optimal solution is at most $(\Delta_\mathcal{M} - 1)|\mathcal{W}^\star| + (\Delta_\mathcal{W} - 1)|\mathcal{M}^\star|$ by Theorem 4. Note that for instances of SMC-1, Theorem 8 yields a polynomial-time algorithm already if both $|\mathcal{W}^\star|$ and $\Delta_\mathcal{M}$ are constant.

Theorem 8 $(\star)$. *SMC can be solved in time* $O(|I|^{(\Delta_\mathcal{M}-1)|\mathcal{W}^\star|+(\Delta_\mathcal{W}-1)|\mathcal{M}^\star|+1})$.

Importantly, restricting only three of the values $|\mathcal{W}^\star|$, $|\mathcal{M}^\star|$, $\Delta_\mathcal{W}$, and $\Delta_\mathcal{M}$ to be constant does not yield tractability for SMC, showing that Theorem 8 is tight. Indeed, Theorem 6 implies that restricting the maximum length of the preference lists on only one side still results in a hard problem: SMC remains W[1]-hard with parameter $b + \Delta_\mathcal{M}$, even if $\Delta_\mathcal{W} = 3$, $|\mathcal{W}^\star| = 1$, and $|\mathcal{M}^\star| = 0$. Similarly, Theorem 2 shows that the problem remains hard even if $\Delta_\mathcal{W} = \Delta_\mathcal{M} = 3$ and $|\mathcal{M}^\star| = 0$.

6   SMC with Preference Lists of Length at Most Two

We show that the restriction of SMC where the maximum length of preference lists is bounded by 2 on one side leads to polynomial-time algorithms and fixed-parameter algorithms for various parameterizations.

Let I be an instance of SMC with underlying graph G. Let M_s be a stable matching in I, and let $\mathcal{M}_0^\star$ and $\mathcal{W}_0^\star$ denote the set of distinguished men and women, respectively, unmatched by M_s. Furthermore, let $\mathcal{M}_0$ and $\mathcal{W}_0$ denote the set of all men and women, respectively, unmatched by M_s. A path P in G is called an *augmenting path*, if $M_s \Delta P$ is a matching, and either both endpoints of P are in $\mathcal{M}_0^\star \cup \mathcal{W}_0^\star$, or one endpoint of P is in $\mathcal{M}_0^\star \cup \mathcal{W}_0^\star$, and its other endpoint is not distinguished. We will call an augmenting path P *masculine* or *feminine* if it contains a man in $\mathcal{M}_0^\star$ or a woman in $\mathcal{W}_0^\star$, respectively; if P is both masculine and feminine, then we call it *neutral*. If P is not neutral, we say that it *starts* at the (unique) person from $\mathcal{M}_0^\star \cup \mathcal{W}_0^\star$ it contains, and *ends* at its other endpoint.

Covering Constraints on One Side. We give a polynomial-time algorithm for SMC-1 when each man finds at most two women acceptable, and show NP-hardness of SMC-1 even if each woman finds at most two men acceptable.

Theorem 9. *There is a polynomial-time algorithm for the special case of SMC-1 where each man finds at most two women acceptable.*

The main observation behind Theorem 9 is that if $\Delta_\mathcal{M} \leq 2$, then any two augmenting paths starting from different women in $\mathcal{W}_0^\star$ can only intersect at their endpoints. Thus, we can modify the stable matching M_s by selecting augmenting paths starting from each woman in $\mathcal{W}_0^\star$ in an almost independent fashion: intu-itively, we simply need to take care not to choose paths sharing an endpoint—a task which can be managed by finding a bipartite matching in certain auxiliary graph G'. To ensure that the number of blocking pairs in the output is minimized, we will assign costs to the augmenting paths. The cost of an augmenting path P roughly determines the number of blocking pairs introduced when modifying M_s along P (though certain special edges need not be counted); hence, our problem reduces to finding a min-weight bipartite matching in G'.

To present the algorithm of Theorem 9 in detail, we start with the following properties of augmenting paths which are easy to prove assuming that $\Delta_{\mathcal{M}} \leq 2$:

Proposition 1. *Let P_1 and P_2 be augmenting paths starting at w_1 and w_2, resp. If $w_1 \neq w_2$, then P_1 and P_2 are either vertex-disjoint, or they both end at some $m \in \mathcal{M}_0$, with $V(P_1) \cap V(P_2) = \{m\}$. If there is an edge $\{m, w\}$ of G (with $m \in \mathcal{M}$ and $w \in \mathcal{W}$) connecting P_1 and P_2, then $m \in \mathcal{M}_0$ and P_1 or P_2 must end at m. If $w_1 = w_2$ and P is the maximal common subpath of P_1 and P_2 starting at w_1, then either $V(P_1) \cap V(P_2) = V(P)$, or P_1 and P_2 both end at some $m \in \mathcal{M}_0$ and $V(P_1) \cap V(P_2) = V(P) \cup \{m\}$.*

With a set P of edges (typically a set of augmenting paths) where $M_s \triangle P$ is a matching, we associate a *cost*, which is the number of blocking pairs that $M_s \triangle P$ admits. A pair $\{m, w\}$ for some $m \in \mathcal{M}$ and $w \in \mathcal{W}$ is *special*, if $m \in \mathcal{M}_0$ and w is the second (less preferred) woman in $L(m)$. As it turns out, such edges can be ignored during certain steps of the algorithm; thus, we let the *special cost* of P be the number of non-special blocking pairs in $M_s \triangle P$.

Lemma 1 ($\star$). *For vertex-disjoint augmenting paths P_1 and P_2 with costs c_1 and c_2, resp., the cost of $P_1 \cup P_2$ is at most $c_1 + c_2$. Further, if the cost of $P_1 \cup P_2$ is less than $c_1 + c_2$, then the following holds for $\{i_1, i_2\} = \{1, 2\}$: there is a special edge $\{m, w\}$ with P_{i_1} ending at m and w appearing on P_{i_2}; moreover, $\{m, w\}$ is blocking in $M_s \triangle P_{i_2}$, but not in $M_s \triangle (P_1 \cup P_2)$.*

We are ready to provide the algorithm, in a sequence of four steps.

- **Step 1: Computing all augmenting paths.** By Proposition 1, if we delete $\mathcal{M}_0$ from the union of all augmenting paths starting at some $w \in \mathcal{W}_0^\star$, then we obtain a tree. Furthermore, these trees are mutually vertex-disjoint for different starting vertices of $\mathcal{W}_0^\star$. This allows us to compute all augmenting paths in linear time, e.g., by an appropriately modified version of the DFS algorithm (so that only augmenting paths are considered). During this process, we can also compute the special cost of each augmenting path in a straightforward way.

- **Step 2: Constructing an auxiliary graph.** Using the results of the computation of Step 1, we construct an edge-weighted single bipartite graph G_{path} as follows. The vertex set of G_{path} is the union of $\mathcal{W}_0^\star$ and $\mathcal{M}_0 \cup \{w' \mid w \in \mathcal{W}_0^\star\}$, so for each woman $w \in \mathcal{W}_0^\star$ we create a corresponding new vertex w'. We add an edge between $w \in \mathcal{W}_0^\star$ and $m \in \mathcal{M}_0$ with weight c if there exists an augmenting path with endpoints w and m having special cost c (and no such path with lower special cost exists). Further, for each $w \in \mathcal{W}_0^\star$ we compute the minimum special cost $c_w^{\min}$ of any augmenting path starting at w and *not* ending in $\mathcal{M}_0$, and add an edge between w and w' with weight $c_w^{\min}$ in G_{path}.

- **Step 3: Computing a minimum weight matching.** We compute a matching M_P in G_{path} covering $\mathcal{W}_0^\star$ and having minimum weight. Observe that such a matching corresponds to a set of augmenting paths $\mathcal{P} = \{P_w \mid w \in \mathcal{W}_0^\star\}$ that are mutually vertex-disjoint by Proposition 1. Recall that the special cost of P_w is the weight of the edge in M_P incident to w.

- **Step 4: Eliminating blocking special edges.** In this step, we modify $\mathcal{P}$ iteratively. We start by setting $\mathcal{P}_{\text{act}} = \mathcal{P}$. At each iteration we modify $\mathcal{P}_{\text{act}}$ as follows. We check whether there exists a special edge $\{m^*, w^*\}$ that is blocking in $M_s \triangle \mathcal{P}_{\text{act}}$. If yes, then notice that m^* is not matched in $M_s \triangle \mathcal{P}_{\text{act}}$, because $\{m^*, w^*\}$ is special and thus $m^* \in \mathcal{M}_0$. Let P be the path of $\mathcal{P}_{\text{act}}$ containing w^*. We modify $\mathcal{P}_{\text{act}}$ by truncating P to its subpath between its starting vertex and w^*, and appending to it the edge $\{m^*, w^*\}$. This way, $\{m^*, w^*\}$ becomes an edge of the matching $M_s \triangle \mathcal{P}_{\text{act}}$. The iteration stops when there is no special edge blocking $M_s \triangle \mathcal{P}_{\text{act}}$. Note that once a special edge ceases to be blocking in $M_s \triangle \mathcal{P}_{\text{act}}$, it cannot become blocking again during this process, so the algorithm performs at most $|\mathcal{M}_0|$ iterations. For each $w \in \mathcal{W}_0^\star$, let P_w^* denote the augmenting path in $\mathcal{P}_{\text{act}}$ covering w at the end of Step 4; we define $\mathcal{P}^* = \{P_w^* \mid w \in \mathcal{W}_0^\star\}$ and output the matching $M_s \triangle \mathcal{P}^*$.

This completes the description of the algorithm; we now provide its analysis.

Lemma 2 ($\star$). $M_{\text{sol}} := M_s \triangle \mathcal{P}^*$ *is a feasible matching for I, and the number of blocking pairs for M_{sol} is at most the weight of M_P.*

To show that our algorithm is correct and M_{sol} is optimal, by Lemma 2 it suffices to prove that the weight of M_P is at most the number of blocking pairs in M^{opt}, an optimal solution in I. To this end, we define a matching covering $\mathcal{W}_0^\star$ in G_{path} whose weight is at most the number of blocking pairs in M^{opt}.

Clearly, $M_s \triangle M^{\text{opt}}$ contains an augmenting path Q_w covering w for each $w \in \mathcal{W}_0^\star$. If some Q_w ends at a man $m \in \mathcal{M}_0$, then clearly no other path in $M_s \triangle M^{\text{opt}}$ can end at m. Take the matching M_Q in G_{path} that includes all pairs $\{m, w\}$ where Q_w ends at $m \in \mathcal{M}_0$ for some $w \in \mathcal{W}_0^\star$. Also, we put $\{w, w'\}$ into M_Q if Q_w does not end at a man of $\mathcal{M}_0$. Note that M_Q is indeed a matching.

It remains to show that the weight of M_Q is at most the number of blocking pairs in M^{opt}. By definition, the weight of M_Q is at most the sum of the special costs of the paths Q_w for every $w \in \mathcal{W}_0^\star$. By Lemma 1, any non-special blocking pair in $M_s \triangle Q_w$ remains a blocking pair in $M_s \triangle (\bigcup_{w \in \mathcal{W}_0^\star} Q_w)$, and hence in M^{opt} as well. Hence, there is a matching in G_{path} with weight at most the number of blocking pairs in an optimal solution, implying the correctness of our algorithm. As the algorithm runs in polynomial time, Theorem 9 follows.

Contrasting Theorem 9, if men may have preference lists of length 3, SMC-1 (and hence SMC) is NP-hard even if each woman finds at most two men acceptable.

Theorem 10 ($\star$). *SMC-1 is* NP-*hard even if* $\Delta_{\mathcal{W}} = 2$ *and* $\Delta_{\mathcal{M}} = 3$.

Covering Constraints on Both Sides. If $\max(\Delta_{\mathcal{W}}, \Delta_{\mathcal{M}}) \leq 2$, the graph underlying the instance is a collection of paths and cycles, and therefore:

Observation 11. *SMC with* $\max(\Delta_{\mathcal{W}}, \Delta_{\mathcal{M}}) \leq 2$ *is polynomial-time solvable.*

Recall that the case where $\Delta_{\mathcal{W}} = 2$ and $\Delta_{\mathcal{M}} = 3$ is NP-hard by Theorem 10, even if there are no distinguished men to be covered. However, switching the role of men and women, Theorem 9 shows that if there are no women to be covered, then $\Delta_{\mathcal{W}} \leq 2$ guarantees polynomial-time solvability for SMC. This raises the natural question whether SMC with $\Delta_{\mathcal{W}} \leq 2$ can be solved efficiently if the number of distinguished women is bounded. Next we show that this is unlikely, as the problem turns out to be NP-hard for $|\mathcal{W}^\star| = 1$.

Theorem 12 ($\star$). *SMC is* NP-*hard, even if* $\Delta_{\mathcal{W}} = 2$, $\Delta_{\mathcal{M}} = 3$ *and* $|\mathcal{W}^\star| = 1$.

Contrasting Theorem 12, we establish fixed-parameter tractability of the case $\Delta_{\mathcal{W}} \leq 2$. The relevant cases (whose tractability or intractability does not follow from our results obtained so far) are as follows (assuming $\Delta_{\mathcal{W}} \leq 2$ throughout). First, we can take the number of distinguished persons as parameter (note that we know NP-hardness of the cases where $|\mathcal{W}^\star| = 1$ or $|\mathcal{M}^\star| = 0$). Second, we can consider the number of blocking pairs as the parameter. We show the following:

Theorem 13 ($\star$). *There is a fixed-parameter algorithm for the special case of* SMC *where each woman finds at most two men acceptable (i.e., $\Delta_{\mathcal{W}} \leq 2$), with parameter the number $|\mathcal{W}_0^\star| + |\mathcal{M}_0^\star|$ of distinguished men and women left unmatched by some stable matching (and hence by any stable matching).*

As each augmenting path contains at least one edge that blocks M^{opt}, the number of blocking pairs admitted by M^{opt} is at least $(|\mathcal{W}_0^\star| + |\mathcal{M}_0^\star|)/2$. Thus:

Corollary 2 ($\star$). *There is a fixed-parameter algorithm with parameter b for the special case of* SMC *where each woman finds at most two men acceptable.*

References

1. Arulselvan, A., Cseh, Á., Groß, M., Manlove, D.F., Matuschke, J.: Matchings with lower quotas: algorithms and complexity. Algorithmica (2016)
2. Aziz, H., Seedig, H.G., von Wedel, J.K.: On the susceptibility of the deferred acceptance algorithm. In: Proceedings of AAMAS 2015, pp. 939–947 (2015)
3. Bartholdi III, J.J., Tovey, C.A., Trick, M.A.: How hard is it to control an election? Math. Comput. Modell. **16**(8–9), 27–40 (1992)
4. Biró, P., Fleiner, T., Irving, R.W., Manlove, D.F.: The college admissions problem with lower and common quotas. Theoret. Comput. Sci. **411**(34–36), 3136–3153 (2010)
5. Biró, P., Irving, R.W., Schlotter, I.: Stable matching with couples: an empirical study. ACM J. Exp. Algorithmics **16**, 1–2 (2011)

6. Biró, P., Manlove, D.F., McDermid, E.J.: "Almost stable" matchings in the room-mates problem with bounded preference lists. Theoret. Comput. Sci. **432**, 10–20 (2012)
7. Cechlárová, K., Fleiner, T.: Pareto optimal matchings with lower quotas. In: Proceedings of COMSOC 2016 (2016)
8. Cseh, Á., Manlove, D.F.: Stable marriage and roommates problems with restricted edges: complexity and approximability. Discret. Optim. **20**, 62–89 (2016)
9. Cygan, M., Fomin, F.V., Kowalik, Ł., Lokshtanov, D., Marx, D., Pilipczuk, M., Pilipczuk, M., Saurabh, S.: Parameterized Algorithms. Springer, Cham (2015)
10. Dias, V.M., da Fonseca, G.D., de Figueiredo, C.M., Szwarcfiter, J.L.: The stable marriage problem with restricted pairs. Theoret. Comput. Sci. **306**(1–3), 391–405 (2003)
11. Ehlers, L., Hafalir, I.E., Yenmez, M.B., Yildirim, M.A.: School choice with controlled choice constraints: hard bounds versus soft bounds. J. Econ. Theory **153**, 648–683 (2014)
12. Fleiner, T., Irving, R.W., Manlove, D.F.: Efficient algorithms for generalised stable marriage and roommates problems. Theor. Comput. Sci. **381**, 162–176 (2007)
13. Fleiner, T., Kamiyama, N.: A matroid approach to stable matchings with lower quotas. Math. Oper. Res. **41**(2), 734–744 (2016)
14. Fragiadakis, D., Iwasaki, A., Troyan, P., Ueda, S., Yokoo, M.: Strategyproof matching with minimum quotas. ACM Trans. Econ. Comput. **4**(1), 6 (2016)
15. Fragiadakis, D., Troyan, P.: Improving matching under hard distributional constraints. Theoret. Econ. **12**, 864–908 (2017)
16. Gale, D., Shapley, L.S.: College admissions and the stability of marriage. Amer. Math. Mon. **69**(1), 9–15 (1962)
17. Gale, D., Sotomayor, M.: Some remarks on the stable matching problem. Discret. Appl. Math. **11**(3), 223–232 (1985)
18. Goto, M., Iwasaki, A., Kawasaki, Y., Kurata, R., Yasuda, Y., Yokoo, M.: Strategyproof matching with regional minimum and maximum quotas. Artif. Intell. **235**, 40–57 (2016)
19. Gusfield, D., Irving, R.W.: The Stable Marriage Problem: Structure and Algorithms. MIT Press, Cambridge (1989)
20. Hamada, K., Iwama, K., Miyazaki, S.: The hospitals/residents problem with lower quotas. Algorithmica **74**(1), 440–465 (2016)
21. Huang, C.-C.: Classified stable matching. In: Proceedings of SODA 2010, pp. 1235–1253 (2010)
22. Immorlica, N., Mahdian, M.: Marriage, honesty, and stability. In: Proceedings of SODA 2005, pp. 53–62 (2005)
23. Impagliazzo, R., Paturi, R., Zane, F.: Which problems have strongly exponential complexity? J. Comput. System Sci. **63**(4), 512–530 (2001)
24. Irving, R.W., Manlove, D., O'Malley, G.: Stable marriage with ties and bounded length preference lists. J. Discret. Algorithms **7**, 213–219 (2009)
25. Kamiyama, N.: A note on the serial dictatorship with project closures. Oper. Res. Lett. **41**(5), 559–561 (2013)
26. Knuth, D.E.: Mariages stables et leurs relations avec d'autres problèmes combinatoires. Les Presses de l'Université de Montréal, Montreal (1976)
27. Kojima, F., Pathak, P.A., Roth, A.E.: Matching with couples: stability and incentives in large markets. Q. J. Econ. **128**(4), 1585–1632 (2013)
28. Kominers, S.D., Sönmez, T.: Matching with slot-specific priorities: theory. Theoret. Econ. **11**(2), 683–710 (2016)

29. Manlove, D.F.: Algorithmics of Matching Under Preferences. Series on Theoretical Computer Science, vol. 2. World Scientific, Singapore (2013)
30. Marx, D., Schlotter, I.: Parameterized complexity and local search approaches for the stable marriage problem with ties. Algorithmica **58**(1), 170–187 (2010)
31. Marx, D., Schlotter, I.: Stable assignment with couples: parameterized complexity and local search. Discret. Optim. **8**, 25–40 (2011)
32. Mnich, M., Schlotter, I.: Stable marriage with covering constraints: a complete computational trichotomy (2017). https://arxiv.org/abs/1602.08230
33. Monte, D., Tumennasan, N.: Matching with quorums. Econ. Lett. **120**(1), 14–17 (2013)
34. Roth, A.E., Peranson, E.: The redesign of the matching market for american physicians: some engineering aspects of economic design. Amer. Econ. Rev. **89**, 748–780 (1999)
35. Veskioja, T.: Stable marriage problem and college admission. Ph.D. thesis, Department of Informatics, Tallinn University of Technology (2005)
36. Westkamp, A.: An analysis of the German University admissions system. Econ. Theory **53**(3), 561–589 (2013)
37. Yokoi, Y.: A generalized polymatroid approach to stable matchings with lower quotas. Math. Oper. Res. **42**(1), 238–255 (2017)

Fairly Allocating Contiguous Blocks
of Indivisible Items

Warut Suksompong[(✉)]

Department of Computer Science, Stanford University,
353 Serra Mall, Stanford, CA 94305, USA
warut@cs.stanford.edu

Abstract. In this paper, we study the classic problem of fairly allocating indivisible items with the extra feature that the items lie on a line. Our goal is to find a fair allocation that is *contiguous*, meaning that the bundle of each agent forms a contiguous block on the line. While allocations satisfying the classical fairness notions of proportionality, envy-freeness, and equitability are not guaranteed to exist even without the contiguity requirement, we show the existence of contiguous allocations satisfying approximate versions of these notions that do not degrade as the number of agents or items increases. We also study the efficiency loss of contiguous allocations due to fairness constraints.

1 Introduction

We consider the classic problem in economics of *fair division*: How can we divide a set of resources among interested agents in such a way that the resulting division is fair? This is an important issue that occurs in a variety of situations, including students splitting the rent of an apartment, couples dividing their properties after a divorce, and countries staking claims in disputed territory. The fair division literature often distinguishes between two types of resources. Some resources, such as cake and land, are said to be *divisible* since they can be split arbitrarily among agents. Other resources, like houses and cars, are *indivisible*—each house or car must be allocated as a whole to one agent.

To reason about fairness, we must define what it means for an allocation of resources to be fair. Several notions of fairness have been proposed, three of the oldest and best-known of which are proportionality, envy-freeness, and equitability. An allocation is said to be *proportional* if the utility that each agent gets from the bundle she receives is at least a $1/n$ fraction of her utility for the whole set of resources, where n is the number of agents among whom we divide the resources. The allocation is called *envy-free* if every agent thinks that her bundle is at least as good as the bundle of any other agent, and *equitable* if all agents have the same utility for their own bundle. It turns out that there is a significant distinction between the two types of resources with respect to these notions. On the one hand, when resources are divisible, allocations that satisfy

---

The full version of this paper is available at http://arxiv.org/abs/1707.00345.

© Springer International Publishing AG 2017
V. Bilò and M. Flammini (Eds.): SAGT 2017, LNCS 10504, pp. 333–344, 2017.
DOI: 10.1007/978-3-319-66700-3_26

the three notions simultaneously always exist [1]. On the other hand, a simple example with two agents who both positively value a single item already shows that the existence of a fair division cannot be guaranteed for any of the notions when we deal with indivisible items.

In this paper, we study the problem of allocating indivisible items with the added feature that the items lie on a line. We are interested in finding a fair allocation that moreover satisfies the requirement of *contiguity*, i.e., the bundle that each agent receives forms a contiguous block on the line. Several practical applications fit into this model. For instance, when we divide offices between research groups on the same floor, it is desirable that each research group get a contiguous block of offices in order to facilitate communication within the group. Likewise, when we allocate retail units on a street, the retailers often prefer to have a contiguous block of units in order to operate a larger store. The contiguity condition can also be interpreted in the temporal sense, as opposed to the spatial sense described thus far. An example is a situation where various organizers wish to use the same conference center for their conferences. Not surprisingly, the organizers typically want to schedule a conference in a contiguous block of time rather than during several separate periods.

Since allocations that satisfy any of the three fairness notions do not always exist in general, the same is necessarily true when we restrict our attention to contiguous allocations. Nevertheless, we show that in light of the contiguity requirement, the existence of allocations that satisfy approximate versions of the notions can still be guaranteed. More precisely, for each notion we define an approximate version that depends on an additive factor $\epsilon \geq 0$. An allocation is said to be ϵ-proportional if the utility of each agent is at most ϵ away from her "proportional share", ϵ-envy-free if each agent envies any other agent by at most ϵ, and ϵ-equitable if the utilities of any two agents differ by at most ϵ. Denoting the maximum utility of an agent for an item by $u_{\max}$, we establish the existence of a contiguous $u_{\max}$-proportional allocation and a contiguous $u_{\max}$-equitable allocation for any number of agents, a contiguous $u_{\max}$-envy-free allocation for two agents, and a contiguous $2u_{\max}$-envy-free allocation for any number of agents. Importantly, the approximation factors do not degrade as the number of agents or items grows. We also prove that our approximation factor is the best possible for proportionality and equitability with any number of agents as well as for envy-freeness with two agents. Finally, for proportionality the factor can be improved to $\frac{n-1}{n} \cdot u_{\max}$ if we know the number n of agents, and we show that this is again tight.

Our results suggest that adding the contiguity requirement does not entail extra costs in terms of the approximation guarantees. Indeed, the approximation factors for proportionality and equitability with any number of agents and for envy-freeness with two agents remain tight even if we allow arbitrary allocations. This can be seen as somewhat surprising, since the space of contiguous allocations is significantly smaller than that of arbitrary allocations. Indeed, when there are n agents and m items, the number of arbitrary allocations is n^m, while the number of contiguous allocations for a fixed order of items on a line is at most $\binom{m+n-1}{n-1}n!$. The latter quantity is much less than the former if m is large compared to n.

In addition, we investigate the efficiency loss of contiguous allocations due to fairness constraints using the *price of fairness* concept initiated by Caragiannis et al. [8]. The price of fairness quantifies the loss of social welfare that is necessary if we impose a fairness constraint on the allocation. A low price of fairness means that we can get fairness at virtually no extra cost on social welfare, while a high price of fairness implies that even the most efficient "fair" allocation has social welfare far below that of the most efficient allocation overall. Caragiannis et al. studied the price of fairness for the three notions of fairness using utilitarian welfare for both divisible and indivisible items. Later, Aumann and Dombb [2] focused on contiguous allocations of divisible items and considered both utilitarian and egalitarian welfare. In this paper, we complete the picture by providing tight or almost tight bounds on the price of fairness for contiguous allocations of indivisible items, again for all three classical notions of fairness and with respect to both utilitarian and egalitarian welfare. Our results are summarized in Table 1 along with a comparison to results from previous work.

Table 1. Comparison of our results on the price of fairness to previous results in [2,8]. The bounds with an asterisk hold for infinitely many values of n.

Indivisible	Contiguous (this work)			Non-contiguous ([8])	
	Utilitarian		Egalitarian	Utilitarian	
	Lower	Upper		Lower	Upper
Proportionality	$n-1+\frac{1}{n}$		1	$n-1+\frac{1}{n}$	
Equitability	$\frac{3}{2}$ for $n=2$		1 for $n=2$	2 for $n=2$	
	∞ for $n>2$		∞ for $n>2$	∞ for $n>2$	
Envy-freeness	$\frac{\lfloor\sqrt{n}\rfloor}{2}$	$\frac{\sqrt{n}}{2}+1-o(1)$	$\frac{n}{2}$	$\frac{3n+7}{9}-O\left(\frac{1}{n}\right)$	$n-\frac{1}{2}$
Divisible	Contiguous ([2])			Non-contiguous ([8])	
Proportionality	$\frac{\sqrt{n}}{2}$*	$\frac{\sqrt{n}}{2}+1-o(1)$	1	$\Omega(\sqrt{n})$	$O(\sqrt{n})$
Equitability	$n-1+\frac{1}{n}$	n	1	$\frac{(n+1)^2}{4n}$	n
Envy-freeness	$\frac{\sqrt{n}}{2}$*	$\frac{\sqrt{n}}{2}+1-o(1)$	$\frac{n}{2}$	$\Omega(\sqrt{n})$	$n-\frac{1}{2}$

1.1 Related Work

The contiguity condition has been studied with respect to the three classical fairness notions in the context of divisible items, often represented by a cake, with the motivation that one wants to avoid giving an agent a "union of crumbs". In particular, Dubins and Spanier [13] exhibited a moving-knife algorithm that guarantees a contiguous proportional allocation. Cechlárová et al. [10] showed that for any ordering of the agents, a contiguous equitable allocation that assigns contiguous pieces to the agents in that order exists. Stromquist [17,18] proved that a contiguous envy-free allocation always exists, but cannot be found by a finite algorithm. Su [19] used techniques involving Sperner's lemma to establish the existence of a contiguous envy-free allocation and moreover considered the related problem of rent partitioning.

Recently, Bouveret et al. [7] studied the allocation of indivisible items on a line with the contiguity condition and showed that determining whether a contiguous fair allocation exists is NP-hard when the fairness notion considered is either proportionality or envy-freeness. They also considered a more general model of the relationship between items where the items are vertices of an undirected graph. Aumann et al. [3] investigated the problem of finding a contiguous allocation that maximizes welfare for both divisible and indivisible items. They showed that while it is NP-hard to find the optimal contiguous allocation, there exists an efficient algorithm that yields a constant factor approximation. Bei et al. [5] and Cohler et al. [11] also considered the objective of maximizing welfare, but under the additional fairness constraint of proportionality and envy-freeness, respectively.

Additively approximating fairness notions using $u_{\max}$, the highest utility of an agent for an item, has been studied before. Lipton et al. [15] showed that without the contiguity requirement, a $u_{\max}$-envy-free allocation exists even for general monotone valuations. Caragiannis et al. [9] used the term "envy-freeness up to one good" (EF1) to refer to a closely related property of an allocation.

Besides the allocation of goods, the price of fairness has also been investigated for the allocation of chores. In particular, Caragiannis et al. [8], who initiated this line of research, studied the notion for both divisible and indivisible chores. Heydrich and van Stee [14] likewise considered the setting of divisible chores but, similarly to our work and that of Aumann and Dombb [2], focused on contiguous allocations. Finally, Bilò et al. [6] applied this concept to machine scheduling problems.

2   Preliminaries

Let $N = \{1, 2, \ldots, n\}$ denote the set of agents, and $M = \{1, 2, \ldots, m\}$ the set of items to be allocated. We assume that the items lie on a line in this order.

Each agent $i \in N$ has some nonnegative utility $u_i(j)$ for item $j \in M$. For an agent i, define $u_{i,\max} := \max_{j \in M} u_i(j)$ to be the highest utility of i for an item. Let $u_{\max} := \max_{i \in N} u_{i,\max}$ be the highest utility of any agent for an item. As is very common, we assume for most of the paper that utilities are *additive*. Additivity means that $u_i(M') = \sum_{j \in M'} u_i(j)$ for any agent i and any subset of items $M' \subseteq M$. An *allocation* $\mathcal{M} = (M_1, \ldots, M_n)$ is a partition of all items into bundles for the agents so that agent i receives bundle M_i. The *utilitarian welfare* of $\mathcal{M}$ is $\sum_{i \in N} u_i(M_i)$ and the *egalitarian welfare* of $\mathcal{M}$ is $\min_{i \in N} u_i(M_i)$. We call the allocation *contiguous* if each bundle M_i forms a contiguous block of items on the line. Furthermore, we refer to a setting with agents, items, and utility functions as an *instance*.

We are now ready to define the fairness notions that we will consider in this paper. We use additive versions of approximation; this is much stronger than multiplicative versions as the number of items grows.

Definition 1. *An allocation* $\mathcal{M} = (M_1, \ldots, M_n)$ *is said to be* proportional *if* $u_i(M_i) \geq \frac{1}{n} \cdot u_i(M)$ *for all* $i \in N$. *For* $\epsilon \geq 0$, *the allocation is said to be* ϵ-proportional *if* $u_i(M_i) \geq \frac{1}{n} \cdot u_i(M) - \epsilon$ *for all* $i \in N$. *We refer to* $\frac{1}{n} \cdot u_i(M)$ *as* the proportional share *of agent* i.

Definition 2. *An allocation* $\mathcal{M} = (M_1, \ldots, M_n)$ *is said to be* envy-free *if* $u_i(M_i) \geq u_i(M_j)$ *for all* $i, j \in N$. *For* $\epsilon \geq 0$, *the allocation is said to be* ϵ-envy-free *if* $u_i(M_i) \geq u_i(M_j) - \epsilon$ *for all* $i, j \in N$.

Definition 3. *An allocation* $\mathcal{M} = (M_1, \ldots, M_n)$ *is said to be* equitable *if* $u_i(M_i) = u_j(M_j)$ *for all* $i, j \in N$. *For* $\epsilon \geq 0$, *the allocation is said to be* ϵ-equitable *if* $|u_i(M_i) - u_j(M_j)| \leq \epsilon$ *for all* $i, j \in N$.

There is a strong relation between proportionality and envy-freeness, as the following proposition shows.

Proposition 1. *Any* ϵ-envy-free allocation is ϵ-proportional.

The proof of Proposition 1 can be found in the full version of this paper. In particular, when $\epsilon = 0$, the proposition reduces to the well-known fact that any envy-free allocation is proportional. When there are two agents, proportional allocations are also envy-free (and in fact, more generally, ϵ-proportional allocations are 2ϵ-envy-free.) This is, however, not necessarily the case if there are at least three agents. An example is when an agent values her own bundle $1/n$ of the whole set of items and values the bundle of another agent the remaining $(n - 1)/n$ of the whole set of items. On the other hand, equitability neither implies nor is implied by proportionality or envy-freeness.

For each of the three (non-approximate) fairness notions defined above, the set of instances for which a contiguous allocation satisfying the notion exists is strictly smaller than the corresponding set when contiguity is not required. Indeed, suppose that there are three items and two agents who share a common utility function u with $u(1) = u(3) = 1$ and $u(2) = 2$. An allocation in which one agent gets items 1 and 3 while the other agent gets item 2 is proportional, envy-free, and equitable. In contrast, no contiguous allocation satisfies any of the three properties.

We end this section by giving the definition of the various forms of the price of fairness.

Definition 4. *Given an instance (along with a set of allocations considered), its* utilitarian price of proportionality *(resp.,* utilitarian price of equitability, utilitarian price of envy-freeness*) is defined as the ratio of the utilitarian welfare of the optimal allocation over the utilitarian welfare of the best proportional (resp., equitable, envy-free) allocation. If a proportional (resp., equitable, envy-free) allocation does not exist, the utilitarian price of proportionality (resp., equitability,* *envy-freeness) is not defined for that instance. The (overall) utilitarian* price of proportionality *(resp., utilitarian price of equitability, utilitarian price of envy-freeness) is then the supremum utilitarian price of proportionality (resp., utilitarian price of equitability, utilitarian price of envy-freeness) over all instances.*

The egalitarian price of proportionality, egalitarian price of equitability, *and* egalitarian price of envy-freeness *are defined analogously.*

3   Proportionality

We begin with proportionality. Our first result shows the existence of a contiguous allocation in which every agent receives at least her proportional share less $\frac{n-1}{n}$ times her utility for her highest-valued item.

Theorem 1. *Given any instance, there exists a contiguous allocation $\mathcal{M}$ such that $u_i(M_i) \geq \frac{1}{n} \cdot u_i(M) - \frac{n-1}{n} \cdot u_{i,max}$ for all agents $i \in N$. In particular, there exists a contiguous $\frac{n-1}{n} \cdot u_{max}$-proportional allocation.*

Proof. We process the items from left to right using the following algorithm.

1. Set the current block to the empty block.
2. If the current block yields utility at least $\frac{1}{n} \cdot u_i(M) - \frac{n-1}{n} \cdot u_{i,\text{max}}$ to some agent i, give the block to the agent. (If several agents satisfy this condition, choose one arbitrarily.)
   - If all agents have received a block as a result of this, allocate the leftover items arbitrarily and terminate.
   - Otherwise, if some agent receives a block in this step, remove that agent from consideration and return to Step 1.
3. Add the next item to the current block and return to Step 2.

If an agent i receives a block of items from this algorithm, she obtains utility at least $\frac{1}{n} \cdot u_i(M) - \frac{n-1}{n} \cdot u_{i,\text{max}}$. Hence it suffices to show that the algorithm allocates a block to every agent. To this end, we show by (backward) induction that when there are k agents who have not been allocated a block, each agent i among them has utility at least $\frac{k}{n} \cdot u_i(M) - \frac{n-k}{n} \cdot u_{i,\text{max}}$ for the remaining items. This will imply that the last agent has utility at least $\frac{1}{n} \cdot u_i(M) - \frac{n-1}{n} \cdot u_{i,\text{max}}$ left, which is enough to satisfy our condition.

The base case $k = n$ trivially holds. For the inductive step, assume that the statement holds when there are $k + 1$ agents left, and consider an agent i who is *not* the next one to receive a block. When there are $k + 1$ agents left, her utility for the remaining items is at least $\frac{k+1}{n} \cdot u_i(M) - \frac{n-k-1}{n} \cdot u_{i,\text{max}}$. Since she does not receive the next block, her utility for the block excluding its last item is less than $\frac{1}{n} \cdot u_i(M) - \frac{n-1}{n} \cdot u_{i,\text{max}}$. This means that her utility for the block is less than $\frac{1}{n} \cdot u_i(M) + \frac{1}{n} \cdot u_{i,\text{max}}$. Hence her utility for the remaining items is at least $\left(\frac{k+1}{n} \cdot u_i(M) - \frac{n-k-1}{n} \cdot u_{i,\text{max}}\right) - \left(\frac{1}{n} \cdot u_i(M) + \frac{1}{n} \cdot u_{i,\text{max}}\right)$, which is equal to $\frac{k}{n} \cdot u_i(M) - \frac{n-k}{n} \cdot u_{i,\text{max}}$, as desired. □

The algorithm in Theorem 1 is similar to the Dubins-Spanier algorithm for proportional cake-cutting [13] and runs in time $O(mn)$, which is the best possible since the input also has size $O(mn)$. It can also be implemented as a mechanism that does not elicit the full utility functions from the agents, but instead asks them to indicate when the value of the current block reaches their threshold. While the mechanism is not truthful, a truthful agent always receives no less than her proportional share minus $\frac{n-1}{n}$ times her utility for the item she values most.

As the next example shows, the additive approximation factor $\frac{n-1}{n} \cdot u_{\max}$ is the best possible in the sense that the existence of a contiguous $\alpha \cdot \frac{n-1}{n} \cdot u_{\max}$-proportional allocation is not guaranteed for any $\alpha < 1$. In fact, this is the case even if we remove the contiguity requirement.

Example 1. Suppose that there are $m = n - 1$ items any of which each agent has a utility of 1. The proportional share of every agent is $\frac{n-1}{n}$. On the other hand, in any (not necessarily contiguous) allocation, some agent does not receive an item and therefore has a utility of 0. For any fixed $\alpha < 1$, the utility of this agent is less than her proportional share minus $\alpha \cdot \frac{n-1}{n} \cdot u_{\max}$.

Even though a contiguous $u_{\max}$-proportional allocation always exists, in some cases we might also want to choose the ordering on the line in which the agents are allocated blocks of items, in addition to imposing the contiguity requirement. For instance, the owner of a conference center could have a preferred lineup of the conferences, and a building manager might want to assign offices in certain parts of the floor to certain research groups. Nevertheless, the following example shows that for approximate proportionality, the ordering cannot always be chosen arbitrarily.

Example 2. Suppose that there are two agents and $m \geq 6$ items. The first agent has utility 1 for the last three items and 0 otherwise, while the second agent has utility 1 for every item. The proportional share of the two agents less half of the utility for their highest-valued item is 1 and $\frac{m-1}{2}$, respectively. If we want to give a left block to the first agent and the remaining right block to the second agent, the left block needs to include up to item $m - 2$. But this means that the second agent gets utility at most 2, which is less than the required $\frac{m-1}{2}$.

By increasing m and the number of items for which the first agent has utility 1, we can extend the example to show that the existence of a contiguous allocation with a fixed ordering of agents is not guaranteed even if we weaken the approximation factor $u_{\max}$ to $ku_{\max}$ for any $k > 1$.

4 Equitability

We next consider equitability. As with proportionality, we show that a contiguous allocation in which the values of different agents for their own block differ by no more than $u_{\max}$ always exists. Unlike for proportionality, however, for equitability we can additionally choose the order in which the agents receive blocks on the line.

Theorem 2. *Given any instance and any ordering of the agents, there exists a contiguous u_{max}-equitable allocation in which the agents are allocated blocks of items on the line according to the ordering.*

Note that in order to ensure that all agents are treated equally, one should normalize the utilities across agents before applying Theorem 2, for example by rescaling the utilities so that $u_{i,\max} = 1$ for all $i \in N$.

Proof. Assume without loss of generality that the required ordering of agents is $1, 2, \ldots, n$ from left to right. Start with an arbitrary allocation satisfying the ordering. For any allocation $\mathcal{M}$, let $\max(\mathcal{M}) = \max_{i=1}^{n} u_i(M_i)$ and $\min(\mathcal{M}) = \min_{i=1}^{n} u_i(M_i)$. In each iteration, as long as the allocation does not satisfy the approximate equitability, we will move an item at the end of a block to the block that the item is adjacent to. Here is the description of the algorithm.[1]

1. Choose a block M_i such that $u_i(M_i) = \max(\mathcal{M})$. If there are many such blocks, choose one arbitrarily.
2. If $\max(\mathcal{M}) \leq \min(\mathcal{M}) + u_{\max}$, stop and return the current allocation.
3. Choose a block M_j such that $u_j(M_j) = \min(\mathcal{M})$. If there are many such blocks, choose arbitrarily from the ones that minimize $|j - i|$.
4. Let M_k be the block between M_i and M_j that is next to M_j, i.e., $k = j - 1$ if $j > i$ and $k = j + 1$ if $j < i$. (It is possible that $k = i$.) The block M_k must be non-empty; otherwise we would have chosen M_k instead of M_j. Move the item in M_k that is adjacent to M_j to M_j.
   (a) If $k = i$ and the moved item has nonzero utility for agent i, go to Step 1.
   (b) Else, go to Step 2.

If the algorithm terminates, then as the ordering of the agents for the blocks never changed, the algorithm returns a $u_{\max}$-equitable allocation with the desired ordering. Hence it remains to show that the algorithm terminates.

To this end, observe that when an item is moved in Step 4 of the algorithm, no new block with utility $\max(\mathcal{M})$ or more to the block owner is created. Indeed, the block that gets an additional item now yields utility at most $\min(\mathcal{M}) + u_{\max}$ to its owner, which is less than $\max(\mathcal{M})$ since the condition in Step 2 is not yet satisfied. Moreover, since items are only being moved farther away from the main block M_i, we must eventually reach a point where $k = i$; formally, the quantity $\sum_{z=1}^{n} |i - z||M_z|$ strictly increases, where $|M_z|$ denotes the number of items in M_z. Since the quantity is bounded from above, after a finite number of moves we will have $k = i$ and meet the condition of Step 4(a).

The argument in the previous paragraph shows that the number of blocks with utility $\max(\mathcal{M})$ decreases during the course of the algorithm. When this number reaches zero, the value $\max(\mathcal{M})$ decreases. Since there are only a finite number of allocations, the algorithm must terminate, as claimed. □

---

[1] The algorithm is inspired by work on block partitions of sequences [4].

The algorithm in Theorem 2 runs in time $O(n^2 m^4)$. For each iteration, computing the maximum and minimum blocks takes $O(m)$. There are $O(m^2)$ possible blocks. Each block cannot be used as the block M_i in Step 1 more than once for each of the n agents, since no new blocks with utility $\max(\mathcal{M})$ is created during an execution of the algorithm. Finally, once the block M_i is fixed, the quantity $\sum_{z=1}^n |i-z||M_z|$ can increase at most $O(mn)$ times, yielding the claimed running time.

Example 1 shows that for any number of agents, the approximation factor $u_{\max}$ for equitability cannot be improved even if we remove the contiguity requirement (and hence also the ordering). On the other hand, using the same algorithm, we can generalize Theorem 2 to any monotonic, not necessarily additive utility function with zero utility for the empty set. In particular, a $u_{\max}$-equitable allocation can be found when agents are endowed with such utility functions, where the generalized definition of $u_{\max}$ is the highest marginal utility of any agent for a single item, i.e., $u_{\max} = \max_{i \in N, j \in M, S \subseteq M}(u_i(S \cup \{j\}) - u_i(S))$.

Although the algorithm in Theorem 2 guarantees that an approximate equitable allocation exists, such an allocation can be "equally bad" rather than "equally good" for the agents. Indeed, if we start with an allocation that yields zero utility to every agent, then the algorithm will terminate immediately despite the possible existence of an equitable allocation with positive utility for all agents. If we insist on choosing the ordering of the agents, then the next example shows that a situation that leaves some agent unhappy may be unavoidable.

Example 3. Suppose that there are two items and two agents with $u_1(2) = u_2(1) = 1$ and $u_1(1) = u_2(2) = 0$. The allocation that gives item 1 to agent 2 and item 2 to agent 1 yields a utility of 1 to both agents. If we require that agent 1 receive a left block and agent 2 a right block, however, some agent is necessarily left with no utility.

Nevertheless, we show next that if we allow the freedom of choosing the ordering of the agents, then an allocation with a better efficiency guarantee for the agents can always be found. In particular, we can find an allocation whose egalitarian welfare equals the highest egalitarian welfare over all contiguous allocations of the instance. The proof mirrors that of the analogous result for divisible items by Aumann and Dombb [2] and can be found in the full version of this paper.

Theorem 3. *Given any instance, there exists a contiguous u_{max}-equitable allocation whose egalitarian welfare equals the highest egalitarian welfare over all contiguous allocations of the instance.*

As is the case for Theorem 2, the result can be generalized to monotonic, not necessarily additive utility functions with zero utility for the empty set, where $u_{\max}$ is again defined as the highest marginal utility of any agent for a single item. A similar argument also shows that for any ordering of the agents, there exists a contiguous $u_{\max}$-equitable allocation whose egalitarian welfare equals the highest egalitarian welfare over all contiguous allocations *with that ordering of the agents* for the instance. However, the proof of Theorem 3 does not give rise to an efficient algorithm for computing a desired allocation.

5   Envy-Freeness

We now turn to envy-freeness. If we remove the contiguity requirement, it is well-known that a simple algorithm yields an ϵ-envy-free allocation for any number of agents and items: Let the agents pick their favorite item in a round-robin manner from the remaining items until all items are allocated. We show in this section that an ϵ-envy-free allocation exists when there are two agents, and a 2ϵ-envy-free allocation exists for an arbitrary number of agents.

For two agents, Theorem 1 directly implies the following.

Theorem 4. *Given any instance with two agents, there exists a contiguous allocation such that agent i has envy at most $u_{i,max}$ toward the other agent. In particular, there exists a contiguous u_{max}-envy-free allocation.*

Tightness of the approximation factor $u_{\max}$ follows from Example 1 with $n = 2$. Moreover, an example similar to Example 2 shows that the result does not hold if we fix the ordering of the agents, even when we replace $u_{\max}$ by $ku_{\max}$ for some $k > 1$.

To tackle the general setting with an arbitrary number of agents, we model the items as divisible items. Since a contiguous envy-free allocation always exists for divisible items [17], we can round such an allocation to obtain an approximate envy-free allocation for indivisible items.

Theorem 5. *Given any instance, there exists a contiguous allocation such that agent i has envy at most $2u_{i,max}$ toward any other agent. In particular, there exists a contiguous $2u_{max}$-envy-free allocation.*

Proof. Consider a cake represented by the interval $[0, m]$. For $j \in M$, agent i has uniform utility $u_i(j)$ for the interval $[j - 1, j]$. Take any contiguous envy-free allocation of the cake. We round the allocation as follows: Allocate each item j to the agent who owns point j of the cake.

The resulting allocation is contiguous; we show that each agent i has envy at most $2u_{i,\max}$. The agent has no envy before the rounding. As a result of the rounding, she loses utility at most $u_{i,\max}$, and any other agent gains utility at most $u_{i,\max}$ from her point of view. Hence agent i has envy at most $2u_{i,\max}$, as claimed.   □

6   Price of Fairness

In this section, we quantify the price of fairness for contiguous allocations of indivisible items with respect to the three notions of fairness and the two types of welfare. We derive tight or almost tight bounds for each of the six resulting combinations. Previous work has studied the problem for the setting of arbitrary (i.e., not necessarily contiguous) allocations of divisible and indivisible items [8] as well as contiguous allocations of divisible items [2]; our results therefore close the remaining gap. The comparison of our results to previous work is shown in Table 1. In fact, for several of the results we will be able to adjust arguments from previous work to our setting. We state our results below and leave the proofs to the full version of this paper.

Theorem 6. *The utilitarian price of proportionality for contiguous allocations of indivisible items is $n - 1 + \frac{1}{n}$.*

Theorem 7. *The utilitarian price of equitability for contiguous allocations of indivisible items is $\frac{3}{2}$ for $n = 2$ and infinite for $n > 2$.*

Theorem 8. *The utilitarian price of envy-freeness for contiguous allocations of indivisible items is in the interval $\left(\frac{\lfloor \sqrt{n} \rfloor}{2}, \frac{\sqrt{n}}{2} + 1 - o(1) \right)$.*

Theorem 9. *The egalitarian price of proportionality for contiguous allocations of indivisible items is 1.*

Theorem 10. *The egalitarian price of equitability for contiguous allocations of indivisible items is 1 for $n = 2$ and infinite for $n > 2$.*

Theorem 11. *The egalitarian price of envy-freeness for contiguous allocations of indivisible items is $\frac{n}{2}$.*

7   Conclusion and Future Work

In this paper, we study the problem of fairly allocating indivisible items on a line in such a way that each agent receives a contiguous block of items. This can be used to model a variety of practical situations, including allocating offices to research groups, retail units to retailers, and time slots for using a conference center to conference organizers. We show that we can find contiguous allocations that satisfy approximate versions of classical fairness notions. Notably, these approximation guarantees do not degrade as the number of agents or items grows. We also quantify the loss of efficiency that occurs when we impose fairness constraints on contiguous allocations.

We conclude the paper by presenting some directions for future work.

– For envy-freeness with an arbitrary number of agents, can we close the approximation factor gap between $u_{\max}$ and $2u_{\max}$? Can we obtain similar guarantees if we also require Pareto optimality?
– Can we show the asymptotic existence or non-existence of contiguous allocations satisfying proportionality or envy-freeness if we assume that the utilities are drawn from certain distributions? This has been shown for non-contiguous allocations [12, 16, 20].
– Does there exist an efficient algorithm that computes an approximate equitable allocation with a nontrivial welfare guarantee?
– How do the prices of fairness change if we define them with respect to approximate fair allocations (which always exist) instead of non-approximate fair allocations (which do not always exist)?

References

1. Alon, N.: Splitting necklaces. Adv. Math. **63**(3), 247–253 (1987)
2. Aumann, Y., Dombb, Y.: The efficiency of fair division with connected pieces. ACM Trans. Econ. Comput. **3**(4), 23 (2015)
3. Aumann, Y., Dombb, Y., Hassidim, A.: Computing socially-efficient cake divisions. In: Proceedings of the 12th International Conference on Autonomous Agents and Multiagent Systems, pp. 343–350 (2013)
4. Bárány, I., Grinberg, V.S.: Block partitions of sequences. Isr. J. Math. **206**(1), 155–164 (2015)
5. Bei, X., Chen, N., Hua, X., Tao, B., Yang, E.: Optimal proportional cake cutting with connected pieces. In: Proceedings of the 26th AAAI Conference on Artificial Intelligence, pp. 1263–1269 (2012)
6. Bilò, V., Fanelli, A., Flammini, M., Monaco, G., Moscardelli, L.: The price of envy-freeness in machine scheduling. Theor. Comput. Sci. **613**, 65–78 (2016)
7. Bouveret, S., Cechlárová, K., Elkind, E., Igarashi, A., Peters, D.: Fair division of a graph. In: Proceedings of the 26th International Joint Conference on Artificial Intelligence (2017, forthcoming)
8. Caragiannis, I., Kaklamanis, C., Kanellopoulos, P., Kyropoulou, M.: The efficiency of fair division. Theory Comput. Syst. **50**, 589–610 (2012)
9. Caragiannis, I., Kurokawa, D., Moulin, H., Procaccia, A.D., Shah, N., Wang, J.: The unreasonable fairness of maximum Nash welfare. In: Proceedings of the 17th ACM Conference on Economics and Computation, pp. 305–322 (2016)
10. Cechlárová, K., Doboš, J., Pillárová, E.: On the existence of equitable cake divisions. Inf. Sci. **228**, 239–245 (2013)
11. Cohler, Y.J., Lai, J.K., Parkes, D.C., Procaccia, A.D.: Optimal envy-free cake cutting. In: Proceedings of the 25th AAAI Conference on Artificial Intelligence, pp. 626–631 (2011)
12. Dickerson, J.P., Goldman, J., Karp, J., Procaccia, A.D., Sandholm, T.: The computational rise and fall of fairness. In: Proceedings of the 28th AAAI Conference on Artificial Intelligence, pp. 1405–1411 (2014)
13. Dubins, L.E., Spanier, E.H.: How to cut a cake fairly. Am. Math. Mon. **68**(1), 1–17 (1961)
14. Heydrich, S., van Stee, R.: Dividing connected chores fairly. Theor. Comput. Sci. **593**, 51–61 (2015)
15. Lipton, R.J., Markakis, E., Mossel, E., Saberi, A.: On approximately fair allocations of indivisible goods. In: Proceedings of the 5th ACM Conference on Economics and Computation, pp. 125–131 (2004)
16. Manurangsi, P., Suksompong, W.: Asymptotic existence of fair divisions for groups. Math. Soc. Sci. (Forthcoming)
17. Stromquist, W.: How to cut a cake fairly. Am. Math. Mon. **87**(8), 640–644 (1980)
18. Stromquist, W.: Envy-free cake divisions cannot be found by finite protocols. Electron. J. Comb. 15, 11 (2008)
19. Su, F.E.: Rental harmony: Sperner's lemma in fair division. Am. Math. Mon. **106**(10), 930–942 (1999)
20. Suksompong, W.: Asymptotic existence of proportionally fair allocations. Math. Soc. Sci. **81**, 62–65 (2016)

Author Index

Printed in the United States
By Bookmasters